Mechanical Vibration

Analysis, Uncertainties, and Control

MECHANICAL ENGINEERING
A Series of Textbooks and Reference Books

Founding Editor

L. L. Faulkner

*Columbus Division, Battelle Memorial Institute
and Department of Mechanical Engineering
The Ohio State University
Columbus, Ohio*

1. *Spring Designer's Handbook*, Harold Carlson
2. *Computer-Aided Graphics and Design*, Daniel L. Ryan
3. *Lubrication Fundamentals*, J. George Wills
4. *Solar Engineering for Domestic Buildings*, William A. Himmelman
5. *Applied Engineering Mechanics: Statics and Dynamics*, G. Boothroyd and C. Poli
6. *Centrifugal Pump Clinic*, Igor J. Karassik
7. *Computer-Aided Kinetics for Machine Design*, Daniel L. Ryan
8. *Plastics Products Design Handbook, Part A: Materials and Components; Part B: Processes and Design for Processes*, edited by Edward Miller
9. *Turbomachinery: Basic Theory and Applications*, Earl Logan, Jr.
10. *Vibrations of Shells and Plates*, Werner Soedel
11. *Flat and Corrugated Diaphragm Design Handbook*, Mario Di Giovanni
12. *Practical Stress Analysis in Engineering Design*, Alexander Blake
13. *An Introduction to the Design and Behavior of Bolted Joints*, John H. Bickford
14. *Optimal Engineering Design: Principles and Applications*, James N. Siddall
15. *Spring Manufacturing Handbook*, Harold Carlson
16. *Industrial Noise Control: Fundamentals and Applications*, edited by Lewis H. Bell
17. *Gears and Their Vibration: A Basic Approach to Understanding Gear Noise*, J. Derek Smith
18. *Chains for Power Transmission and Material Handling: Design and Applications Handbook*, American Chain Association
19. *Corrosion and Corrosion Protection Handbook*, edited by Philip A. Schweitzer
20. *Gear Drive Systems: Design and Application*, Peter Lynwander
21. *Controlling In-Plant Airborne Contaminants: Systems Design and Calculations*, John D. Constance
22. *CAD/CAM Systems Planning and Implementation*, Charles S. Knox
23. *Probabilistic Engineering Design: Principles and Applications*, James N. Siddall
24. *Traction Drives: Selection and Application*, Frederick W. Heilich III and Eugene E. Shube
25. *Finite Element Methods: An Introduction*, Ronald L. Huston and Chris E. Passerello
26. *Mechanical Fastening of Plastics: An Engineering Handbook*, Brayton Lincoln, Kenneth J. Gomes, and James F. Braden
27. *Lubrication in Practice: Second Edition*, edited by W. S. Robertson
28. *Principles of Automated Drafting*, Daniel L. Ryan

WITHDRAWN

LIVERPOOL
JOHN MOORES UNIVERSITY
AVRIL ROBARTS LRC
TITHEBARN STREET
LIVERPOOL L2 2ER
TEL. 0151 231 4022

LIVERPOOL JMU LIBRARY

3 1111 01156 8316

29. *Practical Seal Design*, edited by Leonard J. Martini
30. *Engineering Documentation for CAD/CAM Applications*, Charles S. Knox
31. *Design Dimensioning with Computer Graphics Applications*, Jerome C. Lange
32. *Mechanism Analysis: Simplified Graphical and Analytical Techniques*, Lyndon O. Barton
33. *CAD/CAM Systems: Justification, Implementation, Productivity Measurement*, Edward J. Preston, George W. Crawford, and Mark E. Coticchia
34. *Steam Plant Calculations Manual*, V. Ganapathy
35. *Design Assurance for Engineers and Managers*, John A. Burgess
36. *Heat Transfer Fluids and Systems for Process and Energy Applications*, Jasbir Singh
37. *Potential Flows: Computer Graphic Solutions*, Robert H. Kirchhoff
38. *Computer-Aided Graphics and Design: Second Edition*, Daniel L. Ryan
39. *Electronically Controlled Proportional Valves: Selection and Application*, Michael J. Tonyan, edited by Tobi Goldoftas
40. *Pressure Gauge Handbook*, AMETEK, U.S. Gauge Division, edited by Philip W. Harland
41. *Fabric Filtration for Combustion Sources: Fundamentals and Basic Technology*, R. P. Donovan
42. *Design of Mechanical Joints*, Alexander Blake
43. *CAD/CAM Dictionary*, Edward J. Preston, George W. Crawford, and Mark E. Coticchia
44. *Machinery Adhesives for Locking, Retaining, and Sealing*, Girard S. Haviland
45. *Couplings and Joints: Design, Selection, and Application*, Jon R. Mancuso
46. *Shaft Alignment Handbook*, John Piotrowski
47. *BASIC Programs for Steam Plant Engineers: Boilers, Combustion, Fluid Flow, and Heat Transfer*, V. Ganapathy
48. *Solving Mechanical Design Problems with Computer Graphics*, Jerome C. Lange
49. *Plastics Gearing: Selection and Application*, Clifford E. Adams
50. *Clutches and Brakes: Design and Selection*, William C. Orthwein
51. *Transducers in Mechanical and Electronic Design*, Harry L. Trietley
52. *Metallurgical Applications of Shock-Wave and High-Strain-Rate Phenomena*, edited by Lawrence E. Murr, Karl P. Staudhammer, and Marc A. Meyers
53. *Magnesium Products Design*, Robert S. Busk
54. *How to Integrate CAD/CAM Systems: Management and Technology*, William D. Engelke
55. *Cam Design and Manufacture: Second Edition; with cam design software for the IBM PC and compatibles*, disk included, Preben W. Jensen
56. *Solid-State AC Motor Controls: Selection and Application*, Sylvester Campbell
57. *Fundamentals of Robotics*, David D. Ardayfio
58. *Belt Selection and Application for Engineers*, edited by Wallace D. Erickson
59. *Developing Three-Dimensional CAD Software with the IBM PC*, C. Stan Wei
60. *Organizing Data for CIM Applications*, Charles S. Knox, with contributions by Thomas C. Boos, Ross S. Culverhouse, and Paul F. Muchnicki
61. *Computer-Aided Simulation in Railway Dynamics*, by Rao V. Dukkipati and Joseph R. Amyot
62. *Fiber-Reinforced Composites: Materials, Manufacturing, and Design*, P. K. Mallick
63. *Photoelectric Sensors and Controls: Selection and Application*, Scott M. Juds
64. *Finite Element Analysis with Personal Computers*, Edward R. Champion, Jr., and J. Michael Ensminger

65. *Ultrasonics: Fundamentals, Technology, Applications: Second Edition, Revised and Expanded*, Dale Ensminger
66. *Applied Finite Element Modeling: Practical Problem Solving for Engineers*, Jeffrey M. Steele
67. *Measurement and Instrumentation in Engineering: Principles and Basic Laboratory Experiments*, Francis S. Tse and Ivan E. Morse
68. *Centrifugal Pump Clinic: Second Edition, Revised and Expanded*, Igor J. Karassik
69. *Practical Stress Analysis in Engineering Design: Second Edition, Revised and Expanded*, Alexander Blake
70. *An Introduction to the Design and Behavior of Bolted Joints: Second Edition, Revised and Expanded*, John H. Bickford
71. *High Vacuum Technology: A Practical Guide*, Marsbed H. Hablanian
72. *Pressure Sensors: Selection and Application*, Duane Tandeske
73. *Zinc Handbook: Properties, Processing, and Use in Design*, Frank Porter
74. *Thermal Fatigue of Metals*, Andrzej Weronski and Tadeusz Hejwowski
75. *Classical and Modern Mechanisms for Engineers and Inventors*, Preben W. Jensen
76. *Handbook of Electronic Package Design*, edited by Michael Pecht
77. *Shock-Wave and High-Strain-Rate Phenomena in Materials*, edited by Marc A. Meyers, Lawrence E. Murr, and Karl P. Staudhammer
78. *Industrial Refrigeration: Principles, Design and Applications*, P. C. Koelet
79. *Applied Combustion*, Eugene L. Keating
80. *Engine Oils and Automotive Lubrication*, edited by Wilfried J. Bartz
81. *Mechanism Analysis: Simplified and Graphical Techniques, Second Edition, Revised and Expanded*, Lyndon O. Barton
82. *Fundamental Fluid Mechanics for the Practicing Engineer*, James W. Murdock
83. *Fiber-Reinforced Composites: Materials, Manufacturing, and Design, Second Edition, Revised and Expanded*, P. K. Mallick
84. *Numerical Methods for Engineering Applications*, Edward R. Champion, Jr.
85. *Turbomachinery: Basic Theory and Applications, Second Edition, Revised and Expanded*, Earl Logan, Jr.
86. *Vibrations of Shells and Plates: Second Edition, Revised and Expanded*, Werner Soedel
87. *Steam Plant Calculations Manual: Second Edition, Revised and Expanded*, V. Ganapathy
88. *Industrial Noise Control: Fundamentals and Applications, Second Edition, Revised and Expanded*, Lewis H. Bell and Douglas H. Bell
89. *Finite Elements: Their Design and Performance*, Richard H. MacNeal
90. *Mechanical Properties of Polymers and Composites: Second Edition, Revised and Expanded*, Lawrence E. Nielsen and Robert F. Landel
91. *Mechanical Wear Prediction and Prevention*, Raymond G. Bayer
92. *Mechanical Power Transmission Components*, edited by David W. South and Jon R. Mancuso
93. *Handbook of Turbomachinery*, edited by Earl Logan, Jr.
94. *Engineering Documentation Control Practices and Procedures*, Ray E. Monahan
95. *Refractory Linings Thermomechanical Design and Applications*, Charles A. Schacht
96. *Geometric Dimensioning and Tolerancing: Applications and Techniques for Use in Design, Manufacturing, and Inspection*, James D. Meadows
97. *An Introduction to the Design and Behavior of Bolted Joints: Third Edition, Revised and Expanded*, John H. Bickford

98. *Shaft Alignment Handbook: Second Edition, Revised and Expanded*, John Piotrowski
99. *Computer-Aided Design of Polymer-Matrix Composite Structures*, edited by Suong Van Hoa
100. *Friction Science and Technology*, Peter J. Blau
101. *Introduction to Plastics and Composites: Mechanical Properties and Engineering Applications*, Edward Miller
102. *Practical Fracture Mechanics in Design*, Alexander Blake
103. *Pump Characteristics and Applications*, Michael W. Volk
104. *Optical Principles and Technology for Engineers*, James E. Stewart
105. *Optimizing the Shape of Mechanical Elements and Structures*, A. A. Seireg and Jorge Rodriguez
106. *Kinematics and Dynamics of Machinery*, Vladimír Stejskal and Michael Valásek
107. *Shaft Seals for Dynamic Applications*, Les Horve
108. *Reliability-Based Mechanical Design*, edited by Thomas A. Cruse
109. *Mechanical Fastening, Joining, and Assembly*, James A. Speck
110. *Turbomachinery Fluid Dynamics and Heat Transfer*, edited by Chunill Hah
111. *High-Vacuum Technology: A Practical Guide, Second Edition, Revised and Expanded*, Marsbed H. Hablanian
112. *Geometric Dimensioning and Tolerancing: Workbook and Answerbook*, James D. Meadows
113. *Handbook of Materials Selection for Engineering Applications,* edited by G. T. Murray
114. *Handbook of Thermoplastic Piping System Design*, Thomas Sixsmith and Reinhard Hanselka
115. *Practical Guide to Finite Elements: A Solid Mechanics* Approach, Steven M. Lepi
116. *Applied Computational Fluid Dynamics*, edited by Vijay K. Garg
117. *Fluid Sealing Technology*, Heinz K. Muller and Bernard S. Nau
118. *Friction and Lubrication in Mechanical Design*, A. A. Seireg
119. *Influence Functions and Matrices*, Yuri A. Melnikov
120. *Mechanical Analysis of Electronic Packaging Systems*, Stephen A. McKeown
121. *Couplings and Joints: Design, Selection, and Application, Second Edition*, Revised and Expanded, Jon R. Mancuso
122. *Thermodynamics: Processes and Applications*, Earl Logan, Jr.
123. *Gear Noise and Vibration*, J. Derek Smith
124. *Practical Fluid Mechanics for Engineering Applications*, John J. Bloomer
125. *Handbook of Hydraulic Fluid Technology*, edited by George E. Totten
126. *Heat Exchanger Design Handbook*, T. Kuppan
127. *Designing for Product Sound Quality*, Richard H. Lyon
128. *Probability Applications in Mechanical Design*, Franklin E. Fisher and Joy R. Fisher
129. *Nickel Alloys*, edited by Ulrich Heubner
130. *Rotating Machinery Vibration: Problem Analysis and Troubleshooting*, Maurice L. Adams, Jr.
131. *Formulas for Dynamic Analysis*, Ronald L. Huston and C. Q. Liu
132. *Handbook of Machinery Dynamics*, Lynn L. Faulkner and Earl Logan, Jr.
133. *Rapid Prototyping Technology: Selection and Application*, Kenneth G. Cooper
134. *Reciprocating Machinery Dynamics: Design and Analysis*, Abdulla S. Rangwala
135. *Maintenance Excellence: Optimizing Equipment Life-Cycle Decisions*, edited by John D. Campbell and Andrew K. S. Jardine
136. *Practical Guide to Industrial Boiler Systems*, Ralph L. Vandagriff

137. *Lubrication Fundamentals: Second Edition, Revised and Expanded*, D. M. Pirro and A. A. Wessol
138. *Mechanical Life Cycle Handbook: Good Environmental Design and Manufacturing*, edited by Mahendra S. Hundal
139. *Micromachining of Engineering Materials*, edited by Joseph McGeough
140. *Control Strategies for Dynamic Systems: Design and Implementation*, John H. Lumkes, Jr.
141. *Practical Guide to Pressure Vessel Manufacturing*, Sunil Pullarcot
142. *Nondestructive Evaluation: Theory, Techniques, and Applications*, edited by Peter J. Shull
143. *Diesel Engine Engineering: Thermodynamics, Dynamics, Design, and Control*, Andrei Makartchouk
144. *Handbook of Machine Tool Analysis*, Ioan D. Marinescu, Constantin Ispas, and Dan Boboc
145. *Implementing Concurrent Engineering in Small Companies*, Susan Carlson Skalak
146. *Practical Guide to the Packaging of Electronics: Thermal and Mechanical Design and Analysis*, Ali Jamnia
147. *Bearing Design in Machinery: Engineering Tribology and Lubrication*, Avraham Harnoy
148. *Mechanical Reliability Improvement: Probability and Statistics for Experi-mental Testing*, R. E. Little
149. *Industrial Boilers and Heat Recovery Steam Generators: Design, Applications, and Calculations*, V. Ganapathy
150. *The CAD Guidebook: A Basic Manual for Understanding and Improving Computer-Aided Design*, Stephen J. Schoonmaker
151. *Industrial Noise Control and Acoustics*, Randall F. Barron
152. *Mechanical Properties of Engineered Materials*, Wolé Soboyejo
153. *Reliability Verification, Testing, and Analysis in Engineering Design*, Gary S. Wasserman
154. *Fundamental Mechanics of Fluids: Third Edition*, I. G. Currie
155. *Intermediate Heat Transfer*, Kau-Fui Vincent Wong
156. *HVAC Water Chillers and Cooling Towers: Fundamentals, Application, and Operation*, Herbert W. Stanford III
157. *Gear Noise and Vibration: Second Edition, Revised and Expanded*, J. Derek Smith
158. *Handbook of Turbomachinery: Second Edition*, Revised and Expanded, edited by Earl Logan, Jr., and Ramendra Roy
159. *Piping and Pipeline Engineering: Design, Construction, Maintenance, Integrity, and Repair*, George A. Antaki
160. *Turbomachinery: Design and Theory*, Rama S. R. Gorla and Aijaz Ahmed Khan
161. *Target Costing: Market-Driven Product Design*, M. Bradford Clifton, Henry M. B. Bird, Robert E. Albano, and Wesley P. Townsend
162. *Fluidized Bed Combustion*, Simeon N. Oka
163. *Theory of Dimensioning: An Introduction to Parameterizing Geometric Models*, Vijay Srinivasan
164. *Handbook of Mechanical Alloy Design*, edited by George E. Totten, Lin Xie, and Kiyoshi Funatani
165. *Structural Analysis of Polymeric Composite Materials*, Mark E. Tuttle
166. *Modeling and Simulation for Material Selection and Mechanical Design*, edited by George E. Totten, Lin Xie, and Kiyoshi Funatani

167. *Handbook of Pneumatic Conveying Engineering*, David Mills, Mark G. Jones, and Vijay K. Agarwal
168. *Clutches and Brakes: Design and Selection, Second Edition*, William C. Orthwein
169. *Fundamentals of Fluid Film Lubrication: Second Edition*, Bernard J. Hamrock, Steven R. Schmid, and Bo O. Jacobson
170. *Handbook of Lead-Free Solder Technology for Microelectronic Assemblies*, edited by Karl J. Puttlitz and Kathleen A. Stalter
171. *Vehicle Stability*, Dean Karnopp
172. *Mechanical Wear Fundamentals and Testing: Second Edition, Revised and Expanded*, Raymond G. Bayer
173. *Liquid Pipeline Hydraulics*, E. Shashi Menon
174. *Solid Fuels Combustion and Gasification*, Marcio L. de Souza-Santos
175. *Mechanical Tolerance Stackup and Analysis*, Bryan R. Fischer
176. *Engineering Design for Wear,* Raymond G. Bayer
177. *Vibrations of Shells and Plates: Third Edition, Revised and Expanded*, Werner Soedel
178. *Refractories Handbook*, edited by Charles A. Schacht
179. *Practical Engineering Failure Analysis*, Hani M. Tawancy, Anwar Ul-Hamid, and Nureddin M. Abbas
180. *Mechanical Alloying and Milling*, C. Suryanarayana
181. *Mechanical Vibration: Analysis, Uncertainties, and Control, Second Edition, Revised and Expanded*, Haym Benaroya
182. *Design of Automatic Machinery*, Stephen J. Derby
183. *Practical Fracture Mechanics in Design: Second Edition, Revised and Expanded*, Arun Shukla
184. *Practical Guide to Designed Experiments*, Paul D. Funkenbusch

Additional Volumes in Preparation

Mechanical Engineering Software

Spring Design with an IBM PC, Al Dietrich

Mechanical Design Failure Analysis: With Failure Analysis System Software for the IBM PC, David G. Ullman

Mechanical Vibration

Analysis, Uncertainties, and Control

Second Edition

Haym Benaroya
Mechanical and Aerospace Engineering
Rutgers University
New Brunswick, New Jersey, U.S.A.

 MARCEL DEKKER

NEW YORK

The first edition of this book was published by Pearson Education, Inc.

Although great care has been taken to provide accurate and current information, neither the author(s) nor the publisher, nor anyone else associated with this publication, shall be liable for any loss, damage, or liability directly or indirectly caused or alleged to be caused by this book. The material contained herein is not intended to provide specific advice or recommendations for any specific situation.

Trademark notice: Product or corporate names may be trademarks or registered trademarks and are used only for identification and explanation without intent to infringe.

Library of Congress Cataloging-in-Publication Data
A catalog record for this book is available from the Library of Congress.

ISBN: 0-8247-5380-1

This book is printed on acid-free paper.

Headquarters
Marcel Dekker, Inc., 270 Madison Avenue, New York, NY 10016, U.S.A.
tel: 212-696-9000; fax: 212-685-4540

Distribution and Customer Service
Marcel Dekker, Inc., Cimarron Road, Monticello, New York 12701, U.S.A.
tel: 800-228-1160; fax: 845-796-1772

World Wide Web
http://www.dekker.com

The publisher offers discounts on this book when ordered in bulk quantities. For more information, write to Special Sales/Professional Marketing at the headquarters address above.

Copyright © 2004 by Marcel Dekker, Inc. All Rights Reserved.

Neither this book nor any part may be reproduced or transmitted in any form or by any means, electronic or mechanical, including photocopying, microfilming, and recording, or by any information storage and retrieval system, without permission in writing from the publisher.

Current printing (last digit):

10 9 8 7 6 5 4 3 2 1

PRINTED IN THE UNITED STATES OF AMERICA

To my wife Shelley, for her kindness and encouragement, and to my children Ana Faye and Adam Nathaniel, who have developed into fine and talented people.

Preface

The decision that the profession needs another textbook on any subject must be made with great humility. That I have come to such a conclusion is in no way meant to be a rejection of other such books. In some respects, other books offer ideas and context that I do not. I have chosen to write this book in the format, content, and depth of description that I would have liked when I learned the subject for the first time. I continue to learn the subject. Writing this book has been an exceptional privilege and an enormous learning experience.

This book is intended to be covered at two course levels, one undergraduate and one graduate. The material included is, and must be, a condensed version of what I would have liked to include in such a book. Had it been all–inclusive, the book would have been so large and expensive that you, and your instructor, likely would not have purchased it.

There are several distinguishing features in this vibration text. Uncertainties modeling and control has been presented as an integral part of vibration. In dealing with the subject of vibration, the engineer must also consider the option of vibration control as well as the effects of uncertainties in the analysis. As part of the integration of vibration, uncertainty and control, some notions on these are introduced prior to the technical details. This plants the seeds before the subject matter has to be studied mathematically. Of course, this is a text on vibration, and for extensive and in–depth studies on randomness and control, specialized texts should be sought.

This text includes example problems, end–of–chapter problems, and an up-to–date set of potential mini–projects based mostly on recently published papers that can guide the student and the instructor to interesting and accessible journal papers, either as part of the course in project form, or for independent study. These, plus the footnotes to the literature, have a double purpose. The first is to provide proper citation and expanded discussion; there is no separate list of references in this text, and these footnotes serve as such attribution. The second is to introduce the student to the relevant journal literature and to some of the very useful texts. In no way is this meant to be the definitive word, but only a starting point.

A small number of biographical summaries of some of the founders of this discipline are included at appropriate locations. Of course, we can never do justice here, in a historical sense, or be as comprehensive as we think is necessary. Rather, the intent is to add for the readers the essential human connection to this subject. The history of the subject provides valuable insights into the subject, even though we rarely spend much time on it. These biographies, downloaded from the world wide web, are included here as a courtesy of Professors E.F. Robertson and J.J. O'Connor, School of Mathematical and Computer Sciences, University of St. Andrews, St. Andrews, Scotland. The web site is at

$$http://www\text{-}groups.dcs.st\text{-}andrews.ac.uk/history/index.html.$$

This text is essentially self–contained. The student may start at the beginning and continue to the end with rare need to refer to other works, except to find additional perspectives on the subject. But then, no one text can cover all aspects of a subject as broad as vibration. Where more details become necessary, other works are cited where the reader will find additional information.

The instructor may choose a variety of options for the use of this text. It is generally possible to skip sections that do not fit with the philosophy of the instructor. A first course is likely to omit the more advanced subjects such as stochastic processes in Chapter 5 and the variational approaches of Chapter 7. A logical sequence of material has been presented in the chapters so that the instructor can leave out sections that do not fit into the particular syllabus. These omitted topics can be studied in a second course, where more advanced topics will provide a broader perspective on vibration.

In particular, an undergraduate course could cover most of the introductory and background Chapters 0 and 1, the single degree of freedom Chapters 2–4, and Chapter 8 on multi–degree of freedom systems. Chapter 5 on randomness and Chapter 6 on feedback control provide the instructor with resources that permit a customized syllabus.

The second, usually graduate, course could briefly review Chapters 0, 2–4, introduce the subjects of randomness and control in Chapters 5 and 6, and spend the most time on the variational techniques of Chapter 7, and the multi–degree and continuous systems of Chapters 9 and 11. In this second edition, Chapter 12 introduces concepts of nonlinear vibration and stability. The choices and emphasis will depend on the level of preparation of the students and the curricular philosophy of the institution. A two–semester sequence can cover all the material contained in this book.

PREFACE

As a computational resource, a MATLAB primer is included on CD. This document, written by Dr. Stephen Kuchnicki, provides original programs that cover the range of applications introduced in the text. These programs can be used to solve complex problems as well as to test one's solutions.

Readers' comments are most welcome, as are any suggestions and corrections. All of these would be most appreciated. I may be most easily reached at *benaroya@rci.rutgers.edu* and all messages will be acknowledged.

Haym Benaroya

Acknowledgments

No project of this magnitude can be completed without the explicit and implicit assistance of others. My father Alfred, one of the best engineers I know, spent many hours helping to make this book useful and readable to engineers, and provided valuable insights and suggestions. What I am today is, to a large measure, a result of his efforts and those of my mother, Esther. I am very grateful.

Professor Mark Nagurka dedicated many hours, that he did not have to spare, to help improve the manuscript. Mark is an exceptional teacher and an honest researcher.

Several students very graciously undertook the task of reading the manuscript. Andrew Mosedale volunteered to read the initial chapters and came to me with pages of corrections and suggestions. Professor Ronald Adrezin and Professor Seon Han spent time preparing solutions to some intricate example and homework problems, as well as a group of figures. Joseph Callahan assisted with a set of intricate figures. Dr. Stephen Kuchnicki wrote the m–files for the MATLAB primer that accompanies this text. Eric Doshna helped with some homework problem solutions. Raymond Essig, Michael Pelardis and Donald Zellman, Jr. all read portions of the manuscript and helped with some of the figures and some of the homework problem solutions. They all also prepared some of the computer generated figures. Thanks go to my former student, Dr. Patrick Bar–Avi, for introducing me to the important problem of the moving continua, and especially to the problem of the flow in an elastic pipe that concludes this book.

This work would not have been possible without the supportive environment provided by the Department of Mechanical and Aerospace Engineering and Rutgers University, for which I am sincerely grateful.

I also am pleased to thank Dr. Thomas Swean, of the Office of Naval Research, for his continued support of my research activity. Such research permits the refinement and development of concepts necessary for the advancement of engineering in particular and society in general.

I am indebted to the authors of the many books from which I began to learn and understand the concepts needed for a career in engineering. In particular, my exposure to vibration began as a student with the text **Introduction to Structural Dynamics** by J.M. Biggs, McGraw–Hill, 1964, and with the first edition of **Dynamics of Structures** by R.W. Clough and J. Penzien, McGraw–Hill, 1975. As a teacher, my learning continued with the following texts: the second edition of **Elements of Vibration Analysis**, by L. Meirovitch, McGraw–Hill, 1986, **Analytical Methods in Vibrations**, by L. Meirovitch, Macmillan, 1967, the fourth edition of **Vibration Problems in Engineering**, by S. Timoshenko, D.H. Young, and W. Weaver, Jr., John Wiley & Sons, 1974, and Engineering Vibration, by D.J. Inman, Prentice–Hall, 1994. All these authors have set a standard for technical thoroughness, style, notation, and content against which I measure the success of my efforts. In particular, I have adopted certain notation and approaches that I felt best provides the readers with an understanding and an appreciation of the subject. Since vibration is considered a classical subject, it is impossible for me to fully attribute the work of each author who has had a hand in the development of the discipline, but I have done so as much as possible, and regret any unintentional slight.

Finally, I would like to thank my editor for this Second Edition at Marcel Dekker, Inc., John Corrigan, for his interest and support, and to Marcel Dekker for taking a chance with this Second Edition. For the Second Edition, I thank Professor Seon Han again for help with the updating of the text and recreating most of the figures, and Mangala Gadagi for her serious reading of the earlier edition and pointing out errors and suggesting clarifications. I also appreciate the corrections and suggestions that my students Jason Florek, Rene David Gabbai and again Mangala Gadagi provided on my new nonlinear oscillations chapter.

Contents

0 Introduction and Background   **1**
    0.1   Qualitative Systems Concepts 4
           0.1.1   Qualitative Structures Concepts 4
    0.2   Qualitative Probability Concepts 5
    0.3   Qualitative Vibration, Uncertainties, and Control 5
    0.4   Analysis and Design: Interrelationships 6
    0.5   Computational Modeling and Design 6
    0.6   Modeling for Vibration 7
           0.6.1   Problem Idealization and Formulation 7
           0.6.2   Concepts of Stiffness, Damping, and Inertia 11
           0.6.3   Statics and Equilibrium 13
           0.6.4   The Equations of Motion 14
           0.6.5   Types of System Models 14
    0.7   Newton's Second Law of Motion 17
    0.8   Units 18

1 Some Mathematics   **21**
    1.1   Taylor Series and Linearization 21
    1.2   Ordinary Differential Equations 23
           1.2.1   Solution of Linear Equations 25
           1.2.2   Homogeneous Solution 25
           1.2.3   Particular Solution 29
    1.3   Matrices 32
           1.3.1   Matrix Operations 34
           1.3.2   Determinant and Matrix Inverse 35
           1.3.3   Eigenvalues and Eigenvectors of a Square Matrix . . . 37
    1.4   Transition 38

2 Single Degree of Freedom Vibration: An Introduction to Discrete Models — 39

- 2.1 Example Problems and Motivation — 40
  - 2.1.1 Transport of a Satellite — 40
  - 2.1.2 Rocket Ship — 40
- 2.2 Mathematical Modeling: Deterministic — 44
  - 2.2.1 Problem Idealization and Formulation — 44
  - 2.2.2 Mass, Damping, and Stiffness — 46
  - 2.2.3 Sources of Deterministic Approximation — 49
  - 2.2.4 Dimensional Analysis — 49
  - 2.2.5 Equations of Motion: Newton's Second Law — 52
  - 2.2.6 Equations of Motion: Energy Formulation — 59
  - 2.2.7 The Rotating Vector Approach to the Equation of Motion — 64
  - 2.2.8 Solution of the Equations of Motion — 66
- 2.3 Free Vibration With No Damping — 67
  - 2.3.1 Alternate Formulation — 69
  - 2.3.2 Phase Plane — 70
- 2.4 Harmonic Forced Vibration With No Damping — 72
  - 2.4.1 Resonance — 74
  - 2.4.2 Vibration of a Structure in Water — 79
- 2.5 Concepts Summary — 81
- 2.6 Problems — 81

3 Single Degree of Freedom Vibration: Discrete Models with Damping — 93

- 3.1 Damping — 94
- 3.2 Free Vibration With Damping — 94
  - 3.2.1 Some Time Constants — 103
  - 3.2.2 Phase Plane — 104
- 3.3 Forced Vibration With Damping — 104
- 3.4 Harmonic Excitation and Damped Response — 108
  - 3.4.1 Harmonic Excitation in Complex Notation — 114
- 3.5 Periodic but Not Harmonic Excitation — 129
- 3.6 Arbitrary Loading: Laplace Transform — 135
- 3.7 Step Loading — 143
- 3.8 Impulsive Excitation — 146
- 3.9 Arbitrary Loading: Convolution — 148
- 3.10 Concepts Summary — 155
- 3.11 Problems — 156

CONTENTS xiii

4 Single Degree of Freedom Vibration: Advanced Topics 169
4.1 Introduction to Lagrange's Equation 169
4.2 Notions of Randomness 172
4.3 Notions of Control 174
4.4 The Inverse Problem 175
4.5 A Self–Excited System and Stability 175
4.6 Solution Analysis and Design Techniques 176
4.7 Concepts Summary 188
4.8 Problems 188
4.9 Mini–Projects 190

5 Single Degree of Freedom Vibration: Probabilistic Forces 195
5.1 Introduction 195
5.2 Example Problems and Motivation 200
    5.2.1 Random Vibration 200
    5.2.2 Fatigue Life 201
    5.2.3 Ocean Wave Forces 203
    5.2.4 Wind Forces 205
    5.2.5 Material Properties 206
    5.2.6 Statistics and Probability 207
5.3 Random Variables 208
    5.3.1 Probability Distribution 208
    5.3.2 Probability Density Function 209
5.4 Mathematical Expectation 211
    5.4.1 Variance 213
5.5 Probability Densities Useful in Applications 215
    5.5.1 The Uniform Density 215
    5.5.2 The Exponential Density 217
    5.5.3 The Normal (Gaussian) Density 218
    5.5.4 The Lognormal Density 220
    5.5.5 The Rayleigh Density 221
5.6 Two Random Variables 222
    5.6.1 Covariance and Correlation 224
5.7 Random Processes 228
    5.7.1 Basic Random Process Descriptors 228
    5.7.2 Ensemble Averaging 229
    5.7.3 Stationarity 233
    5.7.4 Power Spectrum 236
5.8 Random Vibration 242
    5.8.1 Formulation 243
    5.8.2 Derivation of Equations 244
    5.8.3 Response Correlations 245

	5.8.4 Response Spectral Densities	246
5.9	Concepts Summary	250
5.10	Problems	250
5.11	Mini–Projects	258

6 Vibration Control   261
- 6.1 Introduction 262
- 6.2 Feedback Control 263
- 6.3 Performance of Feedback Control Systems 267
  - 6.3.1 Poles and Zeros of a Second Order System 269
  - 6.3.2 Gain Factor 270
  - 6.3.3 Stability of Response 272
- 6.4 Automatic Control of Transient Response 272
  - 6.4.1 Control Actions 272
  - 6.4.2 Control of Transient Response 274
- 6.5 Sensitivity to Parameter Variations 279
- 6.6 State Variable Models 282
  - 6.6.1 Matrix Derivatives and Integrals 286
- 6.7 Concepts Summary 287
- 6.8 Problems 287
- 6.9 Mini–Projects 291

7 Variational Principles   295
- 7.1 Introduction 295
- 7.2 Virtual Work 297
  - 7.2.1 Work and Energy 297
  - 7.2.2 Virtual Work 300
  - 7.2.3 d'Alembert's Principle 303
- 7.3 Lagrange's Equation 308
  - 7.3.1 Lagrange's Equation for Small Oscillations 317
- 7.4 Hamilton's Principle 318
- 7.5 Lagrange's Equation with Damping 322
- 7.6 Concepts Summary 323
- 7.7 Problems 324
- 7.8 Additional Readings 329

8 Multi Degree of Freedom Vibration   333
- 8.1 Example Problems and Motivation 333
  - 8.1.1 Periodic Structures 333
  - 8.1.2 Inverse Problems 334
- 8.2 Stiffness and Flexibility 334
  - 8.2.1 Influence Coefficients 335

	8.3	Derivation of Equations of Motion 343
		8.3.1 $[m]$ and $[k]$ Matrix Properties 349
	8.4	Undamped Vibration 351
		8.4.1 Two Degree of Freedom Motion: Solution by the Direct Method 351
		8.4.2 Forced Vibration by the Direct Method 363
		8.4.3 Coupled Pendulum – Beating 368
	8.5	Direct Method: Free Vibration with Damping 376
	8.6	Modal Analysis 386
		8.6.1 Modal Orthogonality 387
		8.6.2 Modal Analysis with Forcing 393
		8.6.3 Modal Analysis with Proportional Damping 402
		8.6.4 Modal Analysis Compared to the Direct Method . . . 407
	8.7	Concepts Summary 410
	8.8	Problems 410
	8.9	Mini–Projects 419

9 Multi Degree of Freedom Vibration: Advanced Topics   423

	9.1	Overview 423
	9.2	Generalization to n Degrees of Freedom 423
		9.2.1 Modal Matrix $[P]$ 425
	9.3	Unrestrained Systems 427
		9.3.1 Repeated Frequencies 435
	9.4	The Geometry of the Eigenvalue Problem 435
		9.4.1 Repeated Frequencies 438
	9.5	Periodic Structures 439
		9.5.1 Perfect Lattice Models 440
		9.5.2 Effects of Imperfection 442
	9.6	Inverse Vibration: Estimate Mass and Stiffness 443
		9.6.1 Deterministic Inverse Vibration Problem 446
		9.6.2 Effect of Uncertain Data 449
	9.7	Sloshing of Fluids in Containers 454
	9.8	Stability of Motion 458
	9.9	Stochastic Response of a Linear MDOF System 461
	9.10	Rayleigh's Quotient 465
	9.11	Monte Carlo Simulation 470
		9.11.1 Random Number Generation 472
		9.11.2 Generation of Random Variates 473
		9.11.3 Generating a Time–History for a Random Process Defined by a Power Spectral Density 475
	9.12	Concepts Summary 477
	9.13	Problems 479

9.14 Mini–Projects 482

10 Continuous Models for Vibration 487
10.1 Continuous Limit of a Discrete Formulation 487
10.2 Vibration of Strings 489
    10.2.1 Wave Propagation Solution 491
    10.2.2 The Wave Equation via Hamilton's Principle 495
    10.2.3 The Boundary Value Problem for the String 498
    10.2.4 Modal Solution for Fixed–Fixed Boundary Conditions 499
10.3 Longitudinal (Axial) Vibration of Beams 505
    10.3.1 Newton's Approach to the Governing Equation 505
    10.3.2 Hamilton's Approach to the Governing Equation . . . 507
    10.3.3 Simplified Eigenvalue Problem 509
    10.3.4 Orthogonality of the Normal Modes 512
10.4 Torsional Vibration of Shafts 514
    10.4.1 Torsion of Shaft with Rigid Disk at End 516
10.5 Transverse Vibration of Beams 518
    10.5.1 Derivation of the Equations of Motion for the Beam with Shear Distortion: The *Timoshenko* Beam 519
    10.5.2 Boundary Conditions 523
    10.5.3 Simplified Eigenvalue Problem 524
    10.5.4 Orthogonality of the Normal Modes 531
10.6 Beam Vibration: Special Problems 541
    10.6.1 Transverse Vibration of Beam with Axial Force 541
    10.6.2 Transverse Vibration of Beam with Elastic Restraints 544
    10.6.3 Transverse Vibration of Beam on Elastic Foundation . 546
    10.6.4 Response of a Beam with a Moving Support 548
    10.6.5 Response of a Beam to a Traveling Force 549
10.7 Concepts Summary 551
10.8 Problems 551
10.9 Mini–Projects 561

11 Continuous Models for Vibration: Advanced Models 575
11.1 Vibration of Membranes 575
    11.1.1 Rectangular Membranes 575
    11.1.2 Circular Membranes 582
11.2 Vibration of Plates 588
    11.2.1 Derivation of the Equation of Motion for Rectangular Plates 589
    11.2.2 The Eigenvalue Problem 593
11.3 Random Vibration of Continuous Structures 596
11.4 Approximate Methods 600

	11.4.1 Rayleigh's Quotient 600
	11.4.2 Rayleigh–Ritz Method 602
	11.4.3 The Galerkin Method 608
11.5	Where Variables Do Not Separate 613
	11.5.1 Response to Nonharmonic, Time–Dependent Boundary Conditions 613
	11.5.2 Flow in a Pipe with Constant Tension 624
11.6	Concepts Summary 630
11.7	Problems 631
11.8	Mini–Projects 634

12 Nonlinear Oscillation   641

- 12.1 Examples of Nonlinear Vibration 642
- 12.2 Fundamental Nonlinear Equations 645
- 12.3 The Phase Plane 646
  - 12.3.1 Stability of Equilibria 651
- 12.4 Perturbation Methods 654
  - 12.4.1 Lindstedt–Poincaré Method 658
  - 12.4.2 Forced Oscillations of Quasi–Harmonic Systems 662
  - 12.4.3 Jump Phenomenon 666
  - 12.4.4 Periodic Solutions of Non–Autonomous Systems . . . 666
- 12.5 Subharmonic and Superharmonic Oscillations 672
  - 12.5.1 Subharmonics 673
  - 12.5.2 Combination Harmonics 674
- 12.6 The Mathieu Equation 675
- 12.7 The van der Pol Equation 681
  - 12.7.1 Limit Cycles 682
  - 12.7.2 The Forced van der Pol Equation 685
- 12.8 Motion in the Large 687
- 12.9 Advanced Topics 691
  - 12.9.1 Random Duffing Oscillator 692
  - 12.9.2 The Nonlinear Pendulum Using a Galerkin Method . . 694
- 12.10 Concluding Summary 696
- 12.11 Problems 696
- 12.12 Mini–Projects 697

Index   701

Chapter 0

Introduction and Background

"For it is wise to start simply."

You are starting to read a book on the subject of mechanical and structural vibration. When referring to mechanical systems, we mean items such as pumps, compressors and other machines. A discussion of structures implies buildings, aircraft, space structures, and other large–scale objects. Although not universal, a mechanical system is one that is used by humans and a structural system is one that humans can physically enter. Of course, there are objects that fall into both categories.

The subject of vibration has a long history. Its modern study may be viewed as beginning with the two part monograph by Rayleigh.[1] While acoustics and vibration may not initially be viewed as part of the same subject, they do overlap considerably. Many aspects are classical, pursued by physicists and mathematicians during the past four centuries,[2] and yet there is intensive academic and industrial research activity on vibration.

---

[1] Lord Rayleigh, **The Theory of Sound.** This is a Dover book published in 1945 that is still in print. Two interesting Dover books on the rich history of mechanics that include details on vibration are the following: **A History of Mechanics**, R. Dugas, Dover, 1968, and **History of Strength of Materials**, S.P. Timoshenko, Dover, 1983. The reader who is interested in obtaining inexpensive copies of older and out of print scientific and mathematical works is encouraged to contact Dover Publications in Long Island, NY.

[2] A very interesting short article on the relation between the natural sciences, mathematics, and engineering is *Science and the Engineer* by M.A. Biot, **Applied Mechanics Reviews**, Vol. 16, No. 2, Feb. 1963, 89–90. It is timely to this day and also very interesting.

The reason is that, while engineers understand the subject in its basic form, new applications force us to reconsider earlier and simpler formulations which were suitable for less demanding applications. This is particularly true for structures designed for extreme environments such as the ocean, atmosphere and space, and seismically active regions. These require the use of statistical methods for their design due to complexities and uncertainties that cannot be reduced to definitive design constraints. For example:

- Large offshore structures are now being designed for ocean depths of well over 1000 ft. More advanced concepts use cables and tethers that extend through several thousand feet of ocean. These structures must be designed to withstand very large forces due to ocean waves, currents, and winds for many years of operational life. The analysis and design of such structures requires an in–depth understanding of structural behavior, especially vibration.

- Aircraft and spacecraft are amongst our most complex structures. They provide a dual challenge to the engineer. That is, they must be minimized in weight but maximized in strength to withstand the severe forces of the atmosphere and space, respectively. Aircraft contain thousands of mechanical, hydraulic, and electronic components which must work in unison, within the design specifications, to provide a safe haven for air travelers. Spacecraft have appeared more recently on the scene, primarily since the 1950s.[3] These structures, which are designed to operate in space, must be protected from the Earth's environment during their route into orbit. They must also be able to withstand severe temperature gradients in space. Therefore, the engineer must examine all aspects and paths of the spacecraft on its way to the launchpad and beyond.

- Structures and machines designed for the Earth's seismic regions must operate safely when subjected to earthquakes. For the most severe quakes, they must not fail in a way which will harm occupants. This is a tall order for the designer since one does not know in advance when an earthquake will occur or what its magnitude will be. Time of occurrence is of importance for structures that have different occupancy levels for different times of day, such as office buildings. Therefore, for an economical and safe design, statistical methods are needed. Historical records are used to estimate these characteristics for a particular region, and they must be accounted for in any earthquake–resistant design.

---

[3] **Dream Machines: An Illustrated History of the Spaceship in Art, Science, and Literature**, R. Miller, Krieger Publishers, Melbourne, FL, 1993.

The challenges are many. The goal of this book is to provide the reader a basic understanding of the methods of engineering vibration analysis. A variety of example problems will help in fixing these ideas more firmly. With such a grounding, more advanced problems become accessible.

While there are a number of very good texts on vibration,[4,5,6,7] this book is written with a unifying perspective. The study of vibration here includes, in addition to classical vibration analysis, two disciplines which are much younger, but have become intimately part of the study of vibration: *uncertainty modeling* and *vibration control*. As mentioned earlier, classical vibration is based on many simplifying assumptions, one being determinism. It is commonly assumed that parameters and system models are known exactly. Such an approach works in a majority of engineering applications, but cannot be used in problems where complex behavior exists, such as for seismic analysis, or for the estimation of aerodynamic loads. In such instances, statistical scatter of data precludes a deterministic analysis. We do not know when an earthquake will hit or when a peak gust will appear. The best we can hope for is to be able to estimate a high probability of occurrence.

With structures being designed for extreme environments and for new applications, it has become necessary to consider how to minimize excessive vibratory motion. Thus, the need to *control* the vibrational behavior of the structure has become increasingly important. Structural control is a relatively new field. The control of structures and processes has become an extremely sophisticated discipline, transcending many fields, with all levels of intricacies. Much of the development of structural control stems from the earlier developments in the aerospace and space communities, where there was no choice but to integrate control studies into the analysis of aircraft and rocket structures in order to meet the required performance.

First, we will study simple deterministic models with one or just a few degrees of freedom. This permits us to define and understand the rudimentary nature of vibratory systems. These essentials will be rediscovered or refined each time additional complexities are added to the system or environmental models. For example, we will begin to study the simple harmonic motion of a structure that can be modeled using a single degree of freedom. Later, it will be of interest to examine the effects of random loading. Finally, the need to control the vibratory behavior (actually rather complicated) of

---

[4]L. Meirovitch, **Elements of Vibration Analysis**, Second Edition, McGraw–Hill, 1990.

[5]W. Weaver, S.P. Timoshenko, D.H. Young, **Vibration Problems in Engineering**, Fifth Edition, Wiley–Interscience, 1990.

[6]D.J. Inman, **Engineering Vibration**, Prentice–Hall, 1994.

[7]W.T. Thompson, **Theory of Vibration with Applications**, Fourth Edition, Prentice–Hall, 1993.

this oscillator requires consideration of control concepts.

In this manner, insight and intuition will be built up methodically, resulting in a sophisticated understanding of vibratory motion. The remainder of this chapter provides a qualitative introduction to the topics that we will study in the following chapters.

0.1 Qualitative Systems Concepts

Sometimes the terms *systems* and *structures* are distinguished as follows: *structures* can be used represent the range from particular elements such as beams and rods, or mechanical components such as rotors to a large–scale assemblage of components. The term *systems* is meant to be more abstract and general. All structures are systems, but all systems are not structures. Examples of non–structural systems are those based on fluidic or electromagnetic governing principles. We will use both terms to represent structures or more general assemblages in order to signify that simple structural models can be utilized in studies of complex interconnected groups of components.

Implicit in these definitions is the separation of the system from its environment. An *environment* is viewed as external to the system, interacting with it and driving its behavior. For *linear models*, the development of system models proceeds independently of environmental models. Once the vibration characteristics of the system are understood, it is necessary to examine how the system behaves in various environments of differing characteristics. For simple models, Chapters 2, 3 and 4 introduce us to these studies.

0.1.1 Qualitative Structures Concepts

Since all structures are systems, whatever is valid for a system is also valid for a structure. In addition to this, aspects unique to structures must also be examined. For example, it may be said that a system is linear in its behavioral characteristics. Such a statement is significant and implies a number of key concepts which we will study in subsequent chapters.

The assumption (which must be verified) that a structure behaves linearly and elastically defines many characteristics about that structure. Thus, in most instances, the system will be introduced and defined with general concepts, and then the structural interpretations expanded upon.

0.2 Qualitative Probability Concepts

Probability is a much misunderstood and maligned[8] subject that practitioners try to avoid at all costs. To some extent this is a reasonable reaction to a discipline that tends to be very formal and mathematical with connections rarely made to experience. The language is strange, and the outcomes are, after all, not certain. The solutions to problems that have some degree of uncertainty in them will have as much or more uncertainty. For example, if there is 10% uncertainty in the data on the loading, then we expect *at least* 10% uncertainty in the calculated structural displacement.

However, the probabilistic approach is the more general framework for systems and structural analyses since it explicitly takes into account all the possible loads and material properties using the *probability density function*, a concept that we will learn in some detail in Chapter 5. The deterministic case is but one of countless cases built into the probabilistic model. The probabilistic framework provides the analyst and designer a way by which uncertainties can be formalized and quantified so that vibration responses that reflect these can be calculated. It provides a measure of the uncertainties, and, very importantly, helps the analyst and designer ascertain the need for additional information or data before progressing with the design. In particular, one can develop confidence bounds on the values of parameters, and then the structural response.

0.3 Qualitative Vibration, Uncertainties, and Control

We will learn to quantify idealized vibrating systems. Also, to make our mathematical models more realistic, we will include the uncertainties inherent in all models. Then we will begin to understand how to modify the behavior of the system, or control its vibratory characteristics, so that it can perform under stricter margins.

For example, what if our analysis of structural response amplitudes shows that it regularly hits the surrounding structure. The analyst or designer has two options. One is to redesign the structure with new dimensions, masses and stiffnesses. The other option is to generate forces within the structure that will balance the external ones, resulting in an acceptable total displacement. These generated forces are known as control forces and are a function of a structural response such as displacement. Generally, a larger displacement leads to a larger control force. These concepts will be

---

[8] Perhaps rightly so, since it is rarely introduced without much jargon and legalistic discussion.

introduced in the early chapters, and discussed in more detail in Chapter 6 on feedback control.

0.4 Analysis and Design: Interrelationships

As analysts and designers,[9] our interests are with *real* structures and systems. The key word here is real, in the sense that the complicating aspects of the structure should not be ignored. In the design office, we are confronted with a task which requires us to first conceptualize the necessary structure to be built. A preliminary concept is stipulated by the designer that appears to satisfy the needs of the *customer*. Then some rough analyses are performed to get a sense of how well the structure satisfies the major constraints we have placed on its design, constraints which force it to be viable in its intended mission. If the conceptual structural design is still acceptable, then a full–scale analysis and design is required. Here, as many realistic and important aspects of the problem as possible are retained.

If the conceptual design points to some poor characteristics that may compromise the mission of the structure, then a redesign is called for, with appropriate changes made to alleviate any such shortcomings. Experiments on scale models are usually necessary to gather data as well as to gain confidence that our physical understanding matches reality.

0.5 Computational Modeling and Design

Engineers are called upon to tackle some of society's most difficult technical problems, each of which has considerable economic constraints. The complexities of such problems require engineers of many disciplines to work together to build very complex structures. So many factors and so much information must be tracked and integrated that much of our analysis, design, and management of such projects require computational assistance.

This book will provide the conceptual foundations to vibration so that the engineer will be able to utilize computational tools, and then be able to interpret the computational results. We will occasionally hint at computational aspects of the subject under study. Certain computational tools

---

[9] An analyst is traditionally one who performs some mathematical analysis of a structure without necessarily specifying how the structure should be dimensioned or built. The designer has traditionally been the one who makes use of analytical results, whether from a computer analysis or by utilizing a design code, to specify dimensions and construction sequence. Sometimes analysis has taken on an air of being more sophisticated, and design more routine, but those familiar with both recognize the challenges of both. One may actually make a case that analysis is a subset of design.

which are readily available will be utilized[10] to generate some analytical results and some of the numerically–generated results.

0.6 Modeling for Vibration

We have only hinted at how an engineer approaches a new problem. Here we continue this discussion. One thing that is certain is that many of the problems presented in this book do not reflect those initially encountered by the engineer. Generally, the problem as first viewed in engineering practice is vague and ill–defined, and often formulated by a non–technical person who has a need for a product or a process. It is the engineer's task to take the ill–defined problem and turn it into a specific, mathematical formulation to be solved using known or newly developed techniques.

0.6.1 Problem Idealization and Formulation

The process of understanding how the physical world behaves and how it can be modeled generally begins with an understanding of how much simpler systems and components behave. Such simpler systems are called *idealized systems* and the process by which they are derived and formulated requires a sophisticated understanding of the actual system. Usually, the idealized system is used to better understand the intricacies of the real system.

Idealization of the *real system* is required because the real system is usually too complex to be modeled and solved, either mathematically or computationally. The analyst and designer must distill the essential properties of the original system in the creation of the idealized model. This model will then be representative of the original physical characteristics and behavior. Thus, one develops an understanding of the behavior of the real system based on the analysis and observation of the idealized system.

The following four examples are presented in order to provide the reader some intuitive view on the process of model idealization. In all these examples, the full structure is shown first, and then some possible simplified schematic representations are drawn alongside. Each simplified model can be used to better understand some aspects of the behavior of the actual structure. Note that the full model and the actual structure are the end product. They are a result of an understanding that *begins* with the study of the simpler models.

---

[10] In particular, we will utilize the commercial codes *MATLAB* and *MAPLE*. Both are very powerful and somewhat complementary computational tools that every technical person will find of great value. *MATLAB* and *MAPLE* are registered trademarks. A tutorial on the use of *MATLAB* for simple vibration problems is bundled with this text.

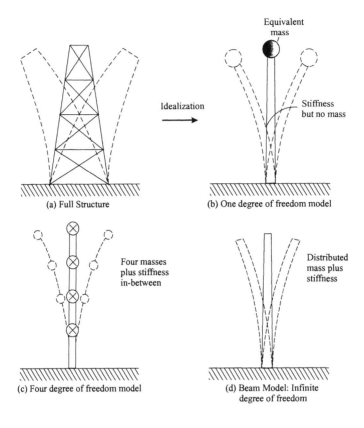

Figure 1: Tower structure and its idealized models.

In Figure 1, a tower structure is shown as it might appear in full, and then is idealized as any of three simplified models. The reference to degree of freedom signifies the number of coordinates needed to define the position of all the masses in the idealization. Also in these simplified models are schematic elements that represent structural stiffness and damping, concepts that are introduced in this chapter. Models *(b)* and *(c)* are studied in Chapters 2, 3, 4, 8 and 9. Model *(d)* is studied in Chapters 10 and 11.

Figure 2 shows a schematic of an automobile, as well as four possible idealized models that are suitable for a preliminary analysis. Models *(b)*, *(c)* and *(d)* include separate modeling of the tire–suspension system. Model *(b)* is studied in Chapters 2–4. Models of the types in *(c)* and *(d)* are studied in Chapters 8 and 9, and models such as in *(e)* are studied in Chapter 10.

A rocket structure along with possible idealizations are shown in Fig-

0.6. MODELING FOR VIBRATION

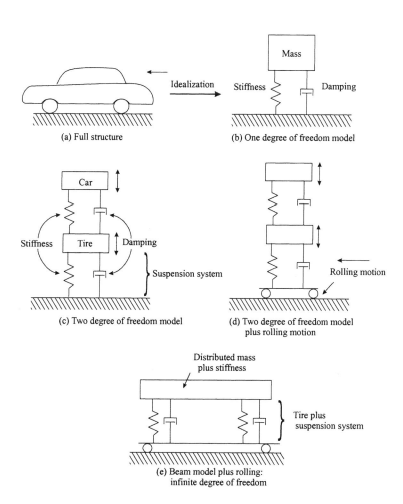

Figure 2: Automobile and its idealized models.

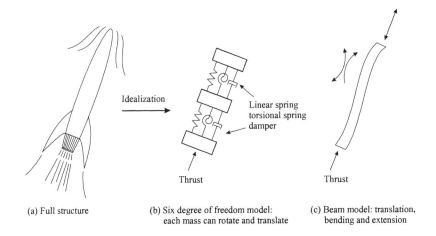

Figure 3: Rocket and its idealized models.

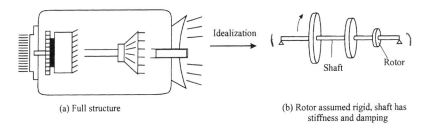

Figure 4: Turbine and idealized model.

ure 3. These simplified models demonstrate the ability to represent axial extension as well as a bending motion of the rocket structure. Models of the type shown in *(b)* are studied in Chapters 8 and 9, and those such as *(c)* are studied in Chapters 10 and 11.

Finally, a schematic of a turbine is presented with its idealized model in Figure 4. The idealized model is conceived of as an elastic, rotating shaft supporting several rigid rotors. Chapters 3, 4, 8, and 9 explore such models.

There are many steps involved in the development of an idealized model that is suitable for studying the behavior of the full–scale structure. It is emphasized that in engineering practice, simple models are first created in order to understand the general behavioral characteristics of the full structure that is yet to be designed. Once the simpler models are understood, more complex and realistic models are used to begin conceptual designs of

0.6. MODELING FOR VIBRATION

the full structure. The process is one of tackling more realistic designs, until the actual structure or machine is being designed.

Nonlinear Models and Stability

Most of this text is focused on problems that can be formulated, and solved, using linear models. Such models are extremely powerful, and carry with them a tremendous body of mathematical analysis for their solution. However, there are important instances where a linear model will not be able to predict some significant behavior characteristics. In such cases, nonlinearities must be retained in the problem formulation, and not simplified. Chapter 12 is devoted to these types of systems and their analysis. Additionally, the issue of stability is examined this chapter as well.

Component and System Modeling

A *system* may be defined as a group of integrated items, behaving as a unit. Examples include structures around us, such as automobiles, airplanes, machinery, and computers. Systems may be engineered or they may exist naturally. Humans, plants, and animals are all natural systems, and are much more complex than engineered systems. Understanding natural systems requires knowledge of biology, chemistry, and physics.

System *modeling* is the process of translating physical characteristics into a mathematical representation, generally consisting of one or more equations. The power of mathematical analysis can then be brought to bear to "solve" the equations.[11] Solving the equations eventually results in numbers. The engineer's task is then to study these numbers, make sure they make sense physically, and to conclude general and specific behavior characteristics so that a design can be created for manufacture or construction.

0.6.2 Concepts of Stiffness, Damping, and Inertia

The dynamic behavior of structures results from the exchange and dissipation of energies. Dynamic forces transfer their energy to the structure, which then responds via several mechanisms, such as bending or extension. Dynamic behavior can be modeled in several ways, the best known is Newton's second law of motion. If the external force is static or quasi–static,

---

[11] It is *very* important to note that system modeling is generally challenging and requires creativity and intuition. It is *not* a cookbook! Modeling is as much an art developed after much practice and experience as it is a science.

structural stiffness forces develop to create an equilibrium. External dynamic forces are balanced in a more complex way with inertial and damping forces.

Inertia is that property of matter which manifests itself as a resistance to any change in its momentum. For a body of mass m, the change in linear momentum p is of the form $\Delta p = m\Delta v$, where Δv is the change in velocity. The change in momentum per unit time is related to the acceleration of the (constant mass) body

$$\frac{\Delta p}{\Delta t} = m\frac{\Delta v}{\Delta t} = ma(t), \qquad (1)$$

where $a(t)$ is the average acceleration during Δt. Thus, inertia must be considered for any body undergoing a change in momentum.

Stiffness, often schematically and conceptually represented by a spring, denotes the capacity of a system to store strain energy. The stiffness force follows *Hooke's law*,[12] $F_s = k_s \Delta x$, where the stiffness constant k_s is expressed in units of force per unit length. This is a linear model that assumes the spring displacement is measured from the rest length. More complicated laws exist, for example, the nonlinear relation (see Section 0.6.5), $F = k(x)\Delta x^a$, where the parameter a would depend on the particular material being modeled, and the stiffness parameter $k(x)$ is a function of how much the spring has been elongated.

Damping defines the ability of a system or structure to dissipate energy.[13] For an oscillatory system, damping is a measure of how much energy is dissipated by the system during an oscillation cycle. For example, structural connections between components add damping to a structure.

System Uncertainty

All systems have associated uncertainties, that is, there is a lack of complete information. Such uncertainties result from the imprecision of our manufacturing tools, instruments that measure data, and of our incomplete understanding of the laws governing the behavior of natural phenomena such as wind and earthquakes.

---

[12] Hooke's law is a simple *constitutive law*, a mathematical model defining the relationship between material characteristics and, in this instance, force and displacement. It is valid for linear behavior.

[13] It should be noted that certain nonlinear systems exhibit negative damping, where energy is actually supplied to the system. Nonlinear systems will not be discussed in this text. An excellent first book for further study of the subject is by J.J. Stoker, **Nonlinear Vibrations in Mechanical and Electrical Systems**, originally published in 1950, and now available through Wiley–Interscience in a 1992 edition. A more modern introduction is offered by A.H. Nayfeh and D.T. Mook, **Nonlinear Oscillations**, John Wiley & Sons, 1979.

0.6. MODELING FOR VIBRATION

Two basic approaches are available to the analyst when confronted by a lack of information. The first approach assumes a deterministic model, where all parameters and dimensions are known exactly. One then takes into account the known imperfections by *overdesigning*, for example, by assuming less material strength than there actually exists.

The second approach requires the use of probabilistic models to explicitly incorporate imprecision into the mathematical models that are to be utilized in the analysis and design. This approach will be explored in more detail in subsequent chapters, in particular Chapter 5.

Both approaches have their place in engineering practice. By far, the deterministic approach is most widely used, but the need for very complex structures that must operate in severe environments has necessitated the introduction of probabilistic tools in analysis and design. Earthquake design codes, for example, are based on probabilistic criteria of structural behavior.

Deterministic Approximations

For many applications, some of the material or geometric uncertainties or imprecision are extremely small when compared to their average value. In these instances, it is reasonable to apply the average material modulus or average dimension. Errors resulting from ignoring any deviations from the true value will not appreciably affect the results of the analysis. In fact, most of the structures we utilize on a daily basis have been designed in this way.

Of course, sometimes it is not possible to distinguish in advance which uncertainties can be ignored and which can be included. In such a case, testing is required to provide us with a better understanding.

0.6.3 Statics and Equilibrium

A *static* system is one which does not experience change as a function of time. Many engineering problems begin by considering the static behavior of the system. The concept of *equilibrium* generally implies a static situation, although dynamic equilibrium[14] is sometimes a useful concept. A static structure resists external forces by deforming itself and/or by transmitting forces through connections to the boundary. Dynamic implies time–dependent behavior.

---

[14] d'Alembert's principle, which we explore in Chapter 7, is essentially a dynamic equilibrium form of Newton's second law of motion.

0.6.4 The Equations of Motion

Formulation, or derivation, of the equations of motion is the step which follows the idealization of a system. The simplicity of the governing equations and the ease of their solution are directly related to the system complexity and the effectiveness of its idealization. Thus, a delicate balance exists between physical reality and mathematical solvability. Often, small changes in the formulation of the idealized model result in significantly more difficult governing equations.

We will focus on two distinct approaches to the derivation of the system governing equations: *Newton's second law of motion* and methods based on the consideration of *kinetic and potential energies*. Each has its advantages. The first method is useful for simpler problems that permit the visualization of the interaction between the forces acting on a system and the resulting internal stiffness and damping reaction forces. *Energy methods* are generally simpler to apply, especially for complex systems with many components that interact in an intricate fashion. Both of these approaches are used and discussed in detail in the following chapters.

0.6.5 Types of System Models

System models will vary depending on the way a physical problem is idealized and the information required for the analysis. Models may be discrete or continuous, linear or nonlinear. The nonlinear model will govern a broader range of dynamic behavior, but can require quite intricate analytical techniques that are beyond the scope of this book. The continuous model is more realistic than the discrete one, but much more difficult to solve. In this text, we will only study linear discrete and linear continuous models. Chapters 10 and 11 introduces linear continuous vibrating systems such as strings, beams, membranes and plates.

Linear Models Approximating Nonlinear Behavior

This book will utilize linear theory as the basis for modeling systems and structures. Much of engineering analysis and design is based on linear models. Two very good reasons why this is so are expressed by *the principle of linear superposition* and *linear systems theory*.

The principle of linear superposition[15] provides analysts with the tools to solve complex linear problems by breaking them down into simpler components. Linearity then permits us to add the simple solutions in order to obtain the complete solution.

---

[15] This principle serves as the basis for the theory of linear differential equations. See Chapter 1 for a review.

0.6. MODELING FOR VIBRATION

Linear systems theory is the general framework for the analysis of linear systems and structures. Thus, as soon as we are able to claim and justify that a system, whether electrical, mechanical, or otherwise, behaves linearly, then all the tools of linear systems theory become available for analysis.

Our link to reality forces us to immediately state the fact that linearity is generally a local phenomenon, meaning that only small displacements or stresses can be viewed as approximately linear. Thus, Hooke's linear law is valid as long as the spring is not stretched beyond a certain fraction of its initial length. "Approximate" linearity is, however, very problem dependent. The error that one can accept depends on the application. In some instances only a nonlinear model is appropriate. One difficult aspect in the analysis of nonlinear systems is that, while there are unifying principles, many solution techniques are required, each for a particular problem. Linear solution techniques are applicable to any linear problem, regardless of its origin.

Dimensionality

One must always be aware that any model contains implicit (unstated) as well as explicit (stated and justified) assumptions. Thus, when a one-dimensional model is utilized, implicit assumptions are made that one coordinate is sufficient to describe the motion and that additional coordinates are not needed to understand the behavior of interest.

For example, the deformation of a rod in its axial direction is accompanied by a contraction in its cross section in a proportion which is given by Poisson's ratio. If one is primarily interested in tensile stresses within the elastic range, the secondary effects, such as the change in rod cross section, can be safely ignored. For larger stresses, the reduction in cross section becomes a significant factor in the calculation of axial stresses.

The important conclusion is that one must always be aware of any defined and *hidden* assumptions in any formulation.

Discrete Models

Many engineering models incorporate physical characteristics such as mass, stiffness, and damping using discrete elements along with discrete mathematical expressions.

In most engineering applications, discrete models are used because of the eventual need for computational models. Computers operate only in discrete time and therefore require discretized models. Common methods used for such analysis are the *finite and boundary element* methods. Our initial studies focus on discrete models which help us begin to understand modeling, and structural characteristics such as inertia and damping.

Single Degree of Freedom Models

A single degree of freedom model implicitly stipulates that one coordinate is sufficient to describe the response of a system to its environment. It reflects the dominance of one response parameter over all others. While such a system rarely exists in application, it proves to be a useful idealized model for learning the concepts of vibration as well as for gathering initial insights into the character of a more complicated dynamic system. Single degree of freedom models help us define and understand many of the key characteristics of a vibrating system and therefore prove to be extremely useful in preliminary studies. We will begin our studies of vibration utilizing basic single degree of freedom models in Chapter 2, and more advanced models in Chapters 3 and 4.

Multi–Degree of Freedom Models

Where more than one coordinate is needed to define the behavior of a system, multiple degrees of freedom are required. Generally, it is necessary to include as many degrees of freedom as there are distinct motions in a system. A continuous system has an infinite number of degrees of freedom, but even these are eventually modeled using a finite number of degrees of freedom.

In many applications, it generally suffices to model the first several to dozen degrees of freedom to help understand the system. Other times, where very detailed behavior is required, hundreds or thousands of degrees of freedom are required. The determining factor will be whether the analysis is used for preliminary studies or for a detailed design. For example, aircraft structures have hundreds of thousands of components and, therefore, hundreds of thousands of degrees of freedom. To gain a basic understanding of aircraft structural vibration, less than ten degrees of freedom are required, but to design such a structure for manufacture will require the information from thousands of degrees of freedom. We will study such systems in Chapters 8 and 9.

Continuous Models

Other engineering systems can only be defined with parameters that are distributed in space. Distributed parameter, or continuous, system models are necessary when the physical characteristics of a structure cannot be accurately lumped at discrete locations. Distributed parameter models of more than one variable are generally in the form of partial differential equations. However, it should be noted that if one utilizes computers for the solution of such models, these will have to be discretized via specialized numerical techniques. We will also discover that many of our solutions for continuous models utilize results from discrete solutions. Such models are studied in Chapters 10 and 11.

Nonlinear Models

This text focuses almost completely on the modeling of system vibration where materials and geometry remain within the linear range. Generally this means that oscillation amplitudes remain "small" and materials are within linear elastic bounds. But some applications cannot be approximated as linear. Some classes of dynamic behavior, such as limit cycles, hysteresis, and chaotic oscillations, do not exist even as approximations in the linear domain. These phenomena require at least a "small" amount of nonlinearity in their governing equations. Such problems are introduced and discussed in Chapter 12.

0.7 Newton's Second Law of Motion

Newton's second law of motion is one of the most important physical laws because it is fundamental to many sciences. Even though it is so important a principle, there is nothing sacred about the relation is states, the relation between the force and acceleration. The equation $F = ma$ is verified only by way of experiments. Consider the following ideal experiment[16] in which a mass particle is subjected to a force F and the acceleration a is measured. Assume that all measurements are exact. The experiment is carried out in an *inertial* setting, that is, the experimental apparatus is not accelerating. If this experiment is carried out on the same mass n times, we find the following,

$$\frac{F_1}{a_1} = \frac{F_2}{a_2} = \cdots = \frac{F_n}{a_n} = c.$$

All the experiments for the same mass particle yield the same constant quantity c for the ratio between force and acceleration. This property of the particle is called is its *inertia*. It can be physically understood as the resistance of the particle to a rate change in velocity. For a given force, a particle with a large inertia has a small acceleration. For the same force, but for a particle with a small inertia, the acceleration is larger. The quantitative measure of inertia is the mass m. Another conclusion drawn from

---

[16] This discussion is based on the excellent and fuller introduction in **Engineering Mechanics: Dynamics**, Fifth Edition, J.L. Meriam, L.G. Kraige, John Wiley & Sons 2002.

the experiments is that the acceleration is always in the direction of the applied force. Therefore, it is necessary that the equation relating force and acceleration be a vector relation.

Therefore, Newton's second law of motion takes the form

$$\mathbf{F} = m\mathbf{a},$$

where the bold notation is the indication that the variable is a vector. Appropriate units must be used. In SI units, the units of force (newtons, N) are derived using Newton's second law from the base units of mass (kilograms, kg) times the acceleration (meters per second per second, m/sec^2). Thus the newton has the units N = kg m/sec^2. In the U.S. or English system of units, the units of mass (slugs) are derived from the units of force (pounds force, lb$_f$) divided by the acceleration (feet per second per second, ft/sec^2). Thus, the mass units are slug = lb sec^2/ft.

If we apply a force to a body sitting on a frictionless horizontal plane, it will accelerate. A force of 1 N on a body of mass 1 kg accelerates at 1 m/sec^2. A force of 1 lb$_f$ on a body of mass 1 slug accelerates at 1 ft/sec^2. A force of 1 lb$_f$ on a body of mass 1 pound mass lb$_m$ accelerates at 32.2 ft/sec^2. A review of units is appropriate and crucial.

0.8 Units

All physical parameters have units that tie them to a particular system. There are two primary systems of units, the *English System* and the *SI System*, where SI stands for *System International*. The *SI* units are considered modern and more appropriate. In this book, both systems will be used since both are used in practice in the United States. In Table 0.1, the English and SI system of units are shown for certain key physical parameters that we will encounter in this text. In Section 2.2.4, we will learn how the dimensions of parameters relevant to a particular problem can be used to derive equations that relate those parameters, as well as for the design of experiments.

Table 1: *English* and *SI* Units for Key Physical Parameters

Parameter	English	SI
Force	1 lb	4.448 N (Newton = kg m/sec^2)
Mass	1 slug (lb sec^2/ft)	14.59 kg (kilogram)
Length	1 ft	0.3048 m (meter)
Acceleration	1 ft/sec^2	0.3048 m/sec^2
Spring constant	1 lb/in	175.12 N/m
Torsional spring constant	1 lb in/rad	0.1130 N m/rad
Damping constant	1 lb sec/in	175.12 N sec/m
Mass moment of inertia	1 lb in sec^2	0.1130 kg m^2
Angle	1 deg	0.0175 rad

Chapter 1

Some Mathematics

"We are reminded of the essential rules."

While we assume that the reader has previously studied the topics in this chapter, a brief introduction and review is developed here, identifying key results of particular interest to those beginning a study of vibration. It is recommended that the example problems be attempted first before reading the solutions.

1.1 Taylor Series and Linearization

The Taylor series[1] of a function is a very useful mathematical identity which we will be using when simplifying certain nonlinear functions. For functions of one variable, the expansion about point a is given by

$$f(x) = f(a) + f'(a)(x-a) + f''(a)\frac{(x-a)^2}{2!} + \cdots, \qquad (1.1)$$

where the primes denote differentiation with respect to the variable x. For a function of two variables, the expansion about the coordinate (a, b) is given by

$$\begin{aligned} f(x,y) &= f(a,b) + (x-a)f_x(a,b) + (y-b)f_y(a,b) \\ &+ \frac{1}{2!}\{(x-a)^2 f_{xx}(a,b) + 2(x-a)(x-b)f_{xy}(a,b) \\ &+ (y-b)^2 f_{yy}(a,b)\} + \cdots, \end{aligned}$$

---

[1] An extremely useful general reference on many topics in calculus, including Taylor series, is by F.B.Hildebrand, **Advanced Calculus for Application**, Prentice–Hall, 1976. (The chapter on matrices is excellent!)

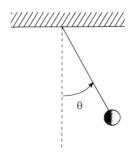

Figure 1.1: Configuration of a pendulum.

where the subscripts denote partial differentiation with respect to the subscripted variable. Thus, $f_{xy}(a,b)$ is the second partial derivative of $f(x,y)$ with respect to x and y evaluated at (a,b).

Such expressions are of use to us when we need to *linearize* functions. For example, some of the equations of motion that we will be deriving include the term $\sin\theta$. The equation of motion for the simple pendulum is

$$\frac{d^2\theta}{dt^2} + \sin\theta = 0, \qquad (1.2)$$

where θ represents the angle of the pendulum with respect to the vertical, as in Figure 1.1.

This is a nonlinear equation due to the sinusoidal function. For small angles, we can *linearize* Equation 1.2. Suppose we expand $\sin\theta$ about point a via a Taylor series. Then, by using Equation 1.1 we find the first four terms in the series to be

$$\sin\theta = \sin a + (\theta - a)\cos a - \frac{(\theta-a)^2}{2!}\sin a - \frac{(\theta-a)^3}{3!}\cos a + \cdots. \qquad (1.3)$$

Generally, we are interested in the behavior about the equilibrium position. In this case, the Taylor series is about $a = 0$, and Equation 1.3 becomes

$$\sin\theta = \theta - \frac{\theta^3}{3!} + \cdots,$$

where θ is in radians. Assume that any oscillation about equilibrium is small, that is $\theta \ll 1$, then

$$\sin\theta \simeq \theta.$$

This can be easily verified numerically and observed when the assumption breaks down. With the above simplification, governing Equation 1.2 be-

Table 1.1: Some Useful Taylor Series Expanded about 0

$e^x = 1 + x + \frac{x^2}{2!} + \frac{x^3}{3!} + \cdots$	$-\infty < x < \infty$
$\ln(1+x) = x - \frac{x^2}{2} + \frac{x^3}{3} - \cdots$	$-1 < x < 1$
$\sin\theta = \theta - \frac{\theta^3}{3!} + \frac{\theta^5}{5!} - \frac{\theta^7}{7!} + \cdots$	$-\infty < \theta < \infty$
$\cos\theta = 1 - \frac{\theta^2}{2!} + \frac{\theta^4}{4!} - \frac{\theta^6}{6!} + \cdots$	$-\infty < \theta < \infty$
$\tan\theta = \theta + \frac{\theta^3}{3} + \frac{2\theta^5}{15} + \frac{17\theta^7}{315} + \cdots$	$|\theta| < \pi/2$
$\sinh\theta = \theta + \frac{\theta^3}{3!} + \frac{\theta^5}{5!} + \frac{\theta^7}{7!} + \cdots$	$-\infty < \theta < \infty$
$\cosh\theta = 1 + \frac{\theta^2}{2!} + \frac{\theta^4}{4!} + \frac{\theta^6}{6!} + \cdots$	$-\infty < \theta < \infty$

comes

$$\frac{d^2\theta}{dt^2} + \theta = 0, \tag{1.4}$$

which is recognized as a simple harmonic oscillator. This equation is only valid for small oscillations about the equilibrium position, $\theta = 0$. To summarize, for small oscillations about the equilibrium, we set $\theta \ll 1$ and retain the first term in the Taylor series of the nonlinear function. This first term of the series is always linear. Some useful series about $a = 0$ are listed in Table 1.1. Oscillations governed by equations such as Equation 1.4 are studied in Chapter 2.

1.2 Ordinary Differential Equations

Much of our study of vibration depends on the solution of linear ordinary differential equations of the form

$$a_2\ddot{y} + a_1\dot{y} + a_0 y = f(t), \quad \text{given } y(0), \ \dot{y}(0), \tag{1.5}$$

where coefficients a_2, a_1 and a_0 are constants, $f(t)$ is a given function of time and overdots denote differentiation with respect to time. Coefficients a_2, a_1 and a_0 may be functions of t in more complex applications. Where $f(t) \neq 0$, the differential equation is called *non-homogeneous*. Otherwise, it is *homogeneous*. Equation 1.5 is known as an *initial-value problem* since its solution depends on the given initial values (or conditions) $y(0)$ and $\dot{y}(0)$.

If the dependent variable y is a function of one independent parameter, in this case time t, then the equation is called an *ordinary* differential equation.

On the other hand, if the dependent variable is a function of two or more parameters, for example time and coordinate, then the differential equation is called a *partial* differential equation. Such equations govern the vibration of continuous systems, and are formulated and solved in Chapter 9.

The *order* of a differential equation is defined to be equal to the highest derivative. Equation 1.5 is of the second order. To solve this equation, the number of initial conditions must equal the order of the equation. Otherwise, there will be arbitrary constants in the solution.

A differential equation is called *linear* if the dependent variable and its derivatives are all to the first power. Of great importance in solving a differential equation is the following theorem:

Theorem 1 *If a_i and $f(t)$ are continuous over a time interval Δt and the initial time t_0 is a point in that interval, then the solution $y(t)$ for the linear initial–value problem is unique.*

The importance of this statement is not yet obvious since we have not applied differential equations to problems of vibration. But what this means is that if *any* solution to linear Equation 1.5 is found, *it is the only solution*. Therefore, when we solve the differential equation governing a linear vibration, we will know that we have the only possible solution. This is not true for nonlinear problems, where there are many valid solutions.

Example 1.1 Verify that
$$y = 3e^{2t} + 3e^{-2t} - 3t$$
is a solution to
$$\ddot{y} - 4y = 12t, \quad y(0) = 6, \quad \dot{y}(0) = -3.$$

Note that the coefficients are constants and that the function $f(t) = 12t$ is continuous. To verify, take the solution y and its second derivative $\ddot{y}$ and substitute these into the differential equation. If the solution is the correct one, then it should satisfy the differential equation. In addition, the initial conditions $y(0)$ and $\dot{y}(0)$ should be equal to the respective given values. ∎

Example 1.2 Verify that $y = ct^2 + t + 3$ is a solution of the linear differential equation
$$t^2 \ddot{y} - 2t\dot{y} + 2y = 6, \quad y(0) = 3, \quad \dot{y}(0) = 1,$$
for any c. Proceed as in the previous example. Note here that the coefficients are functions of time, not constants. ∎

1.2. ORDINARY DIFFERENTIAL EQUATIONS

1.2.1 Solution of Linear Equations

When solving linear differential equations, advantage will be taken of the *principle of linear superposition*. What this means is that the problem is solved in smaller and simpler components. The complete solution is then the sum of these components, as the theorem below explains.

Theorem 2 (Principle of Superposition) *Consider a kth order linear differential equation. If $y_1(t), y_2(t), \ldots, y_k(t)$ are solutions of the equation over an interval, then the linear combination*

$$y = c_1 y_1(t) + c_2 y_2(t) + \cdots + c_k y_k(t)$$

is also a solution.

This can be proven as follows for the case where $k = 2$. Let y_1 and y_2 be solutions to $a_2 \ddot{y} + a_1 \dot{y} + a_0 y = 0$. Define the combination $y = c_1 y_1 + c_2 y_2$. If this sum is also a solution, then it must satisfy the governing equation. Proceed from the governing equation, substitute, and combine like terms, as shown in the following sequence of equations,

$$a_2 \ddot{y} + a_1 \dot{y} + a_0 y = 0$$
$$a_2(c_1 \ddot{y}_1 + c_2 \ddot{y}_2) + a_1(c_1 \dot{y}_1 + c_2 \dot{y}_2) + a_0(c_1 y_1 + c_2 y_2) = 0 \quad (1.6)$$
$$c_1(a_2 \ddot{y} + a_1 \dot{y} + a_0 y) + c_2(a_2 \ddot{y} + a_1 \dot{y} + a_0 y) = 0, \quad (1.7)$$

where the sum within each pair of parentheses in Equations 1.6 and 1.7 is zero. Thus, the principle of superposition is shown to be true.

Superposition has an important extension where the right hand side of the differential equation is non–homogeneous, that is $f(t) \neq 0$. In this instance, the complete solution is equal to the sum of the solutions to the homogeneous equation $y_h(t)$ and the non–homogeneous equation $y_p(t)$:

$$y(t) = y_h(t) + y_p(t).$$

The latter is sometimes called the *particular* solution since it is particular to the specific function on the right hand side of the differential equation.

1.2.2 Homogeneous Solution

Consider the solution of the second order homogeneous governing equation

$$a\ddot{y} + b\dot{y} + cy = 0.$$

We can assume the solution to be of the form $y = e^{rt}$, where the values of r are determined by having the assumed solution satisfy the governing equation. Substituting $y, \dot{y}$ and $\ddot{y}$, find

$$ar^2 e^{rt} + br e^{rt} + c e^{rt} = 0$$
$$e^{rt}(ar^2 + br + c) = 0.$$

Since $e^{rt} \neq 0$, it must be that $ar^2 + br + c = 0$. This quadratic equation for r is very important, as we will often see, and is given the name *characteristic equation*. The values of r must be found using the quadratic formula,

$$r_{1,2} = \frac{-b \pm \sqrt{b^2 - 4ac}}{2a}.$$

There are three possible pairs of values for $r_{1,2}$:

- Case I: There are two real and unequal values, r_1 and r_2. In this case the two solutions are $y_1 = e^{r_1 t}$ and $y_2 = e^{r_2 t}$, and therefore,

$$y(t) = c_1 e^{r_1 t} + c_2 e^{r_2 t}.$$

- Case II: There are two real and equal values, $r_1 = r_2$. In this case the two (independent) solutions are $y_1 = e^{r_1 t}$ and $y_2 = t e^{r_2 t}$, with the complete solution,

$$y(t) = c_1 e^{r_1 t} + c_2 t e^{r_2 t}.$$

- Case III: This is the most interesting case for vibration studies. Here the two values are a *complex conjugate pair*, $r_{1,2} = \alpha \pm i\beta$, with the complete solution given by

$$\begin{aligned} y(t) &= c_1 e^{(\alpha + i\beta)t} + c_2 e^{(\alpha - i\beta)t} \\ &= e^{\alpha t}(c_1 e^{i\beta t} + c_2 e^{-i\beta t}), \end{aligned} \quad (1.8)$$

where $i = \sqrt{-1}$ is the *imaginary number*. Using Euler's formula, $e^{\pm i\theta} = \cos\theta \pm i\sin\theta$, Equation 1.8 becomes[2]

$$y(t) = e^{\alpha t}(\bar{c}_1 \cos\beta t + \bar{c}_2 \sin\beta t). \quad (1.9)$$

Note that c_1 and c_2 are complex numbers, and $\bar{c}_1$ and $\bar{c}_2$ are real numbers. In both Equations 1.8 and 1.9, $y(t)$ is a real value time–history, as required by the physics of the problem.

---

[2] We will see much more detail on this case in Chapters 2 and 3 for single degree of freedom systems.

1.2. ORDINARY DIFFERENTIAL EQUATIONS

Another way of writing Equation 1.8, which is convenient in vibration studies, is
$$y(t) = Ae^{\alpha t}\sin(\beta t + \phi),$$
where $A = \sqrt{\bar{c}_1^2 + \bar{c}_2^2}$, and $\phi = \arctan(\bar{c}_1/\bar{c}_2)$. A is found to be the vibration amplitude and ϕ the phase angle. This last form is convenient from a physical perspective because it separates the exponential part (which is usually a decaying envelope) from the purely oscillatory part. More details are provided in Chapters 2–3.

While it has not been shown, all arbitrary constants are determined by satisfying the given initial conditions, as will be demonstrated in the examples below.

Example 1.3 Solve
$$2\ddot{y} - 5\dot{y} - 3y = 0, \quad y(0) = 1, \quad \dot{y}(0) = 10.$$

Assume $y(t) = e^{rt}$. Differentiating and substituting into the governing equation leads to the characteristic equation $2r^2 - 5r - 3 = 0$, with roots $r_{1,2} = 3, -1/2$. The solution is then
$$y(t) = c_1 e^{3t} + c_2 e^{-t/2}.$$

Constants c_1 and c_2 are evaluated by satisfying the initial conditions, which results in the two algebraic equations
$$\begin{aligned} y(0) = 1 &= c_1 + c_2 \\ \dot{y}(0) = 10 &= 3c_1 - \frac{c_2}{2}. \end{aligned}$$

Solving these leads to $c_1 = 3$ and $c_2 = -2$. ∎

Example 1.4 Solve
$$\ddot{y} - 10\dot{y} + 25y = 0.$$
As above, find the characteristic equation $r^2 - 10r + 25 = 0$, with two equal roots $r_{1,2} = 5$. The solution is then
$$y(t) = c_1 e^{5t} + c_2 t e^{5t}.$$

Note that $y(t)$ increases without bounds because of the second term in the solution. ∎

Example 1.5 Solve

$$\ddot{y} + \dot{y} + y = 0, \quad y(0) = 1, \quad \dot{y}(0) = 0.$$

The characteristic equation is $r^2 + r + 1 = 0$, with two complex conjugate roots $r_{1,2} = (-1 \pm i\sqrt{3})/2$. The complete solution is then

$$\begin{aligned} y(t) &= c_1 \exp\left(-0.5 + i\frac{\sqrt{3}}{2}\right)t + c_2 \exp\left(-0.5 - i\frac{\sqrt{3}}{2}\right)t \\ &= e^{-0.5t}[\bar{c}_1 \cos(\frac{\sqrt{3}}{2}t) + \bar{c}_2 \sin(\frac{\sqrt{3}}{2}t)]. \end{aligned}$$

Evaluate the constants of integration by satisfying the given initial conditions:

$$\begin{aligned} y(0) &= 1 = e^0[\bar{c}_1 \cos(0) + \bar{c}_2 \sin(0)] \to \bar{c}_1 = 1 \\ \dot{y}(0) &= 0 = -0.5\bar{c}_1 + \frac{\sqrt{3}}{2}\bar{c}_2 \to \bar{c}_2 = \frac{\sqrt{3}}{3}. \end{aligned}$$

The homogeneous solution is then

$$y(t) = e^{-0.5t}\left[\cos\left(\frac{\sqrt{3}}{2}t\right) + \frac{\sqrt{3}}{3} \sin\left(\frac{\sqrt{3}}{2}t\right)\right].$$

∎

Homogeneous Solution: An Aside on the Assumed Solution

Before proceeding with solving the particular solution, it is worth summarizing for the reader how we solve differential equations. Where did the assumption come from that *sines*, *cosines*, or the *exponential* are valid solutions to harmonic equations? The key is to remember that these functions are but a shorthand notation for *power series expansions*. Consider the differential equation

$$\ddot{x}(t) + n^2 x(t) = 0.$$

Not knowing the solution, one may generally assume that $x(t)$ can be written as a power series in t:

$$x(t) = a_0 + a_1 t + a_2 t^2 + a_3 t^3 + \cdots.$$

Then,

$$\ddot{x}(t) = 2a_2 + 6a_3 t + 12a_4 t^2 + \cdots.$$

1.2. ORDINARY DIFFERENTIAL EQUATIONS

Substitute these into the ordinary differential equation to derive expressions for the constants a_i,

$$(2a_2 + 6a_3 t + 12 a_4 t^2 + \cdots) + n^2(a_0 + a_1 t + a_2 t^2 + a_3 t^3 + \cdots) = 0,$$

where the equality must be valid for all t. This equality can only hold if the coefficient of every power in t equals zero,

$$2a_2 + n^2 a_0 = 0$$
$$6a_3 + n^2 a_1 = 0$$
$$\cdots .$$

We can solve for the constants defining the power series solution,

$$a_2 = -n^2 a_0 / 2$$
$$a_3 = -n^2 a_1 / 6$$
$$a_4 = -n^2 a_2 / 12 = n^4 a_0 / 24,$$

with as many terms as necessary. Note that the only undetermined coefficients are a_0 and a_1, and combining like terms, we find our general solution

$$x(t) = a_0 \left(1 - \frac{n^2 t^2}{2} + \frac{n^4 t^4}{24} - \frac{n^6 t^6}{720} + \cdots \right)$$
$$+ a_1 \left(t - \frac{n^3 t^3}{6} + \frac{n^5 t^5}{120} - \cdots \right),$$

where a_0 and a_1 are the arbitrary constants required for a second order equation. The series in the first parenthesis is called $\cos nt$, and in the second pair of parenthesis $\sin nt$. We use the shorthand assumption that the homogeneous solution is the sum of a sine and a cosine term. Similarly, the series expansion called the *exponential* is

$$e^z = 1 + z + \frac{z^2}{2} + \frac{z^3}{6} + \frac{z^4}{24} + \cdots,$$

which is equal to the sum of the sine and cosine expansions, where $z = nt$.

The power series is the assumed solution to many important differential equations.

1.2.3 Particular Solution

Consider the following non–homogeneous equation

$$a\ddot{y} + b\dot{y} + cy = f(t),$$

where $f(t)$ may be any of the following functions;

- a constant, k
- a polynomial in t, for example, $c_n t^n + \cdots + c_1 t + c_0$
- an exponential function, $\exp(at)$
- a trigonometric function, $\sin \beta t$, $\cos \beta t$,

or a finite sum and/or product of these functions. As a general rule of thumb, in these instances the particular solution $y_p(t)$ has the same form as the forcing function;

- If $f(t)=$ constant, then $y_p(t) = A$, a constant.
- If $f(t)=$ polynomial, then $y_p(t) =$ a polynomial of the same order.
- If $f(t) = \exp(at)$, then $y_p(t) = c \exp(at)$.
- If $f(t) = \sin \beta t$ or $\cos \beta t$, then $y_p(t) = d_1 \sin \beta t + d_2 \cos \beta t$.

Example 1.6 Consider the following ordinary differential equation in x,

$$y''(x) + 3y'(x) + 2y(x) = 4x^2.$$

The particular solution alone is of the form $y_p(x) = ax^2 + bx + c$. To evaluate constants a, b and c, substitute y_p, y_p' and y_p'' into the governing equation,

$$2a + 3(2ax + b) + 2(ax^2 + bx + c) \equiv 4x^2;$$
$$\text{then} \quad (2a + 2c + 3b) + (6a + 2b)x + (2a)x^2 \equiv 4x^2.$$

By comparing the coefficients of equal powers of x, we have the three equations necessary for solving for the three unknowns:

$$2a + 2c + 3b = 0$$
$$6a + 2b = 0$$
$$2a = 4.$$

Solving these equations yields: $a = 2$, $b = -6$, and $c = 7$. The particular solution is then $y_p(x) = 2x^2 - 6x + 7$. ∎

Example 1.7 In the same manner as in the previous example, solve for the particular solution of

$$y''(x) - 3y'(x) = 4 \sin x.$$

1.2. ORDINARY DIFFERENTIAL EQUATIONS

Assume a particular solution of the form $y_p(x) = a \sin x + b \cos x$. Differentiate and substitute as before; equate sine and cosine terms:

$$-3a - b = 0$$
$$-a + 3b = 4$$

to find $a = -2/5$, $b = 6/5$, and particular solution $y_p(x) = -(2/5)\sin x + (6/5)\cos x$. ∎

Example 1.8 Next consider a differential equation with two terms on the right hand side,

$$y''(x) + 8y(x) = 5 + 2e^{-x}.$$

The forcing function consists of a constant plus an exponential function. For this linear equation, proceed using superposition by assuming a similar particular solution $y_p(x) = A + Be^{-x}$. Following previous procedure, the two equations governing the values of A and B are $9B = 2$, $8A = 5$, with the result

$$y_p(x) = \frac{5}{8} + \frac{2}{9}e^{-x}.$$

∎

Example 1.9 This example involves a product non–homogeneous term,

$$y''(x) - 2y'(x) + y(x) = 10e^{-2x}\cos x.$$

Assume the solution

$$\begin{aligned} y_p(x) &= Ae^{-2x}(B\cos x + C\sin x) \\ &= e^{-2x}(a_1\cos x + a_2\sin x). \end{aligned}$$

The mechanics of substituting this assumed solution into the governing equation is identical to what we have already seen several times. ∎

Example 1.10 As a final example, consider the very important case where the question of linearly independent solutions arises. The differential equation

$$y''(x) + y(x) = \cos x$$

appears simple enough so that previous solution procedures would be directly applicable. Try

$$y_p(x) = A \sin x + B \cos x,$$

and differentiate and substitute as before to find

$$-A\sin x - B\cos x + A\sin x + B\cos x = \cos x$$
$$0 = \cos x,$$

which is not valid over all x. The problem lies with the fact that the chosen particular solution is not linearly independent of the homogeneous solution. This is similar to Case II above, and by analogy, the correct choice for the particular solution here is

$$y_p(x) = Ax\sin x + Bx\cos x.$$

When substituted and solved, the coefficients are found to be $A = 1/2$, $B = 0$, and

$$y_p(x) = \frac{1}{2}x\sin x.$$

An important conclusion here is that one solves the homogeneous equation first, and then solves the particular one. If the assumed particular solution contains a term that is present in the homogeneous solution, $y_p(x)$ is then multiplied by x until no similar terms exist. This problem will be revisited when resonance is studied in Chapter 2. ∎

1.3 Matrices

In this section, we review those aspects of matrix theory that are of use in vibration studies. A *matrix* **A** is a rectangular array of numbers generally arising as short–hand notation for a set of equations. If the *elements* of the matrix are *real* numbers, then the matrix is called a *real matrix*. In a matrix the number of columns is not generally the same as the number of rows. For our vibration applications, the number of rows n always equals the number of columns, resulting in a *square* n x n matrix of *order n*.

A general (n rows, m columns) matrix **A** is written

$$\mathbf{A} = \begin{bmatrix} a_{11} & a_{12} & \cdots & a_{1m} \\ a_{21} & a_{22} & \cdots & a_{2m} \\ \vdots & \vdots & & \vdots \\ a_{n1} & a_{n2} & \cdots & a_{nm} \end{bmatrix}, \quad (1.10)$$

where the matrix is n x m. This matrix is closely related to a *linear transformation* as given by the following n algebraic equations relating

1.3. MATRICES

$x_1, x_2, \ldots, x_m$ to $y_1, y_2, \ldots, y_n$;

$$a_{11}x_1 + a_{12}x_2 + \cdots + a_{1m}x_m = y_1 \qquad (1.11)$$
$$a_{21}x_1 + a_{22}x_2 + \cdots + a_{2m}x_m = y_2 \qquad (1.12)$$
$$\cdots \quad \cdots$$
$$a_{n1}x_1 + a_{n2}x_2 + \cdots + a_{nm}x_m = y_n. \qquad (1.13)$$

A square $n \times n$ matrix having nonzero elements in its main diagonal and a zero in each other location is called a *diagonal* matrix:

$$\mathbf{A} = \begin{bmatrix} a_{11} & 0 & \cdots & 0 \\ 0 & a_{22} & \cdots & 0 \\ \vdots & \vdots & \ddots & \vdots \\ 0 & 0 & \cdots & a_{nn} \end{bmatrix}.$$

The *identity* matrix $\mathbf{I}$ is a square matrix which has a zero in all locations except for the main diagonal, which are all unity:

$$\mathbf{A} = \begin{bmatrix} 1 & 0 & \cdots & 0 \\ 0 & 1 & \cdots & 0 \\ \vdots & \vdots & \ddots & \vdots \\ 0 & 0 & \cdots & 1 \end{bmatrix}.$$

In many matrix computations it is necessary to perform the operation of *matrix transpose*. For $n \times m$ matrix $\mathbf{A}$, if the rows and columns are interchanged, the resulting $m \times n$ matrix $\mathbf{A}^T$ is called the *transpose* of $\mathbf{A}$, and is given by

$$\mathbf{A}^T = \begin{bmatrix} a_{11} & a_{21} & \cdots & a_{n1} \\ a_{12} & a_{22} & \cdots & a_{n2} \\ \vdots & \vdots & \ddots & \vdots \\ a_{1m} & a_{2m} & \cdots & a_{nm} \end{bmatrix}.$$

Several important identities can be established using the transpose operation:

$$\begin{aligned} (\mathbf{A}^T)^T &= \mathbf{A} \\ (\mathbf{A} + \mathbf{B})^T &= \mathbf{A}^T + \mathbf{B}^T \\ (\mathbf{AB})^T &= \mathbf{B}^T \mathbf{A}^T. \end{aligned}$$

A *symmetric* matrix is one which is equal to its transpose,

$$\mathbf{A}^T = \mathbf{A},$$

or $a_{ji} = a_{ij}$.

1.3.1 Matrix Operations

We review the rules of some basic matrix operations. The *sum* of two n x m matrices $\mathbf{A}$ and $\mathbf{B}$ is a matrix where each element is the sum of the corresponding elements of the original matrices,

$$\mathbf{A} + \mathbf{B} = \begin{bmatrix} a_{11} + b_{11} & a_{12} + b_{12} & \cdots & a_{1m} + b_{1m} \\ a_{21} + b_{21} & a_{22} + b_{22} & \cdots & a_{2m} + b_{2m} \\ \vdots & \vdots & & \vdots \\ a_{n1} + b_{n1} & a_{n2} + b_{n2} & \cdots & a_{nm} + b_{nm} \end{bmatrix}.$$

Note that this operation is valid only if both matrices are of the same dimension. One can define the *difference of two matrices* in the same manner as above by replacing all sums by differences.

The *multiplication of a matrix by a scalar* results in a new matrix, $k\mathbf{A}$, where each element a_{ij} is multiplied by k.

The *multiplication of two matrices* is a more intricate operation. Let us multiply the n x m matrix $\mathbf{A}$ by the m x r matrix $\mathbf{B}$,

$$\mathbf{C} = \mathbf{AB} = \begin{bmatrix} a_{11} & a_{12} & \cdots & a_{1m} \\ a_{21} & a_{22} & \cdots & a_{2m} \\ \vdots & \vdots & & \vdots \\ a_{n1} & a_{n2} & \cdots & a_{nm} \end{bmatrix} \begin{bmatrix} b_{11} & b_{12} & \cdots & b_{1r} \\ b_{21} & b_{22} & \cdots & b_{2r} \\ \vdots & \vdots & & \vdots \\ b_{m1} & b_{m2} & \cdots & b_{mr} \end{bmatrix}.$$

The operations required to obtain the element in row 1, column 1 of the new matrix are:

$$(a_{11} \times b_{11}) + (a_{12} \times b_{21}) + (a_{13} \times b_{31}) + \cdots + (a_{1m} \times b_{m1}). \quad (1.14)$$

In order for such a set of operations to work, the matrices must be *commensurate*. This means that the first matrix has as many *columns* as the second matrix has *rows*. Performing all the multiplications and additions in Equation 1.14 results in a number represented by c_{11}. Similarly, the remaining elements of matrix $\mathbf{C}$ are obtained from the general relation

$$c_{ik} = \sum_{j=1}^{m} a_{ij} b_{jk}.$$

1.3. MATRICES

The resulting matrix $\mathbf{C}$ will be of size $(n \times m) \times (m \times r) = n \times r$, where

$$\mathbf{AB} \neq \mathbf{BA}.$$

Note the following on the *cancellation of matrices*. Consider the two *singular* matrices,[3]

$$\mathbf{A} = \begin{bmatrix} 2 & 1 \\ 6 & 3 \end{bmatrix} \neq \begin{bmatrix} 0 & 0 \\ 0 & 0 \end{bmatrix} = \mathbf{0}$$

$$\mathbf{B} = \begin{bmatrix} 1 & -2 \\ -2 & 4 \end{bmatrix} \neq \mathbf{0}.$$

But

$$\mathbf{AB} = \begin{bmatrix} 2 & 1 \\ 6 & 3 \end{bmatrix} \begin{bmatrix} 1 & -2 \\ -2 & 4 \end{bmatrix} = \begin{bmatrix} 0 & 0 \\ 0 & 0 \end{bmatrix} = \mathbf{0}.$$

Therefore, we see that $\mathbf{AB} = \mathbf{0}$ implies neither $\mathbf{A} = \mathbf{0}$ nor $\mathbf{B} = \mathbf{0}$. It is possible that either matrix is zero, but if they are not zero, then both must be singular.

1.3.2 Determinant and Matrix Inverse

The determinant of the square matrix $\mathbf{A}$

$$\det \mathbf{A} = |\mathbf{A}| = \begin{vmatrix} a_{11} & a_{12} & \cdots & a_{1n} \\ a_{21} & a_{22} & \cdots & a_{2n} \\ \vdots & \vdots & & \vdots \\ a_{n1} & a_{n2} & \cdots & a_{nn} \end{vmatrix}, \qquad (1.15)$$

is defined as *a number obtained as the sum of all possible products in each of which there appears one and only one element from each row and each column, each such product being prefixed by a plus or minus sign*.[4] $|\mathbf{A}|$ is said to be of *order n*. By following the rules given below for evaluating the determinant of a matrix, we obtain a unique value, or number, if the elements of the matrix are numbers.

Denote $|M_{ij}|$ as the *minor determinant* of $\mathbf{A}$, which is obtained by taking the determinant of $\mathbf{A}$ after removing row i and column j. The order of $|M_{ij}|$ is $(n-1)$. Attaching a sign (plus or minus) to this minor determinant,

$$|C_{ij}| = (-1)^{i+j} |M_{ij}|,$$

---

[3] A singular matrix is one with a zero determinant, which is defined in the following section.

[4] F.B. Hildebrand, **Methods of Applied Mathematics**, Prentice–Hall, 1965, pg. 10.

results in the term known as the *signed minor determinant*. This corresponds to the matrix element a_{ij}, otherwise known as the *cofactor* of a_{ij}. With these definitions, the value of the determinant can be obtained by "expanding" the determinant in terms of the cofactors for any row r

$$|\mathbf{A}| = \sum_{s=1}^{n} a_{ij}|C_{ij}|.$$

A similar expansion is possible for any column. These expansions are called *Laplace expansions*. For example, Equation 1.15 can be evaluated as follows for the case where $n = 3$ by expanding across row 1:

$$\begin{aligned}
\begin{vmatrix} a_{11} & a_{12} & a_{13} \\ a_{21} & a_{22} & a_{23} \\ a_{31} & a_{32} & a_{33} \end{vmatrix} &= a_{11}|C_{11}| + a_{12}|C_{12}| + a_{13}|C_{13}| \\
&= a_{11} \begin{vmatrix} a_{22} & a_{23} \\ a_{32} & a_{33} \end{vmatrix} - a_{12} \begin{vmatrix} a_{21} & a_{23} \\ a_{31} & a_{33} \end{vmatrix} \\
&\quad + a_{13} \begin{vmatrix} a_{21} & a_{22} \\ a_{31} & a_{32} \end{vmatrix} \\
&= a_{11}(a_{22}a_{33} - a_{23}a_{32}) - a_{12}(a_{21}a_{33} - a_{23}a_{31}) \\
&\quad + a_{13}(a_{21}a_{32} - a_{22}a_{31}).
\end{aligned} \quad (1.16)$$

Computations are greatly reduced by choosing the row containing the most zeros. Looking carefully at these operations, it can be seen that

$$|\mathbf{A}| = |\mathbf{A}^T|.$$

If the value of $|\mathbf{A}|$ is equal to zero, then matrix $\mathbf{A}$ is said to be *singular*. It is easier to evaluate the determinant than it is to explain how to do it.

By definition, the *adjoint* of square matrix $\mathbf{A}$ is the transpose of the matrix obtained from $\mathbf{A}$ by replacing each element of $\mathbf{A}$ by its cofactor. It can be shown that

$$\mathbf{A}(\text{adj } \mathbf{A}) = (\text{adj } \mathbf{A})\mathbf{A} = |\mathbf{A}|\mathbf{I}. \quad (1.17)$$

For example, the 2 x 2 determinants in Equation 1.16 are the *first column* of the adjoint of $\mathbf{A}$. We need to go through the same procedure for rows two and three of $\mathbf{A}$ in order to derive the second and third *columns* of the

1.3. MATRICES

adjoint matrix. For matrix

$$\mathbf{A} = \begin{bmatrix} 1 & 2 & 0 \\ 3 & -1 & -2 \\ 1 & 0 & -3 \end{bmatrix},$$

$$\text{adj } \mathbf{A} = \begin{bmatrix} 3 & 6 & -4 \\ 7 & -3 & 2 \\ 1 & 2 & -7 \end{bmatrix},$$

and

$$\mathbf{A}(\text{adj } \mathbf{A}) = 17 \begin{bmatrix} 1 & 0 & 0 \\ 0 & 1 & 0 \\ 0 & 0 & 1 \end{bmatrix},$$

where $|\mathbf{A}| - 17$.

Another very important matrix operation is known as the *inverse* $\mathbf{A}^{-1}$ of matrix $\mathbf{A}$. The inverse can be interpreted as the matrix counterpart to dividing by a number. Without proof, using Equation 1.17, we have the expression for the inverse of a nonsingular matrix, $\mathbf{A}^{-1} = \text{adj } \mathbf{A}/|\mathbf{A}|$.

Cramer's Rule

As an aside, we refer back to Equations 1.11–1.13, which are a set of n equations with m unknowns, x_i, $i = 1, \ldots, m$. If the number of equations equals the number of unknowns, then matrix $\mathbf{A}$ in Equation 1.10 becomes square, and a unique solution is possible. Many numerical approaches can be used to solve such a set of equations. *Cramer's rule*, for example, can be used, but more efficient numerical approaches exist.

1.3.3 Eigenvalues and Eigenvectors of a Square Matrix

We will be discussing *eigenvalues* and *eigenvectors* very often in subsequent chapters. There, we will be more interested in the physical interpretations for these matrix properties.[5] Here they are viewed formally as part of matrix theory.

For the $n \times n$ matrix $\mathbf{A}$, the determinant

$$|\mathbf{A} - \lambda \mathbf{I}|$$

---
[5] The eigenvalues will be found in Chapter 8 to be the squares of the frequencies of vibration, and the eigenvectors the respective modes of vibration.

is called the *characteristic polynomial* of $\mathbf{A}$, and is an nth order polynomial. The *characteristic equation* is obtained by setting this polynomial to zero:

$$|\mathbf{A} - \lambda\mathbf{I}| = \begin{vmatrix} a_{11} - \lambda & a_{12} & \cdots & a_{1n} \\ a_{21} & a_{22} - \lambda & \cdots & a_{2n} \\ \vdots & \vdots & & \vdots \\ a_{n1} & a_{n2} & \cdots & a_{nn} - \lambda \end{vmatrix}$$

$$= (-1)^n (\lambda^n + c_1\lambda^{n-1} + c_2\lambda^{n-2} + \cdots$$
$$+ c_{n-2}\lambda^2 + c_{n-1}\lambda + c_n) = 0. \tag{1.18}$$

This equation has n *characteristic roots* that are the eigenvalues of the square matrix $\mathbf{A}$. Even though the matrix has real elements, polynomial Equation 1.18 with real coefficients c_i may have *complex conjugate* pairs of roots $\alpha \pm i\beta$. These complex conjugate roots are of primary interest in a vibration analysis.

Note that for a diagonal matrix, the n diagonal elements are the n eigenvalues of the matrix. Also of interest is the fact that *the eigenvalues of matrix $\mathbf{A}$ are equal to the eigenvalues of the inverse matrix $\mathbf{A}^{-1}$*.

Once the eigenvalues λ_i are found, then any nonzero vector $\mathbf{v}_i$ satisfying

$$\mathbf{A}\mathbf{v}_i = \lambda_i\mathbf{v}_i \tag{1.19}$$

is an eigenvector. There is generally one eigenvector for each eigenvalue, but there are degenerate cases. From Equation 1.19, we see that if $\mathbf{v}_i$ is an eigenvector, then for any nonzero scalar α, $\alpha\mathbf{v}_i$ is also an eigenvector since $(\mathbf{A} - \lambda_i)\mathbf{v}_i = 0$.

1.4 Transition

With the conclusion of this chapter, the introductory and review discussions have ended. We are now ready to begin the study of vibratory systems and structures. Our primary goal will always be the physical understanding that allows us to formulate the simplest mathematical model that includes the key system characteristics. The solution of the governing equations always leads to an examination of the physical nature of system behavior. Such an understanding is absolutely necessary in order that the analyst/designed develops an intuition regarding how the structure should behave. Such intuition is critical and cannot be replaced by or relegated to a computer.

Chapter 2

Single Degree of Freedom Vibration: An Introduction to Discrete Models

"The simplest model, exactly known."

We begin our exploration of dynamical systems at the simplest level, where one coordinate completely describes the motion of the structure. This is called a single degree of freedom model.

The study of single degree of freedom systems is very useful since many of the key principles developed here are of use throughout all of dynamics and vibration. One can explore many concepts with such simple models, and learn how to work with their governing equations.

For example, the concept of oscillation frequency is one which will be visited many times in the subsequent chapters. Here it will be introduced for the first time as the most important measure of vibratory motion, and applied to the simplest of all structures. In this chapter, we provide background and learn how to model and solve for the response of very simple oscillators where there is no loss of energy through damping. In Chapter 3 damping is introduced, as are more realistic force models and more complex vibratory environments. This chapter and the next one can be really viewed as parts one and two of the study of single degree of freedom oscillators. The split has been created for pedagogical reasons. First, a few application examples are introduced to motivate the subsequent reading.

2.1 Example Problems and Motivation

Interesting and important vibrating systems are the *bridge, satellite transport, rotating machinery,* and the *rocket ship*; two of these are introduced below in detail. Our studies begin with relatively simple models. The examples are single degree of freedom idealizations. The student may wonder how it is possible to derive anything of use about such complex systems utilizing only a single degree of freedom. While it is true that for a complete detailed analysis one coordinate will not do, it is however possible to estimate the gross behavior of such systems for preliminary design purposes.

2.1.1 Transport of a Satellite

Vibration studies have as their primary goal the isolation of machine or structure from its potentially destructive effects. Since isolation is only partially successful, the secondary goal is to design structures so that they are less susceptible to oscillation. Both goals are typically part of any analysis and design.

A specific application is the analysis, design, and testing of shipping containers. Satellites cannot withstand the rigors of transport from the manufacturing site to the launch pad, and so it becomes necessary to design a container that, among many other purposes, isolates its contents from a spectrum of forcing. This problem is considered in more detail at the end of this chapter. Satellites are designed to orbit the Earth in a microgravity, atmosphere–free environment. However, the designer must also consider what the satellite will experience from its point of departure at the factory until it is operational in orbit. The satellite cannot be efficiently designed to simultaneously survive shipment as well as being shot into orbit for space operations. Competing design constraints govern. Thus, the design requirement for a shipping container is to deliver the satellite in perfect condition to the launch pad. In this way, the satellite can be optimally designed for its final destination in orbit around Earth.

2.1.2 Rocket Ship

A rocket ship is an example of a system with a *variable mass*; mass varies with time. This is because the rocket is propelled into space by *thrust* created by exhausting its fuel, resulting in a decreasing fuel mass within the fuel tanks. Since most of us have not seen a rocket, much less been on one during flight, an analogy with a firing cannon is useful to help understand how thrust is generated. As the cannon is fired, it recoils in the direction opposite of the cannonball, satisfying the *principle of conservation of linear*

2.1. EXAMPLE PROBLEMS AND MOTIVATION 41

momentum. If it is fired rapidly, it picks up speed with each shot. The cannon expels mass in one direction and moves in the opposite direction, moving faster when it shoots out more mass over a relatively short period of time.

The rocket is a container within which fuel is burned and ejected at high speed. The rocket accelerates in a direction opposite that of the fuel ejection. The acceleration is a function of the rate of fuel consumption and the speed with which combustion gases are ejected. The total mass of the rocket varies (decreases) with time.

When we consider this problem quantitatively in a later section, we will use the momentum form of Newton's second law of motion to derive the variable mass equation of motion, and we will briefly discuss how to evaluate the thrust.

Newton

Sir Isaac Newton lived during 1643–1727 and was born in Woolsthorpe, England. He died in London, England. Newton's life can be divided into three quite distinct periods. The first is his boyhood days from 1643 up to his graduation in 1669. The second period, from 1669 to 1687, was the highly productive period in which he was Lucasian professor at Cambridge. The third period (nearly as long as the other two combined) saw Newton as a highly paid government official in London with little further interest in mathematics.

Isaac Newton was born in the manor house of Woolsthorpe, near Grantham in Lincolnshire. Although he was born on Christmas Day 1642, the date given on this card is the Gregorian calendar date. (The Gregorian calendar was not adopted in England until 1752.) Newton came from a family of farmers but never knew his father who died before he was born. His mother remarried, moved to a nearby village, and left him in the care of his grandmother. Upon the death of his stepfather in 1656, Newton's mother removed him from grammar school in Grantham where he had shown little promise in academic work. His school reports described him as 'idle' and 'inattentive'. An uncle decided that he should be prepared for the university, and he entered his uncle's old College, Trinity College, Cambridge, in June 1661. Newton's aim at Cambridge was a law degree.

Instruction at Cambridge was dominated by the philosophy of Aristotle but some freedom of study was allowed in the third year of study. Newton studied the philosophy of Descartes, Gassendi, and Boyle. The new algebra and analytical geometry of Viete, Descartes, and Wallis, and the mechanics of the Copernican astronomy of Galileo attracted him. Newton's talent began to emerge on the arrival of Barrow to the Lucasian chair at Cambridge.

Newton's scientific genius emerged suddenly when the plague closed the University in the summer of 1665 and he had to return to Lincolnshire. There, in a period of less than two years, while Newton was still under 25 years old, he began revolutionary advances in mathematics, optics, physics, and astronomy. While Newton remained at home he laid the foundation for differential and integral calculus several years before its independent discovery by Leibnitz. The 'method of fluxions', as he termed it, was based on his crucial insight that the integration of a function is merely the inverse procedure to differentiating it. Taking differentiation as the basic operation, Newton produced simple analytical methods that unified many separate techniques previously developed to solve apparently unrelated problems such as finding areas, tangents, the lengths of curves, and their maxima and minima. Newton's *De Methodis Serierum et Fluxionum* was written in 1671 but Newton failed to get it published and it did not appear in print until John Colson produced an English translation in 1736.

Barrow resigned the Lucasian chair in 1669 recommending that Newton (still only 27 years old) be appointed in his place. Newton's first work as Lucasian Professor was on optics. He had reached the conclusion during the two plague years that white light is not a simple entity. Every scientist since Aristotle had believed this but the chromatic aberration in a telescope lens convinced Newton otherwise. When he passed a thin beam of sunlight through a glass prism Newton noted the spectrum of colors that was formed. Newton argued that white light is really a mixture of many different types of rays which are refracted at slightly different angles, and that each different type of ray produces a given spectral color. Newton was led by this to the erroneous conclusion that telescopes using refracting lenses would always suffer chromatic aberration. He therefore proposed and constructed a reflecting telescope. Newton was elected a fellow of the Royal Society in 1672 after donating a reflecting telescope.

Also in 1672 Newton published his first scientific paper on light and color in the Philosophical Transactions of the Royal Society. Newton's paper was well received but Hooke and Huygens objected to Newton's attempt to prove by experiment alone that light consists in the motion of small particles rather than waves. Perhaps because of Newton's already high reputation, his corpuscular theory reigned until the wave theory was revived in the 19th C. Newton's relations with Hooke deteriorated and he turned in on himself and away from the Royal Society. He delayed the publication of a full account of his optical researches until after the death of Hooke in 1703. Newton's *Opticks* appeared in 1704. It dealt with the theory of light and color and with *(i)* investigations of the colors of thin sheets, *(ii)* 'Newton's rings,' and *(iii)* diffraction of light. To explain some of his observations he had to use a wave theory of light in conjunction to his corpuscular theory.

Newton's greatest achievement was his work in physics and celestial mechanics, which culminated in the theory of universal gravitation. By 1666 Newton had early versions of his three laws of motion. He had also discovered the law giving the centrifugal force on a body moving uniformly in a circular path. However, he did not have a correct understanding of the mechanics of circular motion. Newton's novel idea of 1666 was to imagine that the Earth's gravity influenced the Moon, counter–balancing its centrifugal force. From his law of centrifugal force and Kepler's third law of planetary motion, Newton deduced the inverse–square law.

In 1679 Newton applied his mathematical skill to proving a conjecture of Hooke's,

2.1. EXAMPLE PROBLEMS AND MOTIVATION

showing that if a body obeys Kepler's second law then the body is being acted upon by a centripetal force. This discovery showed the physical significance of Kepler's second law. In 1684, Halley, tired of Hooke's boasting, asked Newton whether he could prove Hooke's conjecture and was told that Newton had solved the problem five years before but had now mislaid the proof. At Halley's urging, Newton reproduced the proofs and expanded them into a paper on the laws of motion and problems of orbital mechanics. Halley persuaded Newton to write a full treatment of his new physics and its application to astronomy. Over a year later (1687) Newton published the *Philosophiae Naturalis Principia Mathematica* or *Principia* as it is always known. The *Principia* is recognized as the greatest scientific book ever written. Newton analyzed the motion of bodies in resisting and non–resisting media under the action of centripetal forces. The results were applied to orbiting bodies, projectiles, pendula, and free–fall near the Earth. He further demonstrated that the planets were attracted toward the Sun by a force varying as the inverse square of the distance and generalized that all heavenly bodies mutually attract one another. Further generalization led Newton to the law of universal gravitation: *all matter attracts all other matter with a force proportional to the product of their masses and inversely proportional to the square of the distance between them*. Newton explained a wide range of previously unrelated phenomena: the eccentric orbits of comets; the tides and their variations; the precession of the Earth's axis; and motion of the Moon as perturbed by the gravity of the Sun.

After suffering a nervous breakdown in 1693, Newton retired from research to take up a government position in London becoming Warden of the Royal Mint (1696) and Master (1699). In 1703 he was elected president of the Royal Society and was re–elected each year until his death. He was knighted in 1708 by Queen Anne, the first scientist to be so honored for his work.

Leibnitz

Gottfried Wilhelm von Leibnitz lived during 1646–1716 and was born in Leipzig, Saxony, now Germany. He died in Hanover, now Germany. Leibnitz developed the present day notation for the differential and integral calculus.

Leibnitz was the son of a professor of moral philosophy at Leipzig. He taught himself Latin and some Greek by age 12 so that he might read his father's books. From 1661 to 1666 he studied law at the University of Leipzig. In 1666 his continuation to a doctorate course was refused and he went to the University of Altdorf. He received doctorate in law in 1667. Leibnitz declined a chair at Altdorf because he had "very different things in view." He continued a law career in residence at the courts of Mainz (until 1672). He visited Paris in 1672 to try to dissuade Louis XIV from attacking German areas. Leibnitz remained in Paris until 1676, where he continued to practice law. However, in Paris he studied mathematics and physics under Christian Huygens. It was during this period that the basic features of his version of the calculus were developed. The rest of his life was spent at Hanover from 1676 until his death.

By 1673 he was still struggling to develop a good notation for his calculus and his first calculations were clumsy. On November 21, 1675, he wrote a manuscript using the $\int f(x)dx$ notation for the first time. In the same manuscript the product rule for differentiation is given and the quotient rule first appeared in July 1677. By November, 1676 he discovered the familiar $d(x^n) = nx^{n-1}dx$ for both integral and fractional n. Newton was to claim, with justification, that 'not a single previously unsolved problem was solved' here, but the formalism of Leibnitz's approach was to prove vital in the development of the calculus. Leibnitz never thought of the derivative as a limit. This first appears in the work of d'Alembert.

In 1684, Leibnitz published details of his differential calculus in "Nova Methodus pro Maximis et Minimis, Itemque Tangentibus" in *Acta Eruditorum*, a journal established in Leipzig two years earlier. The paper contained the familiar d notation for derivative, the rules for computing the derivatives of powers, products and quotients. However it contained no proofs and the Bernoullis called it an enigma rather than an explanation. In 1686 Leibnitz published, in *Acta Eruditorum*, a paper dealing with the integral calculus with the first appearance in print of the $\int$ notation. Newton's *Principia* appeared the following year. Newton's 'method of fluxions' was written in 1671, but Newton failed to get it published and it did not appear in print until John Colson produced an English translation in 1736. This resulted in a dispute with Leibnitz.

Leibnitz founded the Berlin Academy in 1700 and was its first president. He became more and more of a recluse in his later years.

2.2 Mathematical Modeling: Deterministic

Mathematical modeling is the process of attaching a mathematical framework to the physical behavior of a component or system, as discussed in Chapter 0. Such a framework is essential since it provides the analyst with a systematic approach for specifying constitutive behavior, deriving equations of motion, and classifying forces.

2.2.1 Problem Idealization and Formulation

Most problems of engineering interest are exceptionally complicated if too many aspects of the problem are considered and included in the formulation. However, not all aspects are of equal importance to how the system responds to its environment. Only a few aspects will generally turn out to be important to the system response, while many others can be safely ignored in a preliminary modeling effort.

A judicious choice of those few key parameters of a system and its environment will enable the analyst to derive an adequate physical model of much simpler mathematical complexity. This process of *idealization* is one

2.2. MATHEMATICAL MODELING: DETERMINISTIC

which cannot be taught with a set of rules. In a sense, idealization is an art with which one becomes more proficient given more practice. The idealized model is a good one if it can replace the full model in a study. One is confident of the quality of the idealized model if it has a predictive capability, that is, if it can be used to duplicate experimental data. The range of situations that it can predict are a direct result of the nature and number of idealizations made in its derivation. The accuracy of the model is directly related to our understanding of the physical processes being modeled. The process of experimental verification and idealization is iterative and generally requires several stages of model refinement. For example, if we know nothing about how a structure will behave, we test it. From the test, information or data are gathered that helps the analyst develop a mathematical model of the structure. This model is then used to predict structural behavior for which data do not exist. To gain confidence in the model, additional tests are run for additional comparisons. Only after a series of such iterations between testing and analysis can we be confident that the model is representative of the actual structure and its environment.

The Linear Approximation

The single most important consideration regarding how we expect a structure to behave is whether it is linear or nonlinear. We discussed the concept of *linearization* in Section 1.1, with the idea of expanding a nonlinear function in a Taylor series, and retaining only the linear components.

Most physical systems behave linearly over some range, even if it is a small range. As an example, for a mass hanging on a spring, a small vertical displacement results in an oscillation about the equilibrium. The spring, having been stretched a *small* amount, behaves linearly so that the plot in Figure 2.1 of force versus displacement about equilibrium is a straight line. However, if the mass undergoes a large displacement, then the spring no longer responds linearly with displacement, but rather *nonlinearly*, as shown.

Why is linearity of such importance? Primarily because it leads to the *principle of linear superposition*. This principle allows us to add solutions. Suppose the structure is loaded by force $F_1(t)$ and responds as $x_1(t)$. Subsequently the structure is loaded by force $F_2(t)$ with response $x_2(t)$. This information, along with superposition, implies that if the structure is loaded with the combined force $F_1(t) + F_2(t)$, it will respond as $x_1(t) + x_2(t)$. This remarkable result is due to the linearity of the structure, even if $F_1(t)$ and $F_2(t)$ are nonlinear functions. If the structure possessed nonlinear stiffness or damping, then superposition would not be valid.

Where a governing equation is nonlinear, it is sometimes *linearized* it

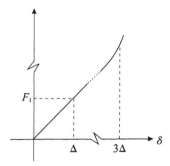

Figure 2.1: Force-displacement curve for spring.

according to the procedures discussed earlier using the Taylor series. This is done with the recognition that any results obtained using the linearized equation are valid only while the structure undergoes small displacements and oscillations.

Taylor

Brook Taylor lived during 1685–1731 and was born in Edmonton, England. He died in London, England. In 1708, Taylor produced a solution to the problem of the center of oscillation which, since it went unpublished until 1714, resulted in a priority dispute with Johann Bernoulli. Taylor's *Methodus Incrementorum Directa et Inversa* (1715) added to mathematics a new branch now called the "calculus of finite differences" and he invented integration by parts. It also contained the celebrated formula known as Taylor's expansion, the importance of which remained unrecognized until 1772 when Lagrange proclaimed it the basic principle of the differential calculus. Taylor also devised the basic principles of perspective in *Linear Perspective* (1715). In addition, the first general treatment of the vanishing points are given. Taylor gives an account of an experiment to discover the law of magnetic attraction (1715) and an improved method for approximating the roots of an equation by giving a new method for computing logarithms (1717). Taylor was elected a Fellow of the Royal Society in 1712 and was appointed in that year to the committee for adjudicating the claims of Newton and of Leibnitz, who both claimed to be the inventors of the calculus.

2.2.2 Mass, Damping, and Stiffness

How a mechanical system responds to its loading environment depends on three key parameters that define it in a dynamic sense. These are its *mass*, *damping*, and *stiffness*. Of the three, damping is especially difficult to

2.2. MATHEMATICAL MODELING: DETERMINISTIC

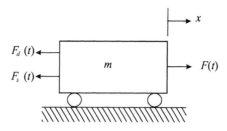

Figure 2.2: Forces on a free body mass. x is shown in the assumed positive direction of motion.

quantify, even for relatively simple structures.

In Section 0.6.2 preliminary ideas on mass, damping, and stiffness were initiated. A mathematical form is necessary to translate physical understanding into a quantitative, predictive tool. Newton's second law of motion relates the forces acting on a body to its acceleration, with the constant of proportionality being the body's mass. The forces acting on a body can be characterized as *external* or *internal* forces. External forces are the obvious ones, such as those due to wind, ocean, or impact. Internal forces are due to stiffness or damping. Refer to Figure 2.2. Stiffness forces are related to the elongation of a representative "spring" element: $F_s(t) = k_s x(t)$, where the initial deformation is taken to be zero for ease of notation. Damping forces are related to the relative velocity between a body and its attachment: $F_d(t) = c\dot{x}(t)$, where the "overdot" notation implies differentiation with respect to time, and where the velocity of the attachment point is assumed zero. The *damping constant c* has units of force per unit velocity. More realistic damping "laws" exist for complicated materials and systems. The difficulty with damping is that, unlike mass and stiffness, it is not generally an inherent characteristic of the system. Rather, damping forces depend on internal and external factors.[1] Four types of damping are:

- *viscous* damping, where the system vibrates as though in a fluid, resulting in an exponential decay in the amplitude of oscillation.

- *structural* damping, which results from within the structure, due to energy loss in the material or at joints.

---

[1] I recommend that you look up the paper by S.H. Crandall, "The Role of Damping in Vibration Theory", *J. Sound and Vibration*, (1970) **11**(1), 3–18. Another very useful reference is by C.W. Bert, "Material Damping: An Introductory Review of Mathematical Models, Measures and Experimental Techniques", *J. Sound and Vibration*, (1973) **29**(2), 129–153. Both these papers provide a significant discussion of the physical causes of damping and how it can be modeled mathematically.

- *Coulomb* damping, resulting from dry friction between the body and another with which it is in contact. The damping force is nearly constant and the structural behavior is characterized by a linear decay. Because of the discontinuity in this damping force, it is considered nonlinear.

- *negative* damping, where energy is added to the system rather than removed. Such systems can be unstable.

In this book, we completely focus on linear viscous damping models.

Electrical Analogy

Mechanical systems are sometimes studied using equivalent electrical circuits, from which experimental results are more conveniently taken. The equivalent electrical circuit is obtained by comparing the governing equations of the two systems and making them mathematically identical. Of course, the electrical analogy is only as good as the original mathematical model. The following example demonstrates the *current–force analogy*, which is a physical analogy.

Example 2.1 Electrical-Mechanical Analogy
To demonstrate the current–force analogy, consider the following two systems, one mechanical and the other electrical. See Figure 2.3. The equation of motion for the mechanical system is

$$m\ddot{x} + c\dot{x} + kx = f(t),$$

while the equation for the current in the circuit is

$$C\ddot{v} + \frac{1}{R}\dot{v} + \frac{1}{L}v = \frac{di}{dt}. \qquad (2.1)$$

These equations demonstrate an analogy between the integral of the voltage and displacement. If Equation 2.1 is integrated with respect to time, we obtain the current–force analogy,

$$C\dot{v} + \frac{v}{R} + \frac{1}{L}\int v\,dt = i(t).$$

2.2. MATHEMATICAL MODELING: DETERMINISTIC

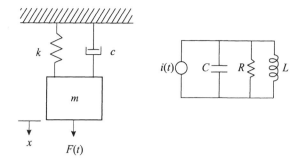

Figure 2.3: Mechanical-electrical analogous systems.

2.2.3 Sources of Deterministic Approximation

A deterministic property or parameter is one that is exact without variance or scatter. Such a variable would possess the same numerical value regardless of how many times it is measured. While slight variability is common and generally ignored for applications, sometimes a significant variability is inherent in some part of the problem and needs to be included in the problem formulation and solution.

Possible sources of uncertainties for a single degree of freedom system are the external forces and the system properties, for example, damping, stiffness, or mass. If all these are of sufficient accuracy, then one may assume that a *deterministic* governing equation will provide accurate results.[2]

In some applications, it is easy to realize that the loading has a *broad frequency band*, that is, it contains energies at many frequencies, or that the amplitudes of the loading are *widely distributed*. Testing is necessary, both to establish that the mathematical model is truly representative of the actual physics of the structure, and also to gather data on the loading frequencies, amplitudes and material properties of the structure. Where such variability exists, *nondeterministic* or probabilistic methods are necessary. We will study these in some detail in Chapter 5 and subsequent chapters.

2.2.4 Dimensional Analysis

Physical variables possess units. Therefore, for any equation to be true it must have consistent units. Just as the numbers on both sides of the equal

---

[2]In fact, it was not until the 1950s, when the aerospace industry began to design aircraft which were to be subjected to essentially unpredictable aerodynamic forces, that uncertainties regarding such forces had to be explicitly included in the vibration analysis. Thus, the discipline of *random vibration* was born.

sign must be equal to each other, so must be the units. *Dimensional analysis* is a technical discipline that seeks to identify significant *dimensionless ratios* or variables in a problem. These dimensionless groups are used because relationships derived for such ratios are independent of a particular set of units. Sometimes the process of identifying these groups leads the analyst to important but unnoticed relationships between particular variables. The *Reynolds number* is one such parameter in fluid mechanics, and the *viscous damping factor* introduced in the next section, is a very important one in vibration studies.

In addition, it is possible to use the units of variables that we expect to be important in a particular analysis to derive analytical relationships between those variables. This may appear remarkable, but as mentioned above, the units must also satisfy the equations. Essentially, the units of the variables provide us with additional equations that must be satisfied. Next we consider an example to demonstrate how the relevant parameters of an oscillator problem are defined on purely dimensional considerations.

Example 2.2 Dimensions of a Vibration

All of the problems we study in this book, and in the physical sciences in general, are formulated in terms of only a few physical variables. These are energy, velocity, density, force, momentum, temperature, and stress, for example. Other disciplines, such as electricity and magnetism, would have their own common variables. Each variable has associated units, and we examine how these units can be used to derive key nondimensional parameters for particular problems. This is called *dimensional analysis* or *similitude theory*. In addition to the kinds of applications we explore in this chapter, such methods are useful in the planning and design of experiments because key parameters are identified and scaling relations can be derived. Therefore, one can focus on the important variables of a problem when designing an experiment and deciding key behavior parameters that need to be measured. For example, in designing an experiment, one can answer questions such as: If the rocket is 1/5–scale, then what scale factors apply for the aerodynamic load and thrust? Variables scale differently according to their units.

We know that stiffness and damping are two of the key parameters that define an oscillatory system's behavior. Use dimensional analysis[3] to derive some key dimensionless parameters considering only units relevant to a vibration.

Solution Consider the undamped oscillator. The following parameters are required to model its behavior: X to represent maximum deflection,

---

[3] W.E. Baker, P.S. Westine, F.T. Dodge, **Similarity Methods in Engineering Dynamics**, Elsevier, 1991 provides good background and interesting applications.

2.2. MATHEMATICAL MODELING: DETERMINISTIC

K to represent spring constant, and M mass. The forcing can be represented by peak magnitude P, with an associated time constant T. The time constant could be representative of frequency or time to peak amplitude.

The basis for this procedure is the *Buckingham–Π Theorem*, which states that any complete physical relation can be expressed in terms of a set of independent dimensionless products composed of the relevant physical parameters, called Π (pi) terms. Because Π terms are products or quotients of zero dimension, a general dimensionless equation expressing is

$$X^a K^b M^c P^d T^e = M^0 L^0 T^0, \tag{2.2}$$

where the equality implies dimensional equality, and superscripts are powers. The right hand side is a product of mass, length, and time, each taken to the zero power, indicating that both sides of the equation must be dimensionless. In order to treat Equation 2.2 as an algebraic equation, we need to substitute appropriate units for the variables on the left hand side, obtaining

$$L^a \left(\frac{M}{T^2}\right)^b M^c \left(\frac{ML}{T^2}\right)^d T^e = M^0 L^0 T^0.$$

Combining like terms on the left hand side results in the simplified equation

$$M^{(b+c+d)} L^{(a+d)} T^{(e-2b-2d)} = M^0 L^0 T^0.$$

This equation is really three equations since the units of each dimensional quantity must satisfy the equality. Therefore, equate the powers of like units:

$$\begin{aligned} b + c + d &= 0 \\ a + d &= 0 \\ e - 2b - 2d &= 0. \end{aligned}$$

These three equations can be solved for any two of the five constants in terms of the others. We solve for a and e to find

$$\begin{aligned} a &= a \\ b &= a + \frac{1}{2}e \\ c &= -\frac{1}{2}e \\ d &= -a \\ e &= e. \end{aligned}$$

Substitute the new expressions for b, c and d into Equation 2.2 to find

$$X^a K^{(a+e/2)} M^{(-e/2)} P^{-a} T^e = M^0 L^0 T^0.$$

Collect terms with the same exponents,

$$\left(\frac{XK}{P}\right)^a \left(\sqrt{\frac{K}{M}}T\right)^e = M^0 L^0 T^0,$$

and identify the Π dimensionless terms

$$\frac{X}{P/K}, \quad \sqrt{\frac{K}{M}}T.$$

The Buckingham–Π theorem further states that there exists a function of these dimensionless terms such as that

$$f\left(\frac{X}{P/K}, \sqrt{\frac{K}{M}}T\right) = 0,$$

meaning that $\frac{X}{P/K}$ is a function of $\sqrt{\frac{K}{M}}T$. These parameters turn out to be important in describing an oscillator. P/K is the static displacement, $\frac{X}{P/K}$ is the maximum displacement normalized with respect to the static displacement, and $\sqrt{K/M}$ is the natural frequency of oscillation. The natural frequency is the single most important parameter for specifying a vibration. In an experiment, data could be gathered to plot $\frac{X}{P/K}$ as a function of $\sqrt{\frac{K}{M}}T$. This is much easier and more useful than plotting all the combinations of the original five parameters with which we started.

These ideas can be used to tackle new and more complex problems in which the key parameters may not even be known. This is a very useful tool. ∎

In the following section, we proceed to formulate the governing equation of an oscillating system.

2.2.5 Equations of Motion: Newton's Second Law

Two general approaches are available for deriving the equations of motion of a dynamic system. The first is based on *Newton's second law of motion*. The second is based on an understanding of the kinetic and potential energies of a system and any dissipation of such energies. For systems with one or only several degrees of freedom, it is possible to use Newton's second law of motion to easily derive the equations of motion. More complicated and realistic systems require energy methods for the derivation of the system of governing equations. This approach is introduced in Section 2.2.6, and in later chapters.

2.2. MATHEMATICAL MODELING: DETERMINISTIC

For a single degree of freedom model, it is relatively straightforward to derive the equation of motion by utilizing an important concept known as the *free body diagram*. Refer again to Figure 2.2. This tool permits us to visualize the *forces* acting on a *free or isolated body*. Newton's second law of motion states that the sum of the external forces on a free body equals the product of the body mass and its resulting acceleration

$$\sum_{i=1}^{n} F_i(t) = ma(t). \tag{2.3}$$

A sign convention needs to be established so that the *vector* sum in Equation 2.3 will properly account for the directions of the forces. For a body in rotational motion, the corresponding equation of motion is $\sum_{i=1}^{n} M_i(t) = I\alpha(t)$, where externally applied moments are being summed. The inertial property is the mass moment of inertia I, and the rectilinear acceleration is replaced by the angular acceleration $\alpha(t)$. Mathematically, though, the result is still a second order linear differential equation.

Based on the above formulation, and the discussion in Section 2.2.2, the general governing equation of motion for a single body, with stiffness and damping properties, undergoing rectilinear motion can now be derived as follows, using Equation 2.3,

$$F_{\text{ext}}(t) - c\dot{x} - k_s x = m\ddot{x}, \tag{2.4}$$

where $F_{\text{ext}}(t)$ is the vectorial sum of external forces in one coordinate direction, an overdot denotes differentiation with respect to time, and[4] $a \equiv \ddot{x}$. In a more standard format, Equation 2.4 becomes[5]

$$m\ddot{x} + c\dot{x} + k_s x = F_{\text{ext}}(t). \tag{2.5}$$

It is customary in structural vibration[6] to work with nondimensional groups. After dividing by the mass m, Equation 2.5 becomes

$$\ddot{x} + 2\zeta\omega_n \dot{x} + \omega_n^2 x = F(t), \tag{2.6}$$

where new parameters ζ and ω are introduced,

$$\zeta = \frac{c}{2m\omega_n}, \quad \omega_n^2 = \frac{k_s}{m}, \quad F(t) = \frac{F_{\text{ext}}(t)}{m};$$

---

[4] We omit the time argument of $x(t)$ and use only x since it is obvious, due to the overdots, that these are time functions.

[5] Suppose positive x is taken in the opposite direction. Then applying Newton's second law of motion yields
$$-F_{\text{ext}}(t) + c\dot{x} + k_s x = -m\ddot{x},$$
which is the same equation of motion, as it should be.

[6] As previously discussed, this is also true in other branches of engineering such as fluid mechanics and soil mechanics.

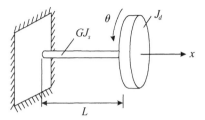

Figure 2.4: Torsional vibration.

ζ is the dimensionless *viscous damping factor*, and ω_n is the *natural frequency* of undamped oscillation with units of *rad/sec*, parameters that will become second nature. This is the equation of motion for a structure idealized by a single degree of freedom model, subjected to external forcing. The subscript s is subsequently omitted from the stiffness constant.

Some ranges of values for the damping factor are listed,

Type of Damping	ζ %
material damping	0.1 – 1
air radiation	0.1 – 2
joints	2 – 5
equipment	2 – 8
special damping materials	1 – 50

Example 2.3 Torsional Vibration
In this example derive the governing equation for torsional oscillation. The governing equation of motion will have the same mathematical form as the translating oscillator, but the parameters have different physical meanings.

The shaft in Figure 2.4 is assumed to be massless and uniform, with a torsional stiffness of GJ_s/L. J_s is the mass moment of inertia of the shaft about the axis of rotation and G is the torsional rigidity. The end disk is assumed to be rigid with mass moment of inertia J_d. Taking the sum of the moments about the x axis of rotation for a small motion θ, and using Newton's second law of motion results in

$$-\frac{GJ_s}{L}\theta = J_d\ddot{\theta},$$

and in standard form

$$\ddot{\theta} + \frac{GJ_s}{J_d L}\theta = 0.$$

2.2. MATHEMATICAL MODELING: DETERMINISTIC

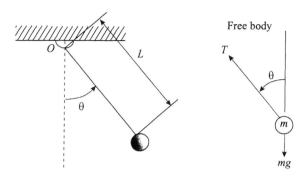

Figure 2.5: Oscillation of a simple pendulum.

This is the governing equation of motion. The solution to this equation is the angular displacement $\theta(t)$. We subsequently learn how to solve this equation. ∎

Example 2.4 Pendulum: Nonlinear and Linearized

The example of a simple pendulum demonstrates how nonlinear motion can be linearized under certain assumptions. Consider Figure 2.5 of the pendulum and its free body. Assume that the pendulum is a point mass so that the rotational inertia can be reasonably ignored.

The forces on the free body suggest two ways by which the equation of motion can be derived, one in terms of forces, the other in terms of moments. Forces can be summed in the instantaneous direction of motion, that is, perpendicular to the string,

$$m(L\ddot{\theta}) = -mg\sin\theta,$$

where the tangential acceleration equals $L\ddot{\theta}$. Otherwise, taking moments about the point of rotation O, we have

$$I_O \alpha = -mg(L\sin\theta),$$

where $L\sin\theta$ is the moment arm for the weight mg and α equals the angular acceleration $\ddot{\theta}$. I_O is the mass moment of inertia about O, which can be expressed as $I_O = I_m + mL^2$, by using the parallel–axis theorem. Based on the point mass assumption made earlier, I_m can be ignored when compared to mL^2. The equation of motion in standard form then becomes

$$m\ddot{\theta} + \frac{mg}{L}\sin\theta = 0. \tag{2.7}$$

This is a nonlinear equation due to the $\sin\theta$ term. (What is the torsional counterpart to this equation?) Following our previous discussions on Taylor series and linearization, we can make the replacement $\sin\theta \approx \theta$ for small θ,

$$m\ddot{\theta} + \frac{mg}{L}\theta = 0. \tag{2.8}$$

As an aside, it is of interest to recast Equation 2.7 in terms of energies. To do this, multiply both sides by $\dot{\theta}$ and integrate as follows,

$$m\int_0^t \dot{\theta}\ddot{\theta}\,dt + \frac{mg}{L}\int_0^t \dot{\theta}\sin\theta\,dt = 0$$

$$m\int_0^t \frac{d}{dt}\left(\frac{1}{2}\dot{\theta}^2\right)dt - \frac{mg}{L}\int_0^t \frac{d}{dt}(\cos\theta)\,dt = 0,$$

or

$$\frac{1}{2}m\dot{\theta}^2 - \frac{mg}{L}\cos\theta = \text{constant}$$
$$T + V = \text{constant},$$

where $T = m\dot{\theta}^2/2$ equals the kinetic energy and $V = -mg\cos\theta/L$ equals the potential energy of the particle. The kinetic plus the potential energy equals the total energy, which is constant for a system without damping. We will solve equations of the form of Equation 2.8 in many applications in the following sections.

Before proceeding, the question of verifying the assumptions made in the formulation is extremely important. *In the modeling, how do we know that reasonable assumptions have been made?* The answer is that we do not know until comparisons are made between the predictions of the mathematical model with experimental data. Reasonable comparisons provide the analyst with assurances that, for similar circumstances, this model is a good predictor. Otherwise, we must go back to the beginning and rethink the assumptions made in the derivation. ■

Example 2.5 System of Variable Mass
We continue with the problem of the rocket that we began in the introduction to this chapter. A schematic of the rocket along with its ejected fuel is shown in Figure 2.6. First derive the equation of motion, and then its solution.

Note that the total mass of the system (rocket + fuel) does not change. But the mass of the rocket decreases by the quantity of fuel that is burned and ejected. The goal is to obtain the equation of motion of the rocket, which includes a term reflecting the variable mass.

2.2. MATHEMATICAL MODELING: DETERMINISTIC

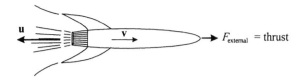

Figure 2.6: Rocket in flight.

Consider the rocket–fuel system at two time instances Δt apart. Assume that initially, at time t, the rocket and all its fuel has mass m and the rocket velocity[7] $\mathbf{v}$. At a later time $t + \Delta t$, a quantity of fuel Δm has been ejected. For this rocket this is a negative quantity,[8] leaving the rocket with less mass: $m + \Delta m$. The rocket is now traveling at an increased velocity, $\mathbf{v} + \Delta \mathbf{v}$. The ejected fuel has mass Δm and is traveling at $-\mathbf{u}$ relative to the rocket. It is reasonable to expect that $\mathbf{u} \gg \mathbf{v}$.

Consider now the momentum of the total system. Newton's second law of motion in momentum form is

$$\mathbf{F}_{external} = \frac{d\mathbf{p}}{dt} = \lim_{\Delta t \to 0} \frac{\Delta \mathbf{p}}{\Delta t}, \qquad (2.9)$$

where $\mathbf{p}$ is the momentum vector, and $\mathbf{F}_{external}$ are external forces such as aerodynamic and thermal loads. During time interval Δt, the change in momentum is

$$\Delta \mathbf{p} = \mathbf{p}_f - \mathbf{p}_i, \qquad (2.10)$$

that is, the final minus the initial momenta. From the previous discussion, these are given by

$$\begin{aligned} \mathbf{p}_i &= m\mathbf{v} \\ \mathbf{p}_f &= (m + \Delta m)(\mathbf{v} + \Delta \mathbf{v}) + \Delta m(-\mathbf{u}). \end{aligned}$$

Equation 2.10 then becomes

$$\begin{aligned} \Delta \mathbf{p} &= (m + \Delta m)(\mathbf{v} + \Delta \mathbf{v}) + (-\Delta m)\mathbf{u} - m\mathbf{v} \\ &= m\Delta \mathbf{v} + \mathbf{v}\Delta m + \Delta m \Delta \mathbf{v} - \mathbf{u}\Delta m. \end{aligned}$$

---

[7] Vector notation is retained initially to remind the reader of the general nature of the problem.

[8] It is customary to denote this loss of mass, a negative quantity, by Δm. In this way for a loss of mass, Δm is negative, and for a mass gain, Δm is positive.

From Equation 2.9

$$\begin{aligned}
\mathbf{F}_{external} &= \lim_{\Delta t \to 0} \frac{\Delta \mathbf{p}}{\Delta t} \\
&= \lim_{\Delta t \to 0} \left[m \frac{\Delta \mathbf{v}}{\Delta t} + (\mathbf{v} - \mathbf{u}) \frac{\Delta m}{\Delta t} + \Delta \mathbf{v} \frac{\Delta m}{\Delta t} \right] \\
&= m \frac{d\mathbf{v}}{dt} + (\mathbf{v} - \mathbf{u}) \frac{dm}{dt},
\end{aligned} \quad (2.11)$$

where $\Delta \mathbf{v} \to 0$ as $\Delta t \to 0$. Using the definition from calculus of the derivative of a product of two functions, Equation 2.11 becomes

$$\mathbf{F}_{external} = \frac{d}{dt}(m\mathbf{v}) - \mathbf{u}\frac{dm}{dt}.$$

In this form, we can easily see that the effect of the variable mass on the equation of motion is the extra term $-\mathbf{u}\, dm/dt$.

The expression $\mathbf{u} - \mathbf{v} \equiv \mathbf{v}_{rel}$ is the velocity of the ejected gases relative to the rocket, and Equation 2.11 can be written as

$$m \frac{d\mathbf{v}}{dt} = \mathbf{F}_{external} + \mathbf{v}_{rel} \frac{dm}{dt}. \quad (2.12)$$

In this form, the second term on the right hand side can be seen to be an equivalent force acting on the system due to the movement of mass. For a rocket, this term is called the thrust. To maximize the thrust, designers make each term, $\mathbf{v}_{rel}$ and dm/dt, as large as possible.[9]

Solution In the solution developed next, it is assumed that the only force acting on the rocket is the thrust. Of course, other forces act on the rocket structure, such as atmospheric aerodynamic forces. These cannot be ignored in an actual design, but we do so here in order to continue with an analytical solution. Making fewer simplifying assumptions would lead us to use computational methods for the solution.

Equation 2.12 becomes

$$m \frac{d\mathbf{v}}{dt} = \mathbf{v}_{rel} \frac{dm}{dt}, \quad (2.13)$$

---

[9] A simple approximation for thrust is given by

$$u_o \frac{dm}{dt} \approx A_o(p - p_o),$$

where u_o is the speed of gas through the orifice of the rocket, A_o is the area of orifice, $p - p_o$ is the pressure difference between the inside of the rocket and the atmospheric pressure just outside the rocket.

2.2. MATHEMATICAL MODELING: DETERMINISTIC

where it is understood that both $\mathbf{v}_{rel}$ and dm/dt are positive quantities. At this point vector notation is dropped to denote motion along one dimension, and the negative sign is introduced to indicate that the direction of the relative velocity is in a direction opposite to the motion of the rocket. Assume fuel is burned at a constant rate. Then, for a period of time dt, we can relate the infinitesimal change in velocity as

$$dv = -v_{rel}\frac{dm}{m}, \tag{2.14}$$

using Equation 2.13.

Assume that m_f fuel was burned during dt. Integrate both sides of Equation 2.14,

$$\int_{v_i}^{v_f} dv = -v_{rel}\int_{m_o}^{m_o-m_f}\frac{dm}{m}$$

$$v_f - v_i = -v_{rel}\ln\left(\frac{m_o - m_f}{m_o}\right),$$

where $m = m_o - m_f$. If the rocket starts from rest, $v_i = 0$ and

$$\frac{m}{m_o} = e^{-v_f/v_{rel}}$$

is the relation between the mass of remaining fuel m and final velocity v_f. ∎

2.2.6 Equations of Motion: Energy Formulation

Newton's second law of motion becomes a cumbersome way to derive the equations of motion for a system having more than a few degrees of freedom. In Chapter 7, we will introduce *Lagrange's* energy approach for the general derivation of governing equations. Here, an early glimpse is provided into the use of energy, which is a scalar.[10]

A vibratory system has several types of energies: *kinetic, potential*, and *strain*. We are very familiar with kinetic and potential energies. Strain energies come into play when work is (temporarily) stored in an element such as a spring. If the system is *conservative*, meaning that energy is conserved without dissipation or losses, the total energy remains constant. The sum of the kinetic T and potential/strain U energies is

$$T + U = constant. \tag{2.15}$$

---

[10] In using Newton's second law of motion for a system with many connected masses, internal forces between masses will appear in the equations of motion. These are generally forces that we do not generally need, but they are amongst the unknowns of the problem.

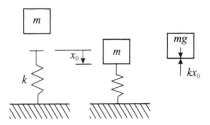

Figure 2.7: Free body diagram for static equilibrium.

For a spring element, the energy stored in the spring due to an elongation x equals the area under the $F - x$ curve. In a linear force–deflection relation of slope k, this area equals $kx^2/2$. If the spring had an initial stretch of x_0, then the total strain energy stored is $k(x_0 + x)^2/2$. A change in the vertical position x of the mass m results in a change in the gravitational potential energy equal to $-mgx$. For a massless spring, the only kinetic energy in the system is due to the motion of the mass and equals $m\dot{x}^2/2$. Substitute all these components into Equation 12.58 to find

$$\frac{1}{2}m\dot{x}^2 + \frac{1}{2}k(x_0 + x)^2 - mgx = \text{constant}.$$

Differentiate this equation with respect to time,

$$(m\ddot{x} + kx)\dot{x} + (kx_0 - mg)\dot{x} = 0,$$

and recognize that, in static equilibrium, $kx_0 = mg$ represents the force balance in the free body diagram of Figure 2.7.

Since $\dot{x} \neq 0$, we are left with

$$m\ddot{x} + kx = 0,$$

the governing equation of motion derived purely on energy considerations. For a system with damping, the procedure is not so simple. Energy loss due to dissipation is discussed in Chapter 7.

Furthermore, for oscillatory conservative systems, the maximum kinetic energy equals the maximum potential energy since no dissipation occurs. Therefore,

$$T_{max} = U_{max}.$$

The displacement of an oscillatory system may be represented by the har-

2.2. MATHEMATICAL MODELING: DETERMINISTIC

monic function $x = A\sin\omega t$. Its velocity is then $\dot{x} = A\omega\cos\omega t$, and

$$U_{max} = \frac{1}{2}kx_{max}^2 = \frac{1}{2}k(A\sin\omega t)|_{max}^2 = \frac{1}{2}kA^2$$

$$T_{max} = \frac{1}{2}m\dot{x}_{max}^2 = \frac{1}{2}m(A\omega\cos\omega t)|_{max}^2 = \frac{1}{2}mA^2\omega^2,$$

and therefore

$$\frac{1}{2}mA^2\omega^2 = \frac{1}{2}kA^2,$$

or $\omega^2 = k/m$, the square of the natural frequency. We will see a more sophisticated and useful version of this idea when multi-degree of freedom systems are studied and *Rayleigh's quotient* is introduced in Chapter 9.

Vertical Motion Compared with Pendulum Motion

The vertical motion of a mass suspended on a spring is governed by the same mathematical equation as the linear oscillation of a pendulum about its equilibrium position. There is a physical difference though. When the mass is in its static equilibrium position, there is energy stored in the spring since the mass' weight is balanced by the spring force kx_{static}. When the pendulum is at static equilibrium, and since the string on which it is suspended is not elastic, there is no stored energy in the system. Gravitational potential energy can be taken as equal to zero at this location.

Example 2.6 Springs with Inertia

Generally, we ignore the mass of a spring when deriving an equation of motion, with the justification that the mass to which the spring is attached will be much larger. Sometimes, such an assumption is unrealistic, requiring us to include the spring mass. If the spring mass is comparable to the mass of the object to which it is attached, then it should be modeled as a continuous system. If the mass of the spring is much smaller than the attached object and cannot be ignored, then the single degree of freedom model can be retained by including the inertia effects of the spring. In effect, we will increase the mass of the object so that it will be representative of both spring and object, as shown schematically in Figure 2.8.

For the actual system, the kinetic energy is $T_{actual} = T_s + (1/2)m\dot{x}^2$, and for the equivalent system, $T_{equiv} = (1/2)m_{equiv}\dot{x}^2$, where T_s is the kinetic energy of spring of mass m_s and m_{equiv} is an equivalent mass representing the inertia of both the spring and the object. Since both kinetic energies are equal by definition, we need the value of T_s in order to evaluate m_{equiv}.

Consider the unstretched spring of length l. An element of this spring of length ds has mass dm. If the object is displaced a distance x, then a

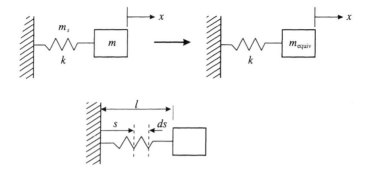

Figure 2.8: Equivalent mass system for spring with inertia.

point on the linear spring at location s will be displaced a distance

$$u(s) = \frac{x}{l} s. \tag{2.16}$$

Suppose the point on the spring is at the fixed end, we have $u(0) = 0$, and for the point at the end connected to the object $u(l) = x$. The kinetic energy of the element ds is $dT_s = (1/2)\, dm\, \dot{u}^2$, where $\dot{u} = (\dot{x}/l)s$ and $dm = (m_s/l)\, ds$. Therefore,

$$\begin{aligned} T_s = \int dT_s &= \int_0^l \frac{1}{2} \frac{m_s}{l} \left(\frac{s}{l}\dot{x}\right)^2 ds \\ &= \frac{1}{2} \frac{m_s}{l} \left(\frac{\dot{x}}{l}\right)^2 \int_0^l s^2 ds \\ &= \frac{m_s \dot{x}^2}{6}. \end{aligned}$$

Setting the actual kinetic energy to the equivalent one gives us

$$\begin{aligned} T_s + \frac{1}{2} m\dot{x}^2 &= \frac{1}{2} m_{equiv} \dot{x}^2 \\ \frac{m_s \dot{x}^2}{6} + \frac{m\dot{x}^2}{2} &= \frac{m_{equiv} \dot{x}^2}{2} \\ \implies m_{equiv} &= m + \frac{m_s}{3}. \end{aligned}$$

Thus, for springs which behave according to the linear displacement model Equation 2.16, we need to add one third of the spring mass to the mass of the vibrating object in order to capture the effect of spring inertia. ∎

2.2. MATHEMATICAL MODELING: DETERMINISTIC

Work Done by Harmonic Loading

It is of interest to calculate the *work* done by a harmonic force $F(t) = F_0 \sin(\omega t + \phi)$ acting on a body that is responding harmonically[11] to this force according to $x(t) = x_0 \sin \omega t$. This could be the work done by a motor driving a shaft or a set of linkages. The work done by the force in moving the body an increment dx in the direction of the force is $F(t)dx$. Considering one cycle of x motion, the argument ωt varies from 0 to 2π. It is necessary to evaluate the work done over one oscillation period. Therefore, let $F(t)dx = F(t)\frac{dx}{dt}dt$, with $0 \leq t \leq 2\pi/\omega$. Then, with the appropriate change of variables, the work is given by

$$\begin{aligned} W &= \int_0^{2\pi/\omega} F(t)\frac{dx}{dt}dt \\ &= \frac{1}{\omega}\int_0^{2\pi} F(t)\frac{dx}{dt}d(\omega t) \\ &= F_0 x_0 \int_0^{2\pi} \sin(\omega t + \phi)\cos\omega t\, d(\omega t) \\ &= F_0 x_0 \int_0^{2\pi} [\sin\omega t \cos\phi + \cos\omega t \sin\phi]\cos\omega t\, d(\omega t) \\ &= F_0 x_0 \cos\phi \int_0^{2\pi} \sin\omega t \cos\omega t\, d(\omega t) \\ &\quad + F_0 x_0 \sin\phi \int_0^{2\pi} \cos^2\omega t\, d(\omega t). \end{aligned} \qquad (2.17)$$

The first integral above equals zero, and the second integral[12] equals π. Therefore,

$$W = \pi F_0 x_0 \sin\phi. \qquad (2.18)$$

We can interpret the above result in the following way. Equation 2.17 contains two parts; in the first part, the force is out of phase with the velocity, whereas in the second part the force is *in phase* with the velocity. Thus, *a force does work only with that component in phase with the structure velocity*.

Example 2.7 Work Done by a Harmonic Force with a Different Frequency[13]

---

[11] A cosine function could have equivalently been chosen.
[12] $\cos^2 x$ varies between 0 and 1, with an average value of $1/2$ over a range of 2π. Therefore, the integral (area) equals $(1/2) \times (2\pi)$, or π.
[13] We follow page 15 of Den Hartog, **Mechanical Vibrations**, Dover, 1984.

As an exercise, let us work through the case where the frequency of the force is different than that of the response motion. Assume that the force is $F(t) = F_0 \sin n\omega t$ and the harmonic response to this force is $x(t) = x_0 \sin(m\omega t + \phi)$. We will determine that the work done by this force during a period is zero. As before, the work done per cycle is $\int F(t)dx$, or

$$\begin{aligned} W &= \int_0^T F(t)\frac{dx}{dt}dt \\ &= F_0 x_0 m\omega \int_0^T \sin n\omega t \cos(m\omega t + \phi)dt, \end{aligned}$$

and when the cosine is expanded, two integrals result of the form

$$\int_0^T \sin n\omega t \sin m\omega t\ dt, \quad \int_0^T \sin n\omega t \cos m\omega t\ dt$$

as before, except that both now equal zero. ∎

To conclude:

- The work done by a harmonic force of one frequency acting upon a harmonic displacement, or velocity, of a different frequency is zero if the time interval includes an integer number of force cycles and a different integer number of velocity cycles.

- The work done by a harmonic force $\pi/2$ radians, or 90 degrees, out of phase with a harmonic velocity of the same frequency is zero during a whole cycle.

- The work done by a harmonic force of amplitude F_0 and frequency ω, in phase with a harmonic velocity $v_0 = x_0 \omega$ of the same frequency, is $\pi F_0 v_0/\omega = \pi F_0 x_0$ over a whole cycle.

2.2.7 The Rotating Vector Approach to the Equation of Motion

The vibrating mass which undergoes harmonic or periodic motion may be easily represented by a rotating vector. Consider $x(t)$ to be represented graphically by a vector, $x(t) = a\cos\omega t$. As ωt increases in the range $0 \leq \omega t \leq 2\pi$, $x(t)$, prescribing a circle of radius a. The angle of the vector with the horizontal is ωt, as shown in Figure 2.9, from which we obtain the term *circular frequency* for variable ω, in units of *radians per second*. Since one *cycle* or revolution is equivalent to 2π radians, the frequency can be equivalently defined as $f = \omega/2\pi$ *cycles per second* or *Hertz (Hz)*.

2.2. MATHEMATICAL MODELING: DETERMINISTIC

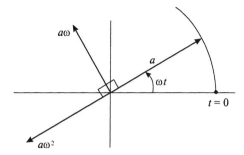

Figure 2.9: Vectorial representation of harmonic motion.

If $x(t)$ is the displacement of the oscillator, then its velocity and acceleration are provided by the following derivatives,

$$\dot{x} = -a\omega \sin \omega t \qquad (2.19)$$
$$\ddot{x} = -a\omega^2 \cos \omega t. \qquad (2.20)$$

The *velocity vector* has magnitude $a\omega$, rotates at the same angular velocity, and is always 90 degrees *ahead* of the displacement vector. Similarly, the *acceleration vector* has magnitude $a\omega^2$, rotates at the same angular velocity, and is always 90 degrees *ahead* of the velocity vector.

Note from Equations 2.19 and 2.20 that the relative magnitudes of the stiffness, damping, and inertia forces are ω-*dependent*.[14] For low–frequency loading, the dominant force is the stiffness, and for high–frequency loading, the inertia dominates. This fact will become important to us when we begin to study the concepts of response magnitude, phase, and resonance.

Complex numbers of the form $a + ib$ can be used to represent vectors, where a and b are real valued constants. These numbers can be plotted as coordinates in the *complex* plane, as shown in Figure 2.10.

If the vector from the origin to the coordinate point has magnitude r, then it can be written as $r(\cos \omega t + i \sin \omega t)$, for an angle of ωt with the horizontal. Thus if $r \cos \omega t$ is the harmonic motion, this is equivalent to the horizontal component of the rotating vector, or *the real part of* the complex number. Note that *Euler's identity*, $e^{i\omega t} = (\cos \omega t + i \sin \omega t)$, can be used to represent harmonic motion in the compact complex exponential form, again implying *the real part of* the expression is retained.

---

[14] Stiffness is related to x, damping to $\dot{x}$, and acceleration to $\ddot{x}$.

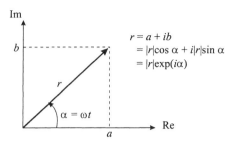

Figure 2.10: A vector represented in the complex plane.

2.2.8 Solution of the Equations of Motion

We now need to solve Equation of motion 2.6, from which we will begin to understand the different classes of vibratory behavior. The character of the response will be a function of the value of the parameter ζ. The next section will begin to draw the basic outline of how this governing equation of motion is to be solved,[15] and, as importantly, how to interpret the possible solutions as functions of damping, loading, and the initial conditions of the problem. In Section 4.6, we will discuss some of the connections between analysis and design.

The reader is reminded that in solving a differential equation, one first solves the homogeneous problem (right hand side set to zero), then the particular problem (solution for the specific right hand side), then adds the two solutions, and finally satisfies the initial conditions in order to specify the arbitrary constants.

The solution of the governing equation of motion will be carried out in several ways because each will provide us with a new perspective on the behavior. The *time domain* solution is simply the structural response as a function of time. This is the most straightforward way to present our solution.

In addition to the time domain solution, we will derive some of the response characteristics, such as amplitude, as functions of the ratio of the driving to the natural frequency. These solutions are in the *frequency domain* because the independent variable is the frequency ratio. The *frequency domain* or *transformed domain* also includes the use of *Laplace transforms*, where the differential equation is transformed into the complex number plane, denoted by $s = a + ib$. One major difference between the two approaches is that the Laplace transform satisfies the initial conditions as part

---

[15] In fact, the solution of this equation is well known to anyone who has had a first course on ordinary differential equations, as we saw in Chapter 1.

2.3 Free Vibration With No Damping

The undamped forced oscillator is representative of a structure where damping plays a small role in the response. It is governed by the equation

$$\ddot{x} + \omega_n^2 x = F(t), \tag{2.21}$$

where the external force is set equal to zero in order to solve for the free vibration. Recall $\omega_n^2 = k/m$. Let $x(0)$ and $\dot{x}(0)$ be, respectively, the initial displacement and initial velocity of this system. The free response depends only on the *initial conditions* of the system.

Since the left hand side of Equation 2.21 includes the response x and its second derivative $\ddot{x}$, the solution, when differentiated twice, is a function that reappears: $\ddot{x} = -(k/m)x$. Functions with this property are the harmonic functions: *sines* and *cosines*. Therefore, assume the solution

$$x(t) = C_1 \sin rt + C_2 \cos rt, \tag{2.22}$$

where r is to be determined and constants C_1 and C_2 are established by satisfying the initial conditions. Substituting the assumed solution into the governing equation results in

$$(-r^2 + \omega_n^2)C_1 \sin rt + (-r^2 + \omega_n^2)C_2 \cos rt \equiv 0,$$

which must be satisfied identically for all t, thus

$$-r^2 + \omega_n^2 = 0$$
$$r = \omega_n \equiv \sqrt{\frac{k}{m}}.$$

Figure 2.11 shows how the natural frequency varies as a function of the ratio of stiffness to mass.

Equation 2.22 becomes

$$x(t) = C_1 \sin \sqrt{\frac{k}{m}} t + C_2 \cos \sqrt{\frac{k}{m}} t. \tag{2.23}$$

As expected, there are two solutions for the second order ordinary differential equation, and the response of the oscillator is at its natural frequency, $\omega_n = \sqrt{k/m}$.

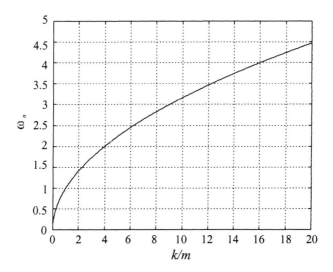

Figure 2.11: ω_n as a function of k/m.

The constants of integration can be evaluated by using the initial displacement $x(0)$ and initial velocity $\dot{x}(0)$. From Equation 2.23, set up equations for $x(0)$ and $\dot{x}(0)$ to find that $C_2 = x(0)$ and $C_1 = \dot{x}(0)/\omega_n$. The general response is then

$$x(t) = x(0)\cos\omega_n t + \frac{\dot{x}(0)}{\omega_n}\sin\omega_n t. \qquad (2.24)$$

We see that for the undamped oscillator there is no decay, meaning that the oscillation continues without a decrease in peak amplitude. The response is a function of the initial conditions, with the frequency of oscillation ω_n.

Example 2.8 Effect of Initial Displacement and Initial Velocity
Figure 2.12 is a time–history for the case $x(0) = 1$, $\dot{x}(0) = 1$, and $\omega_n = 1$ rad/sec= 0.159 Hz. The period is therefore $T = 2\pi$ sec. ∎

Equation 2.24 can be written in terms of an amplitude, frequency, and phase of vibration. To accomplish this, define the equivalent oscillation as

$$x(t) = A\cos(\omega_n t - \phi), \qquad (2.25)$$

where the *amplitude* A and the *phase angle* ϕ need to be evaluated. To do

2.3. FREE VIBRATION WITH NO DAMPING

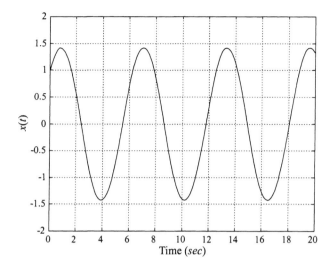

Figure 2.12: Response to initial displacement and velocity.

this, expand the cosine term and equate to Equation 2.24, to find

$$A = \sqrt{[x(0)]^2 + \left(\frac{\dot{x}(0)}{\omega_n}\right)^2}, \quad \phi = \tan^{-1}\left(\frac{\dot{x}(0)}{x(0)\omega_n}\right). \quad (2.26)$$

This is a convenient form, allowing us to see the effects of the initial conditions on the response amplitude and phase, parameters about which we will learn much more in subsequent discussions. Note that care must be taken in calculating ϕ since the arctan will be different depending on the sign of the argument. The phase angle is shown with a negative sign in Equation 2.25 in anticipation of the damped response, where there is a phase *lag* of the structure behind the forcing.

2.3.1 Alternate Formulation

Another approach to solving the equation of motion is to assume $x(t) = Ae^{rt}$ instead of the sinusoidal of Equation 2.22. Physically, A represents amplitude and r the frequency of oscillation. Differentiating this expression and substituting into the governing equation leads to the requirement that

$r^2 + \omega_n^2 = 0$. Or, $r_{1,2} = \pm i\omega_n$, and

$$\begin{aligned} x(t) &= A_1 e^{i\omega_n t} + A_2 e^{-i\omega_n t} \\ &= A_1(\cos\omega_n t + i\sin\omega_n t) + A_2(\cos\omega_n t - i\sin\omega_n t) \\ &= (A_1 + A_2)\cos\omega_n t + i(A_1 - A_2)\sin\omega_n t \\ &= B_1 \cos\omega_n t + B_2 \sin\omega_n t, \end{aligned} \qquad (2.27)$$

which is the same as Equation 2.23. Note that A_1 and A_2 must be complex numbers[16] in order for $x(t)$ to be a real time history, as it must be. The constants of integration, B_1 and B_2 (real numbers), are evaluated by satisfying the initial conditions, in which case

$$\begin{aligned} x(0) &= B_1 \\ \dot{x}(0) &= B_2 \omega_n. \end{aligned} \qquad (2.28)$$

Therefore, the general solution to our problem, in terms of the initial conditions and the system natural frequency, is again Equation 2.24,

$$x(t) = x(0)\cos\omega_n t + \frac{\dot{x}(0)}{\omega_n}\sin\omega_n t.$$

Note that for this undamped structure, the oscillation goes on forever. Next we study the response to a harmonic load, the most important loading case found in practice, because most oscillations are harmonic or sums of harmonics.

2.3.2 Phase Plane

Another way to envision the motion of a harmonic oscillator is in the *phase plane*. We will return to the phase plane in subsequent chapters. The phase plane is created by replacing the second order governing equation $\ddot{x} + \omega_n^2 x = 0$ by an equivalent two first order differential equations

$$\begin{aligned} \dot{x} &= y \\ \dot{y} &= -\omega_n^2 x. \end{aligned}$$

Equilibrium is at $(\dot{x}, \dot{y}) = (0, 0)$, where there is no velocity, and by these equations, at the origin $(x, y) = (0, 0)$. The paths of the particle in the xy

---

[16] Not only must A_1 and A_2 be complex numbers, they must be complex conjugates. Let $A_1 = a_1 + ib_1$ and $A_2 = a_2 + ib_2$. Then, $B_1 = A_1 + A_2$ can be a real number, as it must be, only if $ib_1 + ib_2 = 0$, that is, $b_1 = -b_2$. Similarly, $B_2 = i(A_1 - A_2)$ can only be a real number if $a_1 - a_2 = 0$, that is, $a_1 = a_2$. Thus, we have shown that A_1 and A_2 must be complex numbers. We can easily show that the following equalities hold: $B_1 = 2a_1 = 2a_2$, $B_2 = -2b_1 = 2b_2$.

2.3. FREE VIBRATION WITH NO DAMPING

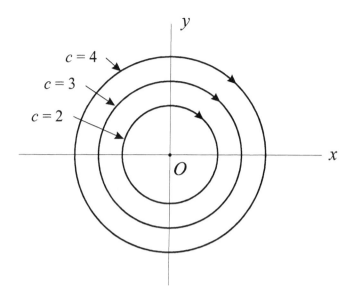

Figure 2.13: Phase plane for $\ddot{x} + \omega_n^2 x = 0$.

plane can be found as follows:

$$\frac{dy}{dx} = -\omega_n^2 \frac{x}{y}$$
$$y^2 + \omega_n^2 x^2 = c,$$

where c is the constant of integration. This is an equation for constant energy, with different curves for different values of c. See Figure 2.13. The arrows show the direction of motion. This can be determined by selecting a quadrant and evaluating the rate of change of a parameter, for example, in the first quadrant, $y = \dot{x} > 0$, y is positive and x increases in value with time.

Such an alternative formulation for oscillatory motions becomes valuable when considering control systems, Chapter 6, and nonlinear systems, Chapter 12. With experience, the phase plane curves provide a qualitative overview of the system dynamics.

2.4 Harmonic Forced Vibration With No Damping

Begin with governing Equation 2.21 where $F(t) = (A/m)\cos\omega t$, ω is the frequency of the driving force $F(t)$ and $\omega \neq \omega_n$. There are two components to any forced vibration response: *transient* or *free vibration*, and *steady state* or *forced vibration*. In mathematical terms, the transient response is the *homogeneous* solution to the differential equation, and the steady state is the *particular* solution. Applying linear superposition, we can solve each problem separately, then add both solutions to obtain the complete response. *It is the complete response that is used to satisfy the initial conditions*. For a damped structure, the free vibration part of the solution is quickly "forgotten," it decays, while the forced vibration response continues as long as the force exists.

For the forced response, assume[17] a solution of the form

$$x(t) = B_1 \cos\omega t.$$

There are two such solutions for the second order differential equation. Differentiate the assumed solution twice and substitute the appropriate expressions into the governing equation $\ddot{x} + \omega_n^2 x = (A/m)\cos\omega t$ to find

$$B_1(-\omega^2 + \omega_n^2)\cos\omega t = \frac{A}{m}\cos\omega t$$

$$B_1 = \frac{A/m}{-\omega^2 + \omega_n^2} = \frac{A/k}{1 - (\omega/\omega_n)^2}. \quad (2.29)$$

In the last equality, both numerator and denominator were divided by ω_n^2. The forced or *steady state* response is then

$$x(t) = \frac{A/k}{1 - (\omega/\omega_n)^2}\cos\omega t. \quad (2.30)$$

The term A/k in the numerator of this solution has the simple physical significance of being the static deflection of the spring in our model under the constant (maximum) force A. Then, with $x_{st} = A/k$, the nondimensional solution becomes

$$\frac{x(t)}{x_{st}} = \frac{1}{1 - (\omega/\omega_n)^2}\cos\omega t. \quad (2.31)$$

---

[17]Note that if $F(t) = A\sin\omega t$, then we would assume a response $x(t) = B_1\sin\omega t$ and arrive at the same value for B_1 as in Equation 2.29. One expects the same response amplitude since the only difference between the sine and cosine is a phase difference of 90 deg. Also, if we would have assumed $x(t) = B_1\cos\omega t + B_2\sin\omega t$, the following analysis would have led to $B_2 = 0$ and B_1 as in Equation 2.29.

2.4. HARMONIC FORCED VIBRATION WITH NO DAMPING 73

The coefficient $1/[1-(\omega/\omega_n)^2]$ represents a *dynamic amplification factor* as a function of the frequency ratio ω/ω_n. Von Kármán called it a *resonance factor*.[18] Physically, this factor shows us that the structure acts as a *filter*, allowing certain frequencies to pass through it and be amplified while blocking others. We will learn more about this, and view this idea in a number of graphs to come.

The complete superposed solution (free vibration plus steady state) is then
$$x(t) = C_1 \sin\omega_n t + C_2 \cos\omega_n t + \frac{x_{st}}{1-(\omega/\omega_n)^2}\cos\omega t,$$

where C_1 and C_2 can be now fixed according to the given initial conditions. Assume that the initial conditions are $x(0) = x_0$ and $\dot{x}(0) = v_0$. Then using Equation ??, for $x(t)$ and its derivative $\dot{x}(t)$, it is straightforward to find the constants of integration to be
$$C_1 = \frac{v_0}{\omega_n}, \quad C_2 = x_0 - \frac{x_{st}}{1-(\omega/\omega_n)^2}.$$

Therefore, the response to harmonic loading with arbitrary initial conditions is
$$\begin{aligned} x(t) &= \frac{v_0}{\omega_n}\sin\omega_n t + \left[x_0 - \frac{x_{st}}{1-(\omega/\omega_n)^2}\right]\cos\omega_n t + \frac{x_{st}}{1-(\omega/\omega_n)^2}\cos\omega t \\ &= \frac{v_0}{\omega_n}\sin\omega_n t + x_0\cos\omega_n t + \frac{x_{st}}{1-(\omega/\omega_n)^2}[\cos\omega t - \cos\omega_n t], \quad (2.32) \end{aligned}$$

where the effects of the initial conditions are discernible from the effect of the harmonic load. Even though the load is harmonic, the response is not harmonic, since it is the difference between two harmonic functions of different frequencies.

Finally, it is of interest to note that the denominator in Equation 2.30, $[1-(\omega/\omega_n)^2]$, is a positive quantity for $\omega/\omega_n < 1$ and a negative quantity for $\omega/\omega_n > 1$. Physically, this means that when the natural frequency is greater than the forcing frequency, the force and motion are in phase, but when the forcing frequency is greater than the natural frequency, they are out of phase. The value of the denominator affects the magnitude of the steady state response.

Example 2.9 Free Plus Forced Vibration Response
Plot Equation 2.32 for the case where $x_0 = 1$ cm, $v_0 = 1$ cm/sec, and $\omega_n = 1$ rad/sec. The static displacement is taken to be $x_{st} = 1$ cm. The

---
[18] Th. von Kármán and M.A. Biot, **Mathematical Methods in Engineering**, McGraw–Hill, 1940, p.134.

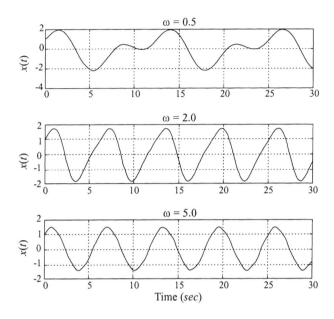

Figure 2.14: Complete response for the cases $\omega = 0.5$, 2.0 and 5.0 rad/sec.

time history for Figure 2.14 is given for the loading frequencies $\omega = 0.5, 2.0$ and 5.0 rad/sec. ∎

2.4.1 Resonance

Vibratory behavior where the forcing frequency ω is close or equal to the natural frequency of the oscillator is a very important case due to the fact that $1/[1 - (\omega/\omega_n)^2]$ in Equation 2.31 becomes very large[19] when $\omega \approx \omega_n$. Therefore, a preliminary engineering design attempts to keep oscillation frequencies as far as possible from the loading frequencies. This is true regardless of the number of degrees of freedom, but becomes almost impossible to achieve for very large structures with many degrees of freedom, since each degree of freedom has an associated frequency.

Consider again the nondimensional solution of Equation 2.31

$$\frac{x(t)}{x_{st}} = \frac{1}{1 - (\omega/\omega_n)^2} \cos \omega t, \qquad (2.33)$$

---

[19] This factor becomes indeterminate, (1/0), when $\omega = \omega_n$, but since all real systems have some damping, the resonance type of behavior is achieved only with zero damping.

2.4. HARMONIC FORCED VIBRATION WITH NO DAMPING 75

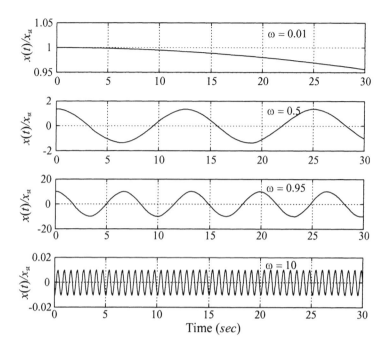

Figure 2.15: (a) Response curves for Equation 2.33.

which is plotted in Figure 2.15(a) for the case $x_{st} = 1$ cm and $\omega_n = 1$ rad/sec. The driving frequencies are $\omega = 0.01, 0.5, 0.95$ and 10 rad/sec. In order to understand how the response amplitude $1/(1 - (\omega/\omega_n)^2)$ is a function of the driving frequency, we will consider three critical[20] cases: *(i)* $\omega = 0^+$, or $\omega/\omega_n \ll 1$, *(ii)* $\omega/\omega_n = 1$, and *(iii)* $\omega/\omega_n \gg 1$. See Figure 2.15(b). It is customary to draw this set of curves so that the absolute value of the displacement is plotted as in Figure 3.21,

$$\left| \frac{1}{1 - (\omega/\omega_n)^2} \right|.$$

Therefore, the negative curve is flipped about the horizontal axis.

For $\omega/\omega_n \ll 1$, the forcing frequency is very slow, and the mass will be displaced essentially to its static deflection with very minor perturbations. In the limit for $\omega = 0$, the static displacement is the response, and $x(t)/x_{st} = 1$ as Figure 2.15(b) shows. On the other hand, for $\omega/\omega_n \gg 1$, the forcing frequency is very high, and the mass cannot follow the rapidly oscillating

---
[20] See Den Hartog, p.44.

force. Therefore, since the average value of the force is zero, in the limit for large ω, the displacement is also zero, and $x(t)/x_{st} = 0$.

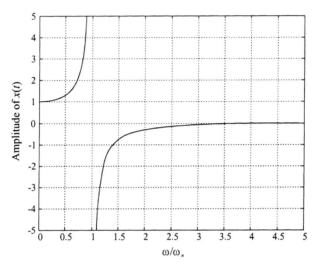

Figure 2.15: (b) Resonance diagram for damped-free oscillator subjected to a constant amplitude force.

The most interesting and important case is when $\omega/\omega_n = 1$. Physically, this means that the forcing frequency coincides exactly (or almost exactly) with the oscillator natural frequency. The force is continuously pushing the mass in the direction of motion, adding energy to the oscillator with each cycle, which then vibrates with indefinitely increasing amplitudes since there is no damping to dissipate any energy. A small force can eventually make the amplitude very large. This phenomenon is called *resonance*.[21] Sometimes the natural frequency is called the *resonant frequency*.

Equation 2.33 is not valid for the case where the driving frequency is equal to the natural frequency, $\omega = \omega_n$. Let us explore this special case here. The equation of motion is

$$\ddot{x} + \omega_n^2 x = \frac{A}{m} \cos \omega_n t. \qquad (2.34)$$

From the theory of linear differential equations, the solution of a nonhomogeneous equation requires the sum of two *linearly independent* solutions, the homogeneous solution plus the particular solution. In assuming a particular solution for Equation 2.34, we cannot choose $x(t) = B_1 \cos \omega_n t$ since

---
[21] Resonance may be understood as a situation where energy can be easily exchanged between systems or structures.

2.4. HARMONIC FORCED VIBRATION WITH NO DAMPING

this will not be linearly independent of the homogeneous solution. In fact, it is the homogeneous solution. Therefore, multiply the nonresonant homogeneous solution, Equation 2.22, by the independent variable t in order to generate the linearly independent solution

$$x(t) = B_1 t \cos \omega_n t + B_2 t \sin \omega_n t,$$

where both sine and cosine terms have been included. To solve, differentiate the assumed solution twice and substitute into the governing equation of motion. Factor common $\cos \omega_n t$ and $\sin \omega_n t$ terms and find

$$\cos \omega_n t [-B_1 t \omega_n^2 + 2 B_2 \omega_n + \omega_n^2 B_1 t]$$
$$+ \sin \omega_n t [-B_2 t \omega_n^2 - 2 B_1 \omega_n + \omega_n^2 B_2 t] \equiv \frac{A}{m} \cos \omega_n t.$$

Simplifying this equation and satisfying the identity leads to two equations for the constants,

$$B_1 = 0$$
$$B_2 = \frac{A}{2\omega_n m} = \frac{A}{2\sqrt{km}}.$$

The response is then

$$x(t) = \frac{A}{2\sqrt{km}} t \sin \omega_n t, \qquad (2.35)$$

where $x(t)$ grows without bound. Expressions where time t appears as a factor are called *secular*. Of course, since all structures possess some damping, one will not find exactly such behavior. However, structures with modest damping can be driven to excessive amplitudes and failure if the driving frequency is close to the natural frequency. If $x(t)$ becomes large enough, the linearity assumption upon which the solution depends will become invalid. Figure 2.16 depicts how a secular oscillation behaves for the special case of $x(t) = t \sin t$. Suppose the strain in the system depends on $x(t)$, then at some instant of time, that strain will exceed the yield strength of the material, and the structure will fail.

Had the force been $(A/m) \sin \omega_n t$, then following the above procedure would result in the response $x(t) = -(A/2\sqrt{km}) t \cos \omega_n t$.

Example 2.10 The Direct Approach to the Resonance Response
Rather than solve the resonance response as a separate problem, consider the general solution, Equation 2.32. Assuming zero initial conditions, the general solution becomes

$$x(t) = \frac{x_{st}}{1 - (\omega/\omega_n)^2} [\cos \omega t - \cos \omega_n t].$$

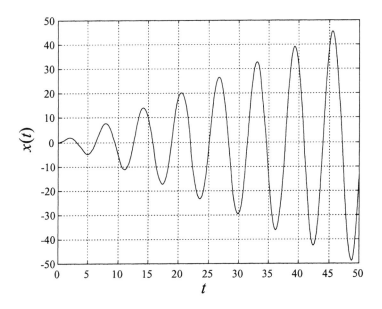

Figure 2.16: *Secular oscillation* for $x(t) = t \sin t$.

Rewrite this equation in a more useful form

$$x(t) = \frac{x_{st}}{1 - (\omega/\omega_n)^2} \left[2 \sin \frac{(\omega_n + \omega)t}{2} \sin \frac{(\omega_n - \omega)t}{2} \right], \qquad (2.36)$$

and define $2\varepsilon = \omega_n - \omega$, where $\varepsilon \ll 1$, to find

$$\begin{aligned} x(t) &= \frac{2x_{st}}{1 - (\omega/\omega_n)^2} \left[\sin(\varepsilon + \omega)t \sin \varepsilon t \right] \\ &= \frac{x_{st}\omega_n^2}{2\varepsilon(\varepsilon + \omega)} \left[\sin(\varepsilon + \omega)t \sin \varepsilon t \right]. \qquad (2.37) \end{aligned}$$

We are interested in two cases: *(i)* resonance, where $\omega = \omega_n$, and *(ii)* beating, where $\omega \approx \omega_n$. In this solution we consider the resonance case. The next example considers case *(ii)*.

Take the limit of Equation 2.37 as $\varepsilon \to 0$, and the result is indeterminate: 0/0. Therefore, use l'Hôpital's rule to find

$$\lim_{\varepsilon \to 0} x(t) = x_{st} \omega_n^2 \left[\frac{t \sin \omega_n t}{2\omega_n} \right] = \frac{A}{2\sqrt{km}} t \sin \omega_n t,$$

where $x_{st} = A/k$ and $\omega \to \omega_n$. This is the same result found before with

2.4. HARMONIC FORCED VIBRATION WITH NO DAMPING

Equation 2.35. ∎

Example 2.11 One Degree of Freedom Beating
Now consider the response when the loading frequency is very close, but not equal, to the natural frequency. In Equation 2.36, note that

$$\frac{1}{1-(\omega/\omega_n)^2} = \frac{\omega_n^2}{(\omega_n+\omega)(\omega_n-\omega)}.$$

When the natural frequency and the driving frequency are close to each other, the factor $(\omega_n - \omega)$ in the denominator becomes very small, equaling 2ε, thus making the ratio very large. This phenomenon is called *beating*, where a harmonic motion with angular frequency $(\omega_n + \omega)/2 \approx \omega_n$ has a periodic but slowly varying amplitude of frequency $(\omega_n - \omega)/2$ rad/sec. The number of beats per unit time equals $|(\omega_n - \omega)/4\pi|$.

At resonance, the period of beating becomes infinite and the buildup is continuous. Therefore, beating may be viewed as a near–resonance phenomenon. We will study beating in more detail for a two mass system in Chapter 8. ∎

l'Hôpital

Guillaume Francois Antoine Marquis de l'Hôpital lived during 1661–1704 and was born in Paris, France. He died in Paris. l'Hôpital wrote the first textbook on calculus in 1696 which was much influenced by the lectures of his teachers Johann Bernoulli, Jacob Bernoulli, and Leibnitz. l'Hôpital served as a cavalry officer but resigned because of nearsightedness. From that time on he directed his attention to mathematics. l'Hôpital was taught calculus by Johann Bernoulli in 1691. l'Hôpital was a very competent mathematician and solved the brachystochrone problem. The fact that this problem was solved independently by Newton, Leibnitz and Jacob Bernoulli puts l'Hôpital in very good company. His fame is based on his book *Analyse des Infiniment Petits pour L'intelligence des Lignes Courbes* (1692) which was the first textbook to be written on the differential calculus. In the introduction, l'Hôpital acknowledges his indebtedness to Leibnitz, Jacob Bernoulli and Johann Bernoulli, but l'Hôpital regarded the foundations provided by him as his own ideas. In this book is found the rule, now known as l'Hôpital's rule, for finding the limit of a rational function whose numerator and denominator tend to zero at a point.

2.4.2 Vibration of a Structure in Water

The technical problem of *fluid-structure interaction* is a very important subject, and also one that is complex. Most of the problems considered

80   CHAPTER 2. SDOF VIBRATION: AN INTRODUCTION

in this text implicitly assume that the medium surrounding the vibrating structure is of such low density, when compared to that of the structure, that it can be completely ignored. Generally this is true. However, if the fluid is water, or air flowing at high speed, its density cannot be ignored. An important engineering application where the fluid characteristics are integral to the problem formulation and solution is in the analysis and design of offshore drilling structures.[22,23,24] These structures can range from several hundred, to well over 1000 ft tall. Forces that such a structure will experience are due to gravity, ocean waves, currents, buoyancy forces, and others.

The question is how to incorporate the *added effects* of the fluid when there is a structure vibrating in its midst. For a static structure, the only fluid force is due to the variation of hydrostatic pressure. This is the *buoyancy force*, which is equal to the weight of the liquid displaced by the body, and acts in the opposite direction of gravity. The buoyancy force balances the weight of a static structure. The structure will sink to an equilibrium location so that there is an equality between buoyancy and gravity forces. Ship structures are designed so that the equilibrium is at an appropriate floating position.

If the structure is displaced from its equilibrium position, then an imbalance of forces leads to a structural oscillation about the equilibrium. Such an oscillation is significantly affected by the water surrounding the structure. When the structure oscillates, it causes motion in the surrounding fluid. In general, the fluid damps the structural motion as a result of drag between it and the structure, and the structure *entrains*, or pulls with it, some of the surrounding fluid as it moves. This means that the structure has an effective mass that is larger than its actual value. Therefore, it is necessary to include this *added mass* in any dynamic analysis. The fluid motion past the structure also creates a drag force that is a function of friction and the shape of the structure. If the fluid flow is an accelerating one, then in addition to the added mass, an inertia force is exerted by the fluid on the submerged structure.

A very important flow effect occurs when *non–aerodynamic* structures, called *bluff* or *blunt bodies*, are placed in steady flows. These bodies may be cylindrical in cross–section or have sharp edges, but are not tailored like a wing. When a fluid flows past such a body, it separates on the far side producing a trailing wake, creating vortices that are shed *alternately*, creating forces perpendicular to the fluid flow. Figure 2.17 depicts this effect

---

[22] O.M. Faltinsen, **Sea Loads on Ships and Offshore Structures**, Cambridge University Press, 1993.
[23] J.F. Wilson, **Dynamics of Offshore Structures**, Wiley–Interscience, 1986.
[24] S. Gran, **A Course in Ocean Engineering**, Elsevier, 1992.

2.5. CONCEPTS SUMMARY

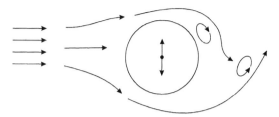

Figure 2.17: Vortices shed alternately from a submerged cylindrical structure.

for a circular cylinder. All fluid interaction effects are very complicated, requiring experimental data for their quantification.

2.5 Concepts Summary

Let us briefly summarize the key concepts we have introduced in this chapter. We have developed the theory of linear elastic vibration of a single degree of freedom system. Linearity and superposition are interchangeable concepts that are the foundation of this theory. This foundation will allow us to formulate and solve problems in Chapter 3 where more complex forcing is studied.

We have considered the formulation of the equation of motion using Newton's second law of motion and using an energy-based approach. The problem of free vibration and forced harmonic vibration have been solved for undamped system. In the next chapter, damping is included.

2.6 Problems

Problems for Section 2.2 – Mathematical Modeling: Deterministic

1. The beam in Figure 2.18 vibrates as a result of some loading. State the necessary assumptions to reduce this problem to a one degree–of–freedom oscillator. Then derive the equation of motion.

2. If a beam is supported continuously on a foundation, as shown in Figure 2.19, damping must be added to an idealized model to represent the viscous effects of the mat foundation. How would you idealize this system as a one degree–of–freedom oscillator. Then derive the equation of motion.

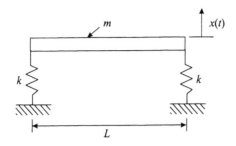

Figure 2.18: Vibrating beam.

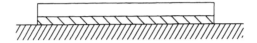

Figure 2.19: Vibrating beam on mat foundation.

3. An idealized one degree–of–freedom model is tested many times in order to estimate its natural frequency. It is relatively straightforward to measure its mass m, but stiffness k only approximately. How would you use the natural frequency data to estimate k ?

4. The cantilever beam in Figure 2.20 undergoes harmonic oscillation, being driven by a force of amplitude A with frequency nominally equal to ω. An examination of a long time–history of the response shows slight fluctuations about an exact harmonic response. If we divide this long time–history into segments of one period $(2\pi/\omega)$ and superpose these, we obtain the set of curves shown in Figure 2.21.

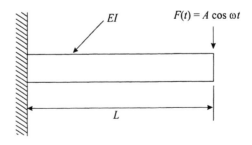

Figure 2.20: Vibrating cantilever beam.

2.6. PROBLEMS

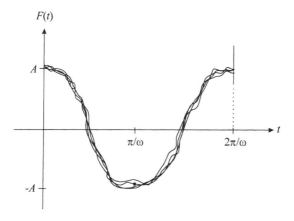

Figure 2.21: Overlapping time-histories.

How serious are such fluctuations in the response of the beam? Is there a way to relate the magnitude of the fluctuation to the maximum response? What would be a reasonable way to specify the value of ω.

5. The following nonlinear equations of motion were derived as part of a larger study. In each case, linearize the equations and include a discussion on the range of validity of the linearized equations. For example, a statement such as the following is necessary: The linearized equation of motion has an $x\%$ error in the nonlinear term $\cos\theta$ for $\theta > \theta_0$. Solve the linearized equations of motion and numerically solve the fully nonlinear equation using a program such as MATLAB. Cross–plot the linear and nonlinear time–histories and discuss the comparisons.

   (i) $\ddot{\theta} + 3\cos\theta = 0$, $\theta(0) = 0.5$, $\dot{\theta}(0) = 0$

   (ii) $\ddot{\theta} + 3\sin\theta = 0$, $\theta(0) = 0.5$, $\dot{\theta}(0) = 0$

   (iii) $\ddot{\theta} + 3\cos^2\theta = 0$, $\theta(0) = 0.5$, $\dot{\theta}(0) = 0$

   (iv) $\ddot{\theta} + 3\sin^2\theta = 0$, $\theta(0) = 0.5$, $\dot{\theta}(0) = 0$.

6. For the idealized models in Figures 2.22–2.25, draw each free body diagram, and derive the respective equations of motion using first Newton's second law of motion and then the energy method. State whether the oscillation is linear or nonlinear. What is the natural frequency of each oscillator?

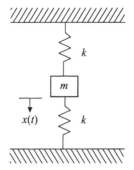

Figure 2.22: Vertical oscillator about equilibrium.

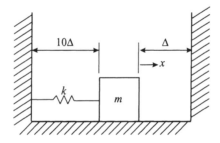

Figure 2.23: Vibration with impact.

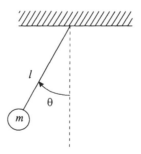

Figure 2.24: Oscillating pendulum.

2.6. PROBLEMS

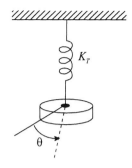

Figure 2.25: Torsional vibration.

7. Derive the equation of motion and natural frequency for the small mass m on the string that is under constant tension T. See Figure 2.26. Assume small displacements and that m is much greater than the mass of the string.

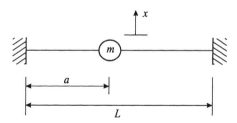

Figure 2.26: Mass on string.

8. Derive the equation of motion for a uniform stiff rod restrained from vertical motion by a torsional spring of stiffness K_T as shown in Figure 2.27. The torsional spring constant is determined by the application of a moment M and the measurement of the rotation θ, that is, $M = K_T \theta$. Calculate the natural frequency of oscillation. Let J define the moment of inertia of the rod about the point of oscillation. State any assumptions you make.

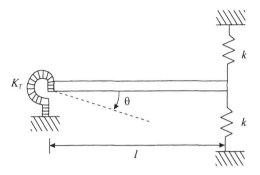

Figure 2.27: Restrained rigid rod.

9. A solid cylinder floating in equilibrium in a body of water is depressed slightly and released into motion, with a schematic shown in Figure 2.28.

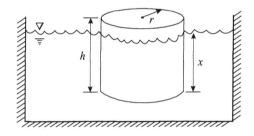

Figure 2.28: Solid body oscillating in a liquid.

Find the equilibrium position and solve for the natural frequency of oscillation assuming the cylinder remains upright at all times. Do your calculations in general so that it will be easy to substitute any liquid into the problem by changing only the specific gravity.

Suppose the assumption that the cylinder remains upright is not reasonable. Then what difficulties do you foresee in the calculations, and how might they be resolved?

10. Derive Equation 2.24.

Problems for Section 2.3 – Free Vibration with No Damping

11. Derive Equation 2.27 beginning with the equation of motion.

2.6. PROBLEMS

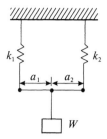

Figure 2.29: Weight hanging from two parallel springs.

12. Show that the period of free vibration of a load weighing W suspended from two parallel springs, as shown in Figure 2.29, is given by T;

$$T = 2\pi\sqrt{\frac{W}{g(k_1 + k_2)}}.$$

Show that the equivalent stiffness is $k = k_1 + k_2$.

Discuss the need to hang the weight asymmetrically, that is $a_1 \neq a_2$, so that the extension of the springs is identical and that the ratio is $a_1/a_2 = k_2/k_1$.

13. For the body suspended between two springs as in Figure 2.30, show that the period of oscillation is

$$T = 2\pi\sqrt{\frac{W}{g(k_1 + k_2)}}.$$

14. Two springs are joined in series as shown in Figure 2.31. If these are to be replaced by an equivalent spring, find the equivalent stiffness as well as the period of oscillation. The solution is

$$k = \frac{k_1 k_2}{k_1 + k_2}, \quad T = 2\pi\sqrt{\frac{w(k_1 + k_2)}{gk_1 k_2}}.$$

15. A *bifilar* pendulum of length $2a$ is suspended with two vertical strings, each of length l, as shown in Figure 2.32. Assuming small rotations of

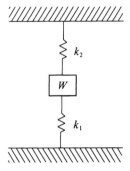

Figure 2.30: Body suspended between two springs.

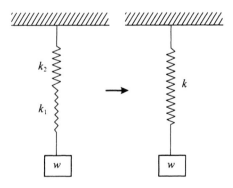

Figure 2.31: Two springs in series.

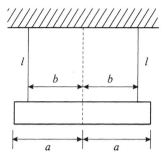

Figure 2.32: A *bifilar* pendulum.

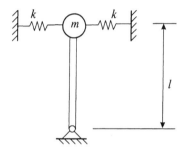

Figure 2.33: Inverted pendulum.

the strings, that the bar is essentially horizontal with half its weight supported by each string, show that the period is given by

$$T = \frac{2\pi a}{b}\sqrt{\frac{l}{3g}}.$$

16. An inverted hinged pendulum with a mass m at the top is suspended between two springs with constants k, as shown in Figure 2.33. The rod can be assumed rigid and massless, and in the vertical position the springs are unstretched. For small motion, the springs can be assumed to remain horizontal. Show that the period of oscillation is

$$T = \frac{2\pi}{\sqrt{\frac{2k}{m} - \frac{g}{l}}}.$$

17. A simple pendulum of initial length l_0 and initial angle θ_0 is released from rest. If length is a function of time according to $l = l_0 + \varepsilon t$, find the position (l, θ) of the pendulum at any time assuming small oscillations. Brush up on your Bessel functions before trying this problem. You should find the governing equation of motion to be

$$(l_0 + \varepsilon)\ddot{\theta} + 2\varepsilon\dot{\theta} + g\theta = 0,$$

and in transformed Bessel form to be

$$x^2\theta'' + 2x\theta' + \frac{xg}{\varepsilon^2}\theta = 0,$$

where $x = l_0 + \varepsilon t$.

18. A spring–mass system is suspended from the ceiling. The governing equation of motion is

$$m\ddot{x} + kx = 0,$$

where there is no external force except gravity and the initial conditions are $x(0) = 1$ in and $\dot{x}(0) = 1$ in/sec. Solve for the response $x(t)$, measured down from the point of suspension, for the two cases: *(i)* spring is assumed massless, and *(ii)* including the inertia effects of the spring as in *Example 2.6*.

Plot both solutions against each other to demonstrate the importance of including such inertia effects. Consider three sets of parameter values: *(a)* $W/g = 1$ lb, $k = 1$ lb/in, *(b)* $W/g = 1$ lb, $k = 10$ lb/in, *(c)* $W/g = 1$ lb, $k = 0.1$ lb/in. What conclusions can be drawn?

Problems for Section 2.4 – Harmonic Forced Vibration with No Damping

19. Derive Equation 2.32.

20. Solve the following governing equation of motion for four loading frequency cases:

$$\ddot{x} + x = F(t),$$

where *(i)* $F(t) = \cos 0.5t$, *(ii)* $F(t) = \cos 0.99t$, *(iii)* $F(t) = \cos t$, and *(iv)* $F(t) = \cos 2t$.

21. For the oscillator that is beating, for example,

$$\ddot{x} + 16x = A\sin(4 - \varepsilon)t,$$

2.6. PROBLEMS

where ε is very close in value but not equal to zero, what do you expect to happen as $\varepsilon \to 0$? Relate the rate of growth of the response amplitude to the value of ε or to the value of $4 - \varepsilon$.

22. Discuss the resonant vibration problem where it is assumed there is no damping. Physically what do you expect to happen as each succeeding amplitude is larger? How would you relate the resonant response to a structural material property such as its Young's modulus? Where does the linearity assumption come into play is this analysis?

23. Solve the following problem and discuss the results physically,

$$\ddot{x} + 9x = 3\sin t + \cos 3t.$$

24. Solve $\ddot{x} + \omega_n^2 = \frac{A}{m} \sin \omega_n t$.

25. Derive Equation 2.37.

Chapter 3

Single Degree of Freedom Vibration: Discrete Models with Damping

"The simplest model, but now with energy dissipation."

We continue the development of single degree of freedom systems started in Chapter 2. More realistic models are used in two ways. The first is with the introduction of damping to the structure, permitting it to respond to loading in a manner that better resembles our intuition and experiments. All motion dissipates energy in one form or another. Vibratory motion is always accompanied by energy loss and amplitude reduction. It is only the external forcing that keeps the system in motion. The second way we approach more realistic models is by examining a variety of loading functions and determining how the structure responds to each differently. We will find that these more complicated models can be built upon the earlier simple models. This is due to the linearity of the system, allowing us to use linear superposition.

3.1 Damping

Damping[1] is defined as the energy dissipation property of materials and structures undergoing time–dependent deformation and/or displacements. Damping is primarily associated with the irreversible transition of mechanical energy into thermal energy. The energy radiation into a surrounding medium is called radiation – or geometric – damping.

Damping can be classified as follows:

- Material Damping – Energy dissipation by deformation in a medium (irreversible intercrystal heat flux, grain boundary viscosity, etc.)

- Structural Damping – Damping in assembled structures including: material damping in members, frictional losses (microslip and macroslip) at contact surfaces (bolted, riveted, damped, welded connections), dissipation in a medium between surfaces in relative motion (gas pumping, squeeze film damping, lubricated bearing)

- Radiation Damping – Energy radiation into surrounding medium

- Active/Passive Damping – Damping with/without external energy and control

- Internal/External Damping – Damping inside/outside defined system boundary.

In the current chapter, as well as subsequent chapters, a relatively simple form of damping is used to represent a mix of material and structural damping. This is called *viscous damping* and is taken to be related to structural velocity, or equal to the structural velocity multiplied by a damping constant that is experimentally determined. Such an approach is approximate and leads the way to more sophisticated and accurate damping models. There are numerous references[2] to damping models and data. A few are referred to in this text.

3.2 Free Vibration With Damping

It is important to consider the effects of viscous damping on the free-vibration of the oscillator governed by

$$\ddot{x} + 2\zeta\omega_n\dot{x} + \omega_n^2 x = 0, \qquad (3.1)$$

---

[1] *The Influence of Damping on Waves and Vibrations*, L. Gaul, *Mechanical Systems and Signal Processing* (1999) **13**(1), 1–30.

[2] **Vibration Damping of Structural Elements**, C.T. Sun, Y.P. Lu, Prentice–Hall 1995.

3.2. FREE VIBRATION WITH DAMPING

where all the notation has already been defined in Chapter 2. All vibration is affected by the dissipation of energy. Such dissipation is modeled as damping. In this book, we use viscous damping exclusively.

To solve this differential equation of motion, a solution of the form $x(t) = Ae^{rt}$ is assumed as in Section 2.3. This equation is differentiated appropriately, and substituted into Equation 3.1. The two roots are determined to be

$$r_{1,2} = [-\zeta \pm \sqrt{\zeta^2 - 1}]\omega_n$$

and the response is then

$$x(t) = A_1 e^{-[\zeta - \sqrt{\zeta^2 - 1}]\omega_n t} + A_2 e^{-[\zeta + \sqrt{\zeta^2 - 1}]\omega_n t}. \tag{3.2}$$

The character of the solution depends on the value of the viscous damping factor ζ. The case where $\zeta = 1$ is known as *critical damping*, since it represents the boundary between aperiodic exponentially decaying motion (*overdamped*, $\zeta > 1$), and exponentially decaying oscillatory motion (*underdamped*, $\zeta < 1$). The critically damped system will approach equilibrium the fastest, but most structures will not have such a high viscous damping factor.

It is of particular interest to examine in more detail the underdamped case, $0 < \zeta < 1$, since the damping in most engineered structures fall in the range $0.01 < \zeta < 0.20$. Equation 3.2 can then be written as

$$x(t) = \left(A_1 e^{i\omega_d t} + A_2 e^{-i\omega_d t}\right) e^{-\zeta \omega_n t},$$

where

$$\omega_d = \omega_n \sqrt{1 - \zeta^2}.$$

ω_d is called the *angular frequency of damped vibration*. Euler's equation, along with a regrouping of terms, yields the more useful form of the solution

$$x(t) = Ce^{-\zeta \omega_n t} \cos(\omega_d t - \phi). \tag{3.3}$$

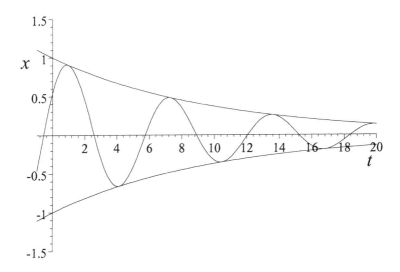

Figure 3.1: Plot of Equation 3.3 for $\zeta = 0.1$, $\omega_n = 1$ rad/sec, $\omega_d = 0.99499$ rad/sec, and $\phi = 1$ rad. $x(t) = \exp(-0.1t)\cos(0.99499t - 1)$.

This equation represents oscillatory motion bounded by an exponentially decaying envelope. See Figure 3.1. The initial displacement is equal to the intercept of the ordinate, x, and the initial velocity is equal to the slope at that point. The oscillation has amplitude $Ce^{-\zeta\omega_n t}$, and vibrates at the constant frequency ω_d, with *phase angle* ϕ. The constants C and ϕ are obtained by satisfying the initial conditions, as follows. Evaluate Equation 3.3 and its derivative at $t = 0$,

$$x(0) = C\cos(-\phi) \tag{3.4}$$
$$\dot{x}(0) = -C(-\omega_d \sin\phi + \zeta\omega_n \cos\phi). \tag{3.5}$$

Solve Equation 3.4 for C, and substitute this into Equation 3.5 (See Figure 3.2) to find

$$C = \frac{x(0)}{\cos\phi}$$
$$\tan\phi = \frac{\dot{x}(0) + x(0)\zeta\omega_n}{x(0)\omega_d}.$$

These equations are explicit functions of the initial displacement and velocity, and therefore,

$$x(t) = \frac{x(0)}{\cos\phi}e^{-\zeta\omega_n t}\cos\left(\omega_d t - \tan^{-1}\left(\frac{\dot{x}(0) + x(0)\zeta\omega_n}{x(0)\omega_d}\right)\right), \tag{3.6}$$

3.2. FREE VIBRATION WITH DAMPING

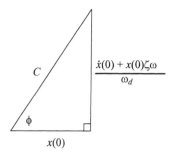

Figure 3.2: Right triangle of intial conditions.

where

$$\cos\phi = \frac{x(0)\omega_d}{\sqrt{[\dot{x}(0) + x(0)\zeta\omega_n]^2 + [x(0)\omega_d]^2}}. \tag{3.7}$$

The phase angle ϕ is a measure of the offset of the first peak from the $t=0$ axis. Specifically, the first peak of the oscillation occurs at $t = \phi/\omega_n$ for an undamped system. Phase is a more important response parameter for forced structures. By substituting Equation 3.7 into Equation 3.6 cancels out the $x(0)$ factor, so there is no way for the denominator to be equal to zero in the factor $x(0)/\cos\phi$.

As an example of the kind of information that can be obtained from Equation 3.6, the time to maximum displacement can be found by differentiating $x(t)$ as follows

$$\frac{dx}{dt} = 0, \tag{3.8}$$

and solving for $t = t_{max}$.

Example 3.1 Numerical Evaluation of Equation 3.6
Graphically examine the response Equation 3.6 more closely to understand the effects of ζ on the response. The plots in Figure 3.3(a) assume $\omega_n = 1$ rad/sec, $x(0) = 1$ cm, $\dot{x}(0) = 1$ cm/sec, and four different values of ζ: 0.01, 0.05, 0.10, and 0.20. We see that damping is of primary importance. Figure 3.3(b) shows the general response for a continuous spectrum of ζ and time, where, for $\zeta \geq 1$, the response turns from oscillatory to purely decaying motion. ∎

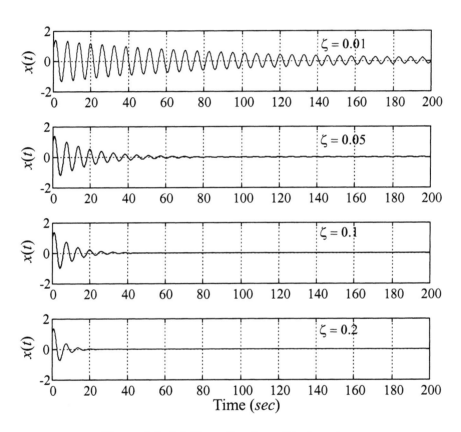

Figure 3.3: (a) Free vibration with damping.

3.2. FREE VIBRATION WITH DAMPING

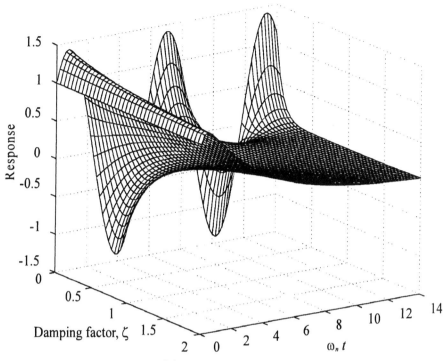

Figure 3.2: (b) Free vibration with damping.

Euler

Leonhard Euler lived during 1707–1783 and was born in Basel, Switzerland. He died in St. Petersburg, Russia. Euler's father wanted his son to follow him into the church and sent him to the University of Basel to prepare for the ministry. However, geometry soon became his favorite subject. Euler obtained his father's consent to change to mathematics after Johann Bernoulli had used his persuasion. Bernoulli became his teacher. He joined the St. Petersburg Academy of Science in 1727, two years after it was founded by Catherine I, the wife of Peter the Great. Euler served as a medical lieutenant in the Russian navy from 1727 to 1730. In St Petersburg he lived with Daniel Bernoulli. He became professor of physics at the academy in 1730 and professor of mathematics in 1733. He married and left Bernoulli's house in 1733. He had 13 children of which 5 survived their infancy. He claimed that he made some of his greatest discoveries while holding a baby on his arm with other children playing round his feet.

The publication of many articles and his book *Mechanica* (1736-37), which extensively presented Newtonian dynamics in the form of mathematical analysis for the first time, started Euler on the way to major mathematical work. In 1741, at the invitation of

Frederick the Great, Euler joined the Berlin Academy of Science, where he remained for 25 years. Even while in Berlin, he received part of his salary from Russia and never got on well with Frederick. During his time in Berlin, he wrote over 200 articles, three books on mathematical analysis, and a popular scientific publication. In 1766 Euler returned to Russia. He had been arguing with Frederick the Great over academic freedom and Frederick was greatly angered at his departure.

Euler lost the sight of his right eye at the age of 31 and soon after his return to St Petersburg he became almost entirely blind after a cataract operation. But because of his remarkable memory, he was able to continue with his work on optics, algebra, and lunar motion. Amazingly, after 1765 (when Euler was 58), he produced almost half his works despite being totally blind. After his death in 1783, the St. Petersburg Academy continued to publish Euler's unpublished work for nearly 50 more years. Euler made large bounds in modern analytic geometry and trigonometry. He made decisive and formative contributions to geometry, calculus and number theory. In number theory he did much work in correspondence with Goldbach. He integrated Leibnitz's differential calculus and Newton's method of fluxions into mathematical analysis. In number theory he stated the prime number theorem and the law of biquadratic reciprocity. He was the most prolific writer of mathematics of all time. His complete works contains 886 books and papers. We owe to him the notations $f(x)$ (1734), e for the base of natural logs (1727), i for $\sqrt{-1}$ (1777), π for pi, $\sum$ for summation (1755), among other symbols. He also introduced the beta and gamma functions, and integrating factors for differential equations. He studied continuum mechanics, lunar theory with Clairaut, the three body problem, elasticity, acoustics, the wave theory of light, hydraulics, and music. He laid the foundation of analytical mechanics, especially in his *Theory of the Motions of Rigid Bodies* (1765).

Example 3.2 The Logarithmic Decrement
Although it is generally assumed that the necessary structural parameter values, such as mass and stiffness, can be easily obtained, this is not true for damping. For complex structures, damping is a very difficult characteristic to establish. A variety of theoretical and experimental techniques have been developed to estimate damping. Here, the concept of the *logarithmic decrement* is introduced and applied to the problem of estimating damping.

Suppose a decaying time–history is obtained experimentally as in Figure 3.4. Consider two successive peaks, or two time instances separated by the damped period $T_d = 2\pi/\omega_d$: $x_1 = x(t_1)$ and $x_2 = x(t_2)$. Then, according to Equation 3.3,

$$\begin{aligned} x_1 = x(t_1) &= Ce^{-\zeta\omega_n t_1}\cos(\omega_d t_1 - \phi) \\ x_2 = x(t_2) &= Ce^{-\zeta\omega_n t_2}\cos(\omega_d t_2 - \phi), \end{aligned}$$

3.2. FREE VIBRATION WITH DAMPING

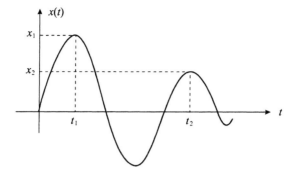

Figure 3.4: Decaying free vibration.

and the ratio of the two displacements is given by

$$\frac{x_2}{x_1} = e^{-\zeta\omega_n(t_2-t_1)} \frac{\cos(\omega_d t_2 - \phi)}{\cos(\omega_d t_1 - \phi)}. \tag{3.9}$$

Since t_2 and t_1 are selected T_d sec apart, we can write

$$\begin{aligned}
\cos(\omega_d t_2 - \phi) &= \cos(\omega_d t_1 - \phi + \omega_d T_d) \\
&= \cos(\omega_d t_1 - \phi + 2\pi) \\
&= \cos(\omega_d t_1 - \phi),
\end{aligned}$$

and Equation 3.9 becomes

$$\frac{x_2}{x_1} = e^{-\zeta\omega_n T_d}.$$

Define the natural log of the amplitude ratio x_1/x_2 as the *logarithmic decrement*, δ, and arrive at

$$\delta = \ln \frac{x_1}{x_2} = \zeta\omega_n T_d = \frac{2\pi\zeta}{\sqrt{1-\zeta^2}}.$$

Using these results, we can show that for a set of measurements n cycles apart,

$$\begin{aligned}
\frac{x_1}{x_{n+1}} &= \frac{x_1}{x_2} \frac{x_2}{x_3} \cdots \frac{x_j}{x_{n+1}} \\
&= \left(e^{\zeta\omega_n T_d}\right)^n.
\end{aligned}$$

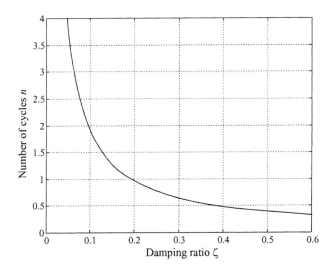

Figure 3.5: Number of cycles n to a 70% reduction in amplitude for small ζ.

Therefore, taking the natural log of each side, and making use of the definition of logarithmic decrement, we have

$$\delta = \frac{1}{n}\ln\frac{x_1}{x_{n+1}}$$

for data points x_1 and x_{n+1}.

Suppose we need to estimate the number of cycles required before a structural amplitude decays 70% of its maximum. Using the above definition,

$$\delta = \frac{1}{n}\ln\left(\frac{1}{0.30}\right),$$

where n is the number of cycles. Substitute for δ in terms of ζ, that is, $\delta = 2\pi\zeta/\sqrt{1-\zeta^2}$. Then evaluating the natural log, we can solve for n, to find

$$n = \frac{1.204\sqrt{1-\zeta^2}}{2\pi\zeta}.$$

This curve can be easily plotted exactly, but assume that ζ is small, say $\zeta < 0.20$. Then $\sqrt{1-\zeta^2} \approx 1$ and $n\zeta = 0.192$, a hyperbolic equation that is graphed in Figure 3.5.

3.2. FREE VIBRATION WITH DAMPING

Using this curve we can answer questions such as: If $0.1 < \zeta < 0.3$ *what is the range for n?* or *What must the value of ζ be so that $n < 3$?* These are important questions in a preliminary design. ∎

3.2.1 Some Time Constants

The concept of *characteristic times* is one that appears in many physical sciences. They are representative of time scales that have intrinsic physical meanings for the structure, fluid, or system in general. We already know one time constant, the frequency of oscillation of a structure,

$$\omega_n = \sqrt{\frac{k}{m}}.$$

The frequency was found, or rather defined, when the ratio of stiffness to mass arose in the equation of motion. In a similar way, when the damping factor is divided by the mass, we obtain

$$\frac{c}{m} = 2\zeta\omega_n \equiv \frac{1}{\tau},$$

where

$$\tau = \frac{1}{2\zeta\omega_n} = \frac{m}{c}$$

is called the *relaxation time* because the product $\zeta\omega_n$ characterizes the decay rate of the amplitude of the freely vibrating damped system. In some applications, the factor 2 is omitted.

Another time parameter, called the *correlation time*, can be defined as a measure of the time the motion can be predicted into the future, linking characteristic time to a probabilistic concept. This will be studied in Chapter 5.

Example 3.3 Some Typical Time Scales
To generate some intuition on time scales, let us consider some representative numbers. Most engineering structures have a damping factor in the range $\zeta \sim 0.01 - 0.20$. For $\omega_n = 1$ rad/sec and $\zeta = 0.05$ we can calculate

$$2\zeta\omega_n = 0.10$$
$$\tau = \frac{1}{2\zeta\omega_n} = 10.0 \text{ sec}$$

Similarly, for $\zeta = 0.15$, $\tau = 3.33$ sec. What if $\omega_n = 3$ rad/sec? Then the above range becomes 3.33 sec−1.11 sec. For $\omega_n = 0.316$ rad/sec the range becomes 31.646 sec -105.48 sec. τ versus $2\zeta\omega_n$ can be easily plotted to show the hyperbolic relation between the expressions. Recall that $2\zeta\omega_n = c/m$, so the time scale is a ratio of inertia to damping, which makes physical sense. ∎

3.2.2 Phase Plane

We revisit the phase plane, this time for the damped oscillator, which can be converted into two first order differential equations,

$$\begin{aligned} \dot{x} &= y \\ \dot{y} &= -\omega_n^2 x - 2\zeta\omega_n \dot{x} \\ &= -\omega_n^2 x - 2\zeta\omega_n y. \end{aligned}$$

The equilibrium position is at the origin, where the velocities $(\dot{x}, \dot{y}) = (0,0)$. The phase paths are given by

$$\frac{dy}{dx} = \frac{-\omega_n^2 x - 2\zeta\omega_n y}{y}.$$

Example paths are shown in Figure 3.6. Note that for the critical damping case $\zeta = 1$, the path goes straight to the equilibrium position without any curving around O. This implies that the particle does not oscillate, but rather decays directly to $(x, y) = (0, 0)$, the equilibrium position. For the underdamped cases $\zeta = 0.1$ and 0.5, the path goes around O, signifying decaying oscillations. The less damping the system posses, the more turns about O.

3.3 Forced Vibration With Damping

Forces on structures are actually very complex functions of time and space. It is necessary to idealize and simplify them so that the structural response to specific loads can be studied and understood. Remarkably, for linear systems, it is possible to study a few loading types and use these cases to build more general and complex loading situations. With this in mind, we will consider the following excitation cases: *(i)* harmonic, *(ii)* periodic but not harmonic, *(iii)* step, *(iv)* impulse, and *(v)* arbitrary excitation.

The logic in this sequence is to build progressively more sophisticated mathematical models for the excitation which more closely approach the

3.3. FORCED VIBRATION WITH DAMPING

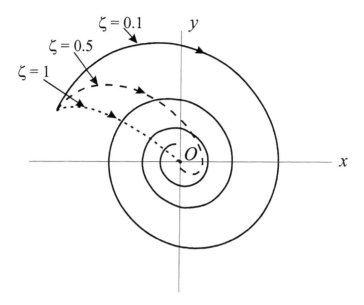

Figure 3.6: Phase paths for the damped oscillator with $x(0) = -10$, $v(0) = 4$, $\omega_n = 1$ rad/sec, and three values of ζ.

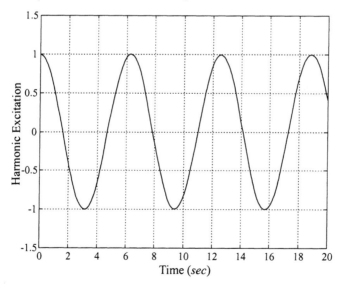

Figure 3.7: Harmonic excitation; $F(t) = A\cos\omega t$, $\omega = 1$ rad/sec, $A = 1$ cm.

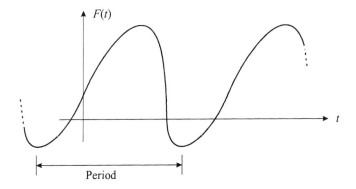

Figure 3.8: Periodic but not harmonic excitation.

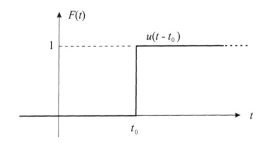

Figure 3.9: Step loading.

reality we perceive.[3] For example, harmonic excitation is an idealized model for loading due to imperfections in rotating machinery. More complicated than these are periodic forces which are not harmonic, such as on/off loads. The introduction of impulse loading prepares the groundwork for the study of the *convolution* integral and the response to arbitrary loads. The step loading is introduced primarily because it is related to the impulse case, and useful in the idealized modeling of impact loads and in control theory.

The governing equation of motion is, in all these instances,

$$\ddot{x} + 2\zeta\omega_n\dot{x} + \omega_n^2 x = F(t), \qquad (3.10)$$

where specific functions will be substituted for $F(t)$, the force per unit mass. The undamped case, $\ddot{x} + \omega_n^2 x = F(t)$, will always be a special case obtained

---

[3]The next more realistic loading model omitted in the above sequence is for a *random* load. This is an extremely important class which requires some prerequisite probability theory; it will be introduced in Chapter 5.

3.3. FORCED VIBRATION WITH DAMPING

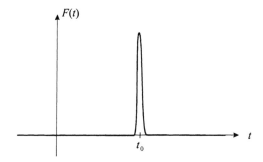

Figure 3.10: Impulse loading.

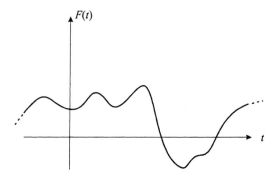

Figure 3.11: Arbitrary loading.

by setting $\zeta = 0$.

3.4 Harmonic Excitation and Damped Response

The governing equation of motion is given by
$$\ddot{x} + 2\zeta\omega_n \dot{x} + \omega_n^2 x = A\cos\omega t, \tag{3.11}$$
where $A\cos\omega t$ is a harmonic load[4] per unit mass with amplitude A and frequency ω. Note that the constant A incorporates a factor $1/m$, that is, A is the maximum force per unit mass, $A = A_1/m$, where A_1 has units of force.

The primary interest is in finding the *amplitude* and the *phase lag* of the structural response. The reason why the introduction of damping causes this response lag is discussed below. In complex notation, the equivalent version of Equation 3.11 is
$$\ddot{x} + 2\zeta\omega_n \dot{x} + \omega_n^2 x = Ae^{i\omega t}, \tag{3.12}$$
where the *real part* of the mathematical solution, $\text{Re}[x(t)]$, will be the physical response.[5]

Analytical approaches to both representations of a harmonic force will be developed since algebraic dexterity with either route is useful. Let us begin with the cosine form of the load. Assume that the (particular) solution to Equation 3.11 has the form[6]
$$x(t) = B_1 \cos\omega t + B_2 \sin\omega t, \tag{3.13}$$
where constants B_1 and B_2 are evaluated by requiring that the assumed solution $x(t)$ satisfies the governing equation, that is,
$$\left(-\omega^2 B_1 + 2\zeta\omega_n\omega B_2 + \omega_n^2 B_1 - A\right)\cos\omega t$$
$$+ \left(-\omega^2 B_2 - 2\zeta\omega_n\omega B_1 + \omega_n^2 B_2\right)\sin\omega t = 0.$$

This equation can be satisfied for all values of t only if the expressions in the parentheses vanish identically,
$$-\omega^2 B_1 + 2\zeta\omega_n\omega B_2 + \omega_n^2 B_1 = A$$
$$-\omega^2 B_2 - 2\zeta\omega_n\omega B_1 + \omega_n^2 B_2 = 0,$$

---

[4] We could have chosen a load equal to $A\sin\omega t$.

[5] Had the right hand side of Equation 3.11 been $A\sin\omega t$, then, with the complex exponential form, the physical solution is given by the *imaginary part* of the solution: $\text{Im}[x(t)]$.

[6] We will need both the cosine and the sine in our solution since the governing equation has first *and* second derivatives. Recall that for the undamped response, only one of the two harmonic terms is non-zero because the governing equation does not contain a first derivative term.

3.4. HARMONIC EXCITATION AND DAMPED RESPONSE

from which

$$B_1 = \frac{A(\omega_n^2 - \omega^2)}{(\omega_n^2 - \omega^2)^2 + (2\zeta\omega_n\omega)^2}$$

$$B_2 = \frac{A(2\zeta\omega_n\omega)}{(\omega_n^2 - \omega^2)^2 + (2\zeta\omega_n\omega)^2}.$$

Therefore, Equation 3.13 becomes

$$x(t) = \frac{A(\omega_n^2 - \omega^2)}{(\omega_n^2 - \omega^2)^2 + (2\zeta\omega_n\omega)^2} \cos\omega t + \frac{A(2\zeta\omega_n\omega)}{(\omega_n^2 - \omega^2)^2 + (2\zeta\omega_n\omega)^2} \sin\omega t. \tag{3.14}$$

This equation is in a form that makes the response difficult to visualize. To obtain a more useful form, Equation 3.14 can be written as

$$x(t) = D\cos(\omega t - \theta), \tag{3.15}$$

where, as in Section 2.3, $D = \sqrt{B_1^2 + B_2^2}$, and $\theta = \tan^{-1}(B_2/B_1)$, or,

$$\begin{aligned} D &= \frac{A}{\sqrt{(\omega_n^2 - \omega^2)^2 + (2\zeta\omega_n\omega)^2}} \\ &= \frac{A/\omega_n^2}{\sqrt{(1 - \omega^2/\omega_n^2)^2 + (2\zeta\omega/\omega_n)^2}}, \end{aligned} \tag{3.16}$$

$$\theta = \tan^{-1}\frac{2\zeta\omega/\omega_n}{1 - \omega^2/\omega_n^2}. \tag{3.17}$$

Note that the phase lag is independent of the loading amplitude. It represents the lag in the peak structure displacement after the load acts. Where $\zeta = 0$, there is no lag and the structure responds instantaneously. Damping results in a delay of structural response to loading.

A *magnification factor* β can be defined as

$$\beta = \frac{1}{\sqrt{(1 - \omega^2/\omega_n^2)^2 + (2\zeta\omega/\omega_n)^2}}, \tag{3.18}$$

so that response Equation 3.15 becomes

$$x(t) = \frac{A}{\omega_n^2}\beta\cos(\omega t - \theta),$$

where A incorporates the factor $1/m$, $(A = A_1/m)$, and therefore

$$D = \frac{A}{\omega_n^2}\beta = \frac{Am}{k}\beta = \frac{A_1}{k}\beta,$$

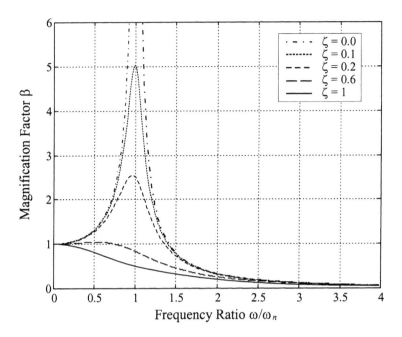

Figure 3.12: Magnification factor β as a function of ω/ω_n.

with A_1/k recognized as the static deflection.

It is important and instructive to plot the magnification factor β and the phase lag θ as functions of the frequency ratio ω/ω_n for various values of the viscous damping factor ζ. The β curves in Figure 3.12 are a family, each curve represents a different value of ζ. An increase in the damping factor results in a decrease in the response amplitude. The maximum amplitude occurs, for $\zeta > 0$, at a frequency ratio which is slightly less than 1, as the following example shows.

Example 3.4 Peak Amplitude vs. Frequency Ratio
To find the peak amplitude or amplification, we need to find the frequency ratio at which β is a maximum. Let

$$\frac{\omega^2}{\omega_n^2} \equiv r.$$

Taking the derivative of β with respect to the frequency ratio r and setting the resulting equation to zero,

$$\frac{d\beta}{dr} = 0,$$

3.4. HARMONIC EXCITATION AND DAMPED RESPONSE

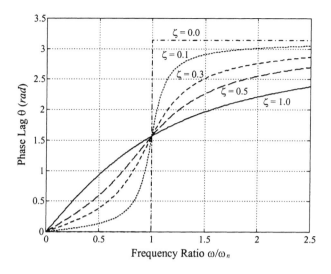

Figure 3.13: Phase lag θ as a function of ζ and ω/ω_n.

leads to the equation for the value of r at which the curve for β is a maximum. Doing this, we find that

$$r = 1 - 2\zeta^2,$$

or

$$\frac{\omega}{\omega_n} = \sqrt{1 - 2\zeta^2}.$$

We see that for ζ values in the range 0.01 to 0.30, the shift from ω_n of the peak amplitude is no more than 10%. ∎

The phase diagram, shown in Figure 3.13 as a function of frequency ratio, is equally interesting. For the case with no damping, for a load with frequency below resonance the phase $\theta = 0$, and above resonance the phase $\theta = \pi$. A discontinuity exists at the resonance point.

When damping is included, the sharp transition observed at resonance for the damped–free case is softened. For all cases regardless of damping, the phase is $\pi/2$ radians or 90° at resonance. This property becomes useful in problems of structural identification and testing. For example, if a vibration test is performed on a machine with phase data plotted as a function of ω/ω_n for a broad band of ω, then at the point where this curve passes through the phase value $\pi/2$ is a resonant frequency, indicating that this is one of

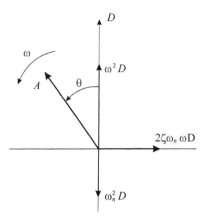

Figure 3.14: Vector diagram for Newton's forces.

the natural frequencies of the structure. Structures have multiple natural frequencies, one for each degree of freedom.

Before adding the steady–state solution to the transient to obtain the complete solution, it is instructive to look at the vector representation of governing Equation 3.11, following Den Hartog (p. 48–50). This discussion will help in better understanding Figure 3.12. Figure 3.14 depicts the magnitude and relative direction of the forces on both sides of Newton's second law of motion, where $x, \dot{x}$ and $\ddot{x}$ are obtained using Equation 3.15. Note that only the magnitude and the directions are shown. Newton's second law requires that the *vector* sum of all inertia, damping, stiffness, and external forces be zero at all times. We know that the stiffness force acts opposite to the displacement; the damping force acts 90° ahead of the stiffness; the inertia force acts 90° ahead of the damping; and the external force acts θ ahead of the displacement.

From Figure 3.14, the vertical and horizontal summations yield, respectively,

$$\omega_n^2 D - \omega^2 D - A\cos\theta = 0$$
$$2\zeta\omega_n\omega D - A\sin\theta = 0.$$

From these two equations, the unknowns D and θ can be solved for directly, now in a new way.

The vector diagram of Figure 3.14 helps us also visualize how the amplitude and phase angle vary with driving frequency. For very slow oscillation ($\omega \approx 0$), the damping and inertia forces are negligible, as discussed earlier, and the stiffness force has a magnitude $A = \omega_n^2 D$, where $\theta = 0$ so that the

3.4. HARMONIC EXCITATION AND DAMPED RESPONSE

forces in the vertical direction balance out. With increasing frequency, the magnitude of the damping force increases according to $2\zeta\omega_n\omega D$, and the inertia force grows even faster as $\omega^2 D$. The phase angle cannot be zero any more since A must have a horizontal component to balance the damping force. The inertia force vector will grow until it becomes as large as the spring force. Then θ must be $\pi/2$ and $A = 2\zeta\omega_n\omega D$. This happens at resonance because $\omega^2 D = \omega_n^2 D$, or $\omega = \omega_n$. Thus, at resonance, the phase angle is $\pi/2$, independent of damping. Above this frequency, $\omega^2 D$ will grow larger than $\omega_n^2 D$, so that A rotates downward and $\theta > \pi/2$. For very high frequencies, stiffness force $\omega_n^2 D$ is insignificant compared to inertia force $\omega^2 D$, so that A is used to balance the inertia force with $\theta = \pi$.

In summary, at low frequencies the spring force overcomes the external force. At high frequencies, the inertia force overcomes the external force. At resonance, the damping force overcomes the external force.

To obtain the complete solution, the particular response, given by Equation 3.13, is added to the homogeneous response, given by Equation 3.3,

$$x(t) = Ce^{-\zeta\omega_n t}\cos(\omega_d t - \phi) + B_1\cos\omega t + B_2\sin\omega t. \quad (3.19)$$

At this point in the analysis, the initial conditions $x(0)$ and $\dot{x}(0)$ can be satisfied and the constants C and ϕ determined. Following the same procedure as in Section 3.2, we find

$$C = \frac{x(0) - B_1}{\cos\phi} \quad (3.20)$$

$$\tan\phi = \frac{\dot{x}(0) + [x(0) - B_1]\zeta\omega_n - B_2\omega}{[x(0) - B_1]\omega_d}. \quad (3.21)$$

The general response, Equation 3.19, has two distinct components. The first, resulting from the initial conditions, is important only during the early stages of the response, while the second, resulting from the forcing, is important as long as the force is acting on the structure. For example, if the load is an explosion or an impact, then we care only about happens in the first few seconds of the response since its largest amplitudes occur during this time. In this case, the harmonic components of the steady state response should be added to the free vibration since the effects of the initial conditions have not yet decayed to insignificance. On the other hand, if the loading is more standard and over long periods of time, then the influence of the initial conditions rapidly becomes small and unimportant when compared to the forced response. The decaying exponential soon erases any memory of the initial conditions. In this case, one may solve only for the forced response, ignoring initial conditions and free vibration, since the long term behavior is of greater interest.

Because of the response characteristics just discussed, the particular solution is often called the *steady state* response, and the homogeneous solution is often called the *transient* response.

3.4.1 Harmonic Excitation in Complex Notation

An alternate approach to derive the steady state response is for the loading to be written in complex exponential form, as in Equation 3.12. Here, two possible forms of the response can be assumed:

$$\begin{aligned} x(t) &= X(i\omega)e^{i\omega t} \\ x(t) &= X(\omega)e^{i(\omega t - \phi)}. \end{aligned} \quad (3.22)$$

In the first, one solves for $X(i\omega)$, a complex quantity that implicitly embodies both amplitude and phase lag information. To see this, recall that any complex number can be written as *modulus* times *phase*,

$$X(i\omega) = |X(i\omega)|e^{-i\phi},$$

where $|X(i\omega)|$ is the modulus, or response amplitude, a real number, and $e^{-i\phi}$ is the phase.[7] Since damped structural response lags behind the load, the negative sign is introduced, thus explaining the term *phase lag* for ϕ. The second formulation for $x(t)$ directly includes amplitude and phase.

Let us proceed with the first. Differentiate assumed solution Equation 3.22 and substitute the resulting expressions for $\dot{x}$ and $\ddot{x}$ into the governing equation. We find

$$\begin{aligned} \dot{x} &= i\omega X(i\omega)e^{i\omega t} \\ \ddot{x} &= -\omega^2 X(i\omega)e^{i\omega t}, \end{aligned}$$

and then

$$X(i\omega)\left[-\omega^2 + 2i\omega\zeta\omega_n + \omega_n^2\right]e^{i\omega t} \equiv Ae^{i\omega t}.$$

Simplifying algebraically, the *frequency response function* is defined as[8]

$$H(i\omega) \equiv \frac{X(i\omega)}{A} = \frac{1/\omega_n^2}{1 - (\omega/\omega_n)^2 + 2i\zeta\omega/\omega_n}. \quad (3.23)$$

---

[7] Let $X(i\omega) = x + iy$, then $|X(i\omega)| = \sqrt{x^2 + y^2}$ and $\phi = \arctan(y/x)$. These are essentially geometric constructs that can be easily written down by drawing the coordinate point (x, y) in the complex plane, and noting that $|X(i\omega)|$ is the position vector to the point, and ϕ is the angle of that vector measured from horizontal.

[8] This expression is sometimes called a *transfer function*, used in control theory to represent the relation between input force and output response. See Chapter 6.

3.4. HARMONIC EXCITATION AND DAMPED RESPONSE

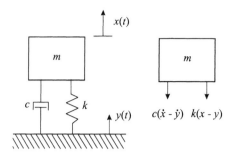

Figure 3.15: Free body diagram of base excited structure.

Introducing $H(i\omega) = |H(i\omega)|e^{-i\phi}$ into Equation 3.23, and substituting the result into Equation 3.22, yields the structural response

$$x(t) = A|H(i\omega)|e^{i(\omega t - \phi)} \quad (3.24)$$

$$\text{where } |H(i\omega)| = \frac{1/\omega_n^2}{\sqrt{(1 - \omega^2/\omega_n^2)^2 + (2\zeta\omega/\omega_n)^2}} \quad (3.25)$$

$$\phi = \tan^{-1}\frac{2\zeta\omega/\omega_n}{1 - \omega^2/\omega_n^2}. \quad (3.26)$$

Note that $|H(i\omega)| = \beta/\omega_n^2$ is the magnification factor of Equation 3.18, and $\phi = \theta$ in Equation 3.17.[9]

There are many applications for the theory just developed for harmonic excitation. Several are presented next. These are the problems of *base excitation, excitation due to rotating unbalance*, and the modeling of *vibration measurement devices*.

Example 3.5 Base Excitation

Base excitation problems have many manifestations in applications. These include the vibration of structures on foundations (such as in earthquake engineering, where the loading is through the base), the response of an automobile to road irregularities and bumps, and the interaction between a machine and its support. We make the assumption here that the excitation is harmonic. This analysis provides us with some of the key characteristics of base excited systems. Therefore, when more realistic excitations such as random loads are considered, we are prepared to examine the general behavioral characteristics.

---

[9]Looking ahead, the above is related to the development in Section 3.6 on Laplace transforms, where if $s = i\omega$ is substituted in Equation 3.44, then $G(s|_{s=i\omega})/\omega_n^2 = H(i\omega)$.

Consider the idealized model in Figure 3.15 of a structure connected to a base. The structure and the base are connected by a substructure with stiffness and damping. We are interested in how the structure responds to the force. The base excitation is assumed to be harmonic and is given the functional form $y(t) = Y \sin \omega_b t$. From the free body diagram, Newton's second law of motion states that[10]

$$m\ddot{x} = -c(\dot{x} - \dot{y}) - k(x - y),$$

or

$$m\ddot{x} + c\dot{x} + kx = c\dot{y} + ky$$
$$\ddot{x} + 2\zeta\omega_n\dot{x} + \omega_n^2 x = 2\zeta\omega_n\omega_b Y \cos\omega_b t + \omega_n^2 Y \sin\omega_b t. \quad (3.27)$$

To solve for the steady state response, proceed as we did for the general solution of a damped oscillator under harmonic excitation. To simplify some of the algebra, let the right hand side of Equation 3.27 be written as $A_1 \cos \omega_b t + A_2 \sin \omega_b t$, where $A_1 = 2\zeta\omega_n\omega_b Y$ and $A_2 = \omega_n^2 Y$. Assume a response of the form

$$x(t) = B_1 \cos \omega_b t + B_2 \sin \omega_b t,$$

differentiate and substitute into governing differential Equation 3.27 to find

$$(-\omega_b^2 B_1 + 2\zeta\omega_n\omega_b B_2 + \omega_n^2 B_1 - A_1) \cos\omega_b t$$
$$+(-\omega_b^2 B_2 - 2\zeta\omega_n\omega_b B_1 + \omega_n^2 B_2 - A_2) \sin\omega_b t = 0.$$

For the sum to be equal to zero for all time t, the expressions in parentheses must be identically zero, that is,

$$-\omega_b^2 B_1 + 2\zeta\omega_n\omega_b B_2 + \omega_n^2 B_1 = A_1 \quad (3.28)$$
$$-\omega_b^2 B_2 - 2\zeta\omega_n\omega_b B_1 + \omega_n^2 B_2 = A_2, \quad (3.29)$$

which must be solved for B_1 and B_2.

As an alternate to this algebraically intensive approach, it would be more convenient to use the general solution, Equation 3.15, along with Equations 3.16 and 3.17. In this way, we solve the response to each harmonic load separately and then add the two solutions to obtain the complete re-

---

[10] We choose relative velocity and relative displacement as $(\dot{x} - \dot{y})$ and $(x - y)$ because we are interested in the motion of mass m and how the damping and stiffness forces, $c\dot{x}$, kx, are altered by the motion of the base, $y, \dot{y}$.

3.4. HARMONIC EXCITATION AND DAMPED RESPONSE

sponse. Doing this we have

$$\begin{aligned}
x(t) &= x_1(t) + x_2(t) \\
&= \frac{2\zeta\omega_n\omega_b Y}{\sqrt{(\omega_n^2 - \omega_b^2)^2 + (2\zeta\omega_n\omega_b)^2}} \cos(\omega_b t - \phi) \\
&\quad + \frac{\omega_n^2 Y}{\sqrt{(\omega_n^2 - \omega_b^2)^2 + (2\zeta\omega_n\omega_b)^2}} \sin(\omega_b t - \phi) \\
\phi &= \arctan\left(\frac{2\zeta\omega_n\omega_b}{\omega_n^2 - \omega_b^2}\right).
\end{aligned}$$

The phase ϕ is the same for both parts of the solution because the phase difference between the two has already been accounted for by using a *sine* and a *cosine* solution. The two harmonic functions can be combined since they have the same arguments, with the steady state response given by

$$x(t) = Y\left[\frac{1 + (2\zeta\omega_b/\omega_n)^2}{(1 - \omega_b^2/\omega_n^2)^2 + (2\zeta\omega_b/\omega_n)^2}\right]^{1/2} \cos(\omega_b t - \phi - \gamma) \quad (3.30)$$

$$\gamma = \arctan\left(\frac{\omega_n}{2\zeta\omega_b}\right).$$

The algebraic expression multiplying the cosine function is a constant once numbers are substituted for all the parameters. Define this constant as X. Then the ratio X/Y, called the *displacement transmissibility*, is given by the expression

$$\frac{X}{Y} = \left[\frac{1 + (2\zeta\omega_b/\omega_n)^2}{(1 - \omega_b^2/\omega_n^2)^2 + (2\zeta\omega_b/\omega_n)^2}\right]^{1/2}, \quad (3.31)$$

and is plotted in Figure 3.16. Comparing this result with Equation 3.18 for the magnification factor β is instructive. The second term in the numerator is due to the base loading that is carried through the damping element. The first term is due to the loading that is carried through the stiffness element.

In Figure 3.16, note that $X = Y$ at $\omega_b/\omega_n = 0$ and at $\omega_b/\omega_n = \sqrt{2}$. Furthermore, the displacement transmissibility at these values is independent of damping. Thus, the parametric curves in ζ all intersect at these two points.

From this figure, it is important to note a key phenomenon of base excited systems. For the frequency ratio range $0 \leq \omega_b/\omega_n \leq \sqrt{2}$, smaller damping results in larger transmissibility values: less damping leads to larger response. On the other hand, for the frequency ratio range $\sqrt{2} \leq \omega_b/\omega_n$, we have smaller amplitudes of response but for *smaller* damping!

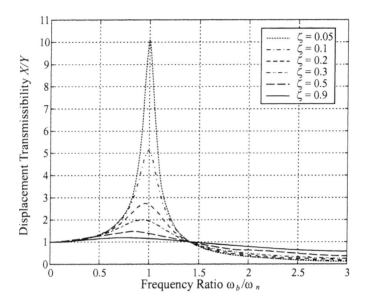

Figure 3.16: Base excitation displacement transmissibility (X/Y) as a function of ζ and ω_b/ω_n.

3.4. HARMONIC EXCITATION AND DAMPED RESPONSE

This is counter-intuitive and needs to be seriously considered in a preliminary design since the frequency ratio in which the system operates will determine whether one designs using larger or smaller damping.

Next, the designer needs to evaluate the force transmitted by the base excitation to the structure using the equation $F_{transmit}(t) = k(x - y) + c(\dot{x} - \dot{y})$, which is the equal and opposite force on the body. See the free body diagram. This is equivalent, from Newton's second law of motion, to $-m\ddot{x}$. Differentiating Equation 3.30 twice and substituting into the term $-m\ddot{x}$ leads to the expression for the transmitted force,

$$F_{transmit}(t) = m\omega_b^2 Y \left[\frac{1 + (2\zeta\omega_b/\omega_n)^2}{(1 - \omega_b^2/\omega_n^2)^2 + (2\zeta\omega_b/\omega_n)^2} \right]^{1/2} \cos(\omega_b t - \phi - \gamma).$$

Rewrite $m\omega_b^2 Y$ as $kY(\omega_b/\omega_n)^2$, and define the magnitude of the transmitted force as

$$F_T = kY(\omega_b/\omega_n)^2 \left[\frac{1 + (2\zeta\omega_b/\omega_n)^2}{(1 - \omega_b^2/\omega_n^2)^2 + (2\zeta\omega_b/\omega_n)^2} \right]^{1/2}.$$

The ratio

$$\frac{F_T}{kY} = \frac{\omega_b^2}{\omega_n^2} \left[\frac{1 + (2\zeta\omega_b/\omega_n)^2}{(1 - \omega_b^2/\omega_n^2)^2 + (2\zeta\omega_b/\omega_n)^2} \right]^{1/2} \quad (3.32)$$

is called the *force transmissibility*. This is a dimensionless relation between maximum base displacement Y and the force magnitude F_T applied to the structure. When Equation 3.32 is plotted in the same way as was Figure 3.16, we find the same qualitative result. That is, for $\omega_b/\omega_n < \sqrt{2}$, less damping results in larger transmitted forces, as expected. But for $\omega_b/\omega_n > \sqrt{2}$, less damping results in smaller transmitted forces, again a counter-intuitive result.

Suppose, for example, a turbomachine on a base imparts a motion of $y(t) = 0.5 \sin \omega_b t$ cm to the floor that surrounds its base. It is necessary to estimate the force that would be experienced by a 6,000 kg compressor that needs to be placed adjacent to the turbomachine. The compressor is connected to the floor via a mat that has stiffness $k = 80,000$ N/m and damping 1,000 N sec/m. The peak force occurs at approximately $\omega_b = \omega_n$, thus simplifying the equation for force transmissibility to

$$\frac{F_T}{kY} = \left[\frac{1 + (2\zeta)^2}{(2\zeta)^2} \right]^{1/2}.$$

The transmitted force is then given by the simplified relation

$$F_T = \frac{kY}{2\zeta}(1 + 4\zeta^2)^{1/2}$$

with ζ given by

$$\zeta = \frac{c}{2\sqrt{km}} = \frac{1000}{2\sqrt{80000 \times 6000}} = 0.023,$$

and $Y = 0.005$ m. The transmitted force is numerically found to be

$$F_T = \frac{80000 \times 0.005}{2(0.023)}(1 + 4(0.023)^2)^{1/2} = 8,705 \text{ N}.$$

This is just under 2000 lb. It is of interest to note that the transmitted force is approximately 15% of the weight of the compressor, given by 6000 kg × 9.81 m/sec^2 = 58,860 N. Recall that this is just an approximation of the peak value of a harmonic base force. If this peak force is too large, the design of the compressor mat must be changed, and the force re-calculated. Otherwise, it may be necessary to stiffen the floor upon which all the machines sit. ∎

Example 3.6 (Excitation due to Rotating Unbalance)
For rotating components, such as turbines and generators, imperfections in component geometry and irregularities in mass distribution create dynamic unbalances. These result in an effective harmonic load on the structure at the frequency of the rotating component. An example we are all familiar with is the car tire. The new tire must be balanced so that no periodic forces develop during operation. The balancing is done on a special machine that rotates the tire at high speed and calculates the magnitude and location of any "deficit mass." At this location, a small lead mass is attached by a mechanic. This process is repeated until the tire is balanced. Tires require re-balancing because of uneven wear. Similar, but much more intricate considerations are part of an internal combustion engine design. A very detailed and applied discussion of engine balance and vibration is provided by Taylor.[11]

To perform a simple analysis based on our understanding of harmonically loaded structures, we replace the imperfection by an eccentrically rotating mass which has an equivalent effect. Thus, the unbalance in the *idealized* model is due to a mass m_0 rotating about a point at some eccentricity e, as per Figure 3.17. The constant frequency of rotation ω_r is that of the eral, the unbalance will result in forces in all directions.

---

he **Internal Combustion Engine in Theory and Practice**, ustion, **Fuels, Materials, Design**, Revised Edition, C.F. Taylor,

3.4. HARMONIC EXCITATION AND DAMPED RESPONSE

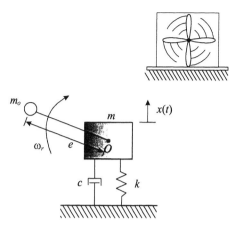

Figure 3.17: Rotating unbalance.

Here, we assume that the machine is constrained to move only in the vertical x direction. The force exerted by the rotating mass is obtained using Newton's second law of motion, in the following way.

The vertical component of the motion of the rotating mass is $x_r(t) = e \sin \omega_r t$, where $\omega_r t$ represents the arc length of the motion in radians. The force generated is then related to its acceleration,

$$F_r = m_0 \ddot{x}_r = -e m_0 \omega_r^2 \sin \omega_r t,$$

which acts at O on mass $(m - m_0)$. The machine has a mass of m. Now, using a free body diagram of the machine, the sum of all external forces can be set equal to the sum of all the respective products of mass and acceleration,

$$-kx - c\dot{x} = (m - m_0)\ddot{x} + m_0 \frac{d^2}{dt^2}(x + e \sin \omega_r t).$$

Simplifying the right hand side and dividing by m leads to

$$\ddot{x} + 2\zeta \omega_n \dot{x} + \omega_n^2 x = \frac{m_0 e \omega_r^2}{m} \sin \omega_r t.$$

This equation is identical to Equation 3.11 with $A = m_0 e \omega_r^2 / m$, and there-

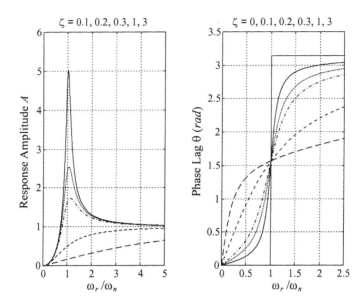

Figure 3.18: Nondimensional magnitude as function of frequency ratio.

fore the steady state response is given by

$$x(t) = A\sin(\omega_r t - \theta), \qquad (3.33)$$
$$A = \frac{m_0 e \omega_r^2 / m \omega_n^2}{\sqrt{(1 - \omega_r^2/\omega_n^2)^2 + (2\zeta \omega_r/\omega_n)^2}},$$
$$\theta = \tan^{-1}\left(\frac{2\zeta \omega_r/\omega_n}{1 - \omega_r^2/\omega_n^2}\right).$$

The physical meaning of these results is best observed in Figure 3.18.

It is easily observed that for high speed machinery, $\omega_r > \omega_n$, the system experiences a resonance with large magnitude response during the powering up or down phase. This is true for all except highly damped structures. In a design, it is important that the rotating component passes through resonance quickly and that for those brief moments of passage the structure vibrates acceptably. ∎

Example 3.7 Rotating Shafts and Whirling
A major problem in the design of machines[12] is the analysis and design of

---

[12] A detailed reference on this subject is by G. Genta, **Vibration of Structures and Machines – Practical Aspects**, Second Edition, Springer–Verlag, 1995.

3.4. HARMONIC EXCITATION AND DAMPED RESPONSE

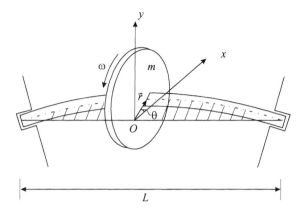

Figure 3.19: A rotating shaft with bending.

rotating shafts, and in particular the behavior known as whirling. We begin by considering the schematic of a rotating shaft in Figure 3.19.

The simplest useful model in the study of the flexural behavior of rotors is the Jeffcott rotor, consisting of a thin disk of mass m attached to a shaft with mass so small when compared to m that it can be safely ignored. Damping is neglected here. The only force acting on m is the stiffness force due to the shaft elastic restoring force. This stiffness is due to the shaft and the support structure. Both can be included in a sophisticated model, but for the simplified model introduced here, the resulting equation of motion will be the same. In addition to the rotation, the shaft will oscillate in bending as a beam. This behavior is ignored here.

How does the originally motionless and straight shaft bend as it rotates to its operating speed? Generally, *imperfections* in the shaft and the disk create an unbalance of forces resulting in a net *centripetal* force and a bent shaft. It is the rotation about the bearing axis of the plane that contains the bent shaft that is known as whirling. That plane is shaded in Figure 3.19. Looking at the disk end–on, we see something like the schematic of Figure 3.20.

Denote C_m as the disk center of mass, and C_s as the center of the shaft. This distance is the eccentricity ε due to imperfections in the disk. If the disk is perfect, the eccentricity equals zero. However, whirling may still occur due to shaft imperfections.

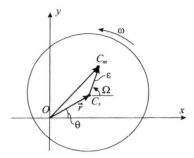

Figure 3.20: Eccentric disk.

Define the following position vectors:

$$\mathbf{r}_{OC_m} = (x + \varepsilon \cos \Omega)\mathbf{i} + (y + \varepsilon \sin \Omega)\mathbf{j}$$
$$\mathbf{r} = x\mathbf{i} + y\mathbf{j}$$
$$|\mathbf{r}_{C_sC_m}| = \varepsilon,$$

where $\dot{\theta}$ is the rotation speed of the plane formed by the bent shaft and its axis, and $\dot{\theta} \neq \omega$ in general, where ω is the rotation speed of the disk and is a general function of time. Define $\Omega = \omega t$. Bold notation denotes vectors, and the unit vectors are $\mathbf{i}$ and $\mathbf{j}$. We follow Genta, p. 218–222, for the following derivations and discussion. To derive the equations of motion in the x and y directions, we use the same procedure as for the rotating unbalance problem. This leads us to the two governing equations,

$$m\frac{d^2}{dt^2}(x + \varepsilon \cos \Omega) + k_x x = 0$$
$$m\frac{d^2}{dt^2}(y + \varepsilon \sin \Omega) + k_y y = 0.$$

Carrying out the derivatives on the left hand side where Ω is a function of time, we find

$$m(\ddot{x} - \varepsilon \dot{\Omega}^2 \cos \Omega - \varepsilon \ddot{\Omega} \sin \Omega) + k_x x = 0 \quad (3.34)$$
$$m(\ddot{y} - \varepsilon \dot{\Omega}^2 \sin \Omega + \varepsilon \ddot{\Omega} \cos \Omega) + k_y y = 0. \quad (3.35)$$

Assuming that ω is constant and that there is *synchronous whirl*, that is $\dot{\Omega} = \omega$, Equations 3.34 and 3.35 become

$$m(\ddot{x} - \varepsilon \omega^2 \cos \omega t) + k_x x = 0$$
$$m(\ddot{y} - \varepsilon \omega^2 \sin \omega t) + k_y y = 0,$$

3.4. HARMONIC EXCITATION AND DAMPED RESPONSE

or in the standard harmonic loading form,

$$\ddot{x} + \omega_{nx}^2 x = \varepsilon \omega^2 \cos \omega t$$
$$\ddot{y} + \omega_{ny}^2 y = \varepsilon \omega^2 \sin \omega t,$$

which can be solved using techniques we already know. Our two equations are uncoupled, meaning that motion in the two directions is independent. This is due to all the prior assumptions and is generally not the case.

For the case where $k_x = k_y = k$, it is possible to combine Equations 3.34 and 3.35 into one complex parameterized equation by letting $z = x + iy$,

$$m\ddot{z} + kz = m\varepsilon(\dot{\theta}^2 - i\ddot{\theta})e^{i\theta}, \quad (3.36)$$

where each original equation can be recovered by taking the real or imaginary parts of Equation 3.36.

With the constant ω assumption, Equation 3.36 becomes

$$\ddot{z} + \omega_n^2 z = \varepsilon \omega^2 e^{i\omega t},$$

since $\ddot{\Omega} = \dot{\omega} = 0$. This equation has the solution $z = z_0 e^{i\omega t}$, with

$$z_0 = \frac{\varepsilon \omega^2}{\omega_n^2 - \omega^2}. \quad (3.37)$$

We see that the critical value of the disk rotation frequency ω is $\omega_{cr} = \omega_n$. The *critical flexural speed* then is defined as the speed at which the frequency of rotation equals the natural frequency of the system.[13] z represents the rotation of the plane of the bent shaft about the bearing axis, and rotates with speed ω in the $x - y$ plane. Considering Equation 3.37 in the form

$$\frac{z_0}{\varepsilon} = \frac{1}{(\omega_n/\omega)^2 - 1},$$

we can plot $|z_0|/\varepsilon$ versus ω_n/ω to explore the response magnitude as a function of ω. Recall that $|z_0| = \sqrt{x^2 + y^2}$.

In the *subcritical range*, $\omega < \omega_n$, the amplitude grows from zero and will, in the limit, approach infinity as $\omega \to \omega_n$ since no damping has been included in the model. The amplitude always remains positive. In the *supercritical range*, $\omega > \omega_n$, the amplitude is always negative, and with increasing ω the magnitude of the amplitude decreases monotonically to $-\varepsilon$ as $\omega \to \infty$. This phenomenon is referred to as *self-centering* since the rotor tends to rotate about its center of mass instead of its geometric center.

---

[13] This is not a general result for all rotors, but only for the simplified model studied here.

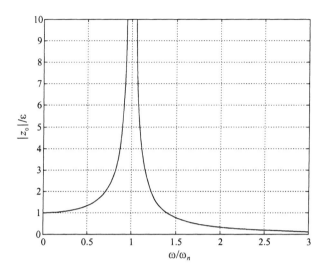

Figure 3.21: Response magnitude $|z_o|/\varepsilon$ versus frequency ratio ω_n/ω in whirling.

With the addition of damping, we can proceed as above with the added possibility that the system may become unstable due to a phenomenon where energy is fed into the system by the damping, rather than the usual removal of energy. In this case the response amplitude will grow exponentially, leading to failure. Such a phenomenon is known as *self-excited whirling*. An introduction to self-excited oscillations is given in Section 4.5.

∎

Example 3.8 Resonance in Electric Motors
This example[14] offers a look at how large amplitude vibration problems may occur in practice. Here, the phenomenon of resonance in rotating equipment, and how it can be alleviated, is explored. Rotating equipment includes pumps, motors, and compressors. Although these systems possess many natural frequencies, we will assume that the fundamental or lowest such frequency is of primary concern.

A designer of such equipment intends that operation be away from the natural frequency where large amplitude motion occurs. If a preliminary design leads to such motion, it is generally necessary to redesign the pump or

---

[14] This example is based on the article by W.R. Campbell, "Practical Solution of Resonance of Electric Motors", *Shock & Vibration Digest*, Vol. 26, No. 1, Jan/Feb 1994.

3.4. HARMONIC EXCITATION AND DAMPED RESPONSE

the compressor impellers, change shaft diameter, bearing stiffness or spans, or the mass of the unit. We know this from our basic studies since frequency is directly proportional to the (square root of) stiffness and inversely proportional to the (square root of) mass. Any change in geometry will alter the stiffness properties of the system. The addition or removal of mass will lower or raise the characteristic frequency of the machine.

The running of the rotor results in a force. If the resonance frequency, or natural frequency, is close to the running speed, one sees an increase in the running speed amplitude, z_0 in the last example. If the resonance cannot be moved away from the running speed, *trim balancing* the rotor will reduce the resonant effect. Such balancing alters the running speed by a change in mass, thus taking it further from the resonant frequency.

Resonances close to the running speed can normally be moved above the running speed by increasing the stiffness of the pedestals, clamping the feet, removing distortion at the feet by shimmying, or using reduced bearing clearances. A particular example offered by Campbell is that of a number of identical vertical pump and motor units whose running speed was amplified by a resonance. The decision was to move the resonance by adding mass (lowering the resonant frequency) to the unit at the motor flange. This was in preference to the difficult process of stiffening the unit. The first step was to decide how far the resonant frequency must be shifted. This allowed for an estimate of the amount of mass that was needed to do this.

In this practical problem, the following parameters are known or measurable: the resonant frequency is measured at approximately 14.8 Hz, the weight of the motor, shroud, drive shaft, and filled pump is 1650 lb, and the mass is obtained by diving the weight by gravitational acceleration $M = W/g = 1650/386$ slug. The frequency in cycles per minute, cpm, is

$$f = \frac{60}{2\pi}\sqrt{\frac{K}{M}},$$

which can be used to evaluate the stiffness of the system,

$$K = \frac{f^2 4\pi^2 M}{60^2} = 36,714 \text{ lb/in.}$$

If the weight is increased by 400 lb, the new mass becomes $(1650+400)/386 = 5.31$ slug, and the measured resonant frequency is 794 cpm = 13.2 Hz, a drop of about 10% or 90 cpm. In practice, when the weight was added physically to the motor flange, the resonant frequency was reduced approximately 90 cpm from the motor's original frequency. The effect was to reduce the vibration amplitude by approximately 80% at the top of the motor. ■

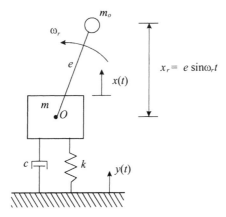

Figure 3.22: Base excited structure with rotating unbalance.

Example 3.9 Base Excited Structure with Rotating Unbalance
We are interested in combining the effects introduced in *Example Problems* 3.5 and 3.6 to learn how, if at all, they interact. A schematic of the structure is given in Figure 3.22. The free body diagram helps in the derivation of the equation of motion. Linearity permits the addition of two effects as distinct loadings on the system.

The governing equation is then

$$m\ddot{x} + c\dot{x} + kx = m_0 e \omega_r^2 \sin \omega_r t + c\dot{y} + ky.$$

As before, assuming the harmonic base motion to be $y(t) = Y \sin \omega_b t$, and dividing through by m leads to

$$\ddot{x} + 2\zeta\omega_n\dot{x} + \omega_n^2 x = \frac{m_0}{m} e \omega_r^2 \sin \omega_r t + 2\zeta\omega_n\omega_b Y \cos \omega_b t + \omega_n^2 Y \sin \omega_b t. \quad (3.38)$$

Assume a response which includes both harmonic components, then substitute this into the governing equation to find the coefficients. A simpler way is to solve for the response to each of the inputs on the right hand side of Equation 3.38, and then add all three components for the complete solution, $x = x_1 + x_2 + x_3$, where x_1, x_2 and x_3 are governed by

$$\ddot{x}_1 + 2\zeta\omega_n\dot{x}_1 + \omega_n^2 x_1 = \frac{m_0}{m} e \omega_r^2 \sin \omega_r t$$
$$\ddot{x}_2 + 2\zeta\omega_n\dot{x}_2 + \omega_n^2 x_2 = 2\zeta\omega_n\omega_b Y \cos \omega_b t$$
$$\ddot{x}_3 + 2\zeta\omega_n\dot{x}_3 + \omega_n^2 x_3 = \omega_n^2 Y \sin \omega_b t.$$

Each of these equations is of the form of Equation 3.11 with solution given by Equation 3.15. Several resonances are now possible. If any of the driving

3.5. PERIODIC BUT NOT HARMONIC EXCITATION

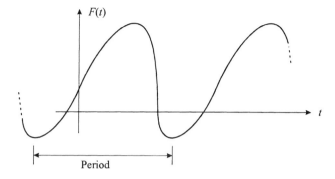

Figure 3.23: Periodic but not harmonic function.

frequencies, ω_r or ω_b, are at or near the resonant frequency ω_n, then a significant magnification of the response amplitude may occur. The study of such possible behavior is an important part of any preliminary analysis and design. A numerical study would be useful to examine dangerous loading combinations. ∎

3.5 Periodic but Not Harmonic Excitation

Structural loading may be periodic but not harmonic. In such a case, there exists a finite period of time after which the waveform of the loading begins to repeat itself. However, the shape of the waveform is not representable by a single sine or cosine function, and the utility of the *Fourier series* becomes apparent, since it transforms such periodic functions into infinite sums of sines and cosines.

The Fourier series of an arbitrary, periodic function, $f(t)$, is given by[15]

$$f(t) = \frac{a_0}{2} + \sum_{p=1}^{\infty}(a_p \cos p\omega_T t + b_p \sin p\omega_T t), \qquad \omega_T = \frac{2\pi}{T},$$

where $p = 1, 2, 3, \ldots$, and T = period of function. The Fourier coefficients a_p and b_p are derivable using the orthogonality property of *sine* and *cosine*

---
[15] Other equivalent definitions exist for the Fourier series.

functions:

$$a_p = \frac{2}{T}\int_{-T/2}^{T/2} f(t)\cos p\omega_T t\, dt \quad p=0,1,2,\ldots$$

$$b_p = \frac{2}{T}\int_{-T/2}^{T/2} f(t)\sin p\omega_T t\, dt \quad p=1,2,\ldots$$

Note that, in principle, an infinite number of harmonic components are required to exactly duplicate the waveform $f(t)$. The Fourier coefficients are a measure of participation of each harmonic component to the final waveform. One can also see from the equation for a_p that $a_0/2$ is the average value of the periodic waveform.

By the principle of superposition, we can solve problems of periodic forcing using the procedures and results of Section 3.4. The governing equation of motion is

$$\ddot{x} + 2\zeta\omega_n \dot{x} + \omega_n^2 x = F(t)$$
$$= \frac{a_0}{2} + \sum_{p=1}^{\infty}(a_p \cos p\omega_T t + b_p \sin p\omega_T t),$$

where the oscillator is loaded by an *infinite* number of sinusoidals, and $F(t)$ is a force per unit mass. In principle then, we would superpose the response due to each of these, with the total sum being the complete response of the oscillator.

Therefore, the steady state response is

$$x(t) = \sum_{p=0}^{\infty} x_p(t)$$
$$= \sum_{p=0}^{\infty} A_p[a_p \cos(p\omega_T t - \theta_p) + b_p \sin(p\omega_T t - \theta_p)],$$

where

$$A_p = \frac{1/\omega_n^2}{\sqrt{(1-p^2\omega_T^2/\omega_n^2)^2 + (2\zeta p\omega_T/\omega_n)^2}}$$

$$\theta_p = \tan^{-1}\frac{2\zeta p\omega_T/\omega_n}{1-(p\omega_T/\omega_n)^2}.$$

Since the oscillator is loaded by harmonics of ω_T, if any *one* of $p\omega_T$ equals ω_n, a resonant-type response occurs. It is sufficient that only one such equality, or very-near equality, occur for a resulting resonance. However,

3.5. PERIODIC BUT NOT HARMONIC EXCITATION

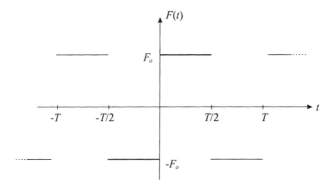

Figure 3.24: Square wave load.

with sufficient damping, the problem may be insignificant. As a practical matter, resonances in the higher frequencies tend to be less important for many applications, since they contain less energy than those in the lower frequencies. The application determines the importance of a resonance.

Example 3.10 Application of the Fourier Series Representation
Consider the simple undamped oscillator $\ddot{x} + \omega_n^2 x = F(t)$, where $F(t)$ is the *square wave* in Figure 3.24. The square wave is the simplest of all non-harmonic, periodic functions. Solve for the response using the Fourier series representation for $F(t)$, assuming zero initial conditions.

Solution Since the forcing is periodic but not harmonic, choose the Fourier series to decompose the periodic function into its harmonic components. This square wave is defined by

$$F(t) = \begin{cases} F_0 & \text{if } 0 < t < T/2 \\ -F_0 & \text{if } -T/2 < t < 0. \end{cases}$$

We need only consider one period of the function in the series expansion. Since $F(t)$ is an odd function the a_p coefficients equal zero. The b_p coefficients are evaluated as follows,

$$\begin{aligned} b_p &= \frac{2}{T}\left[\int_{-T/2}^{0} F(t)\sin p\omega_T t\, dt + \int_{0}^{T/2} F(t)\sin p\omega_T t\, dt\right] \\ &= \frac{4}{T}\int_{0}^{T/2} F(t)\sin p\omega_T t\, dt \\ &= \frac{4}{T}\int_{0}^{T/2} F_0 \sin p\omega_T t\, dt \\ &= \frac{4F_0}{Tp\omega_T}[1 - \cos(p\omega_T T/2)], \end{aligned}$$

where $\omega_T = 2\pi/T$. The cosine term above can be simplified as $\cos(p\omega_T T/2) = \cos p\pi = (-1)^p$. Therefore,

$$b_p = \frac{2F_0}{p\pi}[1 - (-1)^p] = \frac{4F_0}{p\pi}, \quad p \text{ odd},$$

and zero otherwise. The square wave loading is given by the series

$$F(t) = \frac{4F_0}{\pi}\sum_{p=1,3,5,\ldots}^{\infty}\frac{1}{p}\sin p\omega_T t.$$

The equation of motion is then

$$\ddot{x} + \omega_n^2 x = \frac{4F_0}{\pi}\sum_{p=1,3,5,\ldots}^{\infty}\frac{1}{p}\sin p\omega_T t.$$

The response to this infinite number of harmonic loads is

$$\begin{aligned} x(t) &= c_1 \sin\omega_n t + c_2 \cos\omega_n t \\ &\quad + \frac{4F_0}{\pi\omega_n^2}\sum_{p,\text{ odd}}\frac{\sin p\omega_T t}{p[1-(p\omega_T/\omega_n)^2]}. \end{aligned} \qquad (3.39)$$

Applying the initial conditions, $x(0) = 0$ and $\dot{x}(0) = 0$, we find that both constants of integration turn out equal to zero. This is not always true, but let us see how this comes about,

$$\begin{aligned} x(0) &= c_2 = 0 \\ \dot{x}(0) &= \omega_n c_1 + \frac{4F_o}{\pi}\sum_{p=1,3,\ldots}\frac{p\omega_T}{p(\omega_n^2 - p^2\omega_T^2)} \\ c_1 &= -\frac{4F_o}{\pi}\sum_{p=1,3,\ldots}\frac{\omega_T/\omega_n}{(\omega_n^2 - p^2\omega_T^2)} \end{aligned}$$

3.5. PERIODIC BUT NOT HARMONIC EXCITATION

In our case,
$$c_1 = -\frac{4F_o}{\pi\omega_n^2} \sum_{p=1,3,..} \frac{1}{2} \frac{1}{(1-p^2/4)},$$

which converges to zero (shown below). Then,

$$\begin{aligned} c_1 &= -\frac{4F_o}{\pi\omega_n^2} \sum_{p=1,3,..} \frac{4}{2} \frac{1}{(4-p^2)} \\ &= -\frac{8F_o}{\pi\omega_n^2} \sum_{p=1,3,..} \frac{1}{(4-p^2)}. \end{aligned}$$

The following calculation shows that $\sum_{p=1,3,..} 1/(4-p^2) = 0$, so that $c_1 = 0$:

$$\sum_{p=1,3,..} \frac{1}{(4-p^2)} = \frac{1}{4} \sum_{p=1,3,..} \frac{1}{(2-p)} + \frac{1}{(2+p)}$$

where

$$\begin{aligned} \sum_{p=1,3,..} \frac{1}{(2-p)} &= -\sum_{p=1,3,..} \frac{1}{(p-2)} \\ &= -\left(\frac{1}{-1} + \frac{1}{1} + \sum_{p=5,7,..} \frac{1}{(p-2)}\right) \\ &= -\sum_{p=5,7,..} \frac{1}{(p-2)} \\ &= -\sum_{p=1,3,..}^{\infty} \frac{1}{(p+2)}. \end{aligned}$$

Then,

$$\begin{aligned} \sum_{p=1,3,..} \frac{1}{(2-p)} + \frac{1}{(2+p)} &= -\sum_{p=1,3,..}^{\infty} \frac{1}{(p+2)} + \sum_{p=1,3,..}^{\infty} \frac{1}{(p+2)} \\ &= 0. \end{aligned}$$

Therefore, $c_1 = 0$.

It is seen that resonances occur for $p\omega_T = \omega_n$. Figure 3.25 shows the time histories of the loading and the response. These curves utilize only the first four terms in the respective Fourier series representations. The response was obtained for the case $\omega_n = 2\omega_T$. Even though only four terms

134   CHAPTER 3. SDOF VIBRATION: WITH DAMPING

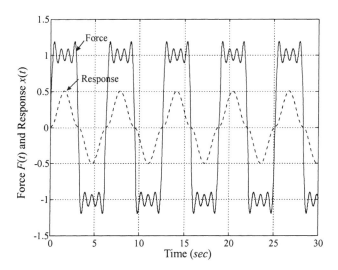

Figure 3.25: Loading and response time histories for the case $\omega_n = 2\omega_T$; four term series are used for both loading and response.

in the series for the square wave were used, the response turns out to be very close to exact. Had there been a response component at $2\omega_T$, this undamped system would have infinite amplification. With damping this would not occur. If $\omega_n = 6\omega_T$, there would be a magnification about that frequency. However, since there is relatively little energy driving the system at the higher frequencies, amplification is not a concern. Response at the lower frequencies will be larger. ∎

Fourier

Jean Baptiste Joseph Fourier lived during 1768–1830 and was born in Auxerre, France. He died in Paris. Fourier trained for the priesthood but did not take his vows. Instead, he took up studying mathematics (1794) and later teaching mathematics at the new École Normale. In 1798 he joined Napoleon's army in its invasion of Egypt as scientific advisor. He helped establish educational facilities in Egypt and carried out archaeological explorations. He returned to France in 1801 and was appointed Prefect of the department of Isère by Napoleon. He published *Theorie Analytique de la Chaleur* in 1822, devoted to the mathematical theory of heat conduction. He established the partial differential equation governing heat diffusion and solved it by using infinite series of trigonometric functions. In this he introduced the representation of a function as a series of sines and

3.6. ARBITRARY LOADING: LAPLACE TRANSFORM

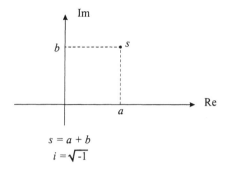

Figure 3.26: Laplace transform domain.

cosines, now known as Fourier series. Fourier's work provided the impetus for later work on trigonometric series and the theory of functions of a real variable.

3.6 Arbitrary Loading: Laplace Transform

The *Laplace transform* is one of a powerful group of techniques known as *transform methods*. The Laplace transform of the function $x(t)$ is defined by

$$\mathcal{L}x \equiv x(s) = \int_0^\infty e^{-st} x(t) dt, \qquad (3.40)$$

where s is a complex number, and notationally, any variable with s as its argument is a Laplace transformed variable.

In vibration analysis, the essential purpose in using the Laplace transform is to derive an equivalent *algebraic* expression for the governing differential equation of motion. This is accomplished by transforming the differential equation from the time domain to the complex number domain. The forward process, transforming from t to s, is relatively straightforward. In the complex s domain, the algebraic relation is solved for the transform of the response. Then it is necessary to transform back to the time domain to obtain the solution in terms of t. The inverse transform can be intricate, requiring *complex variable methods*, but we will rely on tabulated transforms.[16]

---

[16] One should not lose sight of the fundamental nature of problem solving, that is, the *conservation of complexity*: techniques that make one part of the problem easier, will generally make another part more difficult. Anyone of you who finds a method to make all parts easier has found something truly special.

In addition to the algebraic nature of the Laplace transform technique, the solution includes automatic satisfaction of the initial conditions. Consider the generic, damped oscillator, driven by an arbitrary force per unit mass, $F(t)$, governed by Equation 3.10, $\ddot{x} + 2\zeta\omega_n\dot{x} + \omega_n^2 x = F(t)$, with initial conditions $x(0)$ and $\dot{x}(0)$. The procedure begins by taking the Laplace transform of both sides of the governing differential equation. Using the definition of the Laplace transform, the following relations are found,

$$\begin{aligned}
\mathcal{L}x(t) &= x(s) \\
\mathcal{L}\dot{x}(t) &= sx(s) - x(0) \\
\mathcal{L}\ddot{x}(t) &= s^2 x(s) - sx(0) - \dot{x}(0) \\
\mathcal{L}F(t) &= F(s),
\end{aligned}$$

where any initial conditions are understood to be $x(t=0)$ and $\dot{x}(t=0)$. Substituting these transformed variables into the governing equation, and solving for the transform of the solution $x(s)$, one finds

$$x(s) = \frac{F(s) + \dot{x}(0) + (s + 2\zeta\omega_n)x(0)}{(s^2 + 2\zeta\omega_n s + \omega_n^2)}, \qquad (3.41)$$

where the initial conditions are explicit. For any load per unit mass $F(t)$, Equation 3.41 must be inverted to solve for $x(t)$.

Using *Borel's theorem*,[17] and a good table of Laplace transform pairs, we have

$$\mathcal{L}^{-1}\left\{\frac{F(s)}{(s^2 + 2\zeta\omega_n s + \omega_n^2)}\right\}$$
$$= \frac{1}{\omega_d}\int_0^t F(\tau)e^{-\zeta\omega_n(t-\tau)}\sin\omega_d(t-\tau)d\tau$$

$$\mathcal{L}^{-1}\left\{\frac{\dot{x}(0)}{(s^2 + 2\zeta\omega_n s + \omega_n^2)}\right\}$$
$$= \frac{\dot{x}(0)}{\omega_d}e^{-\zeta\omega_n t}\sin\omega_d t$$

$$\mathcal{L}^{-1}\left\{\frac{(s + 2\zeta\omega_n)x(0)}{(s^2 + 2\zeta\omega_n s + \omega_n^2)}\right\}$$
$$= \frac{x(0)}{(1-\zeta^2)^{1/2}}e^{-\zeta\omega_n t}\cos(\omega_d t - \psi),$$

$$\omega_d = (1-\zeta^2)^{1/2}\omega_n \qquad \psi = \tan^{-1}\frac{\zeta}{(1-\zeta^2)^{1/2}}.$$

---

[17]Borel's theorem states that the inverse Laplace transform of the product of two transforms is equal to the convolution of their inverse transforms.

3.6. ARBITRARY LOADING: LAPLACE TRANSFORM

Table 3.1: A Short List of Laplace Transform Pairs.

$F(t)$	$F(s)$
$\delta(0)$	1
1	$1/s$
$\exp(-at)$	$1/(s+a)$
$\sin \omega t$	$\omega/(s^2 + \omega^2)$
$\cos \omega t$	$s/(s^2 + \omega^2)$
$\exp(-at)\sin \omega t$	$\omega/\left((s+a)^2 + \omega^2\right)$
$\exp(-at)\cos \omega t$	$(s+a)/\left((s+a)^2 + \omega^2\right)$
$\frac{1}{\omega_d}\exp(-\zeta\omega_n t)\sin \omega_d t,\ \zeta < 1$	$1/\left(s^2 + 2\zeta\omega_n + \omega_n^2\right)$

Therefore, the inverse transform of Equation 3.41 is

$$x(t) = \frac{1}{\omega_d}\int_0^t F(\tau)e^{-\zeta\omega_n(t-\tau)}\sin\omega_d(t-\tau)d\tau \tag{3.42}$$

$$+\frac{\dot{x}(0)}{\omega_d}e^{-\zeta\omega_n t}\sin\omega_d t + \frac{x(0)}{(1-\zeta^2)^{1/2}}e^{-\zeta\omega_n t}\cos(\omega_d t - \psi). \tag{3.43}$$

We will find the identical solution when we derive the response to arbitrary loading using the convolution integral in Section 3.9. The *steady state* response to $F(t)$ is given by Equation 3.42, and the *transient* response to the initial conditions is given by Equation 3.43. The effects of the initial conditions on the response diminish quickly with time, as witnessed by the decaying exponential. Equations of the form 3.42 are known as *convolution integrals*. We will derive the convolution integral in Section 3.9. In addition to the above inverse transforms, Table 3.1 provides a short list of Laplace transform pairs.

Example 3.11 Steady State and Transient Responses: Equations 3.42 and 3.43

A harmonic forcing is one that has existed over all time to the present. If a sinusoidal force is applied for $t \geq 0$, then it is not harmonic. For the purposes of this example, two cases for $F(t)$ are selected;

$$F_1(t) = A\cos\omega t \text{ for all } t$$
$$F_2(t) = A\cos\omega t,\ t \geq 0.$$

The response $x(t)$ is presented in two parts: first the steady state response to the external force per unit mass $F(t)$, and then the transient response to

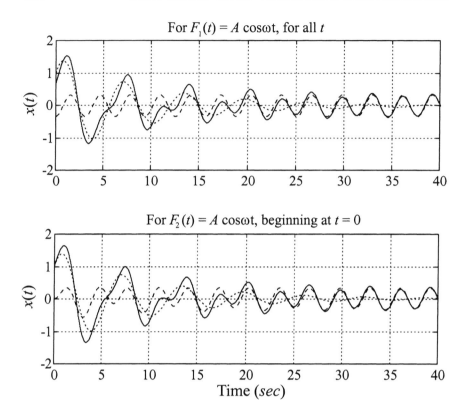

Figure 3.27: Steady state, transient, and complete responses.

the initial conditions. Finally, we plot the sum of the two parts, showing how the complete solution converges to the steady state solution. Our purpose is to show the contribution of each, and how the effects of the initial conditions rapidly decay.

Figure 3.27 provides the steady state (– – –), transient ($\cdots$), and complete (——) responses for each loading function. ∎

Some important definitions are based on ratios of Laplace transformed variables. These ratios will be important in the study of feedback control in Chapter 6. The system *impedance* is defined for zero initial conditions as

$$Z(s) = \frac{F(s)}{x(s)} = s^2 + 2\zeta\omega_n s + \omega_n^2.$$

3.6. ARBITRARY LOADING: LAPLACE TRANSFORM

Its reciprocal is called the *admittance*,

$$Y(s) = \frac{x(s)}{F(s)} = \frac{1}{Z(s)}.$$

In structural applications, the admittance is better recognized by the name *system function* or *transfer function*, and is given the notation $G(s)$,

$$G(s) = \frac{x(s)}{F(s)} = \frac{1}{s^2 + 2\zeta\omega_n s + \omega_n^2}. \tag{3.44}$$

Note that
$$x(s) = G(s)F(s), \tag{3.45}$$

and
$$x(t) = \mathcal{L}^{-1}x(s) = \mathcal{L}^{-1}G(s)F(s).$$

The transfer function, when multiplied by the transform of the loading, yields the transform of the response. With the substitution $s = i\omega$, Equation 3.44 becomes

$$G(s|_{s=i\omega}) = H(i\omega),$$

as in Equation 3.23.

The *characteristic equation* for the system can be obtained by taking the denominator of Equation 3.44 and setting it to zero. The *roots* of this characteristic equation are called the *poles* or *singularities* of the system. The term singularities comes from the fact that $G(s)$ becomes a fraction with a zero in the denominator if s is replaced by a pole. In some problems, there will be a polynomial in the numerator as well.[18] The roots of this polynomial are called the *zeros* of the system, since $G(s) = 0$ if s is replaced by a zero. Poles and zeros are critical frequencies of the system, and their values affect a structure's vibration characteristics. Therefore, poles and zeros are important parameters in a control analysis. Graphically, the complex s–plane plot of poles and zeros portrays the character of the *free* transient response of the system.

Example 3.12 Mechanical Accelerometer for a Rapidly Accelerating Vehicle

In this example, the Laplace transform method is used to analyze the utility of a mechanical *accelerometer*. Such an instrument can be attached to a structure in order to obtain acceleration data. The mechanical version discussed here has been generally superseded by sophisticated electronics.

---
[18] This will be discussed in more detail in Chapter 6 on feedback control.

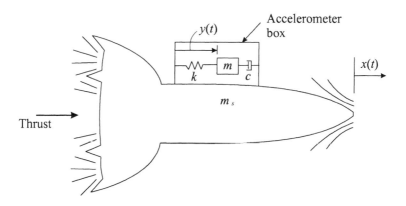

Figure 3.28: Mechanical accelerometer attached to rapidly accelerating vehicle.

It is necessary to evaluate how *quickly* the accelerometer will provide accurate data on the structure, especially if that structure is accelerating at a high rate, as would be a racing car or a missile.

The schematic of Figure 3.28 is taken to be representative of the actual mechanics. We see that the accelerometer is nothing more than a (small) structure with mass, damping, and stiffness properties that is attached to another structure about which we require data.

Here, a missile structure of mass m_s is accelerating due to the thrust force $T(t)$ at a rate of $a(t)$. As expected, there will be a lag in the response time of the accelerometer $y(t)$. Ideally, we would like the accelerometer response to duplicate the structural acceleration: $y(t) = a(t)$ for all time, but this is not possible because there is always a response lag. Generally, this lag will not be a constant, but rather will fluctuate with variations in the acceleration. For this example, assume the relation $y(t) = la(t)$, where the constant l represents this lag. If l is taken to be a time–dependent variable, then the product rule for differentiations must be used in the derivation of the equation of motion below.

To derive the equations of motion for the accelerometer and the missile, free body diagrams are drawn for each, assuming that the accelerometer is rigidly fixed to the missile. The equation of motion for the accelerometer mass m is

$$m\frac{d^2}{dt^2}(y-x) + c\dot{y} + ky = 0, \qquad (3.46)$$

where the only forces on the mass are due to the accelerometer stiffness and

3.6. ARBITRARY LOADING: LAPLACE TRANSFORM

damping, and the inertia force of the accelerating missile. It is assumed that the only force acting on the missile is the rocket thrust $T(t)$. The forces ignored are those due to aerodynamics and the elastic vibration of the missile itself. The thrust–driven equation of motion is then

$$m_s \ddot{x} = T(t). \tag{3.47}$$

Substituting $\ddot{x}$ from Equation 3.47 into 3.46 results in a single equation for the accelerometer response,

$$m\ddot{y} + c\dot{y} + ky = \frac{m}{m_s}T(t). \tag{3.48}$$

Dividing by m, Equation 3.48 becomes

$$\ddot{y} + 2\zeta\omega_n\dot{y} + \omega_n^2 y = f(t),$$

where $T(t)/m_s \equiv f(t)$, and the initial conditions are given as $y(0) = y_0$ and $\dot{y}(0) = v_0$. Taking the Laplace transform of each side,

$$(s^2 y(s) - sy_0 - v_0) + 2\zeta\omega_n(sy(s) - y_0) + \omega_n^2 y(s) = f(s). \tag{3.49}$$

Before continuing, the thrust force needs to be clearly defined. Assuming that the rocket ignites instantaneously, the force can be approximated by a step function of magnitude P. Then,

$$f(t) = Pu(t)$$
$$f(s) = P\frac{1}{s},$$

where $u(t) = 1$ for $t \geq 0$, and zero for $t < 0$. Combining like terms in Equation 3.49 leads to the expression for the transform of the accelerometer response,

$$Y(s) = \frac{P + s(s + 2\zeta\omega_n)y_0 + sv_0}{s(s^2 + 2\zeta\omega_n s + \omega_n^2)}. \tag{3.50}$$

As a numerical example, assume that the initial conditions are zero, $y_0 = 0$, $v_0 = 0$, that $2\zeta\omega_n = 8$ rad/sec and $\omega_n^2 = 15$ (rad/sec)^2. A little algebra shows that the system is overdamped, $\zeta \approx 1.03$. Equation 3.50 becomes

$$y(s) = \frac{P}{s(s+3)(s+5)}.$$

To invert $y(s)$ via transform tables, expand the above into partial fractions,

$$y(s) = \frac{p_1}{s} + \frac{p_2}{s+5} + \frac{p_3}{s+3}.$$

When regrouped and set equivalent to P, the relation becomes

$$y(s) = \frac{P/15}{s} + \frac{P/10}{s+5} + \frac{-P/6}{s+3}.$$

From the tables, the inverse transform of both sides yields

$$y(t) = \frac{P}{15}u(t) + \frac{P}{10}e^{-5t} - \frac{P}{6}e^{-3t}.$$

This equation tells us that as t increases, the second and third terms decay rapidly and $y(t) \to P/15$. These terms represent the transient response. Note that, as expected, for an overdamped system there are no oscillations, just a decay to zero. The first term is the steady state response to the loading. Once $y(t)$ is known, the steady state acceleration of the rocket is found to be $a(t) = y(t)/l = P/(15l)$. ∎

Laplace

Pierre–Simon Laplace lived during 1749–1827 and was born in Beaumont-en-Auge, France. He died in Paris. Laplace attended a Benedictine priory school in Beaumont between the ages of 7 and 16. At the age of 16 he entered Caen University intending to study theology. Laplace wrote his first mathematics paper while at Caen. At the age of 19, Laplace was appointed to a chair of mathematics at the École Militaire in Paris on the recommendation of d'Alembert. In 1773 he became a member of the Paris Academy of Sciences. In 1785, as examiner at the Royal Artillery Corps, he examined and passed the 16 year old Napoleon Bonaparte. During the French Revolution he helped to establish the metric system. He taught calculus at the École Normale and became a member of the French Institute in 1795. Under Napoleon, he was a member, then chancellor, of the Senate, and received the Legion of Honor in 1805. However, Napoleon, in his memoirs written on St. Helene, says he removed Laplace from office after only six weeks "because he brought the spirit of the infinitely small into the government." Laplace became Count of the Empire in 1806 and he was named a marquis in 1817 after the restoration of the Bourbons. In his later years he lived in Arcueil, where he helped to found the Societe d'Arcueil and encouraged the research of young scientists.

Laplace presented his famous nebular hypothesis in *Exposition du Système du Monde* (1796), which viewed the solar system as originating from the contracting and cooling of a large, flattened, and slowly rotating cloud of incandescent gas. Laplace discovered the invariability of planetary mean motions. In 1786 he proved that the eccentricities and inclinations of planetary orbits to each other always remain small, constant, and self-correcting. These results appear in his greatest work, *Traité du Mécanique Céleste*, published in five volumes over 26 years (1799-1825). Laplace also worked on probability and in particular derived the least squares rule. His *Théorie Analytique des Probabilités*

3.7. STEP LOADING

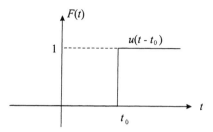

Figure 3.29: Step load.

was published in 1812, putting the theory of mathematical probability on a sound footing. He also worked on differential equations and geodesy. In analysis Laplace introduced the potential function and Laplace coefficients. With Antoine Lavoisier, he conducted experiments on capillary action and specific heat. He also contributed to the foundations of the mathematical science of electricity and magnetism.

3.7 Step Loading

In the last section, the step load and its Laplace transform had been introduced by way of example. Now, we would like to discuss it in greater depth. Consider the response of a single degree of freedom oscillator to a step loading. See Figure 3.29. Such a loading is defined as

$$u(t - t_0) = \begin{cases} 1 & \text{if } t \geq t_0 \text{ sec} \\ 0 & \text{otherwise.} \end{cases}$$

The unit step response $v(t)$ is defined as the response of a system with zero initial conditions to a unit step input at $t_0 = 0$. The step response will be evaluated using the Laplace transform. The governing equation of motion is

$$\ddot{v} + 2\zeta\omega_n \dot{v} + \omega_n^2 v = u(t).$$

Take the Laplace transform of both sides, letting the system transfer function be $G(s)$, as in Equation 3.45, $v(s) = G(s)u(s)$, or

$$v(s) = G(s) \int_0^\infty e^{-st} u(t) dt = \frac{G(s)}{s}. \tag{3.51}$$

The unit step response is then

$$v(t) = \mathcal{L}^{-1} v(s) = \mathcal{L}^{-1} \frac{G(s)}{s},$$

and, for an oscillator with mass, damping and stiffness elements,

$$v(t) = \mathcal{L}^{-1}\left\{\frac{1}{s}\frac{1}{s^2 + 2\zeta\omega_n s + \omega_n^2}\right\}.$$

To transform this equation to the time domain, use Borel's theorem, where

$$\mathcal{L}^{-1}\left\{\frac{1}{s}\right\} = u(t)$$

$$\mathcal{L}^{-1}\left\{\frac{1}{s^2 + 2\zeta\omega_n s + \omega_n^2}\right\} = \frac{1}{\omega_d}e^{-\zeta\omega_n t}\sin\omega_d t,$$

and, using a convolution integral, the response becomes

$$v(t) = \frac{1}{\omega_d}\int_0^t u(\tau)e^{-\zeta\omega_n(t-\tau)}\sin\omega_d(t-\tau)d\tau.$$

The step function $u(\tau)$ can be removed because it "turns on" the response $v(t)$ at $t = 0$. To solve this integral, transform variables $\xi = t - \tau$ and $d\xi = -d\tau$, along with appropriate changes in the limits of integration, to obtain

$$v(t) = -\frac{1}{\omega_d}\int_0^t e^{-\zeta\omega_n \xi}\sin\omega_d \xi\, d\xi. \tag{3.52}$$

A straightforward way to integrate Equation 3.52 is by parts, or to use the identity

$$\sin\omega_d\xi = \frac{e^{i\omega_d\xi} - e^{-i\omega_d\xi}}{2i},$$

simplify the integral, and, after a bit of algebra, obtain the desired solution

$$v(t) = \frac{1}{\omega_n^2}\left[1 - e^{-\zeta\omega_n t}\left(\cos\omega_d t + \frac{\zeta\omega_n}{\omega_d}\sin\omega_d t\right)\right], \quad t \geq 0. \tag{3.53}$$

Physically, what happens a long time after the step load hits the oscillator? The transients die down, and the oscillator only experiences the constant unit load. Thus, we expect the response Equation 3.53 to approach the static response, and indeed it does. As t becomes large, $v \rightarrow 1/\omega_n^2$, which is the static deflection of a mass restrained by a spring of constant k under unit load. (To see this, solve the governing equation of motion where the force per unit mass equals 1.)

For the case where the step is applied at $t = t_0$ sec, Equation 3.51 becomes $v(s) = G(s)e^{-st_0}/s$, and

$$v(t) = \mathcal{L}^{-1}\left\{\frac{e^{-st_0}}{s}\frac{1}{s^2 + 2\zeta\omega_n s + \omega_n^2}\right\}. \tag{3.54}$$

3.7. STEP LOADING

We would then proceed as above to find $v(t)$ explicitly.

Example 3.13 Short Duration Harmonic Forcing
An undamped system is driven by a harmonic function as shown,

$$\ddot{x} + 16x = F(t) = \begin{cases} \cos 4t & \text{if } 0 \leq t < \pi \\ 0 & \text{elsewhere.} \end{cases}$$

The initial conditions are $x(0) = 0$ and $\dot{x}(0) = 1$. We cannot use the usual harmonic solution approach because the loading function does not act over all time. When the term 'harmonic function' is used, implicitly we mean *harmonic over all time*. Solve this problem using Laplace transforms.

Solution To solve this using Laplace transforms, we need to use the second shifting theorem.[19] To do this, rewrite the loading as

$$F(t) = \cos 4t - \cos 4t \, u(t - \pi) = \cos 4t - \cos 4(t - \pi) \, u(t - \pi),$$

where $u(t)$ is the unit step function, and $\cos 4t = \cos(4t - 4\pi)$. Take the Laplace transform of each side of the governing equation,

$$s^2 x(s) - sx(0) - \dot{x}(0) + 16x(s) = \frac{s}{s^2 + 16} - \frac{s}{s^2 + 16} e^{-\pi s}.$$

Substituting the initial conditions and solving for $x(s)$, we find

$$x(s) = \frac{1}{s^2 + 16} + \frac{s}{(s^2 + 16)^2} - \frac{s}{(s^2 + 16)^2} e^{-\pi s},$$

from which

$$\begin{aligned} x(t) &= \frac{1}{4}\mathcal{L}^{-1}\frac{4}{s^2 + 16} + \frac{1}{8}\mathcal{L}^{-1}\frac{8s}{(s^2 + 16)^2} - \frac{1}{8}\mathcal{L}^{-1}\frac{8s}{(s^2 + 16)^2} e^{-\pi s} \\ &= \frac{1}{4}\sin 4t + \frac{1}{8}t\sin 4t - \frac{1}{8}(t - \pi)\sin 4(t - \pi)u(t - \pi). \end{aligned}$$

This solution can be written more clearly in standard form as

$$x(t) = \begin{cases} \frac{1}{4}\sin 4t + \frac{1}{8}t\sin 4t & 0 \leq t < \pi \\ \frac{2+\pi}{8}\sin 4t & t \geq \pi. \end{cases}$$

The plot of this response is shown in Figure 3.30.

---

[19] Two shifting theorems for Laplace transforms are useful. They are provided here without proof:

$$\begin{aligned} \mathcal{L}F(t)e^{at} &= F(s - a) \\ \mathcal{L}F(t - a)u(t - a) &= e^{-as}F(s). \end{aligned}$$

These are also known as *translation theorems*. $u(t)$ is the unit step function.

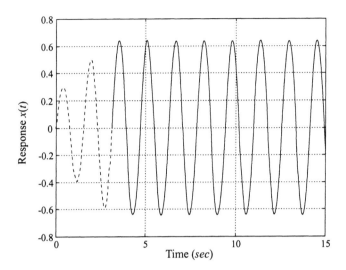

Figure 3.30: Response to short duration harmonic forcing.

Note that the response grows during the loading phase, and then oscillates in steady state after the load disappears. Because there is no damping, the oscillation continues without end. ∎

3.8 Impulsive Excitation

With our understanding of the step response, we can proceed to examine the oscillator response to an "impulse." The impulse is defined as

$$\delta(t - t_0) = 0, \quad t \neq t_0$$
$$\int_0^\infty \delta(t - t_0)dt = 1,$$

where $\delta(t)$ is known as the *Dirac delta function*, defined to have a unit area. This function is also known as the *unit impulse function* because of its usefulness in modeling phenomena of an impulsive nature, where very large forces act over very short intervals of time. Examples are a hammer blow on a structure, an airplane making a hard landing, and a ship being hit by a slamming wave. The delta function is one mathematical way to represent such a force. It is also important for the analysis of a response to an arbitrary force, as we will see in the next section.

3.8. IMPULSIVE EXCITATION

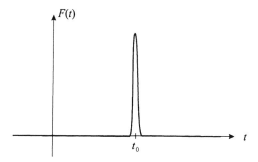

Figure 3.31: Impulsive load.

Let $g(t)$ be defined as the *impulse response* of a system with zero initial conditions to a unit impulse at $t_0 = 0$. See Figure 3.31. The governing equation of motion is then

$$\ddot{g} + 2\zeta\omega_n \dot{g} + \omega_n^2 g = \delta(t). \tag{3.55}$$

As shown in Section 3.6, given a transfer function $G(s)$ and an input force $F(t)$, Equation 3.45 relates the transform of the response to the transform of the force, $x(s) = G(s)F(s)$, and applying this to Equation 3.55 results in $\mathcal{L}g(t) = G(s)\mathcal{L}\delta(t)$, or

$$g(s) = G(s) \int_0^\infty e^{-st}\delta(t)dt = G(s). \tag{3.56}$$

The second equality is due to the fact that the delta function is non–zero only at $t = 0$ and its integral equals 1. The impulse response is found by taking the inverse transform

$$g(t) = \mathcal{L}^{-1}g(s) = \mathcal{L}^{-1}G(s).$$

Comparing the expressions for $g(s)$ and $v(s)$, that is Equation 3.51 and Equation 3.56, we have

$$g(s) = sv(s).$$

The Laplace transform of a derivative of a function equals s times the transform of the function, that is, $\mathcal{L}\dot{x}(t) = sx(s)$ for zero initial displacement. This leads to the conclusion that the unit impulse response equals the time derivative of the unit step response, $g(t) = dv(t)/dt$. Physically, this is intuitive since we can envision that the change or derivative of a unit step response is a discrete change or impulse. But let us show this explicitly with equations that are useful when developing the convolution integral.

By definition,

$$\mathcal{L}\frac{d}{dt}v(t) = sv(s) - v(0)$$
$$= g(s) - v(0),$$

or $g(s) = \mathcal{L}d(v(t))/dt + v(0)$. Taking the inverse transform results in

$$g(t) = \mathcal{L}^{-1}\left(\mathcal{L}\frac{d}{dt}v(t) + v(0)\right) = \frac{dv(t)}{dt} + v(0)\delta(t). \qquad (3.57)$$

Our real interest is to evaluate $g(t)$, which means that the two right-hand terms in Equation 3.57 must be evaluated. Perform the derivative dv/dt using Equation 3.53, use the fact that $v(0) = 0$, and make the substitution $\omega_d^2 = \omega_n^2(1 - \zeta^2)$. Then,

$$g(t) = \frac{1}{\omega_d}e^{-\zeta\omega_n t}\sin\omega_d t\ u(t),$$

since $v(0)\delta(t) = 0$. This impulse response function is a very important relation. We will use it to derive the response of an oscillator of unit mass to an arbitrary load. For an arbitrary mass, the impulse response becomes

$$g(t) = \frac{1}{m\omega_d}e^{-\zeta\omega_n t}\sin\omega_d t\ u(t). \qquad (3.58)$$

Dirac

Paul Adrien Maurice Dirac lived during 1902–1984 and was born in England. He died in the United States. Dirac is famous as the creator of the complete theoretical formulation of quantum mechanics. He studied electrical engineering at the University of Bristol before doing research in mathematics at St John's College, Cambridge. His first major contribution to quantum theory was a paper written in 1925. He published *The Principles of Quantum Mechanics* in 1930 and for this work he was awarded the Nobel Prize for Physics in 1933. Dirac was appointed Lucasian professor of mathematics at the University of Cambridge in 1932, a post he held for 37 years. He was made a fellow of the Royal Society in 1930, was awarded the Royal Society's Royal Medal in 1939, and the Society awarded him the Copley Medal in 1952. Dirac is also well known for a theorem in graph theory on Hamiltonian circuits. This fundamental result was the basis for later work on such circuits. In 1971, Dirac was appointed professor of physics at Florida State University, and was appointed to the Order of Merit in 1973.

3.9 Arbitrary Loading: Convolution

An important goal is to evaluate the response of a structure to an arbitrary but defined load. We have already considered the response to arbitrary

3.9. ARBITRARY LOADING: CONVOLUTION

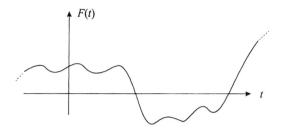

Figure 3.32: Arbitrary deterministic load.

loading via the Laplace transform. Here, we will do so again, but with the convolution integral. In general, arbitrary loads include uncertainties, which must be incorporated into the response analysis. Chapter 5 will develop probabilistic techniques to handle such uncertainties.

Consider Figure 3.32 representing an arbitrary deterministic load $F(t)$. Our approach to deriving the response to such a load is to approximate the arbitrary function by a series of rectangles.[20] The rectangles can be vertical or horizontal as shown in Figure 3.33. In the limit of zero thickness, both will approach the exact curve, but the approximating equations will be different. We develop both approaches and show how they are related to the impulse response and the unit step response. Either approach is correct and independent of the other.

Consider first the case where vertical impulses are used to approximate $F(t)$. As the figure shows, each impulse at time τ has a magnitude $F(\tau)$ and a duration $\Delta \tau$. Given the unit impulse response $g(t)$, the response to an impulsive force of area $F(\tau)\Delta\tau$, applied at $t = \tau$, is

$$\Delta x(t, \tau) = F(\tau)\Delta\tau g(t - \tau). \tag{3.59}$$

Approximating the load by a sum of such impulses,

$$F(t) \approx \sum_{\tau} F(\tau)\Delta\tau,$$

results in the approximate response $x(t) \approx \sum_{\tau} \Delta x(t, \tau)$, or

$$x(t) \approx \sum_{\tau} F(\tau)\Delta\tau g(t - \tau).$$

---

[20] Recall how the integral was derived in calculus and you will have a hint of what is to come.

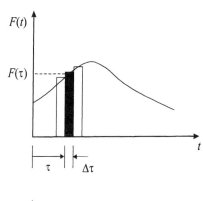
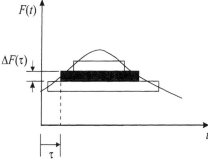

Figure 3.33: Two limiting cases to model the arbitrary load.

3.9. ARBITRARY LOADING: CONVOLUTION

The limit, as $\Delta\tau \to 0$, is

$$x(t) = \int_0^t F(\tau)g(t-\tau)d\tau, \qquad (3.60)$$

an equation known as the *convolution* integral. When we substitute Equation 3.58 for $g(t-\tau)$, Equation 3.60 is called the *Duhamel integral*. It is the principle of superposition for linear systems that validates the idea of the convolution.

Proceed next to the alternate derivation, using a horizontal decomposition of the arbitrary load function. Each of these are horizontal steps, and the response can be modeled by the unit step response $v(t)$ using Equation 3.53. For a step size $\Delta F(\tau)$ initiated at $t = \tau$, the response is

$$\Delta x(t, \tau) = \Delta F(\tau) v(t - \tau),$$

that can be written as

$$\Delta x(t, \tau) = \frac{\Delta F(\tau)}{\Delta \tau} v(t - \tau) \Delta \tau.$$

Let the step size at $t = 0$ be $F(0)$. Then, summing the effects of all the steps to an arbitrary time t gives an approximation of the exact load,

$$x(t) \approx F(0)v(t) + \sum_\tau \frac{\Delta F(\tau)}{\Delta \tau} v(t-\tau) \Delta \tau,$$

and in the limit as $\Delta\tau \to 0$,

$$x(t) = F(0)v(t) + \int_0^t \frac{dF(\tau)}{d\tau} v(t-\tau) d\tau. \qquad (3.61)$$

Since the goal is to derive the convolution integral, integrate Equation 3.61 by parts to find

$$\begin{aligned} x(t) &= F(0)v(t) + v(t-\tau)F(\tau)\big|_0^t - \int_0^t \frac{dv(t-\tau)}{d\tau} F(\tau) d\tau \\ &= v(0)F(t) + \int_0^t v'(t-\tau) F(\tau) d\tau, \end{aligned} \qquad (3.62)$$

where the following equalities were utilized,

$$\frac{dv(t-\tau)}{d\tau} = -\frac{dv(t-\tau)}{d(t-\tau)} \equiv -v'(t-\tau). \qquad (3.63)$$

Using Equation 3.57, repeated here,

$$g(t) = \frac{dv(t)}{dt} + v(0)\delta(t),$$

Equation 3.63 can be written as

$$v'(t-\tau) = g(t-\tau) - v(0)\delta(t-\tau),$$

and Equation 3.62 becomes

$$x(t) = v(0)F(t) + \int_0^t \left[g(t-\tau) - v(0)\delta(t-\tau)\right] F(\tau)d\tau.$$

Noting that $v(0)F(t) = \int_0^t v(0)\delta(t-\tau)F(\tau)d\tau$ leads us again to the convolution integral,

$$\begin{aligned} x(t) &= \int_0^t g(t-\tau)F(\tau)d\tau \\ &= \int_0^t \frac{1}{m\omega_d} e^{-\zeta\omega_n(t-\tau)} \sin\omega_d(t-\tau)F(\tau)d\tau. \end{aligned} \quad (3.64)$$

The complete solution is obtained when Equation 3.64 is added to the free response. Figure 3.34 visually depicts the meaning of the convolution of two functions $g(t)$ and $F(t)$. Note that the time lag can be placed in either function in the integrand. It is equivalent to write

$$\begin{aligned} x(t) &= \int_0^t g(\tau)F(t-\tau)d\tau \\ &= \int_0^t \frac{1}{m\omega_d} e^{-\zeta\omega_n\tau} \sin\omega_d(\tau)F(t-\tau)d\tau. \end{aligned}$$

Example 3.14 Convolution for a General Loading
An underdamped oscillator is subjected to the complex forcing shown in Figure 3.35. Write down the equations which must be solved to obtain the response $x(t)$ over all time.

Solution Since the loading function is of an arbitrary nature, we choose the convolution integral to represent the response of the oscillator. But the load is discontinuous at several time instances, so it is necessary to set up the convolution integral solution for each continuous interval as follows:

3.9. ARBITRARY LOADING: CONVOLUTION

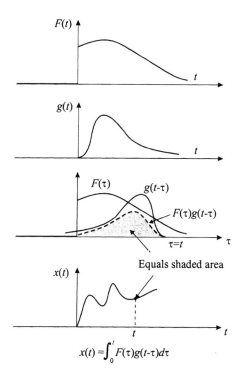

Figure 3.34: The convolution of $g(t)$ and $F(t)$.

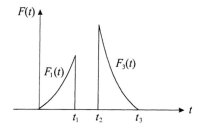

Figure 3.35: Loading history.

$$x(t) = \int_0^t F_1(\tau)g(t-\tau)d\tau, \quad t \leq t_1$$

$$x(t) = \int_0^{t_1} F_1(\tau)g(t-\tau)d\tau, \quad t_1 \leq t \leq t_2$$

$$x(t) = \int_0^{t_1} F_1(\tau)g(t-\tau)d\tau + \int_{t_2}^t F_3(\tau)g(t-\tau)d\tau, \quad t_2 \leq t \leq t_3$$

$$x(t) = \int_0^{t_1} F_1(\tau)g(t-\tau)d\tau + \int_{t_2}^{t_3} F_3(\tau)g(t-\tau)d\tau, \quad t \geq t_3,$$

where $g(t)$ is the oscillator impulse response function, Equation 3.58. In any of the above integrals, we can interchange arguments τ and $t - \tau$ as preferred.

For each time instant, all the loads from earlier time spans up to the present must be added. Due to the discontinuities of the loading function, the convolution integral is evaluated for each continuous segment.

Note that since $F_2(t) = F_4(t) = 0$, there is no contribution to the forcing during these time ranges and the oscillator is in free vibration with initial conditions determined by the final displacement and velocity from the previous loaded range. Therefore, during the time $t_1 \leq t \leq t_2$, the oscillator is in free vibration with behavior set by "initial conditions" $x(t_1)$ and $\dot{x}(t_1)$. After t_3, the free vibration is determined given $x(t_3)$ and $\dot{x}(t_3)$. ∎

Example 3.15 Forcing that is Harmonic Only from t = 0
Suppose that a damped oscillator is governed by the equation

$$\ddot{x} + 2\zeta\omega_n\dot{x} + \omega_n^2 x = F(t),$$

where force per unit mass $F(t) = \sin \omega t$ for $t \geq 0$ and is zero before this time. What is the response?

Solution At first sight, we are tempted to proceed as for any harmonic loading. But our procedure for harmonic loading requires that the forcing function be harmonic over all time. However, the above forcing is not, it turns on at $t = 0$. Assume harmonic response $x(t) = A\sin\omega t + B\cos\omega t$, differentiate twice, and substitute into the governing equation, then

$$(\omega_n^2 - \omega^2)(A\sin\omega t + B\cos\omega t) + 2\zeta\omega_n\omega(A\cos\omega t - B\sin\omega t) \equiv \begin{cases} \sin\omega t & \text{if } t \geq 0 \\ 0 & \text{if } t < 0. \end{cases}$$

There is no way to make this equality valid.

3.10. CONCEPTS SUMMARY

Instead, the Duhamel integral can be used to evaluate the response,

$$\begin{aligned} x(t) &= \frac{1}{\omega_d} \int_0^t F(\tau) e^{-\zeta\omega_n(t-\tau)} \sin\omega_d(t-\tau) d\tau \\ &= \frac{1}{\omega_d} \int_0^t F(t-\tau) e^{-\zeta\omega_n\tau} \sin\omega_d\tau d\tau \\ &= \frac{1}{\omega_d} \int_0^t \sin\omega(t-\tau) e^{-\zeta\omega_n\tau} \sin\omega_d\tau d\tau, \end{aligned}$$

where force per unit mass $F(t)$ has been substituted. This triple product can be evaluated using integration by parts twice. Rather than do this here, assume that damping can be ignored. The last integral above then becomes

$$x(t) = \frac{1}{\omega_n} \int_0^t \sin\omega(t-\tau) \sin\omega_n\tau d\tau.$$

Using the trigonometric identity $\sin\alpha\sin\beta = \frac{1}{2}[\cos(\alpha-\beta) - \cos(\alpha+\beta)]$, the response is

$$x(t) = \frac{m}{k[1-(\omega/\omega_n)^2]} \left(\sin\omega t - \frac{\omega}{\omega_n} \sin\omega_n t \right), \quad t \geq 0.$$

The second term in parenthesis is seen to be the effect of turning the harmonic force on at $t = 0$. ∎

Duhamel

Jean Marie Constant Duhamel lived during 1797–1872 and was born in St Malo, France. He died in Paris. Duhamel was a student at the École Polytechnique and became professor there in 1830. From 1851 he was also professor at the Faculte des Sciences in Paris. Duhamel worked on partial differential equations and applied his methods to the theory of heat and to acoustics. His acoustical studies involved vibrating strings and the vibration of air in cylindrical and conical pipes. His techniques in the theory of heat were mathematically similar to Fresnel's work in optics. His theory of the transmission of heat in crystal structures was based on the work of Fourier and Poisson. The 'Duhamel's principle' in partial differential equations arose from his work on the distribution of heat in a solid with a variable boundary temperature.

3.10 Concepts Summary

This chapter has considered free and forced vibration, with damping. Of particular interest are those parameter combinations that result in under-

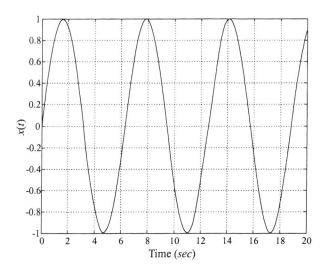

Figure 3.36: Free vibration for Problem 3(*i*).

damped motion. Also of importance are instances where parameter combinations result in the forcing and natural frequencies being equal or nearly equal, leading to resonance and beating behavior, respectively. For arbitrary forcing, the Laplace transform and the convolution integral have been shown to provide the general response.

3.11 Problems

Problems for Section 3.2 -Free Vibration with Damping

1. Solve Equation 3.1 for $\zeta > 1$.

2. Derive Equation 3.3.

3. *(i)* The response of an oscillator with mass $m = 1$ is shown in Figure 3.36. Write down everything you can about the properties of the oscillator and the response. What are the equation of motion and the initial conditions?

   (ii) Now one property is changed, with all others remaining the same. The resulting time history now looks like that in Figure 3.37. Again, write down everything you can about the properties of the oscillator and the response. What are the equation of motion and the initial conditions?

3.11. PROBLEMS

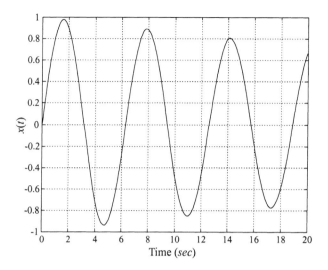

Figure 3.37: Free vibration with one modification for Problem 3(ii).

4. For an mck system in free vibration, derive the equation of motion and solve for the transient response $x(t)$ that is driven by the initial conditions $x(0) = x_0$ and $\dot{x}(0) = v_0$.

   (i) Then make the following substitutions for parameter values: $k = 1$ lb/in, weight $W = 100$ lb, $x_0 = 0$, $v_0 = 10$ in/sec. In this part of the problem, vary the damping constant c so that both underdamped and overdamped responses can be demonstrated. Try values of c such that the cases $\zeta = 0.1$ and $\zeta = 0.9$ are obtained.

   (ii) For the case above with $\zeta = 0.1$, vary initial velocity v_0 and study the variation of the first intercept (zero displacement) as a function of initial velocity.

5. An mck system is tested to determine the value of c. Assume $k = 10$ lb/in and $m = 2$ slug.

   (i) If the vibrational amplitude is observed to decrease to 33% of its initial value after 2 consecutive cycles, what is the value of c?

   (ii) If k, rather than having the exact value given above, has a range of possible values of 8 to 12 lb/in with equal likelihood, then what is the range of possible values of c?

6. Estimate the number of cycles n required for a structural oscillation amplitude decays to $x\%$ of its maximum.

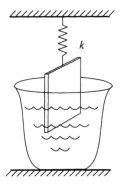

Figure 3.38: Plate suspended in a liquid.

(i) What is the expression for the logarithmic decrement in terms of n and x?

(ii) Solve for n in terms of ζ and x.

(iii) Plot n as a function of ζ for (a) $x = 70\%$, (b) $x = 50\%$, (c) $x = 20\%$ using the exact formula, and then using the approximate formula for $\zeta = 0.1$ and $\zeta = 0.5$.

7. It is usually necessary to test a scale model of a structure to determine some characteristic parameter. Coulomb used the following method to determine the viscosity of liquids. A thin plate of weight W is suspended vertically and set into motion, first in air and then in the liquid, as shown in Figure 3.38. The time duration, t_1 in air and t_2 in the liquid, required for one oscillation is measured. The frictional force between plate and fluid is estimated as $2Acv$, where $2A$ is the area of both sides of the plate, c is the coefficient of viscosity, and v is the plate velocity. The frictional force between plate and air is negligible. Find the value of c in terms of the above parameters.

Problems for Section 3.4 –Harmonic Excitation and Damped Response

8. For the oscillating system in Figure 3.39, solve for the transient response, the steady state response, and the arbitrary constants for the following cases:

(i) $x(0) = 0$ in, $\dot{x}(0) = 5$ in/sec

(ii) $x(0) = 5$ in, $\dot{x}(0) = 0$ in/sec

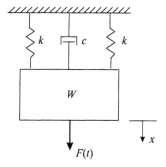

Figure 3.39: Harmonic oscillator.

(iii) $x(0) = 5$ in, $\dot{x}(0) = 5$ in/sec.

The forcing is $F(t) = 10 \sin 15t$ for all the cases. Write down the complete response. Plot transient and steady state responses separately, then cross–plot both of these along with the complete solution. Use the parameter values $W = 40$ lb, $c = 0.85$ lb·sec/in, $k = 12.5$ lb/in. Discuss.

9. For an *mck* system under harmonic loading, $F(t) = \cos \omega t$, solve for the magnification factor β for the case where $k = 20$ lb/in, $W = 40$ lb, and plot for a broad range of driving frequencies ω. If ω ranges between $0 \leq \omega \leq 3\omega_n$, discuss design considerations for the structure as the force is varied from the rest state ($\omega = 0$) to its highest frequency ($\omega = 3\omega_n$). Consider the damping cases: $\zeta = 0.01, 0.1, 0.5$. Discuss parameters of possible concern such as x_{max} and whether material yielding could be of concern.

10. Solve *Problem 9* for $k = 40$ lb/in and $W = 20$ lb.

11. Solve *Problem 9* for $k = 20$ lb/in and $W = 20$ lb.

12. For *Problem 9* where $k = 20$ lb/in and $W = 40$ lb, solve only for the steady state response to the following excitation forces:

    (i) $F(t) = 100 \sin 0.9\omega_n t$

    (ii) $F(t) = 100 \sin \omega_n t$

    (iii) $F(t) = 100 \sin 1.1\omega_n t$

    (iv) $F(t) = 100 \sin 0.5\omega_n t$

    (v) $F(t) = 100 \sin 2\omega_n t$.

    Discuss and plot each case.

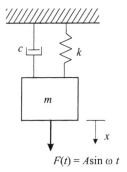

Figure 3.40: Harmonically driven oscillator.

Table 3.2: Cases for t_{max}

$x(0)$ in	$\dot{x}(0)$ in/sec	ζ	ω_n rad/sec
0	5	0.1	2
5	0	0.1	2
5	5	0.1	2
5	5	0.5	2
5	5	1.1	2
5	5	0.1	4
5	5	0.1	10

13. In the idealized model of Figure 3.40, for what range of frequency ratios, ω/ω_n, will the magnification be greater than 1? For what range of frequency ratios will the magnification be greater than 86% of the peak response? Let $m = 1$ slug, $k = 9$ lb/in, and $\zeta = 0.15$. Sketch the response and phase.

14. Derive C and ϕ for Equations 3.20 and 3.21.

15. Solve for $t = t_{max}$ for Equation 3.8 in general for arbitrary variables. Then evaluate t_{max} numerically for the cases in Table 3.3. Plot each case and discuss trends.

16. For the governing equation $\ddot{x} + 2\zeta\omega_n\dot{x} + \omega_n^2 x = A\cos\omega t$, and using solution Equation 3.19, evaluate C and ϕ, and B_1 and B_2. Then plot the free and forced responses separately and then in sum. Let

3.11. PROBLEMS

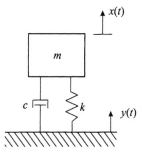

Figure 3.41: Base excited oscillator.

$A = 1$ and do the above for the following parameter cases where $\omega = 1$ rad/sec, and

(i) $\omega_n = 0.5$ rad/sec, $\zeta = 0.1$, $x(0) = 1$ cm, $\dot{x}(0) = 1$ cm/sec,

(ii) $\omega_n = 1$ rad/sec, $\zeta = 0.1$, $x(0) = 1$ cm, $\dot{x}(0) = 1$ cm/sec,

(iii) $\omega_n = 10$ rad/sec, $\zeta = 0.1$, $x(0) = 1$ cm, $\dot{x}(0) = 1$ cm/sec,

(iv) $\omega_n = 0.5$ rad/sec, $\zeta = 0.1$, $x(0) = 0$ cm, $\dot{x}(0) = 1$ cm/sec,

(v) $\omega_n = 0.5$ rad/sec, $\zeta = 0.1$, $x(0) = 10$ cm, $\dot{x}(0) = 1$ cm/sec,

(vi) $\omega_n = 0.5$ rad/sec, $\zeta = 0.1$, $x(0) = 1$ cm, $\dot{x}(0) = 0$ cm/sec,

(vii) $\omega_n = 0.5$ rad/sec, $\zeta = 0.1$, $x(0) = 1$ cm, $\dot{x}(0) = 10$ cm/sec,

(viii) $\omega_n = 0.5$ rad/sec, $\zeta = 0.01$, $x(0) = 1$ cm, $\dot{x}(0) = 1$ cm/sec,

(ix) $\omega_n = 0.5$ rad/sec, $\zeta = 0.95$, $x(0) = 1$ cm, $\dot{x}(0) = 1$ cm/sec.

17. Derive Equations 3.25 and 3.26.

18. Derive B_1 and B_2 in Equations 3.28 and 3.29.

19. Plot Equation 3.32, (F_T/kY) versus ω_b/ω_n for $\zeta = 0.05, 0.10$ and 0.25.

20. The base excited system of Figure 3.41 is driven by a force with frequencies in the range $1.0 \leq \omega_b/\omega_n \leq 2.0$. For a system with mass $m = 1$ N, $k = 9$ N/cm, $Y = 10$ cm, pick a value of c such that the average magnitude equals 1 cm over this frequency range.

21. Solve for the response of the base excited system governed by $\ddot{x} + 2\zeta\omega_n\dot{x} + \omega_n^2 x = 2\zeta\omega_n\dot{y} + \omega_n^2 y$, where $y(t) = A\exp(i\omega_b t)$, using the assumed response $x(t) = X(i\omega)\exp(i\omega t)$.

22. The response of a system excited at its base by $y(t) = Y \sin \omega t$ is given by
$$x(t) = X \sin(\omega_b t - \phi),$$
with
$$\frac{X}{Y} = \left[\frac{1 + (2\zeta\omega_b/\omega_n)^2}{(1 - \omega_b^2/\omega_n^2)^2 + (2\zeta\omega_b/\omega_n)^2}\right]^{1/2}.$$

Use Figure 3.16 for displacement transmissibility when answering the following questions. No substitution of numbers in equations is necessary to answer the questions.

(i) Assume that the maximum amplitude of the base load is $Y = 1$ cm. A design for a rotating machine with the above base load and with natural frequency $\omega_n = 3$ rad/sec is required such that $X \leq 2$ cm across all frequencies ω_b. Which ζ value should be chosen?

(ii) Suppose that instead of the criteria in *(i)*, we require $X \leq 1$ cm, but so that $\zeta < 1.0$. The machine operates at $\omega_b = 9$ rad/sec. What are the options?

(iii) What is the equation for the maximum force transmitted to the base in terms of $\zeta, (\omega_b/\omega_n), k$, and Y? For $\zeta = 1.0$, what is the force transmissibility (F_T/kY) when the machine is running at $\omega_b \approx 0$, $\omega_b = 3\sqrt{2}$ rad/sec, and $\omega_b = 9$ rad/sec?

23. A rotating machine component has an eccentricity of approximately $e = 0.1$ in its center of mass resulting in a harmonic load on the structure. Assume, using the notation of the chapter, that $m_0/m = 0.05$, operating frequency $\omega_r = 100$ rad/sec, $\zeta = 0.10$, and $\omega_n = 10$ rad/sec.

(i) Solve for the amplitude and phase of response in the form of Equation 3.33.

(ii) Suppose amplitude D must be reduced by 15%. What options exist if controlling the structure is not one? Discuss which parameters need to be changed.

24. Derive the general equations of motion, Equations 3.34 and 3.35.

Problems for Section 3.5 –Periodic but Not Harmonic Excitation

25. A machine is loaded by a periodic "sawtooth" force, as per Figure 3.42. The load is assumed to have existed for a very long time. Model this force using the following cases:

(i) use a one–term Fourier series

3.11. PROBLEMS

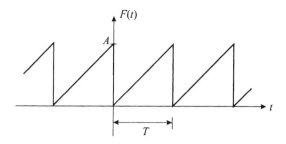

Figure 3.42: Sawtooth loading.

(ii) use a three–term Fourier series

(iii) use a five–term Fourier series, and in each case solve for the response as a function of time. System properties are: $m = 1$ N, $k = 9$ N/cm, $\zeta = 0.15$, $T = 1$ sec, and $A = 1$ cm.

26. Solve *Problem* 25 with (i) $T = 2$ sec, and (ii) $T = 0.5$ sec.

27. Solve *Problem* 25 with $A = 2$ cm.

28. Solve *Problem* 25 with $\zeta = 1.0$.

29. For *Example Problem* 3.10, plot $F(t)$ and $x(t)$ for the case $\omega_n = 6\omega_T$. Compare the results with those of Figure 3.25 and discuss.

30. Solve *Example Problem* 3.10 where the structure has viscous damping, that is, the governing equation of motion is $\ddot{x} + 2\zeta\omega_n\dot{x} + \omega_n^2 = F(t)$, the forcing per unit mass $F(t)$ is the same square wave, and $\omega_n = 2\omega_T$ rad/sec.

31. Solve the above with $\omega_n = 4\omega_T$ rad/sec.

32. A base–excited structure is governed by the equation $\ddot{x} + 2\zeta\omega_n\dot{x} + \omega_n^2 x = 2\zeta\omega_n\dot{y} + \omega_n^2 y$, where $y(t)$ is the base motion. Solve for the response if the base motion is given by the sawtooth function shown in Figure 3.42, where $F(t) = y(t)$.

33. Solve *Problem* 32 where the input is given by the square wave of Figure 3.43.

34. Solve *Problem* 32 where the input is given by the function $\sin t$, $0 \leq t \leq \pi$ repeated periodically, as in Figure 3.44.

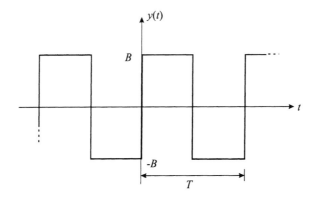

Figure 3.43: Square wave base motion.

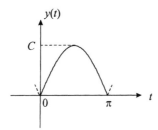

Figure 3.44: Base motion driven by $\sin t$.

3.11. PROBLEMS

Problems for Section 3.6 –Arbitrary Loading: Laplace Transform

35. Solve the following equations of motion using the Laplace transform approach:

    (i) $\ddot{y} + 2\dot{y} + 3y = 5\cos 3t$, $y(0) = 3, \dot{y}(0) = 4$

    (ii) $4\ddot{y} + 5\dot{y} + 5y = 4u(t)$, $y(0) = 1, \dot{y}(0) = 1$, where $u(t)$ is the unit step,

    (iii) $3\ddot{y} + 3\dot{y} + 6y = 3e^{-t} + 2\cos 3t$, $y(0) = 2, \dot{y}(0) = 4$

    (iv) $\ddot{y} + \dot{y} + y = F(t)$, $y(0) = 0, \dot{y}(0) = 0$, where $F(t)$ is given by the square wave function with maximum amplitude 1 and period 1,

    (v) $\ddot{y} + 2\dot{y} + 3y = \cos 3t + \cos 5t$, $y(0) = 0, \dot{y}(0) = 0$.

36. Beginning with Equation 3.57:

$$g(t) = \mathcal{L}^{-1}\left(\mathcal{L}\frac{d}{dt}v(t) + v(0)\right) = \frac{dv(t)}{dt} + v(0)\delta(t),$$

derive Equation 3.58,

$$g(t) = \frac{1}{\omega_d}e^{-\zeta\omega_n t}\sin\omega_d t\, u(t).$$

37. In the accelerometer of *Example Problem* 3.12, the relation between accelerometer response $y(t)$ and structural acceleration $a(t)$ is given a simple form, $y(t) = la(t)$, where l is a constant representing the lag of the component in measuring the structural behavior.

    (i) Do you expect l to be greater than or less than 1?

    (ii) Pick a value for l based on your answer above and discuss any implicit assumptions in your relation $y(t) = la(t)$.

    (iii) We expect that l is better modeled as a time–dependent function that can represent the large acceleration during startup, and the zeroing of the acceleration once the structure approaches a constant velocity. Create such a function.

Problems for Section 3.7 –Step Loading

38. Derive Equation 3.53.

39. Solve Equation 3.54.

40. Using the Laplace transform, solve

$$\ddot{y} + 2\dot{y} + y = F(t)$$
$$\text{where } F(t) = u(t-1) - 2u(t-2) + u(t-3)$$

in terms of unit step functions, and $y(0) = 0$ and $\dot{y}(0) = 0$.

Problems for Section 3.9 –Arbitrary Loading: Convolution

41. Evaluate the response of the undamped oscillator $\ddot{x} + 4x = F(t)$ to the following forces per unit mass using the convolution integral:

    (i) $F(t) = 1 - e^{-t}$, $t \geq 0$
    (ii) $F(t) = \cos 2t$, $0 \leq t \leq 3.1\pi$
    (iii) $F(t) = \cos 2t + 3$, $0 \leq t \leq 3.1\pi$
    (iv) $F(t) = \cos 2t + \cos 3t$, $0 \leq t \leq 3.1\pi$.

42. Evaluate the response of the damped oscillator $\ddot{x} + 2\zeta\omega_n\dot{x} + \omega_n^2 x = F(t)$, with $\omega_n^2 = 4(\text{rad/sec})^2$, to the forces per unit mass listed, and solve for three cases of damping: $\zeta = 0, 0.1$ and $\zeta = 0.9$:

    (i) $F(t) = 1 - e^{-t}$, $t \geq 0$
    (ii) $F(t) = \cos 2t$, $0 \leq t \leq 3.1\pi$
    (iii) $F(t) = \cos 2t + 3$, $0 \leq t \leq 3.1\pi$
    (iv) $F(t) = \cos 2t + \cos 3t$, $0 \leq t \leq 3.1\pi$.

43. From *Example Problem* 3.15, solve for the damped response beginning with the following equation

$$x(t) = \frac{1}{\omega_d} \int_0^t \sin\omega(t-\tau) \exp(-\zeta\omega_n\tau) \sin\omega_d\tau d\tau.$$

Cross–plot the results for the following cases:

$$\omega = 1.0, 1.5, 2.0, 2.5, 3.0 \text{ rad/sec}$$

where $\zeta = 0.1$, $\omega_n = 2$ rad/sec, and $m = 1$ N. Discuss.

44. Structures are sometimes subjected to very rapidly applied loads of extremely short duration. These types of loads are sometimes described as *blast* or *explosive*.

    Consider how such a load time history may look. Figure 3.45(a) is a generic blast load. There is a rapid rise time along with an exponential–like decay, and by time t_1 the load is effectively zero,

3.11. PROBLEMS

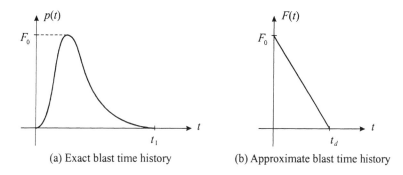

(a) Exact blast time history

(b) Approximate blast time history

Figure 3.45: Exact (left) and approximate (right) blast time histories.

where $t_1 \ll 1$ sec. Such a time history is generally a very complicated function that may not be easily determined because of its sensitivity to many factors, such as the medium through which the load passes, be it the atmosphere or the Earth.

A common approach is to curve–fit the time history with an appropriate combination of functions. However, as a first approximation here, replace the exact curve by the one shown in Figure 3.45(b). To draw such a straight–line approximation, estimate the area under the original curve (this is the impulse) and set it equal to the area under the triangle load, thus fixing the value of t_d. The value of F_0 is the same max for both curves. Although not strictly true for such *high strain–rate* loading, we neglect the effects of damping for the early time response. Solve for the structural response for all time; $0 \leq t \leq t_d$ and $t \geq t_d$. Show that:

$$x(t) = \frac{F_0}{\omega_n^2}\left[1 - \frac{t}{t_d} - \cos\omega_n t + \left(\frac{1}{\omega_n t_d}\right)\sin\omega_n t\right], \quad 0 \leq t \leq t_d$$

$$x(t) = \frac{F_0}{\omega_n^3 t_d}\left[\sin\omega_n t(1 - \cos\omega_n t_d) - \cos\omega_n t(\omega_n t_d - \sin\omega_n t_d)\right],$$

$$t > t_d.$$

45. Solve *Problem 44* except include damping. Discuss the importance of damping by comparing the two results. Plot both sets of results against each other.

46. Solve for the response for all time of an underdamped oscillator that is driven by the forcing function drawn in *(i)* Figure 3.46, and *(ii)* Figure 3.47. Use the convolution integral.

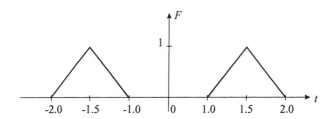

Figure 3.46: Triangular loading function.

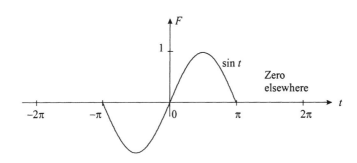

Figure 3.47: Sine loading function.

Chapter 4

Single Degree of Freedom Vibration: Advanced Topics

"Complexities begin to arise."

4.1 Introduction to Lagrange's Equation

In Chapter 7 we will learn the details of *Lagrange's equation*, which is one of the most important and practical principles of dynamics. It is also useful in the derivation of the equations of motion of a discrete or continuous dynamic system. Here we will only state Lagrange's equation and show an application to demonstrate its use. Its effectiveness, however, is best observed when working with systems of several degrees of freedom, where the free body diagrams needed for Newton's second law of motion are difficult to work with because it is necessary to account for the internal forces between the bodies.

Lagrange's equation for a structure with a single degree of freedom $q(t)$ is

$$\frac{d}{dt}\left(\frac{\partial T}{\partial \dot{q}}\right) - \frac{\partial T}{\partial q} + \frac{\partial V}{\partial q} = Q(t),$$

where T equals the system kinetic energy, V equals the system potential energy, and $Q(t)$ includes damping and external forces. It is customary to use the notation $q(t)$ for the coordinate. In the example which follows, we assume no damping so that $Q(t)$ will represent only forces external to the

170   CHAPTER 4. SDOF VIBRATION: ADVANCED TOPICS

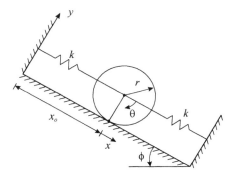

Figure 4.1: Disk on an inclined plane.

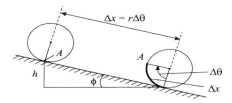

Figure 4.2: Two arbitrary positions of the disk.

body. When considering more than one degree of freedom, there will be one Lagrange's equation for each degree of freedom.

Example 4.1 Application of Lagrange's Equation to a Single Degree of Freedom Oscillator

The disk on the inclined plane of Figure 4.1 is suspended by two massless springs. The disk rolls without slipping, and thus there is no dissipation of energy. The disk oscillates about the equilibrium position indefinitely, and at the equilibrium position has a velocity of $\dot{v}_0$. Use Lagrange's equation to derive the equation of motion. Motion is initiated by an initial displacement or velocity.

Solution Consider first the potential energy of the system. Since the disk is assumed rigid, it cannot possess any deformation strain energy. However, it does store and give up gravitational potential energy V_g as it oscillates about the equilibrium.

To evaluate the change in potential energy of the disk as it rolls down the ramp, consider Figure 4.2. Assume the original orientation to be θ_0,

4.1. INTRODUCTION TO LAGRANGE'S EQUATION

then the change in height is $h = r(\theta - \theta_0)\sin\phi$ and

$$V_g = -mgh = -mgr(\theta - \theta_0)\sin\phi.$$

The two springs store and release potential (strain) energy. Work is done on the springs by the disk as the springs are expanded and compressed. Each spring potential equals the work done on it, with the total spring potential given by

$$V_s = k(\Delta x)^2 = kr^2(\theta - \theta_0)^2,$$

and the total potential energy for the system is then $V_g + V_s$,

$$V = -mgr(\theta - \theta_0)\sin\phi + kr^2(\theta - \theta_0)^2.$$

Next consider the system kinetic energy, due here only to the motion of the disk. There are two ways to visualize this motion. One is the planar motion plus the rotation of the disk. The other is that the disk is in pure rotation about the instantaneous center C. We show both,

$$\begin{aligned} T &= \frac{1}{2}mv_G^2 + \frac{1}{2}I_G\omega^2 \\ &= \frac{1}{2}m(r\dot\theta)^2 + \frac{1}{2}\left(\frac{1}{2}mr^2\right)\dot\theta^2 = \frac{3}{4}mr^2\dot\theta^2, \end{aligned}$$

or

$$T = \frac{1}{2}I_C\omega^2 = \frac{1}{2}\left(\frac{1}{2}mr^2 + mr^2\right)\dot\theta^2 = \frac{3}{4}mr^2\dot\theta^2,$$

as before.

Now, in order to use Lagrange's equation, we need to perform the necessary derivatives of T and V,

$$\begin{aligned} \frac{\partial T}{\partial \dot\theta} &= \frac{3}{2}mr^2\dot\theta \\ \frac{d}{dt}\left(\frac{\partial T}{\partial \dot\theta}\right) &= \frac{3}{2}mr^2\ddot\theta \\ \frac{\partial V}{\partial \theta} &= -mgr\sin\phi + 2kr^2(\theta - \theta_0). \end{aligned}$$

Substitution yields

$$\frac{3}{2}mr^2\ddot\theta - mgr\sin\phi + 2kr^2(\theta - \theta_0) = 0,$$

or

$$\ddot\theta + \frac{4}{3}\frac{k}{m}(\theta - \theta_0) = \frac{2}{3}\frac{g}{r}\sin\phi, \qquad (4.1)$$

the equation of motion of the disk, where the natural frequency is $\sqrt{4k/3m}$. The driver to this motion is a function of ϕ. We could also have used x as the generalized coordinate.

The solution of Equation 4.1 can be used to find the displacement and velocity of point C on the disk via the relations $x = r\theta$ and $\dot{x} = r\dot{\theta}$. ∎

Lagrange

Joseph–Louis Lagrange lived during 1736–1813 and was born Turin, Italy. He died in Paris. Lagrange's interest in mathematics began at a very early age when he read a copy of a book by Halley. Lagrange served as professor of geometry at the Royal Artillery School in Turin from 1755 to 1766 and helped to found the Royal Academy of Science there in 1757. In 1764 he was awarded the first of many prizes when the Paris Academy awarded him for his essay on the libration of the Moon. When Euler left the Berlin Academy of Science, Lagrange succeeded him as director of mathematics 1766. In 1787 he left Berlin to become a member of the Paris Academy of Science, where he remained for the rest of his career. Lagrange survived the French Revolution while others did not. Lagrange said on the death of the chemist Lavoisier, "It took only a moment to cause this head to fall and a hundred years will not suffice to produce its like."

During the 1790s, he worked on the metric system and advocated a decimal base. He also taught at the École Polytechnique, which he helped to found. Napoleon named Lagrange to the Legion of Honor and Count of the Empire in 1808. He excelled in all fields of analysis and number theory, and analytical and celestial mechanics. In 1788 he published *Mécanique Analytique*, which summarized all the work done in the field of mechanics since the time of Newton and is notable for its use of the theory of differential equations. In this work Lagrange transformed mechanics into a branch of mathematical analysis. His early work on the theory of equations was to lead Galois to the idea of a group of permutations. In 1797 Lagrange published the first theory of functions of a real variable, although he failed to give enough attention to matters of convergence.

4.2 Notions of Randomness

Sometimes it is not possible to exactly specify the loading on a structure or machine. How then does one proceed to evaluate an expression such as the convolution integral? Consider what it means when the force is random; if an experiment is run many times to measure this force, the results would be different each time. This difference could be small and insignificant, or large and very important.

One way to approach this dilemma is to average the force, and use this quantity in the convolution integral. This would at least provide us with a *mean value* response, but it would not provide us with a measure of how

4.2. NOTIONS OF RANDOMNESS

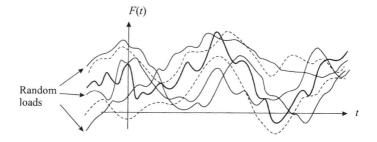

Figure 4.3: Possible Random Loads: Mean Value ± One Standard Deviation

scattered the results might be. In analysis and design, we need to know (in a statistical sense) what are the largest loads and what spread of values is possible. A measure of the scatter in the data is given by the standard deviation of the loading. We will work with these kinds of problems in the next chapter, but here is an example for your subconscious to work with.

Example 4.2 Random Loading
Figure 4.3 depicts three time histories of a series of tests to estimate the statistics of the force acting on a structure. This force is one with much variability. For example, the aerodynamic loads acting on an airplane. Also, the ocean wave forces acting on an oil drilling platform. While each force in the figure is similar in the way it varies with time, the magnitudes are far apart and no two are alike. The question the designer will ask when given such data is: *How do I decide which loading history to use when designing the structure?*

The dark line in the figure is the *average* of all the forces. By itself, it is a useful first measure of the general character of the loading, but not of the *spread* in force magnitudes. This is given by the dashed set of curves above and below the average force. These are the *mean ± one standard deviation* curves. The larger the band between these two curves, the larger the spread in the original test forces. These standard deviation curves, in conjunction with the mean value response, provide a great deal of useful information on the variability of the structural force and how to proceed with a design. One possible approach is to design the structure to resist a force that equals the mean plus one standard deviation. Such questions will be discussed in more detail in the next chapter on probabilistic models. ∎

4.3 Notions of Control

Suppose the convolution integral is solved for the response of a structural component and the displacement $x(t)$ is too large, that is, $x(t_i) > x_{max}$. Because of this, the vibrating component hits an adjacent structure. Another possibility is that the acceleration $\ddot{x}$ is too rapid, $\ddot{x}(t_j) > \ddot{x}_{max}$, resulting in component damage. What can the analyst or designer do to alleviate this problem? There are two choices, one is to create a new system with different mass, damping, and/or stiffness properties so that the response of the structure is acceptable. Sometimes this may be the best approach. Other times, it is not possible to change the structural configuration or properties.

A second possibility is to consider a mechanism to *control* the structure. This would mean the introduction of additional forces or moments that, when added to the existing environmental forces, would result in a structural response that satisfies user–prescribed criteria. We will study such *feedback control* approaches in Chapter 6. Here is a simple example that demonstrates how one can add a force to an existing vibrating system and modify its behavior.

Example 4.3 Control of an Oscillator
A vibrating component within a larger structure is found to be resonating, leading to excessively large amplitude motion. Occasionally, the component will strike the structure within which it sits. Since this is unacceptable behavior, it is concluded that some way must be found to reduce the amplitude of response. Due to the precision nature of the component, modifying its *mass* or *stiffness* properties is not feasible. Therefore, a *feedback force* is used to reduce the response.

Suppose the feedback force is applied to the structure so that its total response $x(t)$ is modified. The governing equation is

$$\ddot{x} + 2\zeta\omega_n\dot{x} + \omega_n^2 x = \frac{1}{m}(F(t)_{ext} + F(t)_{control}), \qquad (4.2)$$

where $F(t)_{control}$ is the feedback control force. The total response is then

$$x(t) = x(t)_{ext} + x(t)_{control},$$

where each response component is a Duhamel integral. With such a solution, the designer can modify c and $F(t)_{control}$ so that $x(t)$ responds in an acceptable way, within certain amplitude and frequency bounds. Here, the analyst acts as the *feedback loop*, modifying the input force and system properties depending on the time history that is being observed. This is usually an automatic process because once a structure or machine is built, automatic control is more effective and efficient. ∎

4.4 The Inverse Problem

For applications such as *system identification, fault detection,* or *nondestructive evaluation*,[1] the system properties are not well known. Thus in the convolution integral, $F(t)$ can be prescribed, but the impulse response $g(t)$ cannot. Usually, the damping of a system is a difficult property to evaluate accurately. One needs to devise a method to estimate such properties.

The inverse problem is a method for estimating $g(t)$ by loading the structure with a prescribed force, and measuring the structural response $x(t)$. Recall the expression for $g(t)$,

$$g(t) = \frac{1}{\omega_d} e^{-\zeta \omega_n t} \sin \omega_d t, \quad t \geq 0, \tag{4.3}$$

which is the response of an oscillator to the unit impulse $\delta(t)$,

$$\ddot{g} + 2\zeta \omega_n \dot{g} + \omega_n^2 g = \delta(t).$$

Equation 4.3 is only a function of system properties, ζ and ω_n.

Another way to look at the inverse problem is by considering the simple oscillator governed by

$$\ddot{x} + 2\zeta \omega_n \dot{x} + \omega_n^2 x = F(t).$$

In this equation, we know the force per unit mass $F(t)$ and the response function $x(t)$. The dimensionless system parameters ζ and ω_n are unknown. A series of experiments must be devised that would provide us with enough data (and equations) to estimate these parameters. One type of inverse problem is introduced in Chapter 9.

4.5 A Self–Excited System and Stability

Our studies in this book focus on stable system behavior, that is, vibration about a stable equilibrium. However, certain classes of loading can result in unstable behavior, by which is meant unbounded growth of the response amplitude.

We have already studied the unstable behavior of a structure that is harmonically forced at or near its natural frequency. *Self-excited* systems also have the possibility of becoming unstable. For example, assume that

---

[1] System identification is a means by which the mass, damping and stiffness properties of a structure are estimated using a known input force and a measured response. Fault detection and nondestructive evaluation are a group of techniques used to evaluate the health of a structure or to locate damage.

the loading on a structure is directly proportional to its velocity with the force per unit mass given by $F(t) = 2A\omega_n \dot{x}$. This can be the case for aerodynamically and fluid loaded structures. For example, the wave force on an offshore structure is related to structural velocity and will be examined in Section 5.2.3.

The equation of motion is then

$$\ddot{x} + 2\zeta\omega_n \dot{x} + \omega_n^2 x = 2A\omega_n \dot{x},$$

which can be re-written as a "free vibration" problem,

$$\ddot{x} + 2(\zeta - A)\omega_n \dot{x} + \omega_n^2 x = 0,$$

with the viscous damping factor being replaced by $\zeta - A$. The response, given by Equation 3.3, becomes

$$x(t) = Ce^{-(\zeta-A)\omega_n t}\cos(\omega_d t - \phi),$$

where ω_d and ϕ are changed accordingly. If $\zeta > A$, the response amplitude decays slower than if $A = 0$, but it still does decay exponentially. If $\zeta < A$, the response amplitude *increases* exponentially without bound. This is an unstable system.

Some applications where self–excited oscillations are possible include[2] engines and governors, mechanical and hydraulic devices, electronic and acoustic devices, rotating shaft phenomena such as whirling, mechanical slip–stick or variable damping phenomena, and aerodynamic phenomena such as the galloping of transmission lines and the flutter of airfoils. Such problems, which are generally nonlinear, are complex and very important in more advanced analysis and design. These tend to be nonlinear phenomena, with which stability becomes a fundamental concern. An introduction is provided in Chapter 12.

4.6 Solution Analysis and Design Techniques

It is appropriate to conclude this chapter with a discussion of how one might use single degree of freedom models for concept definition and preliminary design. The design of a structure or a machine always has two aspects. One is to design the structure so that it meets its function adequately. The pump must move a certain amount of fluid in a specified time. The engine must develop a certain thrust while maintaining specified temperature and

---

[2] R.H. Scanlan and R. Rosenbaum, **Aircraft Vibration and Flutter**, Dover Publications, 1968.

4.6. SOLUTION ANALYSIS AND DESIGN TECHNIQUES

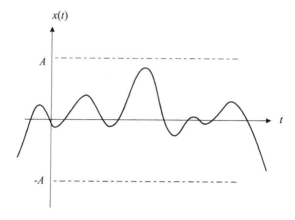

Figure 4.4: Design for $|x(t)| < A$ for all time.

vibration constraints. Vibration constraints can be specified in a variety of ways. As an example, Figure 4.4 represents an amplitude constraint. The vibration amplitude $|x(t)|$ can never exceed the value A. Such a requirement exists because there is no space for larger magnitude deflections, or because larger magnitude deflections will lead to a structural failure.

This leads us to the second aspect of design. Beyond design for function, there is design for structural integrity, which is connected to structural life and reliability. Fatigue life is always a concern when designing vibrating components since it is directly related to the number of cycles that the component undergoes. Of course, design for function is always accomplished in parallel with structural integrity. But to obtain a deeper understanding of the individual aspects of a problem, they should be studied separately.

Figure 4.5 has a similar design goal, except that amplitudes greater than A are permissible, but only after a certain time t_0 sec. Other variations of such design constraints exist.

This chapter is now concluded with two interesting examples of how single degree of freedom systems can provide useful preliminary information for important applications.

Example 4.4 Water Landing of a Space Module

As of this writing, an international space station is in the initial fabrication stage. While the station and its crew will be serviced by the Shuttle and similar Space Transportations Systems (STS), it is necessary to provide the crew with an emergency system that can transport them safely back to

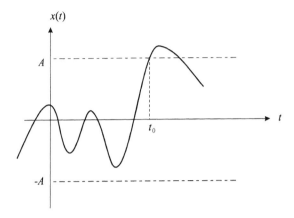

Figure 4.5: Design for $|x(t)| < A$ for t_0 sec.

Earth.[3] This system may be used as a result of a medical emergency, a space station catastrophe, or if the next STS launch is not soon enough.

Current plans call for a water landing space module, a capsule design based on the Apollo command module that was flown to the Moon. There are several concerns regarding the design of this module. It is expected to be used in an emergency, with little warning. Once it ejects from the space station at some random location in orbit and falls to some body of water, it is likely to be floating for a relatively long period of time before it can be rescued. This necessitates that the crew be comfortable and safe. The module needs to parachute to the ocean surface and be maintained in an upright position.

To assist in the design of such a module, a 1/5 scale model has been built and tested. A 1/5 scale model results in a mass scale factor of $(1/5)^3 = 0.008$. It turns out that the time scale factor is $\sqrt{1/5} = 0.4472$. This follows from our earlier discussions on dimensional analysis. Tests are performed for various center of gravity locations (CG) with and without use of flotation or stabilization mechanisms. It is found that the pitch oscillation increased as the CG was offset horizontally. Less significant, pitch oscillation increases were observed with an increasing vertical CG offset. To assist in stabilization, *attitude spheres* are attached at various positions on the model just below the water line. These tests suggest design considerations for a full–scale module. Figure 4.6 shows a schematic of the attitude

---

[3]This example is based on the paper by D.E. Van Sickle and L.A. Anderson, "Pitch Analysis of a Space Module After Water Landing," *AIAA Journal of Spacecraft and Rockets*, Vol.32, No.4, July–August 1995, pp. 601–607.

4.6. SOLUTION ANALYSIS AND DESIGN TECHNIQUES

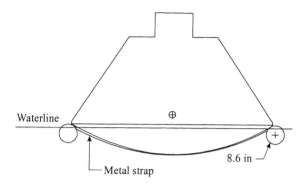

Figure 4.6: Attitude sphere attachment. Copyright © 1995 AIAA. Reprinted with permission.

sphere attachment.

To determine the mass moment of inertia for the module and the equation of motion for the CG, a pendulum test setup was built, as shown in Figure 4.7. From the figure, we see that the test module will oscillate as a pendulum about the support point. If the module mass moment of inertia about its CG is I_G, then by the parallel axis theorem, the inertia about the support point distance a from the CG is $(I_G + ma^2)$, where m is the mass of the module. From our studies in this chapter, we can find the nonlinear pendulum equation of rotational motion to be

$$(I_G + ma^2)\ddot{\theta} + mga\sin\theta = 0. \tag{4.4}$$

If only small angles are allowed in the testing, then Equation 4.4 can be simplified by letting $\sin\theta \approx \theta$. When solved,

$$\theta(t) = A\sin\left(\frac{mga}{I_G + ma^2}t + \gamma\right).$$

A and γ are determined using the initial conditions. Knowing the natural frequency, the period can be evaluated to be

$$T = 2\pi\sqrt{\frac{I_G + ma^2}{mga}} \text{ sec}.$$

The unknowns are a and I_G. The location of the CG is found by swinging the module from its top and its bottom in two tests. In this manner, two sets of equations for the period are used to determine the location of the

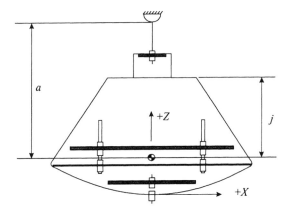

Figure 4.7: Pendulum test setup to determine module dynamic characteristics. Copyright © 1995 AIAA. Reprinted with permission.

CG. The pendulum tests also permit the calculation of the mass moments of inertia about each horizontal axis, I_x and I_y.

The above tests are performed to establish the geometrical and mass properties of the module. Next, it is necessary to model the oscillation of the module as it floats in a variety of sea states, from calm to severe. A floating body will generally have six degrees of motion. As an initial study, only *pitch* motion is considered to be the critical parameter that establishes the stability of the module. Figure 4.8 depicts the coordinate system and forces acting on the module as it pitches or *rolls*.

The pitch equation of motion can be obtained via Newton's second law for the moments acting on the module. From the figure, the restoring force can be seen to be the buoyancy moment due to the angular deflection. The buoyancy force is the weight W and the moment arm is the distance from the CG to the new center of volume of the immersed section. Also taken into consideration is the added inertia term I_w due to the water entrained by the module. Therefore, the total inertia is equal to $I = I_G + I_w$.

Taking the sum of the moments, one finds[4] the nonlinear equation of motion

$$I\ddot{\theta} + c\dot{\theta} + k\sin(\theta - \phi) = 0,$$

where ϕ equals the slope of the wave,

$$\phi(t) = \alpha \sin \omega t.$$

---
[4] For complete details, see the original paper. You will be able to follow it!

4.6. SOLUTION ANALYSIS AND DESIGN TECHNIQUES

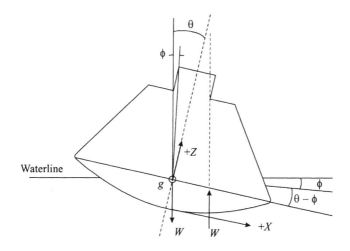

Figure 4.8: Pitch oscillations for module. Copyright © 1995 AIAA. Reprinted with permission.

For small angles, the usual linearization is made, and the wave forcing term is placed on the right hand side of the equation of motion, resulting in

$$\ddot{\theta} + \frac{c}{I}\dot{\theta} + \frac{k}{I}\theta = \frac{k\alpha}{I}\sin\omega t.$$

The linearized equation of motion of free vibration is

$$\ddot{\theta} + 2\zeta\omega_n\dot{\theta} + \omega_n^2\theta = 0,$$

where

$$c = 2\zeta\omega_n I, \quad k = \omega_n^2 I.$$

ζ is obtained by measuring several successive peak amplitudes of a free oscillation and applying the concept of the logarithmic decrement. Once ζ is evaluated, the natural frequency ω_n can be determined using the damped period measurements, $\omega_n = 2\pi/T_d\sqrt{1-\zeta^2}$.

Studies based on this model show that the natural frequency of the module may overlap those of a wave state, resulting in large amplitude pitching as in resonance. This potential problem can be resolved by altering the craft's natural frequency by changing the position of the CG, or by adding attitude spheres. Concluding, we find much value in first solving the simplified problem, and then using our preliminary conclusions to help us develop general design criteria for the actual craft. ∎

182   CHAPTER 4. SDOF VIBRATION: ADVANCED TOPICS

Finally, we consider the very interesting and important problem of protecting manufactured components and structures during shipment from the point of manufacture to the end user. Here, a satellite must be protected when shipped to its launch site over bumpy roads and other obstacles. Similar approaches can be applied to a variety of such container and packaging problems.

Example 4.5 Vibration Analysis of a Satellite in a Shipping Container[5]
A shipping container needs to be designed to transport a satellite from the factory to the launch site. Designing this container is a critical aspect of spacecraft design and manufacture. We will be concerned only with the vibration design criteria. Other considerations include the thermal environment, contamination issues, physical compatibility between the spacecraft and the container, as well as between the container and the transportation system being utilized. Vibration loads are difficult to quantify and, depending on the type of transportation used (truck, rail, or air), test data from the actual route must be used to design the shipping container's suspension system.

Let us consider an example with some representative design criteria. For example, an engineer would receive the design order with some specifications. For truck shipment, the load could be specified as a "$3g$ decaying sinusoid with a decay rate, or logarithmic decrement, δ of 0.5, with load occurring over the frequency range of 2–200 Hz." Figure 4.9 illustrates this loading case for 5Hz. The loading characteristics and the frequency range will depend on the kind of bumps in the road as well as the truck speed. How bumps get transmitted to the satellite is a function of the container properties.

The equation for this load is

$$\ddot{x} = 3ge^{-\zeta\omega t}\sin\omega t \qquad (4.5)$$

$$\zeta = \frac{\delta}{\sqrt{(2\pi)^2 + \delta^2}}, \qquad (4.6)$$

where ω is the driving frequency of the loading in rad/sec. It is common in the study of shock and vibration, for example in earthquake engineering, to specify loading in terms of an acceleration.

Assume that the maximum load the satellite can withstand is $8g$ at the CG at frequencies below 15 Hz. *Above 15 Hz, we can assume that the spacecraft cannot resist any significant acceleration.* Thus the suspension

---

[5]Copyright 1993, 1996, Lockheed Martin Corporation. Used with Permission.

4.6. SOLUTION ANALYSIS AND DESIGN TECHNIQUES

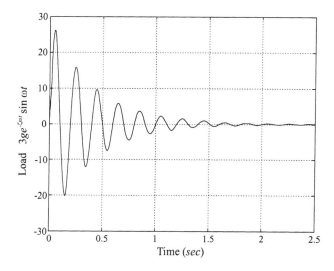

Figure 4.9: Decaying sinusoid load at 5 Hz.

system requirements must be derived for maximum attenuation in the vertical direction at frequencies greater than 15 Hz.

Furthermore, we are told that to prevent *coupling* between the container suspension system and the spacecraft, the suspension system must have a natural frequency of 5 Hz or less.[6] If coupling occurs, a beating–type behavior will result, leading to large amplitude motion. The satellite in its shipping configuration has a first natural frequency of 20 Hz.

The above information is provided to a shipping container subcontractor who then translates the above requirements into a container design proposal. One such proposal is:

The spacecraft will be mounted on a frame assembly which weighs 1550 lb_f. The frame assembly will be mounted on four shock mounts attached to the shipping container base. With the spacecraft installed, 4.0 in of clearance exists between the shipping container insulation and the spacecraft envelope. With an assumed static spring constant of 1275 lb_f/in, dynamic spring constant of 1.3×1275 lb_f/in, and a damping factor of 11% for each shock mount, this system will satisfy suspension performance requirements.

As the satellite owner, you must find some way to verify that the above

---

[6] As we will learn, structures generally vibrate at many frequencies; each frequency is associated with a degree of freedom. Since we have idealized the suspension system as a single degree of freedom oscillator, we characterize its behavior with a single natural frequency.

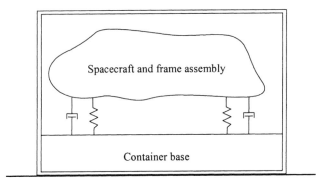

Figure 4.10: Idealized satellite in container.

container will truly meet your needs to safeguard the satellite en route to the launch site. Typically, shipping containers and their suspension systems are verified through drop tests. In this test, an equivalent mock satellite is dropped in such a container, released from a height resulting in a $3g$ decaying sinusoid vibration input. You must now determine this *equivalent drop height* to test the container. This is best estimated analytically.

Solution   *Part One*   To begin to solve this problem, the satellite–container system is idealized as a base–excited, single degree of freedom system. To establish the safety of the satellite in the container, derive and solve the equation governing the satellite oscillation inside the container and show that the maximum response to the $3g$ load is within the allowable space. This will verify the suspension system design. If internal impact occurs, the system must be redesigned. Once the system model is verified, determine a drop height to test the container's suspension system. *Why perform this test?* The testing is necessary because we have used a very simple analytical model for the system. The test will verify that the design can be based on the simplified model.

The base–excited system is shown in Figure 4.10. Further assume that the satellite and the container CGs are centered. Thus, upon impact there are no eccentricities that will induce rotational motion. Also, there is no other motion of the container or of the structure (truck) upon which the container sits. Of course, all these assumptions can be removed if necessary.

From the data provided, additional parameters needed to solve the equation of motion can be evaluated. The total system weighs $W = 4000$ lb$_f$. Thus,

$$m = \frac{W}{g} = \frac{4000}{386.4} = 10.35 \frac{\text{lb sec}^2}{\text{in}}.$$

4.6. SOLUTION ANALYSIS AND DESIGN TECHNIQUES

The input frequency has a range of $f = 2 \to 200$ Hz. Therefore, calculate displacements for a range of frequencies in order to be certain that the maxima have been captured. The forcing frequency is given by $\omega = 2\pi f$ rad/sec. From Equation 4.6, the damping factor $\zeta = 0.079$. Note that a *relaxation time* constant for a frequency of $f = 200$ Hz is $\tau = 1/(\zeta\omega) = 0.01$ sec. This can be physically interpreted as the amount of time the structure is at its peak oscillation. From the shipping container suspension system parameters, the spring constant of all 4 shock mounts is $k = 4 \times 1.3 \times 1275 = 6,630$ lb$_f$/in. The natural frequency is

$$f_n = \frac{\omega_n}{2\pi} = \frac{1}{2\pi}\sqrt{\frac{k}{m}} = 4.028 \text{ Hz},$$

the damped natural frequency is

$$f_d = \frac{\omega_d}{2\pi} = \frac{\omega_n\sqrt{1-\zeta^2}}{2\pi} = 4.015 \text{ Hz},$$

and finally the damping coefficient is

$$c = 2m\zeta\omega_n = 41.395 \text{ lb}_f \text{ sec/in}.$$

The governing equation of motion for a base excited structure is

$$m\ddot{y} + c(\dot{y} - \dot{x}) + k(y - x) = 0, \tag{4.7}$$

where $y(t)$ is the structural response, and $x(t)$ and $\dot{x}$ are the base container input displacement and velocity, respectively. These are derived by twice integrating Equation 4.5, $\ddot{x} = 3ge^{-\zeta\omega t}\sin\omega t = 3ge^{\rho t}\sin\omega t$, where we define $-\zeta\omega = \rho$. Assuming zero initial velocity, the first integration results in

$$\dot{x} = \frac{3g}{\rho^2 + \omega^2}\left[-\omega e^{\rho t}\cos\omega t + \rho e^{\rho t}\sin\omega t + \omega\right]. \tag{4.8}$$

A second integration, assuming zero initial displacement, yields

$$\begin{aligned} x(t) = &-\frac{3g}{\rho^4 + 2\rho^2\omega^2 + \omega^4}\left[2e^{\rho t}\omega\rho\cos\omega t \right.\\ &\left.+ e^{\rho t}(\omega^2 - \rho^2)\sin\omega t - \omega t\rho^2 - \omega^3 t - \omega\rho\right]. \end{aligned} \tag{4.9}$$

Substituting Equations 4.8 and 4.9 into the governing Equation 4.7 results in an equation of motion that is easier to solve numerically,

$$\begin{aligned} m\ddot{y} + c\dot{y} + ky = &\frac{3gc}{\rho^2 + \omega^2}\left[-\omega e^{\rho t}\cos\omega t + \rho e^{\rho t}\sin\omega t + \omega\right]\\ &+ \frac{3gk}{\rho^4 + 2\rho^2\omega^2 + \omega^4}[2e^{\rho t}\omega\rho\cos\omega t + e^{\rho t}(\omega^2 - \rho^2)\sin\omega t\\ &- \omega t\rho^2 - \omega^3 t - \omega\rho]. \end{aligned} \tag{4.10}$$

Table 4.1: Maximum Spacecraft Acceleration and Maximum Relative Displacement $(x - y)$ For Various Driving Frequencies f

f Hz	$\ddot{y}$ g	$(x - y)$ in
2	4.077	1.948
3.6	6.066	3.505
3.7	6.121	3.578
3.9	6.073	3.577
4	5.987	3.553
14	1.027	0.352
20	0.613	0.222
25	0.461	0.179
100	0.106	0.044
200	0.051	0.022

Note that there are secular terms on the right hand sides of the last two equations. The ω term in Equation 4.8 has led to the secular terms and the term $2\omega\rho$ of Equation 4.9. These arise from the requirement that $x(0) = 0$ and $\dot{x}(0) = 0$. Such terms are signs of potential trouble and instability, but they do not cause difficulties here in the numerical solution of the Equation 4.10. We are interested only in the first cycle of the oscillation, that is for small time $t \ll 1$ sec. This is when the peak amplitudes and accelerations are experienced by the satellite.

For a range of frequencies, this equation can be solved for the maximum spacecraft acceleration and for the maximum spacecraft/container relative displacement. This information allows us to determine whether the system requirements are met. Table 4.1 shows how maximum spacecraft acceleration $\ddot{y}$ and maximum relative displacement $(x - y)$ vary as a function of driving frequency f. In the above equations $\omega = 2\pi f$ and therefore $\rho = -\zeta\omega = -2\pi\zeta f$.

The system model meets the requirements that the maximum acceleration be less than $8g$ and the maximum displacement be less than 4 in. From the Table, for the range of frequencies 2 – 4 Hz, there is dynamic coupling between the spacecraft container and the truck suspension, resulting in significant satellite motion. At higher frequencies, for example 14 – 200 Hz, the shock loads are isolated, as required by the design specifications. ∎

Example 4.6 Solution   *Part Two*   The second part of this solution

4.6. SOLUTION ANALYSIS AND DESIGN TECHNIQUES

is the determination of the drop height that results in loads and displacements similar to the $3g$ decaying sinusoid. This drop height will be used to design an experiment to verify compliance with design requirements, and the height will be determined using the single degree of freedom model just developed. Recalling basic physics, when an object is dropped from rest at height h, its original gravitational potential energy is translated to kinetic energy at the instant of impact, $mgh = mv_0^2/2$, from which

$$v_0 = \sqrt{2gh}. \tag{4.11}$$

Once the container strikes the ground, the internal component will oscillate in free vibration with *initial velocity* v_0. It is assumed that upon impact, the container remains in contact with the ground rather than bouncing up and down. During the free fall, the springs are in a zero or unstretched position, but after impact the oscillation will be about the equilibrium. Therefore, there will also be an *initial displacement* y_0 given by $-4ky_0 = mg$, and

$$y_0 = -\frac{mg}{4k}.$$

To make the above equations work, we need to *guess* an initial height h, and then use Equation 4.11 to calculate initial velocity $\dot{y}(0) = v_0$. The initial displacement $y(0) = y_0$ is independent of initial height and can be calculated immediately as $y_0 = -0.794$ in. Governing Equation 4.7 becomes

$$m\ddot{y} + c\dot{y} + ky = 0,$$

since $x = 0$ and $\dot{x} = 0$ at impact. This equation can be solved exactly in closed–form, as in Section 3.2. See Equation 3.1 with solution Equation 3.6, repeated here,

$$y(t) = \frac{y(0)}{\cos\phi} e^{-\zeta\omega_n t} \cos\left(\omega_d t - \tan^{-1}\left(\frac{\dot{y}(0) + y(0)\zeta\omega_n}{y(0)\omega_d}\right)\right),$$

where

$$\cos\phi = \frac{y(0)\omega_d}{\sqrt{[\dot{y}(0) + y(0)\zeta\omega_n]^2 + [y(0)\omega_d]^2}}.$$

Beginning with a guess of $h = 12$ in, we find $(x - y) = 3.474$ in and $\ddot{y} = 5.823\ g$. For a guess of $h = 13$ in, $(x - y) = 3.604$ in and $\ddot{y} = 6.040\ g$. Finally, for a guess of $h = 14$ in, we find $(x - y) = 3.729$ in and $\ddot{y} = 6.248\ g$. We can use the last value of $h = 14$ in as the height for a drop test to verify the protective value of the satellite container. This is because of the requirement that maximum displacement be less than 4 in and the maximum acceleration be less that $8\ g$. Both these conditions are met by the drop test height $h = 14$ in. ∎

These last two examples demonstrate that difficult vibration problems can be tackled using single degree of freedom models.

4.7 Concepts Summary

Lagrange's equation has been introduced as a prelude to more powerful energy based techniques for problem formulation. Some more advanced ideas were touched upon: randomness, control, identification, and stability. Two detailed preliminary design problems were discussed, closing the chapter. Chapter 5 introduces methods to analyze systems with uncertain parameters and forcing.

4.8 Problems

Problems for Section 4.1 –Introduction to the Use of Lagrange's Equation

1. Use Lagrange's equation to derive the equations of motion for the systems shown in Figures 4.11 and 4.12.

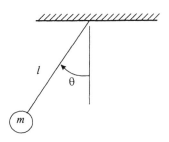

Figure 4.11: Pendulum via Lagrange.

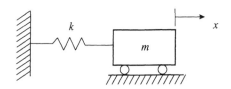

Figure 4.12: Spring mass via Lagrange.

4.8. PROBLEMS

2. Consider *Example Problem* 4.16 and derive the equation of motion using x as the generalized coordinate. Then solve this equation with initial conditions: x_0, v_0 to find
$$x(t) = \frac{v_0}{\omega} \sin \omega t - \frac{mg}{2k} \sin \phi (\cos \omega t - 1) + x_0,$$
where $\omega = \sqrt{4k/3m}$.

3. Derive the equation of motion of *Example Problem* 4.1 using Newton's second law of motion.

Problems for Sections 4.2-4.6

4. For the problem of oscillator control, given by Equation 4.2, consider the specific governing equation
$$\ddot{x} + 2\zeta\omega_n \dot{x} + \omega_n^2 x = A \cos \omega t + F(t)_{control},$$
where ζ, ω_n and $F(t)_{control}$ must be determined so that the maximum amplitude of the response is $x_{max} < \alpha A$, where $\alpha = 0.5$. Since there is no single answer, describe how to proceed and what considerations must be made during the analysis. What is one solution?

5. How does randomness of excitation alter the analyst's ability to evaluate structural response?

6. What difficulties arise if a system parameter such as k or m are only approximately known?

7. If the structure responds in a way that is unacceptable for a particular application, what options exist for the designer?

8. If the analyst decides that a structure requires control in order to fulfill the needs of an application, describe the procedure by which a control force can be derived and then verified that it performs as expected.

9. Consider a damped oscillator where the force is a function of structural velocity,
$$m\ddot{x} + c\dot{x} + kx = F_0 \dot{x}.$$
Discuss the stability of this system in terms of the parameters given: m, c, k, F_0.

10. Consider a damped oscillator where the force is a function of structural displacement,
$$m\ddot{x} + c\dot{x} + kx = F_0 x.$$
Discuss the stability of this system in terms of the parameters given: m, c, k, F_0.

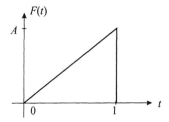

Figure 4.13: Ramp loading.

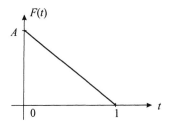

Figure 4.14: Shock loading.

11. For an mck oscillator subjected to harmonic loading, $A\cos\omega t$, conceive a design such that the displacement response $|x(t)| < A$ for all time. Assume $A = 10$ cm, $m = 1$ N, and $\omega = 3$ rad/sec. In your design, specify c, k.

12. For an mck oscillator subjected to harmonic loading, conceive a design such that the displacement response $|x(t)| < A$ for t_0 sec. Assume $A = 10$ cm and $m = 1$ N. In your design, specify c and k.

13. Solve *Problem* 13, but for the load as shown in Figure 4.13.

14. Solve *Problem* 13, but for the load as shown in Figure 4.14.

4.9 Mini–Projects

This is the first chapter that contains a section on *mini–projects*. Let me briefly state the reason for their inclusion here and in the remaining chapters. There is often a valid complaint regarding introductory texts, especially of subjects that tend to be very mathematical, that there are few links made between the material introduced in the text and the proverbial

4.9. MINI–PROJECTS

real world. This text includes realistic examples and homework problems for the student, but in addition offers another connection to the concerns of the profession. That connection is through the literature of the vibration community. Through an examination of these papers, the student will begin to understand that the problems discussed in this book are also of interest to those at the forefront of analysis and design. Another benefit for the student is that the names of these journals will also become familiar.

The instructor may assign any of the following papers for the students to locate and read. An example assignment statement is: *Look up paper number 1 and read it as far as you can. Focus on the problem formulation and especially the statement of assumptions. Try to reproduce the early mathematical development. Study the results and any plots. Try to understand the physical behavior and see if you can link a particular physical characteristic to certain parts of the mathematics. Study the discussion of results. Prepare your study of this paper in written form.*

The mini–projects are intended to be individual or small group projects for the student so that the material of the chapter can be applied or extended to more advanced cases. Generally, the student will be able to begin to understand the papers cited here, but it is unlikely that all the material in the papers will be accessible. Additional mini–project lists appear at the ends of Chapters 6–12.

1. *The Role of Damping in Vibration Theory*, S.H. Crandall, *J. Sound and Vibration* (1970) 11(1), 3–18.

    In many applications of vibration and wave theory the magnitudes of the damping forces are small in comparison with the elastic and inertia forces. These small forces may, however, have very great influence under certain special circumstances. Damping arises from the removal of energy by radiation or dissipation. It is generally measured under conditions of cyclic or near–cyclic motion. Damping is of primary importance in controlling vibration response amplitudes under conditions of steady–state resonance and stationary random excitation.

2. *Resonance, Tacoma Narrows Bridge Failure, and Undergraduate Physics Textbooks*, K.Y. Billah, R.H. Scanlan, *American J. Physics* 59 (2), Feb. 1991, 118–124.

    The dramatic Tacoma Narrows bridge disaster of 1940 is still very much in the public eye today. Notably, in many undergraduate physics texts the disaster is presented as an example of elementary forced resonance of a mechanical oscillator, with the wind providing an external periodic frequency that has matched the natural structural frequency. In the present article the engineers' viewpoint is presented to the physics community to make it clear where substantial disagreement exists.

3. *Modelling of a Hydraulic Engine Mount Focusing on Response to Sinusoidal and Composite Excitations*, J.E. Colgate, C–T. Chang, Y–C. Chiou, W.K. Liu, and L.M. Keer, *J. Sound and Vibration* (1995) 184(3), 503–528.

   Engine mounts serve the principal functions of vibration isolation and engine support. In the past decade, the automotive industry's shift to small, four cylinder engines and transversely mounted front–wheel–drive power trains has made these two functions increasingly incompatible. This paper investigates the frequency response characteristics of a hydraulic engine mount, both experimentally and with analytical models.

4. *A Simple Application of Nonstandard Analysis to Forced Vibration of a Spring–Mass System*, F. Farassat, M.K. Myers, *J. Sound and Vibration* (1996) 195(2), 340–345.

   This paper is of a more mathematical nature, suitable for graduate level study. Nonstandard mathematical analysis, as an application, is used to study resonance in a single degree of freedom forced oscillator.

5. *Dynamic Vibration Absorbers*, R. Smith, *Sound and Vibration*, November 1998, 22–27.

   The design and application of dynamic vibration absorbers to control machinery vibration problems is reviewed. Three case histories are presented to demonstrate the techniques.

6. *Linear Damping Models for Structural Vibration*, J. Woodhouse, *J. Sound and Vibration* (1998) 215(3), 547–569.

   Linear damping models for structural vibration are examined. Simple expressions for damped natural frequencies, complex mode shapes, and transfer functions are obtained.

7. *Silicon–Micromachined Scanning Confocal Optical Microscope*, D.L. Dickensheets, G.S. Kino, *J. of Microelectromechanical Systems*, Vol. 7, No. 1, March 1988, 38–47.

   A miniature scanning confocal optical microscope is constructed from micromachined components using silicon and fused silica. The design, fabrication, and characterization of the components of the microscope as well as the assembly of the system are described.

8. *A Simple Way to Measure Mass Moments of Inertia*, J.B. Andriulli, *Sound and Vibration*, Nov. 1997, 18–19.

4.9. MINI–PROJECTS

A simple rotational pendulum method to measure the radii of gyration or mass moments of inertia of a rotor and other assemblies is described.

9. *Electrostatic Combdrive–Actuated Micromirrors for Laser–Beam Scanning and Positioning*, M.–H. Kiang, O. Solgaard, K.Y. Lau, and R.S. Muller, *J. Microelectromechanical Systems*, Vol. 7, No. 1, March 1998, 27–37.

    The design and fabrication of surface-machined resonant microscanners that have large scan angles and fast scan speeds are described.

10. *Spring's Effective Mass in Spring Mass System Free Vibration*, Y. Yamamoto, *J. Sound and Vibration* (1999) 220(3), 564–570.

    The free longitudinal vibration of a fixed–lumped mass rod is examined numerically in order to estimate the spring's effective mass in free vibration of the fixed–lumped mass spring system.

11. *Vibration and Acoustic Testing of Spacecraft*, T.D. Scharton, *Sound and Vibration*, June 2002, 14–18.

    Spacecraft are subjected to a variety of dynamic environments which may include quasi–static, vibration and acoustic loads at launch, pyrotechnic shocks generated by separation mechanisms, on–orbit jitter, and sometimes planetary landing loads. Some of these are discussed here.

12. *Recent Advances in Vibroacoustics*, W.O. Hughes, M.E. McNelis, *Sound and Vibration*, June 2002, 20-27.

    Numerous vibroacoustics advances and problems in the aerospace industry have occurred over the last 15 years. This article addresses some of these that developed from engineering programmatic task–work at the NASA Glenn Research Center at Lewis Field.

13. *Helpful Guidelines for Single-Axis Shaker Testing*, R. Vinokur, *Sound and Vibration*, October 2002, 16–19.

    This article presents common-sense guidelines for single-axis sweep sine and random vibration testing. Several actual case histories are included as well to demonstrate these techniques.

14. *Viscoelastic Damping*, P. Macioce, *Sound and Vibration*, April 2003, 8–10.

15. *Diagnosing Vibration Problems with Embedded Sensitivity Functions*, C. Yang, D.E. Adams, S.-W. Yoo, *Sound and Vibration*, April 2003, 12–21.

    Embedded sensitivity functions are described. They indicate which changes in mass, damping or stiffness will suppress vibration problems without introducing other problems.

Chapter 5

Single Degree of Freedom Vibration: Probabilistic Forces

"The uncertainties must be confronted."

5.1 Introduction

The main goal of the study of structures under random loading is to predict the response (output) statistics given the loading (input) statistics. *Statistics* is the discipline that organizes data in a form that is meaningful. The governing equation is the second order differential equation for an oscillator, Equation 2.6,

$$\ddot{x} + 2\zeta\omega_n \dot{x} + \omega_n^2 x = F(t), \tag{5.1}$$

except that force per unit mass $F(t)$ is now a random function of time.[1] If $F(t)$ is a deterministic function of time as in the last chapter, we would know how to solve the convolution integral for the vibration response. *But what is to be done when the function oscillates in such a complex manner as in Figure* 5.1? One possibility is to carry out many experiments and gather data on $F(t)$ in the form of time–histories.

---

[1] It is important for us to distinguish between inherently random molecular forces due to Brownian motion, such as those experienced by atoms on a molecular scale, and the environmental forces of concern here. Environmental forces needed for vibration studies are not inherently random, but they undergo very complex cycles. We are unable to model such complex phenomena using deterministic techniques. Therefore, we adopt the tools of probability and statistics to provide us with a way to quantify our uncertainties.

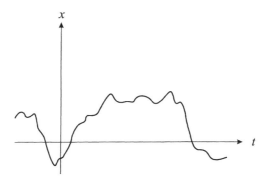

Figure 5.1: Random behavior.

Then, the time–history with the largest amplitudes can be used for the deterministic analysis and design. This would work, but if the largest amplitude force occurs only infrequently, the structure would be over–designed. This means that the structure is stronger than it has to be and is therefore uneconomical. What if all the time–histories were averaged and this *averaged or mean value time–history* is used as a deterministic load that is used in the convolution integral? This would be a good start, but of course, the response calculated in this manner would underestimate the actual response too often! How often depends on the scatter, or standard deviation, of possible time–histories.

The next question to ask is then: *How much scatter is there above and below the mean value response?* Perhaps if we knew the mean value response as well as a measure of the scatter, this information could be used in a safe and economical design. This is indeed an approach that makes sense.

Finally, we need to ask: *How does the engineer know how often a very large amplitude force occurs?* If the very large force, such as an earthquake, occurs only once in 100 *years*, how is that fact used in a design? What the designer needs is a way to give more *weight* to more likely events, without completely ignoring more severe but less likely loads. We will learn that there is a quantitative measure called the *probability density function* that acts as such a weighting function.

All of the ideas described above are actually probabilistic concepts. We need to develop some introductory ideas with important motivating examples. This will set the stage for our effort at random vibration modeling.

Before proceeding, it should be noted that other tools exist for evaluating how parameter variations affect changes in system response. A powerful tool for doing this is known as *sensitivity analysis*, introduced in Section 6.5

5.1. INTRODUCTION

of Chapter 6. Using sensitivity, the analyst can understand whether small variations in certain parameters translate into small or large variations in other parameters of interest or in the response. Sensitivity analysis should be viewed as complementing, not replacing, the probabilistic approach introduced in this chapter.

A Definition for Probability

Let us first define probability. There is dispute about the definition of probability, primarily pitting those who view it as a subjective quantity against others who believe that only with experimentation can a rigorously derived probability be possible. The former will counter that it is often not possible to perform enough experiments to arrive at that rigorous probability and judgment must be used. Fortunately, our purposes here do not require us to resolve this debate. We can assume that in some manner it is possible to obtain the probabilities necessary for our computations, usually based on data analysis.[2]

Think of a randomly vibrating oscillator where random behavior implies unpredictable periods, amplitudes, and frequencies. These all appear to vary from one instant of time to the next. How can we answer a question such as: *What is the probability that the amplitude is greater than a specific number?* Using probability notation,

$$\Pr\{A > A_0\} = ?$$

What we are really asking is how much of the time is the oscillator at amplitudes greater than A_0. This implies a *fraction* or *frequency* interpretation for probability. For the above problem, we look at a long time history of the oscillation and figure the amount of time the amplitude is greater than A_0. That *excursion frequency* is the probability estimate,

$$\Pr\{A > A_0\} = \frac{amount\ of\ time\ > A_0}{total\ time}.$$

For example, if the oscillation time history is 350 hours long and for 37 of those hours $A > A_0$, then $\Pr\{A > A_0\} = 37/350 = 0.106$; the probability that $A > A_0$ is estimated to be 10.6%. This is only an estimate since it is expected that if the test lasts 3,500 hours instead of 350 hours there would be some change in the estimated probability. The key to a good estimate is that the test be long enough so the probability estimate has approximately converged. Figure 5.2 depicts this procedure of estimating probabilities by

---

[2] J.S. Bendat and A.G. Piersol, **Random Data: Analysis and Measurement Procedures**, Second Edition, John Wiley & Sons, 1986. This book provides an excellent development of the theory and techniques of data analysis.

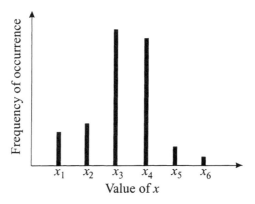

Figure 5.2: Relative frequency.

frequencies of occurrence. This is the most common approach to estimating probabilities for the technical arts, and the one we use here.

While the above discussion of the meaning of probability involved time, this is not always the case. Probabilistic models of mechanical systems are a natural result of the observation that most physical variables may take on a range of possible values. For example, if 100 machine shafts are manufactured, there will be 100 different diameters if enough significant figures are kept. Figure 5.3 depicts a *histogram* of diameter data where three significant figures are kept. As expected, the diameters are very close in value, but not exactly the same.

How do we account for such a spread of values if we are interested in measuring the strength of the shaft in torsion? What numbers should be substituted into the stress–strain relation? Similarly, running ultimate tensile–strength tests a number of times on "identical specimens" will show no two identical results. Small differences in dimensions, material properties, and boundary conditions make it impossible to exactly duplicate experimental results. *There will always be some scatter.* How should this information be utilized?

Randomness is possible for constants as well as functions (of time or space). A constant with a scatter of possible values is called a *random variable*. A function with scatter is generally called a *random*[3] *process*. Random variables are those which can only be prescribed to a certain level of certainty. An important example is material yield characteristics that define the transition from elastic to plastic behavior. *Random processes*

---

[3] or the Greek *stochastic*: στοκος

5.1. INTRODUCTION

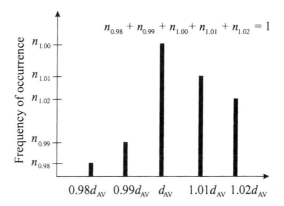

Figure 5.3: Histogram of machine shaft diameters.

are time–dependent (or space–dependent) phenomena that, with repeated observation under essentially identical conditions, do not show the same time histories.

It is thus increasingly important to understand, and be able to model, uncertainties and qualitative information in engineering analysis and design. An example of qualitative information is a verbal description of size or strength. Developing the ability to analyze uncertainties allows the engineer to decide for which applications they are insignificant and may be ignored. However, there will be applications where scatter cannot be ignored because of the resulting variability of response.

Chapter Outline

This chapter is different from the others you will read. Primarily this is because there is a steep learning curve before it is possible to consider even the simplest random vibration problem. A new way of thinking with uncertainty as our *paradigm*[4] must be learned. In some ways the probabilistic paradigm is very uncomfortable for engineers since we are raised to believe that, given enough experiments and theoretical development, any problem is solvable exactly, or at least to within measurement tolerances. We are

---

[4] A paradigm, pronounced "para–dime," is a way of thinking. It may be viewed as the beliefs, values, and techniques shared by a particular group of people. Therefore, a new paradigm in a technical area implies a completely new way of thinking about that area. A recent example of a *paradigm shift* is the development of the field of *chaos* in nonlinear dynamics.

about to learn that certainty exists only in idealized models, not in the physical systems that must be understood and designed. Nothing is exact, but sometimes uncertainty can be ignored for particular applications. Here we begin to learn how to proceed when uncertainty cannot be ignored. To do this, some basic concepts in probability must be learned. Be patient and do not be discouraged at what may at first reading appear to be a major diversion from what you are primarily interested in: *vibration.* This is no more a diversion than was our study of ordinary differential equations. Let us look at some motivating examples first.

5.2 Example Problems and Motivation

In order to demonstrate the importance of uncertainty modeling in mechanical systems, a number of examples are chosen for brief and qualitative discussion. There are a number of texts on random vibration,[5] where the subject is explored fully.

5.2.1 Random Vibration

The discipline of random vibration of structures was born of the need to understand how structures respond to dynamic loads that are too complex to model deterministically. Essentially, the question that must be answered is: *Given the statistics[6] of the loading, what are the statistics[7] of the response?* Generally, for engineering applications the statistics of greatest usefulness are the mean or average value and the variance or scatter. These concepts are discussed in detail in this chapter.

Example 5.1 Aerodynamic Loads
Suppose that we are aircraft designers currently working on the analysis and design of a wing for a new airplane. As engineers, we are very familiar with the mechanics of solids, and can size the wing for static loads. Also, we have vibration experience and can evaluate the response of the wing to a harmonic or impulsive forcing. But this wing will be attached to an airplane flying

---

[5] A very useful one that includes a broad spectrum of theory and application is by I. Elishakoff, **Probabilistic Methods in the Theory of Structures**, Wiley–Interscience, 1983. Two exceptionally clear early books on random vibration are worth reading. The first is **Random Vibration in Mechanical Systems** by S.H. Crandall and W.D. Mark, Academic Press, 1963. The other is **An Introduction to Random Vibration** by J.D. Robson, Elsevier, 1964. My first exposure to the subject was in **An Introduction to Random Vibrations and Spectral Analysis** by D.E. Newland, Longman, 1975, now in its third edition.

[6] *read: uncertainties*

[7] *read: most likely values with bounds*

5.2. EXAMPLE PROBLEMS AND MOTIVATION

through a turbulent atmosphere. Even though we are not fluid dynamicists, we know that turbulence is a very complicated physical process. In fact, the fluid (air) motion is so complicated that probabilistic models are required in order to make any progress. *Here, a plausibly deterministic but very complicated dynamic process is taken to be random for purposes of modeling.*

For wing design, needed are estimates of the forces due to the interaction between fluid and structure. An *averaging* of forces and moments is needed. The question is how to make sense of such intricate motion.

The next step for the analyst and designer is to run some scale model tests. A wing section is set up in the wind tunnel and representative aerodynamic forces are generated. Data on wind forces and structural response are gathered and analyzed. With additional data analysis, we begin to have some measures of the force magnitudes to be expected. Estimates of likely or mean values of these forces are possible, as well as of the range of possible forces.

Now we can begin to study the behavior of the wing under a variety of realistic loading scenarios, using the tools of probability to model this complex physical problem. *This chapter introduces the use of probabilistic information in a vibration analysis.* ∎

5.2.2 Fatigue Life

The fatigue life of mechanical components and structures[8] depends on many factors such as material properties, temperature, corrosion environment, and also vibration history. A first step in estimating fatigue life involves the characterization of the cycles the structure has experienced. Were there many cycles, what were the amplitude ranges, and was the loading harmonic or of broad frequency band?

Fatigue life estimates are extremely important for the proper operation of a modern industrial society. Such estimates are intimately linked to the reliability of machines and structures. They determine how often components need to be replaced, how economical is the operation of the machine, and what will be the insurance rates.

Anyone studying fatigue life data will be immediately struck by the significant scatter. Components normally considered to be identical can have a wide range of lives. As engineers, we are concerned about having a rigorous basis for estimating the fatigue lives of ostensibly identical manufactured components. Eventually, it is necessary to relate the life estimate of the structure to that of its components. This is generally a difficult task, one

---

[8] A very useful book with which to begin the study of fatigue is by V.V. Bolotin, **Prediction of Service Life for Machines and Structures**, ASME Press, 1989.

that requires the ability to evaluate structural and machine response to random forces.

Example 5.2 Miner's Rule for Fatigue Damage
One of the most important early works on the estimation of fatigue life is by Miner,[9] who was a strength test engineer with the Douglas Aircraft Company. *Miner's rule* is a deterministic way to deal with the uncertainties of structural damage and fatigue.

The phenomenon of cumulative damage under repeated loads is assumed to be related to the total work absorbed by a test specimen. The number of loading cycles applied, expressed as a percentage of the number of cycles to failure at a given stress level, would be the proportion of useful structural life expended. When the total damage reaches 100%, the fatigue test specimen should fail. Miner presented experimental verification using aluminum sheets.

At a certain stress level for a specific material and geometry,[10] this rule estimates the number of cycles to failure. Mathematically, this can be written as
$$\frac{n}{N} = 1, \tag{5.2}$$
where n equals the number of cycles undergone by the structure at a specific stress level, and N is the experimentally known number of cycles to failure at that stress level. Since most structures undergo a mixture of loading cycles at different stress levels, Equation 5.2 must be written for each stress level i as follows,
$$\sum_i \frac{n_i}{N_i} = 1, \tag{5.3}$$
where each fraction represents the percentage of life used up at each stress level. Therefore, suppose we have two stress levels, $i = 1, 2$, with corresponding $N_1 = 100$ and $N_2 = 50$. According to Equation 5.3, the following relation holds
$$\frac{n_1}{100} + \frac{n_2}{50} = 1$$
between the number of possible cycles n_1 and n_2 for each stress level. There are numerous combinations that lead to failure. For example: $(n_1, n_2) = (50, 25)$, $(n_1, n_2) = (100, 0)$, $(n_1, n_2) = (0, 50)$, with others easy to find.

Miner realized that these summations were only approximations. His experiments showed that sometimes a component failed before the sum

---

[9] M.A. Miner, *Cumulative Damage in Fatigue*, pp. A-159–A-164, *Journal of Applied Mechanics*, September 1945.

[10] Corners and discontinuities cause high stress concentrations resulting in lower fatigue life.

5.2. EXAMPLE PROBLEMS AND MOTIVATION

totaled one, and other times did not fail until the sum was greater than one. Furthermore, failure by this rule is independent of the ordering of the stress cycles. This means that fatigue life is the same whether high stress cycles precede or follow lower stress cycles. We know, however, that stress history affects fatigue life.

In the 60 years since Miner's paper, a vast amount of work has been done to build on his and other work to better understand fatigue, but Miner's rule and its variants are still a widely utilized practical method. ∎

5.2.3 Ocean Wave Forces

There is similarity between approaches used to model ocean wave forces on structures and those for wind forces. The differences are primarily due to the added mass of the water and to differences in structural types found in the ocean. This topic is a subset of the engineering specialty known as ocean engineering.[11] As might be expected, many engineering disciplines are utilized in ocean engineering. The estimation of wave forces on offshore oil drilling platforms, ships, and other ocean and hydraulic structures such as water channel spillways and dams is one of the very important aspects of the design of such structures. Without these estimates, there is no way to analyze or design the structure. The estimation of loads is always first on the list of tasks for an engineer.

Example 5.3 Wave Forces on an Oil Drilling Platform
The need to drill for oil in the oceans has driven our ability to design ocean structures for sites of ever increasing depths. Today's fixed-bottom ocean structures, when taken with their foundations, are taller than our tallest skyscrapers. As might be expected, the dynamic response of these towers to ocean waves and currents is significant and must be understood and analyzed. Consider the ocean wave force on a simple structure, as shown in Figure 5.4.

The most important single paper[12] on the force exerted by ocean waves on fixed structures, even though it was written over half a century ago, derived what became to be universally known as the *Morison equation.* In

---

[11] Some useful books, among hundreds of volumes, are the following: J.F. Wilson, **Dynamics of Offshore Structures**, Second Edition, John Wiley & Sons, 2003. O.M. Faltinsen, **Sea Loads on Ships and Offshore Structures**, Cambridge University Press, 1990. S. Gran, **A Course in Ocean Engineering**, Elsevier Science, 1992. This is a very thorough book.
[12] J.R. Morison, M.P. O'Brien, J.W. Johnson, and S.A. Schaff, *The Force Exerted by Surface Waves on Piles*, pp.149–154, *Petroleum Transactions*, Vol. 189, 1950. This is the original work.

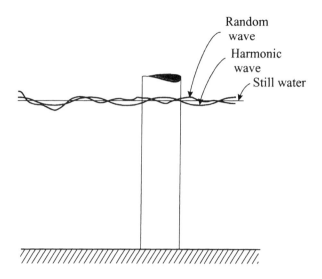

Figure 5.4: Schematic of wave on cylindrical structure.

this paper, after much experimental work, Morison and colleagues came to the conclusion that the force exerted by unbroken surface waves on a circular cylindrical column that extends from the bottom upward above the wave crest is made up of two components:

(i) a drag force proportional to the square of the wave particle velocity, with proportionality represented by a *drag coefficient* having substantially the same value as for steady flow, and

(ii) an inertia force proportional to the horizontal component of the inertia force exerted on the mass of water displaced by the column, with proportionality represented by an *inertia coefficient*.

The drag force on an element of length dx is given by

$$dF_D = C_D \rho D \frac{u|u|}{2} dx,$$

where C_D is the experimentally determined drag coefficient, ρ is the density of water, D is the diameter of the cylinder, and u is the instantaneous horizontal water particle velocity. The term $u|u|$ ensures that the direction of the force is in the direction of the flow.

The inertia force on an element of length dx is given by

$$dF_I = C_I \rho \frac{\pi D^2}{4} \dot{u} dx,$$

5.2. EXAMPLE PROBLEMS AND MOTIVATION

where $\dot{u}$ is the instantaneous horizontal water particle acceleration and C_I is the inertia coefficient. The dimensionless drag and inertia coefficients are functions of flow characteristics, cylinder diameter and fluid density. Depending on the application, the analyst may assume them to be effectively constant, or may need to account for the scatter in their values.

The Morison force equation is the sum of the above drag and inertia components, and this appears as the forcing function in the governing equation of motion. Classical deterministic fluid mechanics is used to derive wave particle velocities and accelerations.

Many tall ocean structures will oscillate appreciably. To take account of this, the relative velocity and acceleration between fluid and structure is used in the Morison equation, where u is replaced by $(u-\dot{x})$ and $\dot{u}$ is replaced by $(\dot{u} - \ddot{x})$, where $\dot{x}$ and $\ddot{x}$ are the structural velocity and acceleration, respectively.

Also, in order to better characterize the complexity of the wave motion, fluid velocity, acceleration, and the resulting force are modeled as random functions of time. We will explore the concept of random forces in detail later in this chapter. More details on the vibration of structures in fluids can be found in numerous books.[13] ∎

5.2.4 Wind Forces

Engineering structures such as cooling towers, aircraft, skyscrapers, rockets and bridges are all exposed to wind and aerodynamic loads. Wind is the natural movement of the atmosphere due to temperature and pressure gradients. Aerodynamic loads are the atmospheric forces resulting from the interaction of wind and structure. While we know how to write an equation for a harmonic force, what does an equation for wind force look like? Due to the complexity of the fluid mechanics of wind, it is generally necessary to approximate the force due to wind. There are various levels of approximate relations, depending on the application. In all instances, the force relation includes at least one experimentally–determined parameter or coefficient. Such *semi–empirical* force equations are very valuable in engineering practice. These will look very much like the Morison equation above.

---

[13] These two books are worth having a look at: R.D. Blevins, **Flow–Induced Vibration**, van Nostrand Reinhold, 1977. There is a second edition. A.T. Ippen, Editor, **Estuary and Coastline Hydrodynamics**, McGraw–Hill, 1966.

5.2.5 Material Properties

While the modeling of randomness in material properties is beyond our scope in this book, it is worthwhile to briefly mention this type of modeling because of the importance of many new materials that have *effective* properties, that is, properties that are an average over a cross–section. These include various *composites* and *tailored materials*, modern materials designed for particular structural applications, especially where high strength and durability, but light weight, are needed. The design requires that a complicated mix of fibers and substrates be organized to obtain particular properties. The difficulty then is to model these materials so that their properties are included in a vibration analysis. Defining stress–strain relations and Young's modulus, for example, is not straightforward. It is sometimes necessary that properties be averaged or effective properties be defined.

The soil is a naturally occurring material that is extremely complex and cannot be modeled in a traditional manner. It is common that two nearby volumes of soil have very different mechanical properties. Therefore, in structural dynamic applications, such as earthquake engineering, the loading is effectively random, in part, because by the time it reaches the structure, the force has traversed through a complex topology of earth.[14]

Data on the variability of material properties are tabulated in numerous references.[15] From Haugen, for example, hot rolled 1035 steel round bars of diameters in the range 1–9 *in* have yield strengths of between 40,000–60,000 psi, with an average yield of just under 50,000 psi. In addition, the variability can change appreciably depending on temperature. A titanium–aluminum–lead alloy has an ultimate shear strength of between 88,000–114,000 psi at $90°F$, but at $1,000°F$ the strength drops to between 42,000–60,000 psi. The obvious conclusion is that variability can be significant and is a function of different causes. In an analysis and design, it is therefore necessary to know the environment where the structure will operate. While temperature and thermal effects are not discussed in an introductory text on vibration, these can be critical factors in many advanced aerospace and machine designs.

---

[14] This area of research is known as *earthquake engineering* and the specific study of how energy propagates through complex materials such as soils is known as the study of *waves in random media*.

[15] One can begin with the text by E.B. Haugen, **Probabilistic Mechanical Design**, Wiley–Interscience, 1980. There are interesting applications of probability to mechanical engineering, primarily based on the Gaussian distribution.

5.2.6 Statistics and Probability

The previous examples of natural forces all have one factor in common. It is that they depend on experimentally determined parameters. Just as linearity depends on small oscillations, these semi–empirical equations are valid only for a particular range in the data. While deterministic models also depend greatly on experimental data for their formulation and ultimately their validity, random models are an attempt to explicitly deal with observed scatter in the data and with very intricate dynamic behavior. Random models also show how data scatter affects response scatter.

Data are always our link to valid probabilistic models, their derivation and validation. While this is not the focus of our efforts here, it is important that the reader is at least aware that this step precedes any valid probabilistic model.

Example 5.4 From Data to Model and Back to Data
As we have emphasized, modeling can be as much an art as a science. Engineers are generally handed a problem that needs to be solved, not an equation, not even a well thought out description of the problem. For example: *We need to go to the Moon in ten years!*

Engineering is predicated on understanding how structures and materials behave under various operating conditions. This understanding is based on theory and *data*. Many experiments have been performed to get us to our current level of understanding and intuition about vibration. The experiments suggest *cause* and *effect* between variables. They provide us with parameter values. Finally, they are the basis for the equations we derive.

Data has scatter, and the significance of the scatter to a particular problem determines whether it can or cannot be ignored. If it cannot be ignored, then data is used to estimate the statistics of the randomness. The resulting probabilistic model is used to study the particular problem at hand, and the model's validity is established by comparing its predictions with available data. Such comparisons help define the limits of model validity.

In this way, a full circle has been achieved. Data gives birth to understanding and parameter values, which lead to governing equations and their predictions, and finally validity is established by comparing model predictions with new data that is not part of the original set. ■

5.3 Random Variables

We begin to explore the properties of *random variables*. Probability affords us a framework for defining and utilizing such variables in the models developed for engineering analysis and design. Mathematical models of physical phenomena are essentially relationships between variables. Where some of these variables have associated uncertainties, there are a multiplicity of possible values for each *random variable*. An example is the set of possible values of Young's moduli determined from a series of experiments on "identical" test specimens. This multiplicity is represented by the *probability density function*, discussed in the following section. A random variable may be *discrete*, *continuous*, or *mixed*. If a parameter is a random variable, its probability density function provides a complete description of its variability.

In the following discussion, we adopt the notation that random variables are represented by capital lettered variable, with exceptions sometimes provided by Greek letter names, and we use lower case letters to denote realizations of a random variable.

5.3.1 Probability Distribution

The likelihood of a random variable taking on a particular range of values is defined by its distribution function. The *probability distribution function*[16] is defined as

$$F(x) = \Pr\{X \leq x\},$$

where $\Pr\{X \leq x\}$ is the probability that random variable X is less than or equal to the number x. See Figure 5.5. This probability is, of course, a function of the particular value x.

Based on the axioms of probability,[17] it can be shown that $F(x)$ is an increasing function of x, and is bound by 0 and 1. The *impossible* event has a zero probability, and the *certain* event has a probability of one. In

---

[16] $F(x)$ is sometimes called the *cumulative distribution function*, since probability is accumulated as x becomes larger.

[17] An axiom is a rule that is assumed to be true and upon which further rules and facts are deduced. For engineering, the deduced facts must conform to reality. An excellent book on the basics of probabilistic modeling is by A. Papoulis, **Probability, Random Variables, and Stochastic Processes**, McGraw–Hill. There are several editions. I find the first edition most readable. I would also encourage the reader to look up some of the other fine texts by Papoulis on probability and stochastic processes; they are amongst the best. A different approach to explaining probability is offered by C. Ash in **The Probability Tutoring Book: An Intuitive Course for Engineers and Scientists (and everyone else!)**, IEEE Press, 1993. It offers an introduction through problem solving.

5.3. RANDOM VARIABLES

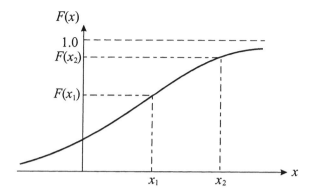

Figure 5.5: Cumulative distribution function.

particular,

$$F(-\infty) = 0$$

since $\Pr\{X < -\infty\} = 0$; all *realizations*[18] of the random variable must be greater than negative infinity. Similarly,

$$F(+\infty) = 1$$

since $\Pr\{X < +\infty\} = 1$; all realizations of the random variable must be less than positive infinity. Thus, bounds on $F(x)$ are $0 \leq F(x) \leq 1$, and if

$$F(x_1) \leq F(x_2) \implies x_1 < x_2,$$

since $\Pr\{X \leq x_1\} \leq \Pr\{X \leq x_2\}$. Note that the probability distribution function is *non-decreasing*.

The cumulative distribution function is one way to probabilistically describe a random variable. But we still have not fully answered the question posed at the beginning of this chapter: *How is the more likely force given added weight in the computation of structural response?* The probability density function gives us the answer.

5.3.2 Probability Density Function

The *probability density function* presents the same information contained in the probability distribution function, but in a more useful form. (See Figure

---

[18] A *realization* is one of many possible values of a random variable.

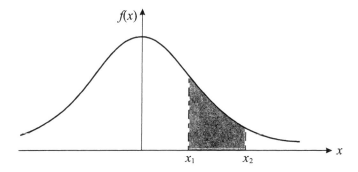

Figure 5.6: Probability density function.

5.6.) Assuming continuity of the distribution,[19] the probability density function $f(x)$ is defined as

$$f(x) = \frac{dF(x)}{dx}.$$

Alternately, by integrating both sides and rearranging,

$$F(x) = \Pr\{X \leq x\} = \int_{-\infty}^{x} f(\xi)d\xi. \tag{5.4}$$

Equation 5.4 provides a useful interpretation of the density function: *the probability that a continuous random variable X has a value less than or equal to the number x is equal to the area under the density function for values less than x.* Similarly, for arbitrary x_1 and x_2, the probability that $x_1 \leq X \leq x_2$ is

$$\Pr\{x_1 \leq X \leq x_2\} = \int_{x_1}^{x_2} f(x)dx. \tag{5.5}$$

Note the important *normalization* property

$$\int_{-\infty}^{+\infty} f(x)dx = 1, \tag{5.6}$$

signifying that the density function is representative of all possible outcomes or realizations of the random variable with the area under the density func-

---

[19] The distribution function does not have to be a continuous function. In many instances it may have discrete jumps where a finite probability exists for a certain realization. It is just easier to work with a continuous function.

5.4. MATHEMATICAL EXPECTATION

tion *normalized* to 1. Since probability is numerically in the range 0 to 1, the density function must be a positive semi–definite[20] function: $f(x) \geq 0$.

It is important to recall that the random variable is a static property. That is, the shape of the density function does not change with time. Where the density function is time–dependent, the variable is called a *random or stochastic process*. This more advanced topic is discussed in Section 5.7.

Example 5.5 Use of the Density Function
Suppose the probability density function of random variable X is $f(x) = ce^{-|x|}$. Evaluate the constant c and then find $\Pr\{-2 \leq X \leq 2\}$.

Solution The constant c must be first evaluated using the normalization property before the density function can be used to derive probabilities of events. Using Equation 5.6,

$$c \int_{-\infty}^{\infty} e^{-|x|} dx = 1$$

$$2c \int_{0}^{\infty} e^{-x} dx = 1$$

$$\Rightarrow c = \frac{1}{2},$$

and $f(x) = \frac{1}{2} e^{-|x|}$. Then,

$$\Pr\{-2 \leq X \leq 2\} = \int_{-2}^{2} \frac{1}{2} e^{-|x|} dx = 1 - e^{-2} \approx 0.86.$$

Such density functions have numerous engineering applications. The exponential density is used for reliability analysis in Section 5.5.2. ∎

Before examining some important densities, we need to define an averaging procedure known as the mathematical expectation for probabilistic variables.

5.4 Mathematical Expectation

The single most important descriptor of a random variable is its *mean* or *expected value*. This defines the most likely value of the variable. However, numerous random variables may have the same mean, but their *spread of possible values* or their *variance* can be considerably different. This explains the variance as the next most important statistical descriptor. The

---
[20] A *positive definite* function is one that has all values greater than zero. If it is positive *semi-definite*, then it may also be equal to zero.

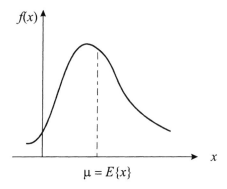

Figure 5.7: Expected value.

mean and variance of a random variable are statistical averages, and can be evaluated using the concept of the *mathematical expectation of a function of random variable* X, defined as

$$E\{g(X)\} = \int_{-\infty}^{\infty} g(x)f(x)dx. \tag{5.7}$$

The *expected* or *mean value* is defined, using Equation 5.7, as

$$\mu = E\{X\} = \int_{-\infty}^{\infty} xf(x)dx. \tag{5.8}$$

The expected value is a constant, *first–order* statistic, and is also known as the *first moment* because the variable x appears to the first power. The term moment is used by analogy to the center of mass in the mechanics of solids. The result $E\{X\}$ is the center of "probability mass." Note that the density function, acting as a probabilistic "weighting" function, is a larger factor in the integral for more probable values of the random variable. It is also clear from the definition of expectation that the expected value of a constant is that constant. See Figure 5.7.

Example 5.6 Expected Value
Equation 5.8 is written for a continuous random variable. There are instances when the variable of interest is discrete. How is a discrete variable analyzed?
    Solution   For a discrete random variable, the integral in the mathematical expectation becomes a summation. Suppose an experiment for the

5.4. MATHEMATICAL EXPECTATION

yield strength of a material is run ten times with the following frequency data:

$$10.0,\ 9.8,\ 11.1,\ 9.1,\ 9.9,\ 9.7,\ 10.3,\ 10.1,\ 9.9,\ 10.0$$

To find the expected value use the discrete counterpart to Equation 5.8,

$$\mu = E\{X\} = \sum_{i=1}^{10} x_i f(x_i),$$

where x_i are the test results and $f(x_i)$ are the probability weights, in this case the fraction of times that a particular value occurred. For the data listed, test results that occur once have a probability of 1/10. Results such as 9.9 and 10.0 that occur twice have a probability 2/10. Note that $\sum_{i=1}^{10} f(x_i) = 1$ signifying that all possible outcomes have been included. Then,

$$\begin{aligned}
\mu &= \left(10.0 \times \frac{2}{10}\right) + \left(9.8 \times \frac{1}{10}\right) + \left(11.1 \times \frac{1}{10}\right) \\
&+ \left(9.1 \times \frac{1}{10}\right) + \left(9.9 \times \frac{2}{10}\right) + \left(9.7 \times \frac{1}{10}\right) \\
&+ \left(10.3 \times \frac{1}{10}\right) + \left(10.1 \times \frac{1}{10}\right) \\
&= \frac{99.9}{10} = 9.99.
\end{aligned}$$

On the other hand, if the yield strength is assumed to be continuous between the values 9.1 and 11.1 with continuous density $f(x) = 1/(11.1-9.1) = 1/2$, then

$$\mu = \int_{9.1}^{11.1} x \cdot \frac{1}{2} dx = 10.1.$$

This result is slightly different than that for the discrete case due to the uneven discrete distribution. ∎

Now that the question of the most likely value of a random variable has been addressed using the expected value, we can derive an equation that provides a measure of the scatter about the mean value.

5.4.1 Variance

The *variance* is a *second-order moment*. It is defined as

$$Var\{X\} = E\{(X - E\{X\})^2\} = \int_{-\infty}^{\infty} (x - \mu)^2 f(x) dx.$$

Expand the squared term, integrate term by term, and find the variance equal to

$$Var\{X\} = E\{X^2\} - (E\{X\})^2, \tag{5.9}$$

or, the difference between the *mean–square value* and the *mean value squared*. Here, the second–moment analogy is with the *mass moment of inertia*.

In order that the measure of dispersion have the same dimensions as the random variable, the *standard deviation* is defined as the positive square root of the variance,

$$\sigma = +\sqrt{Var\{X\}}. \tag{5.10}$$

An important dimensionless parameter is the *coefficient of variation*,

$$\delta = \frac{\sigma}{\mu}.$$

It is used as a non–dimensional measure of the degree of uncertainty in a parameter, that is, the scatter of its data. In engineering practice, one expects a δ value of between 0.05–0.15, or 5–15%. Values larger than this imply that a serious lack of knowledge exists about the system itself and its underlying physics. If this is the case, then a significant program of experiments is necessary before one can consider the analysis and design of such a system.

Which density functions are of use in engineering applications? To be able to engineer a product such as a structure or a machine, one needs to be able to understand the behavior of materials, the characteristics of a vibrating system, and the external forces. Usually, the largest uncertainties are with the loading. Even so, in practice, we expect probability densities to have most of their area about the mean value, that is, with a small variance. Sometimes in engineering, the uncertainty is such that we only know the *high/low* values of a variable. In this instance all intermediate values are equally probable. This leads us to the uniform probability density, studied next. Other times our experience tells us that parameter values significantly different from the mean can happen, even if these are unlikely. This implies the Gaussian density, also studied below. What we see is that data from testing and design experience considerably helps in the decision regarding the choice of the most physically realistic probability density. The next section provides the details.

5.5 Probability Densities Useful in Applications

It turns out that a handful of density functions are sufficient for probabilistic modeling in many engineering applications. Here, five of these are discussed: the uniform, exponential, normal or Gaussian, lognormal, and the Rayleigh densities.

5.5.1 The Uniform Density

The *uniform density* is a good model for a variable with known upper and lower bounds, and *equally likely* values within the range. From Figure 5.8 it can be seen that for any range Δx, the area under the density curve (*a horizontal line*) is the same. According to Equation 5.5, there is equal probability for the variable to be in any range.

Suppose that X is a continuous random variable that can have any value in the interval $[a, b]$, where both a and b are finite. For the probability density function is given by

$$f(x) = \begin{cases} 1/(b-a) & \text{if } a \leq x \leq b \\ 0 & \text{otherwise,} \end{cases} \qquad (5.11)$$

X is called *uniformly distributed*. The probability distribution function for a uniformly distributed random variable is

$$F(x) = \Pr\{X \leq x\} = \int_{-\infty}^{x} f(s)ds$$

$$= \begin{cases} 0 & x < a \\ \frac{(x-a)}{(b-a)} & a \leq x < b \\ 1 & x \geq b. \end{cases}$$

The mean and mean–square values of uniformly distributed random variable X are given by

$$E\{X\} = \int_a^b x \frac{1}{b-a} dx = \frac{b+a}{2}$$

$$E\{X^2\} = \int_a^b x^2 \frac{1}{b-a} dx = \frac{b^2 + ab + a^2}{3}.$$

The standard deviation is then easily found to be

$$\sigma = \frac{b-a}{\sqrt{12}} = 0.289(b-a).$$

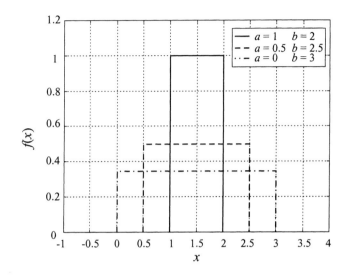

Figure 5.8: Uniform density function.

As expected, the statistics are functions only of the upper and the lower bounds.

Example 5.7 Uniform Density
A constant force F is known to have a value of between 10 lb$_f$ and 25 lb$_f$, but there is no additional information as to where in this range the actual value lies. All values in the range are of equal probability, given existing knowledge. Find the most likely value, and estimate the probability that the force $F > 20$ lb$_f$. Also, find the variance and coefficient of variation.

Solution   For such cases, where any value in a range is equally likely, a uniform density is chosen, $f_F = c$, where c is a constant. Use the normalization property of the density to find the value of c,

$$\int_{10}^{25} f_F dF = \int_{10}^{25} c \, dF = 1,$$

and $c = \frac{1}{15}$. The mean value for a uniform density is just the midpoint between the upper and lower bounds, $\mu_F = (25 + 10)/2 = 17\frac{1}{2}$ lb$_f$. Or use the definition of mean value, $E\{F\} = \int_{10}^{25} F f_F dF$, and come to the same result.

Finally, the probability that force $F > 20$ lb$_f$ is

$$\Pr\{F > 20\} = \int_{20}^{25} \frac{1}{15} dF = \frac{1}{3}.$$

5.5. PROBABILITY DENSITIES USEFUL IN APPLICATIONS

The variance can be evaluated from

$$\sigma_F^2 = E\{F^2\} - \mu_F^2 = 18.75 \text{ lb}_f^2,$$

and the coefficient of variation is then

$$\delta = \frac{\sigma_F}{\mu_F} = \frac{4.3}{17.5} = 0.25,$$

or 25%. This is a relatively large scatter about the mean value. In engineering applications, coefficients of variation greater than 15% or, 0.15, imply a need for further data gathering. ∎

Example 5.8 Quadratic Density
For comparison, suppose that instead of a uniform density, F is distributed according to a quadratic law, $f_F = \alpha F^2$, also with $10 \leq F \leq 25$. Following the above procedure, it is straightforward to find $\alpha = 0.00021$, $\mu_F = 19.98$ lb$_f$, and

$$\Pr\{F > 20\} = \int_{20}^{25} 0.00021 F^2 dF = 0.53,$$

which makes sense, when compared to the uniform density results, since much more of the area under the quadratic density function is located near the upper end of the range. In fact, 53% of the area is in the range $20 \leq F \leq 25$.

Here, the variance is $\sigma_F^2 = 6.76$ lb$_f^2$, a much smaller value than for the uniform density, and the coefficient of variation is $\delta = 0.13$ or 13%, again signifying that the spread of values is much smaller for the quadratic. (Sketch the quadratic superimposed on the uniform to get a good visual!) ∎

5.5.2 The Exponential Density

For mechanical reliability,[21] the exponential distribution is most commonly used to estimate failure times. The (failure) density is

$$f(t) = \lambda e^{-\lambda t}, \quad \lambda > 0, t \geq 0,$$

where λ is a constant (failure) rate per unit time, and $1/\lambda$ is the mean (time to failure).

---

[21] A good starting point for studying reliability is the book by B.S. Dhillon, **Mechanical Reliability: Theory, Models and Applications**, AIAA, 1988.

Example 5.9 Time to Failure
A pump is known to fail according to the exponential distribution with a mean of 1000 hours. Then $\lambda = 1/1000$. Suppose that a critical mission requires the pump to operate for 200 hours. Calculate the failure probability.

Solution For an exponential density, the probability distribution function is in general
$$F(t) = 1 - e^{-\lambda t}.$$
This is the probability that failure will occur during $t \leq t_0$. Here, $t_0 = 200$ hours,
$$F(200) = 1 - e^{-200/1000} = 0.1813.$$
The probability that the pump will fail during the first 200 hours is 0.1813 or 18.13% according to the distribution. Knowing this value will help in making the decision whether a backup pump needs to be on hand. ∎

5.5.3 The Normal (Gaussian) Density

Many physical variables are assumed to be governed by the normal or Gaussian density. There are two reasons for this: the Gaussian is mathematically tractable and tabulated, and the broad applicability of the *central limit theorem*. The central limit theorem states that under very general conditions, as the number of variables in a sum becomes large, the density of the sum of random variables will approach the Gaussian regardless of the individual densities. Examples of variables which arise as the sum of a number of random effects, where no one effect dominates, are noise generated by falling rain, the effects of a turbulent boundary layer, and the response of linear structures to a turbulent environment. Many naturally occurring physical processes approach a Gaussian density.

Random variable X governed by the Gaussian density has the probability density function
$$f(x) = \frac{1}{\sigma\sqrt{2\pi}} \exp\left\{-\frac{1}{2}\left(\frac{x-\mu}{\sigma}\right)^2\right\}, \tag{5.12}$$
where the meaning of μ and σ are found by taking the expected value[22] of X and X^2, respectively,
$$E\{X\} = \frac{1}{\sqrt{2\pi}} \int_{-\infty}^{\infty} (\sigma y + \mu) e^{-y^2/2} dy = \mu \tag{5.13}$$
$$E\{X^2\} = \frac{1}{\sqrt{2\pi}} \int_{-\infty}^{\infty} (\sigma y + \mu)^2 e^{-y^2/2} dy = \mu^2 + \sigma^2. \tag{5.14}$$

---
[22] Make use of the transformation of variables: $y = (x-\mu)/\sigma$ and note that $dx = \sigma dy$.

5.5. PROBABILITY DENSITIES USEFUL IN APPLICATIONS

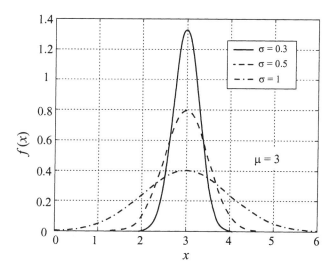

Figure 5.9: The normal or Gaussian density function.

We see then that the mean value is μ and, using Equation 5.10, the standard deviation is σ.

Note that the Gaussian density extends from $-\infty$ to $+\infty$, and therefore, cannot represent any physical variable except *approximately*. Since there are no physical parameters that can take on all possible values on the real number line, we may rightly wonder how good a model is the Gaussian. But the approximation in many instances turns out to be very good. For example, consider a positive–definite random variable X that is modeled as a Gaussian random variable with coefficient of variation $\delta = 0.20$, or $\mu = 5\sigma$. What is the effect of the area under the density function in the negative X region? Integrating for $x < 0$, one finds an area of approximately 24×10^{-8}, a negligible probability for most purposes. Thus, the suitability of the Gaussian model depends on the application, and how much the *tails* extend into physically forbidden regions.

When it is not possible to accept any negative values,[23] the analyst sometimes resorts to a truncated Gaussian with the following density function

$$f(x) = \frac{A}{\sigma\sqrt{2\pi}} \exp\left[-\frac{(x-x_0)^2}{2\sigma^2}\right], \quad 0 \leq x_1 \leq x \leq x_2,$$

---
[23] This is especially true in reliability calculations where the probabilities of failure may be very small, even on the order of 10^{-8}, and extra care must be taken to ensure that the density function is suitable.

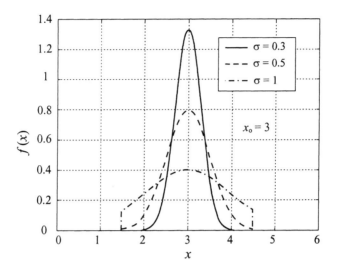

Figure 5.10: The truncated Gaussian density function.

and zero elsewhere. (If $x_1 \to -\infty$ and $x_2 \to +\infty$, then $A \to 1$, and X becomes a Gaussian random variable with $E\{X\} = x_0$ and $Var(X) = \sigma^2$. See Figure 5.10.)

For ease in applications, the Gaussian variable X is sometimes transformed so that the resulting variable S is *zero mean* with *unit variance*,

$$S = \frac{X - \mu_X}{\sigma_X},$$

with the resulting *standard normal density*,

$$f_S(s) = \frac{1}{\sqrt{2\pi}} e^{-s^2/2}. \tag{5.15}$$

The probability distribution is then

$$F_S(s) = \Pr\{S \leq s\} = \int_{-\infty}^{s} \frac{1}{\sqrt{2\pi}} e^{-s^2/2} ds, \tag{5.16}$$

where $F_S(s)$ can be found in tables.

5.5.4 The Lognormal Density

Sometimes it is important to strictly limit possible values of a parameter to the positive range. Then the lognormal density is commonly used. See

5.5. PROBABILITY DENSITIES USEFUL IN APPLICATIONS

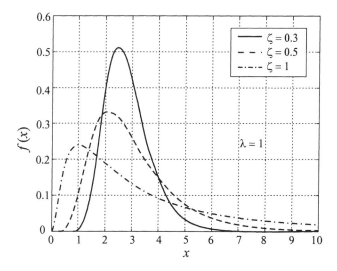

Figure 5.11: The lognormal density function.

Figure 5.11. Applications include material strength, fatigue life, loading intensity, time to the occurrence of an event, and volumes and areas. A random variable X has a *lognormal* probability density function if $\ln X$ is normally distributed, that is,

$$f(x) = \frac{1}{x\zeta\sqrt{2\pi}} \exp\left\{-\frac{1}{2}\left(\frac{\ln x - \lambda}{\zeta}\right)^2\right\}, \quad 0 < x < \infty, \qquad (5.17)$$

where $\lambda = E\{\ln X\}$ is the mean value and $\zeta = \sqrt{Var(\ln X)}$ the standard deviation of $\ln X$.

5.5.5 The Rayleigh Density

The *Rayleigh* density, like the lognormal, is also limited to strictly positive random variables,

$$f(x) = \frac{x}{\sigma^2} \exp\left\{-\frac{x^2}{2\sigma^2}\right\}, \quad x > 0.$$

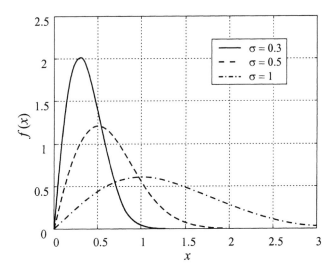

Figure 5.12: The Rayleigh density function.

See Figure 5.12. The first– and second–order statistics can be derived to be

$$E\{X\} = \sqrt{\frac{\pi}{2}}\sigma$$
$$E\{X^2\} = 2\sigma^2$$
$$\sigma_X = \sqrt{\frac{4-\pi}{2}}\sigma \approx 0.655\sigma.$$

As an example of where the Rayleigh density is a good model, consider a random oscillation, or any process that is governed by the Gaussian. This oscillation or process has peaks that are distributed randomly as well. The Rayleigh density is a good model for the distribution of these peak values.

5.6 Two Random Variables

When a problem includes more than one random variable, it becomes necessary to generalize the basic definitions. We will work with two random variables, since this allows us to explore the necessary generalizations without needlessly crowding the concepts with too much algebra. Consider the two random variables X and Y. In general, a *joint distribution function* will completely define their probable values and is denoted by

$$F_{XY}(x,y) = \Pr\{X \leq x, Y \leq y\}.$$

5.6. TWO RANDOM VARIABLES

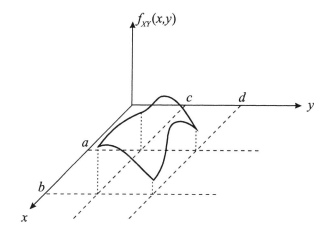

Figure 5.13: Schematic domain of two random variables.

This function defines the probability that random variable X is less than or equal to x *and* random variable Y is less than or equal to y. Generalizing from our study of one random variable, the *joint density function* can be defined as

$$f_{XY}(x,y)dx\,dy = \Pr\{x < X \leq x+dx, y < Y \leq y+dy\},$$

or

$$F_{XY}(x,y) = \int_{-\infty}^{y}\int_{-\infty}^{x} f_{XY}(u,v)du\,dv.$$

Conversely,

$$f_{XY}(x,y) = \frac{\partial^2 F_{XY}(x,y)}{\partial x\, \partial y}$$

and

$$\Pr\{a < X \leq b, c < Y \leq d\} = \int_{c}^{d}\int_{a}^{b} f_{XY}(u,v)du\,dv.$$

The above define the probability that two random variables are *simultaneously* within a certain range, specifically, that $a < X \leq b$ *and* $c < Y \leq d$ See Figure 5.13. Such information is useful when considering problems that have two or more variables where it is necessary to establish how the values of one variable affect values of the other variables. Such questions of correlation are considered next.

5.6.1 Covariance and Correlation

Proceeding from models with one random variable to those with two requires the introduction of the concept of *covariance*, and the related parameter, the *correlation coefficient*. These tell us how *linear* the relationship is between the two random variables. Begin by considering the two random variables X and Y with respective second joint moment,

$$E\{XY\} = \int_{-\infty}^{\infty}\int_{-\infty}^{\infty} xy f_{XY}(x,y)dx\,dy. \quad (5.18)$$

If X and Y are *statistically independent* then the joint density function can be separated into the product of the respective *marginal densities*, $f_{XY}(x,y) = f_X(x)f_Y(y)$, and using Equation 5.18, $E\{XY\} = E\{X\}E\{Y\}$. The covariance is defined as the second joint moment about the mean values μ_X and μ_Y,

$$Cov(XY) = E\{(X-\mu_X)(Y-\mu_Y)\} = E\{XY\} - \mu_X\mu_Y. \quad (5.19)$$

Note that if the variables are independent, $Cov(XY) = 0$. The *correlation coefficient* is defined as the normalized (dimensionless) $Cov(XY)$,

$$\rho = \frac{Cov(XY)}{\sigma_X \sigma_Y}. \quad (5.20)$$

To better understand the correlation coefficient, let us assume that X and Y are linearly related by the equation $X = aY$, where a is a constant. Then $E\{XY\} = aE\{Y^2\}$, and $Cov(XY) = aE\{Y^2\} - aE^2\{Y\}$. Using the definition of variance, the covariance becomes $Cov(XY) = a\sigma_Y^2$, and Equation 5.20 becomes

$$\rho = \frac{a\sigma_Y^2}{\sigma_X \sigma_Y} = \frac{a\sigma_Y}{\sigma_X} = +1, \quad (5.21)$$

since $\sigma_X = a\sigma_Y$. The random variables X and Y are in perfect positive correlation. The last equality in Equation 5.21 is found using the property of the variance of a random variable multiplied by a constant: If Y has variance σ_Y, then aY has variance $a\sigma_Y$. Had we defined $X = -aY$, then we would have found that $\rho = -1$, or X and Y are in perfect negative correlation.[24] We conclude that $-1 \leq \rho \leq +1$.

---

[24] It is important to realize that a high value for ρ may indicate strong correlation, but not direct *cause* and *effect* since X and Y may be correlated by virtue of being related to some third variable. Also, if X and Y are independent, $\rho = 0$, as we see in *Example* 5.10. The converse is not necessarily true. $\rho = 0$ indicates the absence of a *linear* relationship; a random or a nonlinear functional relationship between X and Y is still a possibility.

5.6. TWO RANDOM VARIABLES

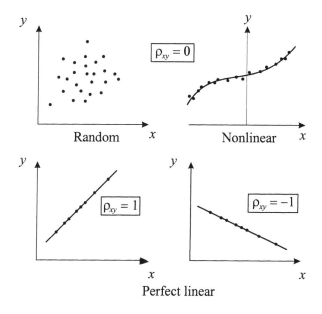

Figure 5.14: Correlation coefficient.

Figure 5.14 depicts representative correlations between data points for random, nonlinear, and perfect linear relationships.

Example 5.10 Jointly Distributed Variables
Two random variables X and Y are jointly distributed according to the joint density $f_{XY}(x,y) = \frac{1}{2}e^{-y}$, $y > |x|$, $-\infty < x < \infty$, as in Figure 5.15. Compute the marginal densities, and the covariance.

Solution   The marginal densities are defined and solved:

$$f_X(x) = \int f_{XY}(x,y)dy = \int_{|x|}^{\infty} \frac{1}{2}e^{-y}dy = \frac{1}{2}e^{-|x|}, \quad -\infty < x < \infty$$

$$f_Y(y) = \int f_{XY}(x,y)dx = \int_{-y}^{y} \frac{1}{2}e^{-y}dx = ye^{-y}, \quad y > 0.$$

Since the joint density is not equal to the product of the two marginal densities, the variables are not statistically independent. Covariance Equa-

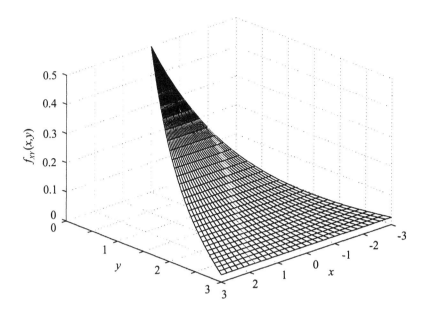

Figure 5.15: $f_{XY}(x,y) = \frac{1}{2}e^{-y}$, $y > |x|$, $-\infty < x < \infty$.

tion 5.19 requires an evaluation of the second joint moment,

$$E\{XY\} = \int_0^\infty \int_{-y}^y \frac{1}{2}xye^{-y}\,dx\,dy$$
$$= \frac{1}{2}\int_0^\infty ye^{-y}\int_{-y}^y x\,dx\,dy = 0,$$

and the respective mean values

$$E\{X\} = \int_{-\infty}^\infty xf_X(x)dx = \int_{-\infty}^0 \frac{1}{2}xe^x dx + \int_0^\infty \frac{1}{2}xe^{-x}dx = 0,$$

and by a similar procedure, $E\{Y\} = 2$. Therefore, $Cov(XY) = 0$, not because the two variables are independent, but because the right hand side equals zero. ∎

Example 5.11 Correlation Coefficient and Reliability
We have drawn a number of our examples from the discipline of reliability, in part because it is very important in all aspects of engineering. Reliability

5.6. TWO RANDOM VARIABLES

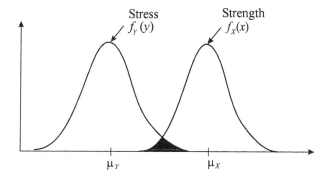

Figure 5.16: Overlap of strength and stress densities for reliability.

is also a function of vibration characteristics, which depend on uncertainties in loading and material parameter values.

Suppose that we have been tasked with the job of estimating the reliability of a component that is to be used in a particular stress environment. Assume that we can have tests run on both component and loading to gather data. How do we proceed?

Solution Define the *strength* of the component as X psi and the **stress** it experiences due to loading as Y psi. The strength may be a yield stress or an ultimate stress. Test a sufficient number of "identical" components in order to establish its strength probability density function, and find it to be $f_X(x)$. Similarly, loading data leads us to a loading or stress density function $f_Y(y)$.

The strength is designed to exceed the stress for all but the most rare of cases, as one would expect. This is shown schematically in Figure 5.16.

The shaded region in the figure represents the realizations where the loading stress is greater than the component strength. *This is defined as a failure of the component.* The probability of failure is equal to $\Pr\{(X-Y) \leq 0\}$. Define $Z = X - Y$ and find $Pr\{Z \leq 0\}$. The reliability of the component is then defined as $R = 1 - \Pr\{Z \leq 0\}$, which is designed to be a very large number.

From Equations 5.19 and 5.20 we have

$$E\{XY\} = \rho \sigma_X \sigma_Y + \mu_X \mu_Y,$$

and using Equation 5.9, the variance of Z is

$$Var\{Z\} = Var\{X\} + Var\{Y\} - 2\rho \sigma_X \sigma_Y. \tag{5.22}$$

Since strength and loading stress are uncorrelated, $\rho = 0$ and the variance of Z equals the variance of X plus the variance of Y, or

$$\sigma_Z = +\sqrt{\sigma_X^2 + \sigma_Y^2}.$$

The mean value of Z is

$$\mu_Z = \mu_X - \mu_Y.$$

We have derived the mean and variance of the probability of failure. If we can establish the density function $f_Z(z)$, then the reliability of the component is given by

$$R = 1 - \int_{-\infty}^{0} f_Z(z) dz. \tag{5.23}$$

Suppose that both X and Y are Gaussian. Then it is known, and not proven here, that Z is also Gaussian. In this instance, $f_Z(z)$ is fully defined given μ_Z and σ_Z, and Equation 5.23 can then be evaluated. ∎

5.7 Random Processes

The study of the response of a single degree of freedom oscillator to random forcing requires the extension of concepts to include time–dependent random variables, or random processes. This will further involve an understanding of the notion of correlation and power spectrum. One is the Fourier transform of the other, and each provides important physical insights to the process under consideration. A random process can informally be thought of as a random variable that varies with time in a probabilistic way. With this in mind, it will be straightforward to extend the earlier definitions for random variables to those for random processes. With these tools, the random force can be probabilistically modeled.

The discussion in the following subsections is necessarily brief and is meant to be introductory. It should be viewed as a glimpse at some of the concepts necessary for further studies of applied random processes. This basic introduction provides us with what we need to start understanding the study of random vibration, initiated in Section 5.8.

5.7.1 Basic Random Process Descriptors

A *random process* may be understood to be a time–dependent random variable. For a specific time t, $X(t)$ is a random variable with distribution function

$$F_{X(t)}(x; t) = \Pr\{X(t) \leq x\}.$$

5.7. RANDOM PROCESSES

This is the *first-order distribution* of the process $X(t)$. It describes how the probability characteristics of the random process change with time. That is, how do the mean value and variance change with time? It is customary to separate the time variable from the other variables by a semicolon.

The corresponding *first-order density* is given by

$$f_{X(t)}(x;t) = \frac{\partial F_{X(t)}(x;t)}{\partial x}.$$

The *second-order distribution* for $X(t_1)$ and $X(t_2)$ is the joint distribution

$$F(x_1, x_2; t_1, t_2) = \Pr\{X(t_1) \leq x_1, X(t_2) \leq x_2\} \quad (5.24)$$

with corresponding density

$$f_{X_1 X_2}(x_1, x_2; t_1, t_2) = \frac{\partial^2 F_{X_1 X_2}}{\partial x_1 \partial x_2}. \quad (5.25)$$

Note that $X_1 \equiv X(t_1)$ and $X_2 \equiv X(t_2)$. Also, subscripts may be dropped if it is clear as to which variables are being addressed.

Equation 5.24 is the probability that at time t_1, random process $X(t)$ will be less than or equal to the value x_1, *and* that at time t_2, it will be less than or equal to the value x_2. The density Equation 5.25 has the same information but in a form that is easier for calculation: the volume under the second–order density equals the respective probability. Equations 5.24 and 5.25 are the starting points for the mathematical modeling of the probabilistic evolution of $X(t)$ in time. For random vibration, this is the random force. The question considered next is: *If the random function is represented by many possible time–histories, how can averages be determined?*

5.7.2 Ensemble Averaging

The random function of time $X(t)$ with density function $f(x;t)$ is representative of many possible time–histories or a *sample population*. Theoretically, there are an infinite number of samples $X_i(t)$ with statistical properties governed by the density function $f(x;t)$. As Figure 5.17 shows, for any time t_1 there will exist an infinite number of possible values: $x_1(t_1), x_2(t_1), x_3(t_1), \ldots$. To account for all these possible values, average them to obtain the most likely value for the function at time t_1. This process is known as *ensemble averaging* since the group of samples is known as an *ensemble*. The averaging procedure uses the same mathematical expectation of Equation 5.8,

$$\mu_{X(t)}(t) = E\{X(t)\} = \int_{-\infty}^{\infty} x f(x;t) dx, \quad (5.26)$$

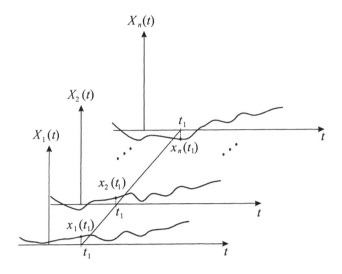

Figure 5.17: Ensemble averaging.

except the statistical parameters $\mu_{X(t)}(t)$ and $f(x;t)$ are now functions of time. Note that x is the dummy variable in the integration. A possible time–dependent probability density function is $f(x;t) = ce^{-xt}$, where c is a normalization constant.

In general, the random process is governed by a time–dependent density function, and has a time–dependent mean value. Similarly, second–order averages that will be very important in random vibration can be evaluated. The motivation for such averages is the question: *How does the value of the process $X(t)$ at $t = t_1$ affect its value at later time $t = t_2$?* Knowing this helps us to understand how rapidly a process varies. For a slowly varying function, it is expected that if t_1 and t_2 are not *too far* apart, then $X(t_2)$ can be estimated given $X(t_1)$, as shown in Figure 5.18(a). The values are correlated. On the other hand, if $X(t)$ varies rapidly, as in Figure 5.18(b), any estimate of future values is not accurate and the values are much less correlated.

To address this question quantitatively, consider the random process $X(t)$ at any two time instances, $X(t_1)$ and $X(t_2)$. The second–order average is given by

$$E\{X(t_1)X(t_2)\} = \int_{-\infty}^{\infty} \int_{-\infty}^{\infty} x_1 x_2 f(x_1, x_2; t_1, t_2) dx_1 dx_2, \quad (5.27)$$

where the joint density function $f(x_1, x_2; t_1, t_2)$ is required. x_1 and x_2 are

5.7. RANDOM PROCESSES

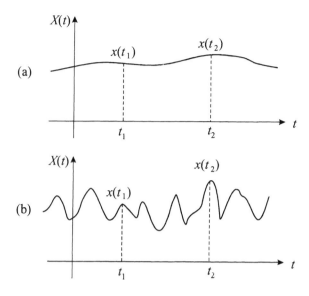

Figure 5.18: Second-order averages for slowly varying and rapidly varying random processes.

dummy variables of integration.

The second order average defined by Equation 5.27, and shown in Figure 5.19, is called the *correlation function* and is given the shorthand notation $R_{XX}(t_1, t_2)$,

$$R_{XX}(t_1, t_2) = E\{X(t_1)X(t_2)\}, \qquad (5.28)$$

where it is clear that the end result of the double integral is a function of t_1 and t_2.

In Equation 5.27, it is necessary to emphasize that at a specific time, a random process is nothing more than a random variable. At each instant of time, the values a random process can take is governed by a probability density. This is exactly what the density $f(x;t)$ implies; the density changes with time, but once a specific time is chosen, the density function is only a function of the realization x, that is, $f(x;t_1) = f(x)$. This is shown in Figure 5.20. For second–order averages, we require the joint density function $f(x_1, x_2; t_1, t_2)$.

The correlation function, similar in purpose to the earlier correlation coefficient, is a measure of the *similarity between the values of one stochastic process at two instants of time*, or *between different stochastic processes at two instances of time*. For different stochastic processes we have a *cross–*

232   CHAPTER 5.   SDOF VIBRATION: PROBABILISTIC FORCES

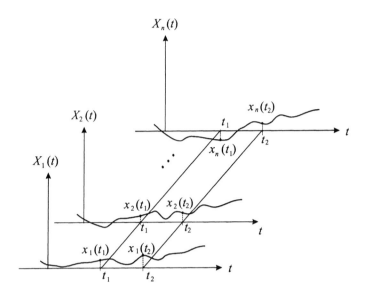

Figure 5.19: Second-order ensemble averaging.

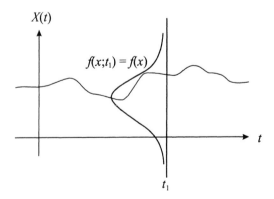

Figure 5.20: The density of a random process at specific time t_1.

5.7. RANDOM PROCESSES

correlation function, $R_{XY}(t_1, t_2) = E\{X(t_1)Y(t_2)\}$.

We can see from Equations 5.27 and 5.28 that the correlation is calculated by multiplying the corresponding values of the functions and then averaging these products using the expectation. If the two functions have similar shapes, then it is expected that a larger correlation will be found, since otherwise some of the products will be smaller or negative, leading to a smaller average. The largest correlation is found for $t_1 = t_2$, $R_{XX}(t_1, t_1) = E\{X(t_1)X(t_1)\}$.

The *autocovariance* of $X(t)$ is defined as

$$C_{XX}(t_1, t_2) = E\{[X(t_1) - \mu_X(t_1)][X(t_2) - \mu_X(t_2)]\},$$

which equals

$$C_{XX}(t_1, t_2) = R_{XX}(t_1, t_2) - \mu_X(t_1)\mu_X(t_2).$$

The *variance* is defined for $t_1 = t_2 = t$ as

$$\sigma^2_{X(t)} = C_{XX}(t, t) = R_{XX}(t, t) - \mu^2_X(t).$$

For a normal random process, for example, the time–dependent probability density function is

$$f_X(x; t) = \frac{1}{\sqrt{2\pi C_{XX}(t, t)}} exp\left[-\frac{(x - \mu_X(t))^2}{2C_{XX}(t, t)}\right].$$

Generally, functions such as $R_{XX}(t_1, t_2)$ are derived experimentally. In practice, preliminary analyses assume a reasonable function for $R_{XX}(t_1, t_2)$ that is representative of the physical process under study. Predictions using this model are then verified with experimental data.

5.7.3 Stationarity

It can be observed that Equations 5.26 and 5.27 are difficult to evaluate, not only mathematically, but also due to the difficulty in obtaining the necessary data to define the joint density function. To *begin* to understand random vibration problems, we make an assumption of *stationarity*. If the statistical properties of a random process are *invariant under translation in time*, the process is called *stationary*. While this assumption may appear to limit the applicability of the following models, in fact, with proper care and understanding, stationarity is a viable assumption for numerous practical applications. To use vibration terminology, *the assumption of stationarity implies steady state behavior in a statistical sense*.

Of particular interest in applications are the mean and correlation. For a stationary random process, the mean value becomes a constant (as for random variables) and the correlation becomes a function of time difference, $\tau = t_2 - t_1$, rather than a function of the specific times t_2 and t_1. For physical processes, the correlation is an even function, $R_{XX}(\tau) = R_{XX}(-\tau)$. Thus,

$$E\{X(t)\} = \mu_X$$
$$R_{XX}(t_1, t_2) = E\{X(t)X(t+\tau)\} = R_{XX}(\tau).$$

Similarly, for two stationary processes, the *cross-correlation* function is $R_{XY}(\tau)$.

It is interesting to note that for $\tau = 0$ the *mean square* value of $X(t)$ is

$$R_{XX}(\tau = 0) = E\{X^2(t)\} = \sigma_X^2 + \mu_X^2,$$

using Equation 5.9. The mean square value is an important quantity in the following physical way. If $X(t)$ is a displacement, then $E\{X^2(t)\}$ is a measure of strain energy. If $X(t)$ is a velocity, then $E\{X^2(t)\}$ is a measure of kinetic energy. The average energy of a stationary process is independent of time and equals the autocorrelation at $\tau = 0$. This knowledge is useful for interpreting the meaning of the *spectral density* in Section 5.7.4. In physical processes, as $\tau \to \infty$, the correlation will approach the mean value squared,

$$R_{XX}(\tau \to \infty) \to \mu_X^2.$$

This means that as time difference τ becomes larger, less correlation exists between the two respective values of the process, and the limit of the correlation becomes the square of the mean value.

Stationarity also implies that the autocovariance is given by

$$C_{XX}(\tau) = R_{XX}(\tau) - \mu_X^2,$$

where, if $\tau = 0$, then $C_{XX}(0) = R_{XX}(0) - \mu_X^2 = \sigma_X^2$. The *correlation coefficient* is defined as

$$r_{XX}(\tau) = C_{XX}(\tau)/C_{XX}(0).$$

The *correlation time* of a stochastic process can be defined as

$$\tau_c = \frac{1}{C_{XX}(0)} \int_0^\infty C_{XX}(\tau) d\tau.$$

The concept of correlation time becomes important in problems which must be solved approximately and simplification must be made on a physical basis.

5.7. RANDOM PROCESSES

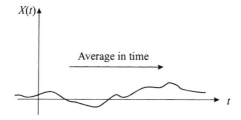

Figure 5.21: Ergodicity.

Ergodicity

As a practical matter, one rarely has the benefit of numerous experiments, but usually must make the best use of *one* trial. This is especially true for expensive testing environments such as space or undersea. Utilizing one trial requires the introduction of the concept of *ergodicity*. A stationary random process is said to be an *ergodic* process if the *time average* of a *single* record is approximately equal to the ensemble average. It is thus possible to average over a long *single* time history, as in Figure 5.21, rather than trying to obtain numerous records over which to perform an ensemble average. The mean value is then given by

$$\mu_X \approx \overline{X} = \lim_{T \to \infty} \frac{1}{2T} \int_{-T}^{T} X(t) dt, \qquad (5.29)$$

where it is assumed that $X(t)$ is one particular realization of the random process.

Such an average makes sense only if μ_X is a constant, otherwise μ_X will be a function of T and the initial assumption is no longer valid. An ergodic process is always stationary, but the opposite is not always true.

The corresponding ergodic definition for the autocorrelation function is

$$R_{XX}(\tau) \approx \lim_{T \to \infty} \frac{1}{2T} \int_{-T}^{T} X(t) X(t + \tau) dt. \qquad (5.30)$$

Now that we have an understanding of the autocorrelation, we proceed to study its Fourier transform, the spectral density.

5.7.4 Power Spectrum

A measure of the "energy" of stochastic process $X(t)$ is given by its *power spectrum*, or *spectral density*, $S_{XX}(\omega)$, which is the *Fourier transform*[25] of its autocorrelation function,

$$S_{XX}(\omega) = \int_{-\infty}^{\infty} R_{XX}(\tau)\exp(-i\omega\tau)d\tau,$$

and thus,

$$R_{XX}(\tau) = \frac{1}{2\pi}\int_{-\infty}^{\infty} S_{XX}(\omega)\exp(i\omega\tau)d\omega. \qquad (5.31)$$

Since $R_{XX}(-\tau) = R_{XX}(\tau)$, $S_{XX}(\omega)$ is not a complex function but a real even function. For $\tau = 0$,

$$\frac{1}{2\pi}\int_{-\infty}^{\infty} S_{XX}(\omega)d\omega = R_{XX}(0) = E\{X(t)^2\} \geq 0,$$

where, without proof, we state that $S_{XX}(\omega) \geq 0$. The integral of the power spectrum equals the "average or mean square power" of the process $X(t)$, confirming our opening statement that it is an energy measure. Where there is no chance of confusion, the subscripts used above can be omitted.

Cross spectral densities are similarly defined;

$$S_{XY}(\omega) = \int_{-\infty}^{\infty} R_{XY}(\tau)\exp(-i\omega\tau)d\tau = S_{YX}(\omega)$$

and

$$R_{XY}(\tau) = \frac{1}{2\pi}\int_{-\infty}^{\infty} S_{XY}(\omega)\exp(i\omega\tau)d\omega.$$

Some examples of $R_{XX}(\tau)$ and $S_{XX}(\omega)$ which have broad physical application are presented below.

---

[25] The Fourier transform of function $q(x)$ is defined as

$$\mathcal{F}\{q(x)\} = Q(\omega) = \int_{-\infty}^{\infty} q(x)e^{-i\omega x}dx.$$

The inverse Fourier transform of $Q(\omega)$ is

$$\mathcal{F}^{-1}Q(\omega) = q(x) = \frac{1}{2\pi}\int_{-\infty}^{\infty} Q(\omega)e^{i\omega x}d\alpha.$$

$Q(\omega)$ and $q(x)$ are known as a Fourier transform pair. It is important to note that sometimes the factor $(1/2\pi)$ appears on the other transform, and sometimes a factor of $(1/\sqrt{2\pi})$ appears with both transforms.

5.7. RANDOM PROCESSES

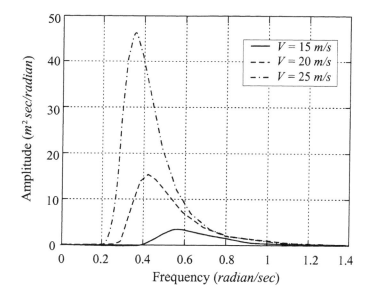

Figure 5.22: The Pierson-Moskowitz ocean wave height spectrum.

Example 5.12 Ocean Wave Spectra
Waves generated by ocean contact with wind are modeled as random processes. A much used spectral density of ocean wave height $\eta(t)$ is the *Pierson–Moskowitz* spectrum,

$$S_{\eta\eta}(\omega) = \frac{8.1 \times 10^{-3} g^2}{\omega^5} \exp -0.74 \left(\frac{g}{V\omega}\right)^4 \quad \text{m}^2\text{sec}, \qquad (5.32)$$

where $\omega > 0$, g is the gravitational constant, and V is the wind speed at a height of 19.5 m above the still water level. Any consistent set of units for g and V can be used, with ω in rad/sec. See Figure 5.22. The Pierson–Moskowitz is an experimentally determined spectrum, as are most if not all spectra used in applications.

Recall that the fluid particle velocities and accelerations are derivable from wave height elevation. There are other spectral densities in use. Each is specific to a particular part of the ocean for a particular time of year. ∎

Example 5.13 Wind Spectra
All ground and mobile structures must be designed to withstand wind forces. One can classify structures subjected to wind as either *streamlined* or *bluff bodies*. Streamlined structures such as wings have a high *aspect ratio* (ratio

of one dimension to another) and are shaped to optimally follow a streamline. They are designed so that the interaction of their shape with the wind results in a desirable configuration of forces. Bluff bodies such as tall buildings or chimneys are generally of low aspect ratio and may have corners or sharp edges. They are designed more for strength than for fitting within the streamlines.

An example of a wind spectrum for horizontal velocity is,[26]

$$S_v(f) = 4\kappa \bar{V}_{10}^2 \frac{L/\bar{V}_{10}}{(2 + \bar{f}^2)^{5/6}} \left(\frac{\text{ft}}{\text{sec}}\right)^2 \text{sec}, \qquad (5.33)$$

where f is the frequency in Hz, $\bar{f} = fL/\bar{V}_{10}$ is a dimensionless frequency, L is a length scale[27] of approximately 4,000 ft, $\bar{V}_{10}$ is the mean wind speed at 10 ft above the ground, and κ is a number in the range $0.005 \leq \kappa \leq 0.05$ that depends on wind profile expected in the region.

As with the wave height spectra, wind characteristics have wide variability depending on location. ■

Example 5.14 Earthquake Spectra
For earthquakes, the difficulty in specifying ground motion spectra for use as input to the structure is due to the significant variability in soil and geologic properties, even for two sites very near each other. We have all examined photos of earthquake related devastation and noted how for two adjacent similar structures, one has little damage and the other has suffered severe destruction. Differences can be due to even minor differences in structures and soil/foundation dynamic characteristics.

The El Centro, California earthquake[28] of 18 May 1940 has been used as input to numerous structural designs to verify that they will survive anticipated ground motion. ■

Units

At first sight, the power spectrum is a strange creation. We claim that it represents a real dynamic physical process, but Equation 5.31 is an integral over negative frequencies! Of course, these negative frequencies do not

---

[26]See A.G. Davenport and M. Novak, in **Shock & Vibration Handbook**, Chapter 29, p. 23, C.M. Harris, Editor, McGraw–Hill, 1988.

[27]A length scale is a dimension that is representative of the process being modeled. For example, if the wavelength of a harmonic force is λ, then this is a length scale of the problem. Similarly, structural dimensions can be used as length scales.

[28]See W.J. Hall, in C.M. Harris, **Shock and Vibration Handbook**, McGraw–Hill, 1987, Chapter 24, p. 5.

5.7. RANDOM PROCESSES

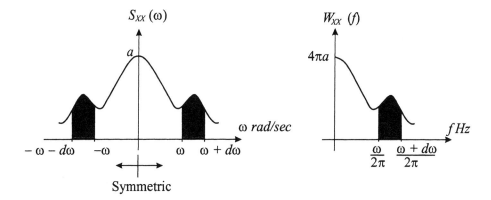

Figure 5.23: Equivalent one-sided spectrum.

exist except mathematically, just as the complex exponential is used for mathematical ease with the understanding that only the real or imaginary parts are retained. The frequency ω has units of rad/sec. In applications, frequencies f with units of cycles/second or Hz (Hertz) are more common.

These two approaches can be simply accommodated by the relation[29] $f = \omega/2\pi$, and defining an equivalent one–sided spectral density function $W_{XX}(f)$. By equating shaded areas in Figure 5.23, we find

$$2S_{XX}(\omega)d\omega = W_{XX}(f)\frac{1}{2\pi}d\omega, \qquad (5.34)$$

or $W_{XX}(f) = 4\pi S_{XX}(\omega)$. In most applications, since these depend on experimental data, the one–sided density function is more common. The wave height and wind spectra are one–sided.

Narrow and Broad Band Processes

In engineering applications, certain types of random processes tend to be frequently found. Since the power spectrum is representative of the distribution of vibratory energy as a function of frequency, then it is possible to define general categories of vibration according to how energy is distributed. A structure vibrating at a single constant frequency ω_0 can be represented in the time and frequency domains as shown in Figure 5.24.

Two very important types of processes are *narrow band* (Figure 5.25) and *broad band* (Figure 5.26) processes. Narrow and broad indicate the spread of the respective frequency bands.

---

[29] 2π radians equals one Hz.

240 CHAPTER 5. SDOF VIBRATION: PROBABILISTIC FORCES

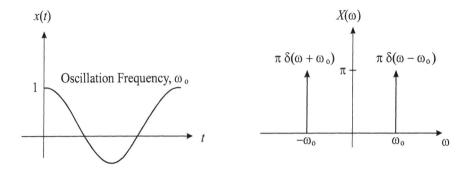

Figure 5.24: Constant frequency process in time and frequency domains.

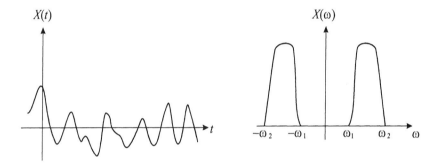

Figure 5.25: Narrow band process in time and frequency domains.

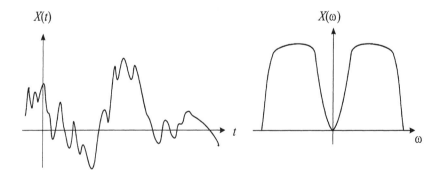

Figure 5.26: Broad band process in time and frequency domains.

5.7. RANDOM PROCESSES

A *narrow band* process is an "almost" harmonic oscillator. Instead of vibrating at one distinct frequency as does a harmonic oscillator, it vibrates with frequencies in a narrow range: $\omega_1 \leq \omega \leq \omega_2$. The spectral density of a narrow band process can be idealized by the spectrum in Figure 5.27. It has a flat spectrum in the frequency band, with constant magnitude S_0.

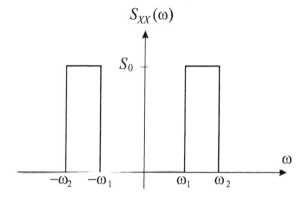

Figure 5.27: Spectral density of an idealized narrow band process.

The autocorrelation function for such a process is evaluated as follows,

$$\begin{aligned}
R_{XX}(\tau) &= \frac{1}{2\pi} \int_{-\infty}^{\infty} S_{XX}(\omega) e^{i\omega\tau} d\omega \\
&= \frac{1}{\pi} \int_{\omega_1}^{\omega_2} S_0 \cos\omega\tau \, d\omega \\
&= \frac{S_0}{\pi\tau} (\sin\omega_2\tau - \sin\omega_1\tau) \quad (5.35) \\
&= \frac{2S_0}{\pi\tau} \cos\left[\frac{\omega_1+\omega_2}{2}\tau\right] \sin\left[\frac{\omega_2-\omega_1}{2}\tau\right], \quad (5.36)
\end{aligned}$$

where the real part of the complex exponential is retained having made use of the symmetry of the power spectrum function.[30] The mean square value can be obtained by taking the limit of the right hand side (rhs) of Equation 5.36,

$$E\{X^2\} = R_{XX}(0) = \lim_{\tau \to 0} \text{rhs} = \frac{S_0}{\pi}(\omega_2 - \omega_1).$$

---

[30] If the random process $X(t)$ is real, then $R_{XX}(\tau)$ is even and real. Therefore, $S_{XX}(\omega)$ is also even, $S_{XX}(-\omega) = S_{XX}(\omega)$, and in this case the exponential can be replaced by the cosine.

Of course, in this case it is very easy to evaluate the mean square by just calculating the area/2π under the spectral density curve,

$$E\{X^2\} = \frac{1}{2\pi}\int_{-\infty}^{\infty} S_{XX}(\omega)d\omega = \frac{S_0}{\pi}(\omega_2 - \omega_1).$$

A *broad band* vibration is one with a significant range of frequencies, and therefore will have a broader frequency band. The above results are also valid, except $\omega_2 - \omega_1$ is a larger range.

White Noise Processes

A *white noise* process is an idealization made for mathematical expediency. Assuming that a process is white noise greatly simplifies the necessary algebra of analysis. The term "white" is adopted from optics to signify that all frequencies are part of such a process, much like white light is composed of the whole color spectrum.

The power spectrum ranges from $-\infty$ to ∞, and the autocorrelation can be evaluated by setting $\omega_1 = 0$ and $\omega_2 \to \infty$ in Equation 5.35,

$$\lim_{\omega_1 \to 0} \frac{S_0}{\pi\tau}(\sin\omega_2\tau - \sin\omega_1\tau) = S_0\frac{\sin\omega_2\tau}{\pi\tau},$$

and

$$R_{XX}(\tau) = \lim_{\omega_2 \to \infty} S_0 \frac{\sin\omega_2\tau}{\pi\tau} = S_0\delta(\tau),$$

with Fourier transform

$$S_{XX}(\omega) = \int_{-\infty}^{\infty} S_0\delta(\tau)e^{-i\omega\tau}d\tau = S_0.$$

This is the flat spectrum anticipated and shown in Figure 5.28. The figure includes the equivalent one–sided spectrum.

A random process that has a band limited spectrum is called, following the optics analogy, *colored noise*.

5.8 Random Vibration

We have arrived at our main goal in the study of structures under random loading, to predict the response statistics given the loading statistics. Such structures, or oscillators, are governed by the second order differential equation,

$$\ddot{X} + 2\zeta\omega_n\dot{X} + \omega_n^2 X = F(t), \tag{5.37}$$

5.8. RANDOM VIBRATION

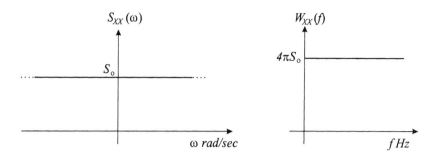

Figure 5.28: White noise spectrum.

where the input force per unit mass is given by stationary random process $F(t)$ and the output displacement by random process $X(t)$.

Before proceeding with the problem development, the reader is urged to review the concepts of *impulse response* and *convolution*. The results of this section are presented *a priori*, the mean value of the response μ_X and the spectral density of the output $S_{XX}(\omega)$,

$$\mu_X = H(0)\mu_F = \frac{1}{\omega_n^2}\mu_F$$

$$S_{XX}(\omega) = |H(i\omega)|^2 S_{FF}(\omega)$$

$$|H(i\omega)|^2 = \left[\frac{1/\omega_n^2}{\sqrt{(1-\omega^2/\omega_n^2)^2 + (2\zeta\omega/\omega_n)^2}}\right]^2,$$

so that the reader will be able to better grasp the mathematical manipulations that follow. The mean value of the response μ_X is proportional to the mean value of the force μ_F, and the response spectral density $S_{XX}(\omega)$ is proportional to the force spectral density $S_{FF}(\omega)$. For both results the proportionality is the complex frequency response function $H(i\omega)$.

5.8.1 Formulation

Consider the linear system defined by Equation 5.37 and assume random process input $F(t)$ to be stationary, with mean μ_F and power spectrum $S_{FF}(\omega)$. The stationarity assumption for the forcing means that transient dynamic behavior cannot be directly considered here.[31] The initial loading

---

[31] There are, however, clever ways by which stationary solutions can be utilized in non-stationary cases. One possibility is to multiply the stationary process by a deterministic time function such that the product is an *evolutionary* or nonstationary process. For

transients of an earthquake, a wind gust, or an extreme ocean wave cannot be considered as stationary. Assuming that the character of the loading does not change, steady state behavior can be assumed to be statistically stationary.

5.8.2 Derivation of Equations

Begin with the equation for the deterministic response of a linear harmonic oscillator, Equation 3.64,

$$X(t) = \int_{-\infty}^{\infty} g(\tau)F(t-\tau)d\tau, \tag{5.38}$$

where $g(t)$ is the impulse response, given by Equation 3.58,

$$g(t) = \frac{1}{\omega_d} e^{-\zeta\omega_n t} \sin \omega_d t,$$

and stationary random load per unit mass $F(t)$ is applied at $t = -\infty$, that is, long before the present time. This ensures stationarity. Beginning with Equation 5.38, take the expected value of both sides, and use the linear property of mathematical expectation to interchange it with the integral:

$$\begin{aligned} E\{X(t)\} &= \int_{-\infty}^{\infty} g(\tau) E\{F(t-\tau)\} d\tau \\ &= E\{F(t)\} \int_{-\infty}^{\infty} g(\tau) d\tau \\ &= \mu_F \int_{-\infty}^{\infty} g(\tau) d\tau \\ &= \mu_F H(0), \end{aligned}$$

where the stationarity of $F(t)$ is utilized in the second and third equations,[32] and $H(0) = H(i\omega)|_{\omega=0}$ using Equation 3.23. This is the first important result,

$$\mu_X = H(0)\mu_F = \frac{1}{\omega_n^2}\mu_F.$$

Since μ_F is time independent, then so must be μ_X.

In order to derive the output spectral density, it is necessary to derive intermediate results involving the correlation function.

---

example, use $A(t)F(t)$ as the forcing function, where $A(t)$ is a deterministic transient function and $F(t)$ is stationary.

[32] The force is stationary and has a constant mean value.

5.8.3 Response Correlations

For a stationary process, the autocorrelation function is given by

$$R_{XX}(\tau) = \int_{-\infty}^{\infty} x(t)x(t+\tau)f_X(x)dx,$$

where $f_X(x)$ is the probability density function of the process $X(t)$. The Fourier transform of this equation cannot be evaluated because the response density function $f_X(x)$ is not known. We can proceed in two other equivalent ways: (i) by utilizing the ergodic definition of the autocorrelation,[33] or (ii) by using Equation 5.38, utilizing available information on $F(t)$ in the following way.

First, derive the cross–correlation between $F(t)$ and $X(t)$. Multiply both sides of Equation 5.38 by $F(t-\alpha_1)$, and take expected values of both sides,

$$E\{X(t)F(t-\alpha_1)\} = \int_{-\infty}^{\infty} g(\tau_1)E\{F(t-\tau_1)F(t-\alpha_1)\}d\tau_1,$$

where the autocorrelation of the force is

$$E\{F(t-\tau_1)F(t-\alpha_1)\} = R_{FF}(\tau_1-\alpha_1)$$
$$\text{and } E\{X(t)F(t-\alpha_1)\} = R_{XF}(\alpha_1),$$

is the cross–correlation between loading $F(t)$ and response $X(t)$. Thus,

$$R_{XF}(\alpha_1) = \int_{-\infty}^{\infty} g(\tau_1)R_{FF}(\alpha_1-\tau_1)d\tau_1, \tag{5.39}$$

and $R_{FF}(\tau)$ is known from experimental data. Next, multiply both sides of Equation 5.38 by $X(t+\alpha_2)$, and take expected values of both sides,

$$E\{X(t+\alpha_2)X(t)\} = \int_{-\infty}^{\infty} g(\tau_2)E\{X(t+\alpha_2)F(t-\tau_2)\}d\tau_2 \tag{5.40}$$

$$R_{XX}(\alpha_2) = \int_{-\infty}^{\infty} g(\tau_2)R_{XF}(\tau_2+\alpha_2)d\tau_2. \tag{5.41}$$

Substitute Equation 5.39 into 5.41 to find[34]

$$R_{XX}(\tau) = \int_{-\infty}^{\infty}\int_{-\infty}^{\infty} g(\alpha)g(\beta)R_{FF}(\tau+\beta-\alpha)d\alpha\,d\beta, \tag{5.42}$$

---

[33] $R_{XX}(\tau) = \lim_{T\to\infty} \frac{1}{2T}\int_{-T}^{+T} X(t)X(t+\tau)dt$

[34] Keep careful track of the *dummy variables* so that appropriate arguments are maintained. Here, let $\alpha_2 \equiv \tau$, $\tau_1 \equiv \alpha$ and $\tau_2 \equiv \beta$ in order to simplify the notation.

which is a double convolution. To evaluate the variance,

$$\begin{aligned} \sigma_X^2 &= E\{X(t)^2\} - E^2\{X(t)\} \\ &= R_{XX}(0) - [H(0)E\{F\}]^2. \end{aligned} \qquad (5.43)$$

Example 5.15 Response Mean and Variance
Following the previous discussion, let us examine how the response mean and variance (or standard deviation) can be very useful in a design. Suppose that an analysis resulted in the following statistics: μ_F and σ_F, where stationarity has been assumed. By stationarity, we recall, is implied that the mean value is not a function of time and that the correlation is only a function of time difference, generally τ.

Equation 5.43 was obtained by setting $\tau = 0$ and by substituting the response mean value. The designer needs both the mean value *and* the variance to establish bounds on the possible response. Example bounds are: $\mu_F \pm \sigma_F$, $\mu_F \pm 2\sigma_F$, or $\mu_F \pm 3\sigma_F$. Of course, the larger the *sigma bounds* the more likely that all possible responses are covered. Along with a higher probability comes this broader band with its vagueness. There is no way around this *uncertainty–type* principle. These upper and lower bounds are used to define the least and most likely range of responses. If designing for strength, then the upper sigma bound can be used to size the structural components.

How wide or narrow the sigma bounds are depends on the underlying density function. For the Gaussian, there is approximately a one–in–three chance of being outside the one sigma bounds, but only a one–in–two hundred chance of being outside the two sigma bounds. Different densities are structured differently.

Therefore, the designer must study the data in order to better understand the underlying density. There is no easy or clear cut answer regarding how many sigma bounds to use in a design. As a practical matter, by retaining larger sigma bounds in the design, the design becomes more conservative, leading to a more costly structure or product. ∎

While this information on the correlation is of interest, the real goal of this section is to evaluate the response spectral density, which is derived in the next section.

5.8.4 Response Spectral Densities

Begin with the Fourier transform relation between power spectrum and correlation function, $S_{XX}(\omega) = \int_{-\infty}^{\infty} \exp(-i\omega\tau) R_{XX}(\tau) d\tau$, substitute Equa-

5.8. RANDOM VIBRATION

tion 5.42 for $R_{XX}(\tau)$, and let $\lambda = \tau + \beta - \alpha$,

$$\begin{aligned} S_{XX}(\omega) &= \int_{-\infty}^{\infty} \exp(-i\omega\tau) \left[\int_{-\infty}^{\infty} \int_{-\infty}^{\infty} g(\alpha)g(\beta)R_{FF}(\lambda)d\alpha d\beta \right] d\tau \\ &= \int_{-\infty}^{\infty} g(\alpha) \exp(-i\omega\alpha)d\alpha \times \int_{-\infty}^{\infty} g(\beta) \exp(+i\omega\beta)d\beta \\ &\quad \times \int_{-\infty}^{\infty} R_{FF}(\lambda) \exp(-i\omega\lambda)d\lambda \\ &= H(i\omega)H^*(i\omega)S_{FF}(\omega), \end{aligned} \qquad (5.44)$$

by definition, where $*$ denotes complex conjugate, and therefore,

$$S_{XX}(\omega) = |H(i\omega)|^2 S_{FF}(\omega). \qquad (5.45)$$

This is the **fundamental** result for random vibration and linear systems theory that allows us to evaluate the output spectral density, given the input spectral density and the system frequency response. It is emphasized here that the derivation of Equation 5.45 made use of the convolution equation, which is valid for linear systems and structures. Any generalization for nonlinear behavior requires problem–specific approaches.[35]

Example 5.16 Oscillator Response to White Noise
Consider a simple application of the above ideas to an oscillator. What is the response of a damped oscillator to a force with white noise density?
Solution   The governing equation of motion is

$$\ddot{X} + 2\zeta\omega_n \dot{X} + \omega_n^2 X = \frac{F(t)}{m},$$

where $F(t)$ is the external force, and the squared magnitude of the system function is given by

$$|H(i\omega)|^2 = \frac{1}{(\omega_n^2 - \omega^2)^2 + (2\zeta\omega_n\omega)^2}.$$

Therefore, given *any* input spectral density $S_{FF}(\omega)$, the response spectral density is

$$S_{XX}(\omega) = |H(i\omega)|^2 S_{FF}(\omega) = \frac{S_{FF}(\omega)}{(\omega_n^2 - \omega^2)^2 + (2\zeta\omega_n\omega)^2}.$$

---

[35] Two widely used techniques for nonlinear stochastic problems are *stochastic linearization* which allows the use of linear theory, and *perturbation methods* which transform a nonlinear equation into an infinite sequence of linear equations, again allowing the use of linear theory.

Suppose, for mathematical simplicity, that the forcing is white noise, $S_{FF}(\omega) = S_0$. Then,
$$S_{XX}(\omega) = \frac{S_0}{(\omega_n^2 - \omega^2)^2 + (2\zeta\omega_n\omega)^2}, \qquad (5.46)$$
and the mean square response is given by
$$E\{X(t)^2\} = \frac{1}{2\pi}\int_{-\infty}^{\infty} S_{XX}(\omega)d\omega = \frac{\pi S_0}{kc} = \frac{\pi S_0}{2m^2\omega_n^3\zeta}.$$

The mean square response can also be written in terms of a one–sided spectrum using Equation 5.34,
$$E\{X(t)^2\} = \frac{\pi S_0}{kc} = \frac{W_0}{4kc},$$
using the one–sided density.

The above integral is not standard, but can be found in texts on random vibration.[36] Even though infinite mean square energy is input to the system,[37] it responds with a finite mean square energy. See Figure 5.29 for plots of the components of Equation 5.46. White noise is frequently used even though it is non–physical because it leads to good approximate results. ∎

Example 5.17 Response to Colored Noise
Suppose the same system as in the last example is subjected to more complex loading, where the spectral density of the forcing is not a constant, but a function of ω. How would the above analysis change?

The output spectral density becomes a more complex function of frequency, for example, if the loading spectral density is, say, the wind load spectrum. Then, the mean square response must be evaluated numerically.

The applied problems are always solved numerically, although hopefully after some significant analytical exposition. ∎

---

[36] For example, the integral of this example problem is a specialized version of
$$\int_{-\infty}^{\infty} \left| \frac{B_0 + i\omega B_1}{A_0 + i\omega A_1 - \omega^2 A_2} \right|^2 d\omega = \frac{\pi(A_0 B_1^2 + A_2 B_0^2)}{A_0 A_1 A_2}.$$

[37] The energy input equals the area under the spectral density, which for white noise is
$$\int_{-\infty}^{\infty} S_0 \, d\omega = \infty.$$

5.8. RANDOM VIBRATION

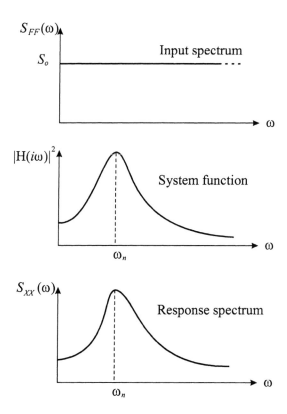

Figure 5.29: $S_{FF}(\omega)$, $|H(i\omega)|^2$, $S_{XX}(\omega)$.

5.9 Concepts Summary

This chapter has introduced the reader to the very basic concepts in applied probability and random processes. In particular, we discovered how a variable that has associated uncertainties in its value can be modeled mathematically through the probability distribution function or the probability density function. We also examined several key density functions that are widely useful in applications.

The concept of the random process is introduced. This led to correlation and the power spectrum. The correlation is physically a measure of how rapidly a function varies, and the power spectrum is a measure of the energy distribution as a function of frequency.

Different random processes are discussed and classified according to their spectrum: broad and narrow band processes are two important cases. The white noise process was introduced as an idealized model that yields acceptable results in applications, and, finally, applied to the example of the random vibration of a damped oscillator.

5.10 Problems

Problems for Section 5.1 –Introduction

1. Twenty-five samples of a steel beam were chosen and tested for the Young's modulus. Eight had a modulus of $E = 30 \times 10^6$ psi. Two had a modulus of $E = 29 \times 10^6$ psi. Fifteen had a modulus of $E = 30.5 \times 10^6$ psi. Estimate the following probabilities:

    (i) $\Pr\{E > 29.5 \times 10^6 \text{ psi}\}$,

    (ii) $\Pr\{E > 30 \times 10^6 \text{ psi}\}$,

    (iii) $\Pr\{E > 28 \times 10^6 \text{ psi}\}$.

    Problems for Section 5.2 – Example Problems and Motivation

2. For the following applications, make a list of parameters and forces that are needed to analyze the problem and distinguish between those that can be assumed deterministic and those that must be described probabilistically: *(i)* airplane design, *(ii)* ship design, *(iii)* turbomachinery design, *(iv)* mechanical watch movements, *(v)* computer hard drive, *(vi)* automobile internal combustion engine, *(vii)* automobile body.

    Problems for Section 5.3 – Random Variables

5.10. PROBLEMS

3. Write the meanings of the following mathematical expressions in words only:

   (i) $F(x)$

   (ii) $F(-\infty)$

   (iii) $F(+\infty)$

   (iv) $\int_{-\infty}^{\infty} f(x)dx = 1$.

4. What does the fact $F(x_1) \leq F(x_2)$ imply about the values of x_1 and x_2? Why?

5. Can the schematics in Figure 5.30 be cumulative distribution functions? Why?

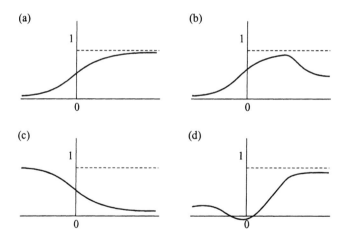

Figure 5.30: Possible cumulative distribution functions.

6. Can the schematics in Figure 5.31 be probability density functions? Why?

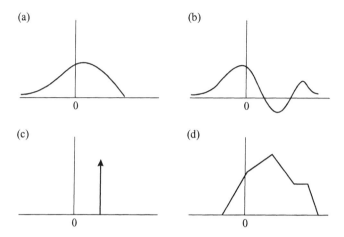

Figure 5.31: Possible probability density functions.

7. If $\Pr\{X \leq x_1\} = 0.1$ and $\Pr\{X \leq x_2\} = 0.2$, sketch the probability density function for random variable X. What is $\Pr\{x_1 \leq X \leq x_2\}$?

8. For the following density functions, evaluate the normalization constant and the respective probability, where c and a are constants:
   (i) $ce^{-|x|}$, $-\infty < X < +\infty$; $\Pr\{X \leq 0\}$
   (ii) ce^{-x}, $4 < X < 10$; $\Pr\{X \leq 5\}$
   (iii) $a/(x_2 - x_1)$, $0 < X < 2$; $\Pr\{X \leq 0\}$.

9. Can the following be densities? a, b, and c are constants:
   (i) $3/(x_2 - x_1)$, $0 \leq X \leq 1$
   (ii) $a \exp x$, $0 \leq X \leq \infty$
   (iii) $c \ln x$, $1 \leq X \leq 2$
   (iv) b, $2 \leq X \leq 5$.

10. Let X have density $f(x) = x^2/9$, $0 \leq x \leq 3$. Is this a legitimate density function? Find the following probabilities:
    (i) $\Pr\{1 \leq X \leq 2\}$
    (ii) $\Pr\{X \leq 1\}$
    (iii) $\Pr\{X \geq 3\}$.

11. What is the probability that a continuous random variable takes on a particular value, that is, $\Pr\{X = x\}$?

5.10. PROBLEMS

12. A density function is given by $f(x) = cx$, $0 \leq x \leq 1$. Evaluate c so that $f(x)$ becomes a probability density function. Then find $\Pr\{X < 0.3\}$.

13. A density function is given by $f(x) = 1 - |x|$, $-1 \leq x \leq 1$. Sketch $f(x)$, show that it is a density, and find $\Pr\{-1/2 \leq X \leq 1/3\}$.

14. The direction at which the wind strikes a tower is a random variable θ with density function $f(\theta) = c\cos\theta$, $-\pi/4 \leq \theta \leq \pi/4$. The angle θ is measured from due East.

    (i) Sketch the density as a function of θ.

    (ii) Evaluate c.

    (iii) Calculate the probability that $\theta > |\pi/4|$.

    (iv) Calculate the probability that $-\pi/10 \leq \theta \leq \pi/10$.

    (v) Sketch the cumulative distribution function.

    Problems for Section 5.4 – Mathematical Expectation

15. Find the mathematical expectation of random variable X with density function $f(x) = a/(x_2 - x_1)$, $x_1 \leq X \leq x_2$. Also evaluate the mean–square value, the variance, and the coefficient of variation.

16. Find the mathematical expectation of random variable X with density function $f(x) = c\ln x$, $1 \leq X \leq 3$, where c is a constant to be determined. Also evaluate the mean–square value, the variance, and the coefficient of variation.

17. We have the following data on the value of a variable: 2.3, 3.5, 3.5, 4.9, 3.7, 0.7, 4.1; find the average value, the mean–square value, the variance, and the coefficient of variation.

18. Compare the mean values of the following two random variables. Do you expect them to have the same means? For the first random variable we have the data: 3.0, 3.5, 4.0, 4.5, 5.0, 5.5, and 6.0. For the second random variable we have a probability density of $f(x) = 1/(x_2 - x_1)$, $3 \leq X \leq 6$.

    Problems for Section 5.5 – Useful Probability Densities

19. The stiffness properties of a new material is established by initial testing. The preliminary results show a uniform scatter of data in the range 9 lb/in $\leq k \leq$ 11 lb/in. There is no additional information. Calculate the mean value, standard deviation, coefficient of variation, and the probability that the stiffness is greater than 9.5 lb/in. The

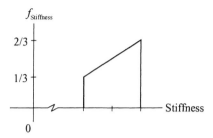

Figure 5.32: Density of stiffness data.

above spread of data is found to be too large, and so additional testing is carried out. A plot of the data shows the density Figure 5.32. Calculate the same values as before, and draw conclusions.

20. Several steel beams delivered to the factory are to be used as columns. The test data for the lot from which these beams were selected are known to have a Young's modulus in the range 29×10^6 psi $\leq E \leq 31 \times 10^6$ psi, with all values equally likely. If the columns have a cross section area of 144 in^2 and are loaded with 100,000 lb, what are the possible stress and strain ranges?

21. Show that the discrete Poisson–distributed random variable X with discrete density

$$\Pr\{X = k\} = \frac{e^{-\lambda}\lambda^k}{k!}, \quad k = 0, 1, 2, \ldots$$

has an expectation $E\{X\} = \lambda$.

22. If a mechanical component fails according to the exponential distribution with a mean value of 5000 hrs, what is the probability that the component will fail by 1000 hrs? How many additional hours before this probability is doubled? Sketch the density function.

23. Derive Equations 5.13 and 5.14.

24. For the Gaussian random variable X, numerically estimate the probability $\Pr\{X < 0\}$ for the two cases:

    (i) mean value $\mu = 10$ and standard deviation $\sigma = 10$

    (ii) mean value $\mu = 10$ and standard deviation $\sigma = 1$.

    Can either of these be used as models for a positive definite variable? Explain.

5.10. PROBLEMS

Problems for Section 5.6 – Two Random Variables

25. Given the joint density function

$$f(x,y) = \frac{1}{(x_2 - x_1)(y_2 - y_1)},$$

$2 \leq x \leq 4$, and $1 \leq y \leq 3$. Find $E\{X\}$, $E\{XY\}$, $Cov(XY)$, and ρ. Note that $f(x,y) = f(x)f(y)$.

26. X and Y have a joint probability density given by

$$f(x,y) = e^{-(x+y)}, \quad x \geq 0, \ y \geq 0.$$

Find $\Pr\{X \geq Y \geq 2\}$ and sketch the region in the x,y plane that defines the region of integration.

27. In *Example* 5.11, verify Equation 5.22.

Problems for Section 5.7 – Random Processes

28. Suppose strain gages are placed at two locations on a wing that is being tested in a wind tunnel. The first gage is near the fixed base of the wing and the other gage is near the free wing tip. Two tests are performed:

    (i) the wing vibration is recorded for low velocity laminar flow and

    (ii) for high velocity turbulent flow.

    For each of these tests, discuss the following:

    (a) Describe and compare the time–histories you would expect to see recorded by each strain gage;

    (b) Discuss the type of cross correlation $R_{X_1 X_2}(\tau)$ you would expect between X_1 and X_2, the locations of the gages.

29. Considering Figure 5.33, discuss whether the processes are stationary.

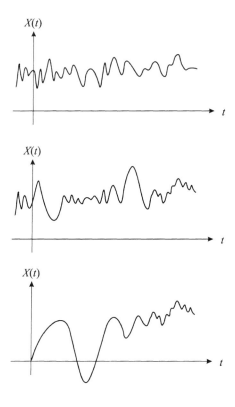

Figure 5.33: Random processes.

30. From the expressions for wind, and ocean wave power spectra, which parameters appear to be important for a proper characterization of the energy distribution?

31. For the Pierson–Moskowitz power spectrum, evaluate numerically the areas under the first 20 frequency bands of width $\Delta\omega = 0.1$ rad/sec. Do this for the cases: *(i)* $V = 10$ m/sec, *(ii)* $V = 20$ m/sec. What conclusions can be drawn regarding the effect of wind speed on the frequency distribution of wind energy for this spectrum? Could similar conclusions have been arrived at by directly studying the equation for the spectrum?

32. Convert the power spectra of Figure 5.34 to one–sided equivalent spectra that are functions of Hz.

5.10. PROBLEMS

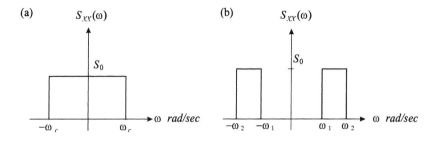

Figure 5.34: Broad band and narrow band spectra.

Problems for Section 5.8 – Random Vibration

33. Beginning with Equation 5.39, derive Equation 5.42.

34. Following *Example 5.16*, we would like to examine the sensitivity of the output spectral density $S_{XX}(\omega)$ and the mean square response $E\{X(t)^2\}$ to various combinations of parameter values, where S_0 is the input spectrum. For the following cases, plot $|H(i\omega)|^2$, and $S_{XX}(\omega)$, and evaluate $E\{X(t)^2\}$ for unit mass, $0 < \omega < 3.0$ rad/sec and S_0 ft^{2}sec:

   (i) $\omega_n = 0.1$ rad/sec and $\zeta = 0.1$
   (ii) $\omega_n = 0.5$ rad/sec and $\zeta = 0.1$
   (iii) $\omega_n = 1.0$ rad/sec and $\zeta = 0.1$
   (iv) $\omega_n = 2.0$ rad/sec and $\zeta = 0.1$
   (v) $\omega_n = 0.1$ rad/sec and $\zeta = 0.5$
   (vi) $\omega_n = 0.5$ rad/sec and $\zeta = 0.5$
   (vii) $\omega_n = 1.0$ rad/sec and $\zeta = 0.5$
   (viii) $\omega_n = 2.0$ rad/sec and $\zeta = 0.5$.

   Compare results and draw general conclusions.

35. Using the fundamental relation between input and output spectra, find the response spectrum for the oscillator governed by

$$\ddot{X} + 2\zeta\omega_n \dot{X} + \omega_n^2 X = F(t),$$

where the spectrum for $F(t)$ is *(i)* the Pierson–Moskowitz (Equation 5.32), *(ii)* the wind spectrum (Equation 5.33). For each, numerically evaluate the mean square response for the following parameters.

For the oscillator, assume $m = 10$ N, $\zeta = 0.15$, $k = 5$ N/m. For the Pierson–Moskowitz, let the wind speed $V = 25$ km/hr. For the wind spectrum, let $\bar{V}_{10} = 20$ mi/hr, and $\kappa = 0.01$. Plot each response spectrum.

5.11 Mini–Projects

The following mini–projects will provide the student with an avenue for further exploration of the applied probabilistic concepts of this chapter. The instructor can assign the student to obtain a copy of any of the following papers, and to begin reading them to the point where they are too difficult to follow. The purpose for such an assignment is to expose the student to some actual examples of research into random vibration and the modeling of uncertainties.

1. *Statistical Aspects of Dynamic Loads*, Y.C. Fung, *J. of the Aeronautical Sciences*, May 1953, 317–330.

   In many dynamic stress problems, the knowledge of the physical phenomena is not precise enough to justify exact predictions with respect to the results of each individual observation. On the one hand, the loading is never exactly known, and on the other, the physical system may be so complicated that detailed calculations for the prediction of dynamic responses are difficult and the numerical results are often uncertain. In such cases the physical data can be regarded as a set of statistical data. In this paper the statistical theory is illustrated by the gust– and landing–load problems.

2. *Transient Response of a Dynamic System Under Random Excitation*, T.K. Caughey, H.J. Stumpf, *J. Applied Mechanics*, Dec. 1961, 563–566.

   This paper analyses the transient response of a simple harmonic oscillator to a stationary random input having arbitrary power spectrum. The application of the results of this analysis to the response of structures to strong–motion earthquakes is discussed.

3. *Mean–Square Response of Simple Mechanical Systems to Nonstationary Random Excitation*, R.L. Barnoski, J.R. Maurer, *J. Applied Mechanics*, Vol. 36, No. 2, June 1969, 221–227.

   This paper concerns the mean–square response of a single degree of freedom system to amplitude modulated nonstationary random noise. Both the unit step and rectangular step functions are used for the amplitude modulation, and both white noise and noise with an exponentially decaying harmonic correlation function are considered.

5.11. MINI-PROJECTS

4. *Mean–Square Response of a Second–Order System to Nonstationary Random Excitation*, L.L. Bucciarelli, Jr., C. Kuo, *J. Applied Mechanics*, Vol. 37, No. 3, Sept. 1970, 612–616.

   The mean–square response of a lightly damped, second–order system to a type of nonstationary random excitation is determined. The forcing function on the system is taken in the form of a product of a well–defined, slowly varying envelope function and a noise function. The latter is assumed to be white or correlated as a narrow band processed.

5. *Random Vibration: A Survey of Recent Developments*, S.H. Crandall, W.Q. Zhu, *J. Applied Mechanics*, Dec. 1983, Vol. 50, 953–962.

   A general overview of the problems, methods, and results achieved in random vibration since its inception as a technical discipline nearly 50 years ago is given with particular emphasis on recent developments.

6. *Structural Dynamics with Parameter Uncertainties*, R.A. Ibrahim, *Applied Mechanics Reviews*, Vol. 40, No. 3, Mar. 1987, 309–328.

   The treatment of structural parameters as random variables has been the subject of structural dynamicists and designers for many years. In this paper, the subject and techniques are reviewed in depth.

7. *Statistical Implications of Methods of Finding Characteristic Strengths*, R.D. Hunt, A.H. Bryant, *J. Structural Engineering*, Vol. 122, No. 2, Feb. 1996, 202–209.

   Various statistical methods of estimating the characteristic strengths of materials or structural components from the results of experimental tests can be assessed in the figures and tables of this paper.

8. *Cassini Spacecraft Force Limited Vibration Testing*, K.Y. Chang, T.D. Scharton, *Sound and Vibration*, March 1998, 16–20.

   This paper discusses the environmental test program for the Cassini spacecraft included a force–limited vertical–axis random vibration testing.

9. *Spacecraft Acoustic and Random Vibration Test Optimization*, J.C. Forgrave, K.F. Man, J.M. Newell, *Sound and Vibration*, March 1999, 28–31.

   Acoustic noise and random vibration tests are key constituents of an effective spacecraft environmental qualification program. This article describes a method for optimizing acoustic and random vibration trials to reduce costs and schedules.

10. *An Inverse Method for the Identification of a Distributed Random Excitation Acting on a Vibrating Structure, Part 1: Theory*, S. Granger, L. Perotin, *J. Mechanical Systems and Signal Processing* (1999) **13**(1), 53–65.

In many practical situations, it is difficult, if not impossible, to perform direct measurements or calculations of the external forces acting on vibrating structures. Instead, vibrational responses can often be conveniently measured. This paper presents an inverse method for estimating a distributed random excitation from the measurement of the structural response at a number of discrete points.

11. *An Inverse Method for the Identification of a Distributed Random Excitation Acting on a Vibrating Structure, Part 1: Flow–Induced Vibration Application*, L. Perotin, S. Granger, *J. Mechanical Systems and Signal Processing* (1999) **13**(1), 67–81.

The experimental setup consists of a straight tube submitted to a complex three-dimensional turbulent water flow.

Chapter 6

Vibration Control

"The response is unacceptable."

Our study of single degree of freedom system models has so far consisted of the structure and the forces acting on it. Given a force and a structure, our analytical purpose is to evaluate the vibration response. We made no judgment whether the response, be it displacement or velocity, was too large or too small. The possibility of modifying the behavior was only briefly hinted.

This chapter on *vibration control*, studies the modification of structural response. Before altering or controlling the response of a structure or machine, criteria needs to be defined that will be used to measure how well the structure is being controlled. Then, if a response exceeds certain tolerances, *appropriate forces*, called *feedback forces*, are needed to bring the response within acceptable levels.

We can only hint at the breadth and depth of structural control here, a vast subject to which hundreds of books are dedicated. A sense is given of how the controls engineer would approach the analysis and design of a structural control system. Performance measures in the time domain are examined. Related issues of stability and sensitivity are introduced.

We do not describe the frequency domain or graphical methods available for the analysis and design of control systems. These include the *Root–Locus* method, the *Bode* plot, the *Nyquist* diagram, and the *Nichols Chart*. To provide a description that would make any sense at all to the uninitiated reader would require a significant increase to the number of pages in this text and is beyond our scope. Just as Chapter 5 provided a brief and focused view of the probabilistic approach, this chapter provides a selective view of control. The reader is referred to several fine books[1] on control and systems that can ably fill in the many gaps that are unavoidable in this introductory chapter.

6.1 Introduction

Machines and structures are designed to behave acceptably. This may require a design where displacement amplitudes remain below a particular value, or that operating frequencies remain far from the natural frequency of the structure. One approach is to design these constraints into the structure. However, it is common for the operating conditions of a machine or structure to change with time so that it becomes impossible to incorporate all the necessary constraints in the design *a priori*.

Also, environmental forces can vary considerably over a single operating cycle. For example, it is not possible to design an efficient solid wing so that an airplane can take off, fly, and land effectively. Wings have a variety of control surfaces that alter flying characteristics needed for turns, decelerations, and varying the lift forces. As another example, we learned in Chapter 3 how a rotating machine has to pass its resonance frequency as it speeds up to its operating speed. At its resonance frequency, amplitudes can become very large. One solution is to add damping so that amplitudes cannot grow too fast. For some applications it may not be possible to do this, and it may be necessary to control the vibration actively, which is the focus of this chapter.

Recall that both time and frequency domains provided useful information and understanding about vibration. The same is true for control problems. The Laplace transform, introduced in Chapter 3, is used to move from the time domain to the (complex) frequency domain in terms of complex variable s, resulting in the *transfer function*. Analytical results from Chapters 2 and 3 for single degree of freedom systems provide a foundation for control studies. *State variable models* are used to provide a framework for the (numerical) solution of systems with many degrees of freedom. In this chapter, the state variable model will be introduced for a single degree of freedom oscillator, but is most useful when extended to multi–degree of freedom systems.

---

[1] J.J. DiStefano III, A.R. Stubberud, and I.J. Williams, **Feedback and Control Systems**, Schaum's Outline Series, McGraw-Hill, 1967. As in all the Schaum's Series, there are many wonderful worked out examples.

R.C. Dorf and R.H. Bishop, **Modern Control Systems**, Ninth Edition, Prentice-Hall, 1995. This is also a very fine book with many examples.

K. Ogata, **System Dynamics**, Third Edition, Prentice-Hall, 1997. Many of the topics we study here are part of the more general discipline called *systems engineering*.

6.2. FEEDBACK CONTROL

Figure 6.1: Block diagram for open loop system.

There are three types of control systems: *(i) engineered control systems* such as electric switches and thermostatic controlled furnaces, *(ii) natural and biological control systems*; a biological example is human motion, in which the brain requires input from vision and touch to effect muscle action; body temperature and blood pressure are kept almost constant by means of physiological feedback. Natural control is demonstrated by planetary temperature maintenance through atmospheric geophysical processes such as wind, clouds and rain, and *(iii) hybrid control systems*, which are a combination of the first two. The example we all have experience with is driving, where human and machine operate as one, with the human acting as a sensor, controller, and actuator, using information provided by the eyes and dashboard instrumentation to effect changes by accelerating, braking, and turning the steering wheel, as necessary. The simplest control systems are those that are engineered. Biological and natural control systems are generally more complex.

6.2 Feedback Control

It is customary to call any object that is to be controlled, such as a piece of equipment or a group of machine components working as a unit with a particular function, the *plant*. A *disturbance* may be a signal, force, or temperature variation that affects the response of a system in an unacceptable manner. Such a disturbance can be internally or externally driven. For applications, specific disturbances, such as impulses, are used to study the behavior characteristics of a particular design.

An *open–loop* control system has a one way logic *flow*, meaning that the control action is independent of the structural response. In such systems, all possible contingencies must be accounted for at the design stage. An open loop system is customarily shown schematically using *block diagrams*, as in Figure 6.1.

As can be observed from the figure, the variables are written in the *Laplace transform* domain; response $Y(s)$ results from input $F(s)$ into the system $G(s)$, all in the transform s domain. $G(s)$ is the same frequency

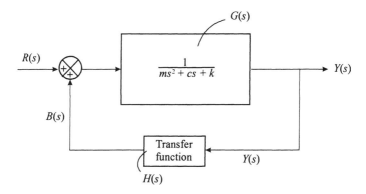

Figure 6.2: Block diagram for feedback control system.

response function of Chapter 3, Equation 3.44, also known as the *transfer function*. For the single degree of freedom oscillator, $G(s) = [ms^2+cs+k]^{-1}$.

When the response $Y(s)$ is not acceptable, the value of the response is used to generate additional forces according to certain *rules* or *laws* such that the modified response behaves according to design and within certain bounds. This results in a *closed–loop* system that incorporates *feedback control*, that is, the response is evaluated by a *sensor* and is fed back to an *actuator* that generates a force or moment. The control action is dependent on the response. The purpose of designing a system with feedback force is to minimize unwanted behavior. Feedback control provides a mechanism for tailoring system behavior to specific standards and needs. Examples include the autopilot in commercial aircraft and the historical mechanical governor of the early steam engine. In block diagram notation, this feedback is depicted in Figure 6.2.

Block diagrams are one way in which the system and its interconnections can be visualized. They depict cause and effect relationships between (input) force and (output) response. Such visualizations provide a very powerful means for understanding and designing complex systems. The block diagram is, however, simply a symbol for the mathematical operation on the input signal. The feedback block diagram is used to demonstrate how the input–output transfer function relations can be derived directly from the diagram.

Referring again to Figure 6.2, at the output juncture, $Y(s)$ is channeled in two directions. One becomes the output response and the other is fed back to a block containing an undefined transfer function. The output or response within the feedback loop is $B(s) = H(s)Y(s)$, and it loops back

6.2. FEEDBACK CONTROL

into the system via the node by which the *reference signal* $B(s)$ is input. Note that the system is loaded by *both* feedback force $B(s)$ and reference signal $R(s)$.

Following the diagram yields the response as

$$\begin{aligned} Y(s) &= G(s)[R(s) + B(s)] \\ &= G(s)[R(s) + H(s)Y(s)], \end{aligned} \quad (6.1)$$

where $Y(s)$ is on both sides of the transfer function relation. While this may seem strange, this relation is essentially a statement of the purpose of a feedback force: *Measure the output, and, if necessary, feedback a force proportional to the output back to the system. Keep doing this until the output is within the required bounds of behavior.* This implies a design procedure.

Equation 6.1 can be solved for $Y(s)$,

$$Y(s) = \left[\frac{G(s)}{1 - G(s)H(s)} \right] R(s), \quad (6.2)$$

where the ratio $G(s)/[1 - G(s)H(s)] \equiv T(s)$ is the new *feedback transfer function*. This type of feedback is known as *positive* feedback because a force is added to the system. A key concern in the design of such feedback is the stability implications for the structure. Stability is affected because the feedback modifies the effective transfer function, and it is the transfer function that characterizes the system eigenvalues. Stability is discussed later in this chapter as well as in Chapter 12. The original transfer function can be recovered by setting $H(s) = 0$. The goal of creating a system such as that modeled by Equation 6.2 is to enable the designer to alter the response $Y(s)$ by changing $H(s)$ appropriately. Of course, life is not so simple, and there are numerous challenges to controlling structural vibration.

Example 6.1 Transfer Function for a Mechanical Accelerometer
The equation of motion for the accelerometer is derived in Chapter 3. Figure 6.3 shows a simple mechanical representation of such a system.

The input signal is the acceleration $\ddot{y}$ of the primary structure, and the governing equation of motion for the accelerometer mass m is

$$m\ddot{x} + c\dot{x} + kx = m\ddot{y}.$$

For zero initial conditions, the Laplace transformed equation is

$$(ms^2 + cs + k)X(s) = mY(s),$$

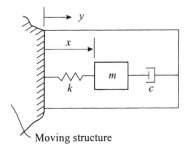

Figure 6.3: Mechanical accelerometer.

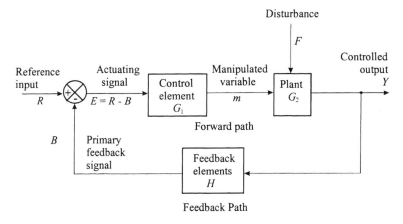

Figure 6.4: Feedback control system with disturbance $F(t)$.

with the transfer function between response and input given by

$$\frac{X(s)}{Y(s)} = \frac{1}{s^2 + (c/m)s + (k/m)}.$$

Figure 6.4 shows a more detailed block diagram of a feedback control system. Here, one can see the *branch* or *takeoff point*, where the signal from a block goes *concurrently* to other blocks or summing points. The *summing point* is denoted by a circle with an X inside, $\otimes$, indicating a summing operation.

The plus or minus sign in each sector in the circle indicates whether that signal is to be added or subtracted. Of course, units of quantities to be

6.3. PERFORMANCE OF FEEDBACK CONTROL SYSTEMS

combined must be identical. When the summing point is as shown in the figure, a difference is taken, indicating that the input and output signals are compared. The difference is an *error* that generates a control action. Also note in Figure 6.4 that the external disturbance (forcing) is shown being input directly to the plant, while the reference input is a signal that is generated internally for purposes of generating a control action once an error signal is generated.

While it appears that control is a concept that should be applied to all structures, there are costs. These include the following:

- There is an increase in the number of components, such as *sensors* and *actuators*. This implies the necessity to build and maintain a more complicated structure as well as a more difficult reliability issue.

- The controlled structure can be more expensive, although an argument can be made that a controlled structure is less expensive in the long run since its behavior is less prone to damage.

- The addition of a feedback force also increases the possibility of instability. This may happen because the additional control force may alter the system vibration characteristics in such a way that, for example, a growing exponential response may result instead of a decaying exponential. Of course, any design considers stability issues, but for very complicated structures such as aircraft and turbines, it is not straightforward to establish stability under all operating conditions, and extensive testing is necessary.

Even so, the advantages outweigh the disadvantages, and for some applications there is no choice but to include feedback control. A major advantage of a feedback control system is that it provides the ability to adjust the transient and steady state performance.

6.3 Performance of Feedback Control Systems

In order to properly design a feedback control system, the desired structural response must be defined in terms of the system specifications. *Performance* will be related to the *poles* and *zeros* of the transfer function, and to the *steady state error*, terms which will be defined soon. Performance must be defined and measured.

To examine some of the properties of a control system, *test input signals* are chosen to establish stability and control system behavior under a variety of operating conditions. Using standard test inputs, such as those listed

Signal Type	Equation	Laplace Transform
Unit Impulse	$F_i(t) = 1/\varepsilon$, $0 \le t \le \varepsilon$	1
Step Function	$F_s(t) = A$, $t > 0$	$F_s(s) = A/s$
Ramp Function	$F_r(t) = At$, $t > 0$	$F_r(s) = A/s^2$
Parabolic Function	$F_p(t) = At^2$, $t > 0$	$F_p(s) = 2A/s^3$
Sine Wave	$A \sin \omega t$	$F_{sine}(s) = \omega/(s^2 + \omega^2)$.

Table 6.1: Some Standard Test Input Signals.

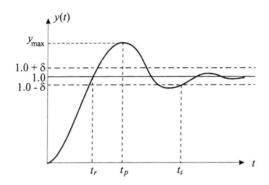

Figure 6.5: Step response performance.

in Table 5.1, gives the analyst and designer a way to compare competing designs. The signals in the table equal zero outside the given range.

Standard performance measures are usually defined in terms of the *step response*, studied in Chapter 3 and shown schematically in Figure 6.5. The general design objectives are *(i)* the speed of response, *(ii)* stability, and *(iii)* accuracy or allowable error. The first objective implies that, in general, it is desirable for a control system to compensate for any disturbance rapidly. Stability considerations are part of any active control system design. The allowable error represents how close to the desired response the control force must bring the structure. These concepts are discussed next with respect to the generic response in Figure 6.5.

The transient response of the system may be described in terms of two factors seen in the figure; *(i)* the swiftness of the response as represented by the *rise time* t_r and *peak time* t_p, and *(ii)* the closeness of the response to the desired response as measured by the *overshoot* ($y_{max} - 1$) and the

6.3. PERFORMANCE OF FEEDBACK CONTROL SYSTEMS

settling time t_s, after which the response is within the required tolerance of 2δ. Others define the same concepts and parameters in slightly different ways, but with the same goals.

6.3.1 Poles and Zeros of a Second Order System

The poles and zeros of a system are defined in terms of the transfer function of the system. The *poles* are those values of s that render the transfer function infinite, and the *zeros* are those values that render the transfer function equal to zero.[2]

The poles then make the denominator equal to zero. We have already studied such equations, known as the characteristic equation for the system. For the second order system $\ddot{y} + 2\zeta\omega_n \dot{y} + \omega_n^2 y = Ax(t)$ with zero initial conditions, the input–output relation in the s domain is

$$Y(s) = \left[\frac{A}{s^2 + 2\zeta\omega_n s + \omega_n^2}\right] X(s),$$

where the transfer function is given by the ratio $Y(s)/X(s)$. The poles of the transfer function are given by the solution of

$$s^2 + 2\zeta\omega_n s + \omega_n^2 = 0,$$

or $s = -\zeta\omega_n \pm \omega_n\sqrt{\zeta^2 - 1}$. These are the roots, or the natural frequencies, of the characteristic equation for the system. In Chapter 3, we studied the various cases that are possible for this system,

- If $\zeta > 1$, both poles are *negative* and *real*.
- If $\zeta = 1$, both poles are *negative, equal,* and *real*; $s = -\omega_n$.
- If $0 < \zeta < 1$, the poles are *complex conjugate* with *negative real parts*; $s = -\zeta\omega_n \pm i\omega_n\sqrt{1-\zeta^2}$ and $\omega_d = \omega_n\sqrt{1-\zeta^2}$.
- If $\zeta = 0$, the poles are *complex conjugate* and *imaginary*; $s = \pm i\omega_n$.
- If $\zeta < 0$, the poles are in the right half of the s–plane, and the system is unstable.

The third case applies to underdamped oscillatory systems, with $s = -\zeta\omega_n \pm i\omega_d$, and where $1/(\zeta\omega_n)$ is the *characteristic time* constant of Chapter 3. Equation 3.3 is the general response for such a system, repeated here,

$$x(t) = Ce^{-\zeta\omega_n t}\cos(\omega_d t - \phi),$$

---

[2] Another way of stating this is that *the roots of the denominator are the system poles* and *the roots of the numerator are the system zeros*.

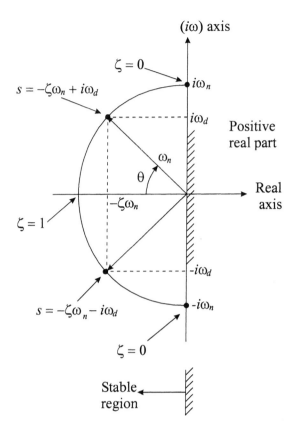

Figure 6.6: Variation of roots with ζ and the region of stability on the left half plane.

showing that when $\zeta < 0$, unbounded growth occurs. How the roots for this case vary with ζ can be examined with the aid of Figure 6.6.

6.3.2 Gain Factor

From the above discussion of zeros and poles, we see that the system transfer function can be specified to within a constant by specifying the system poles and zeros. This constant, usually denoted by K, is known as the *system gain factor*.

Example 6.2 Two Examples

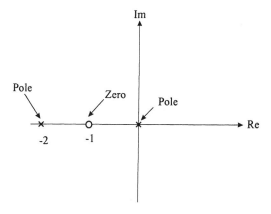

Figure 6.7: Pole-zero map.

(i) Given the transfer function
$$T(s) = \frac{2s+1}{s^2+s+1},$$
the system differential equation is
$$y(t) = \left(\frac{2D+1}{D^2+D+1}\right)x(t),$$
where the shorthand notation $D^n \equiv d^n/dt^n$ is sometimes used. Then,
$$\begin{aligned}(D^2+D+1)y(t) &= (2D+1)x(t)\\ \ddot{y}+\dot{y}+y &= 2\dot{x}+x.\end{aligned}$$
The left hand side represents the plant and the right hand side the forces.

(ii) The transfer function
$$T(s) = \frac{K(s+a)}{(s+b)(s+c)}$$
can be specified by giving the zero location $-a$, the pole locations $-b$ and $-c$, and the gain factor K.

Therefore, if we are given the information that the gain factor of a system is 2, and the pole–zero map is as shown in Figure 6.7 in the s–plane, then the transfer function is
$$T(s) = \frac{2(s+1)}{s(s+2)},$$
from which the system differential equation can be derived as before. ∎

6.3.3 Stability of Response

A key reason for controlling the vibration of a structure or a machine is to make it *more stable*. Stability is always a concern in the design of a dynamic system. There are numerous definitions for stability, two of the most important of which are:

- A system is *stable* if its impulse response approaches zero as time approaches infinity. The impulse response is a useful input since, by the convolution, any load can be represented in terms of the impulse response.

- A system is *stable* if every bounded input produces a bounded output. If an input magnitude is always within a certain finite range, it is called *bounded*.

If the input is random, then there are comparable probabilistic definitions of stability. A necessary condition for the system to be stable is for the roots of the characteristic equation to have *negative real parts*. This ensures that the impulse response will decay exponentially with time. The case $\zeta < 0$ implies that energy is added to the system faster than it can be dissipated, resulting in ever–greater amplitude oscillations and eventual failure.

6.4 Automatic Control of Transient Response

6.4.1 Control Actions

One of the first questions that comes to mind when thinking of controlling structural vibration is, *How can something be controlled automatically?* An automatic controller produces a control signal or *control action* by comparing plant output with the desired value or a desired range of acceptable responses. A deviation can result in the control action.[3]

There are certain control actions that form the basic toolbox of the controls engineer. These are *(i)* on–off, *(ii)* proportional, *(iii)* integral, *(iv)* derivative, and *(v)* combination actions such as proportional–integral *(PI)*,

---
[3] The subject of *sensors* that detect motion and *actuators* that apply the forces, moments or temperature changes on the plant comes under the heading of hardware. This consideration is a very important aspect of the control engineer's task since a design is not complete until appropriate sensors and actuators are chosen. Very often the system dynamics are altered due to the presence of sensors and actuators, especially in lightly damped space structures.

6.4. AUTOMATIC CONTROL OF TRANSIENT RESPONSE

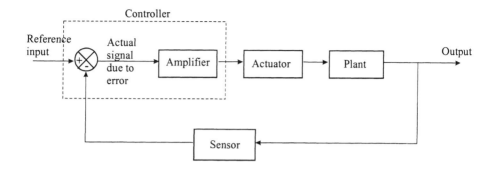

Figure 6.8: Generic control system.

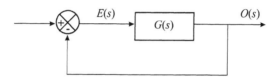

Figure 6.9: Basic control actions.

proportional–derivative *(PD)*, or proportional–integral–derivative *(PID)* controllers. We will briefly discuss some of these control actions and their desirable properties in a design. The generic block diagram of Figure 6.8 depicts schematically how a control system might be logically connected.

Generally, the actuating signal is amplified to activate the actuator that creates a force. The block diagram of Figure 6.8 has no external forces acting on the plant, only feedback control forces. We briefly describe each of the above control actions.

On–off control action: As the name signifies, there are two possible actions here. Generally, one of these actions is *off*, or no action at all. Which of the two actions is active depends on the sign of the error signal. For example, if the error signal is positive there is one control action, and if the error signal is negative there is another action.

The block diagram of Figure 6.9 can be referred to when considering the next three control actions. *Proportional control action* is defined by the proportional gain K_p and the transfer function

$$\frac{O(s)}{E(s)} = G(s) = K_p.$$

Integral control action is defined by the integral gain K_i and the transfer

function
$$\frac{O(s)}{E(s)} = G(s) = \frac{K_i}{s}.$$

Proportional plus Integral control action (PI) is defined by the above gains and the following transfer function

$$\frac{O(s)}{E(s)} = G(s) = K_p\left(1 + \frac{1}{T_i s}\right),$$

where T_i is the *integral time* and $K_p/T_i = K_i$, the integral gain.

Proportional plus Derivative control action (PD) is defined by the above gains and the following transfer function

$$\frac{O(s)}{E(s)} = G(s) = K_p(1 + T_d s),$$

where T_d is the *derivative time*, and $K_p T_d = K_d$, the derivative gain.

Proportional plus Integral plus Derivative control action (PID) is defined by the above gains and time constants and the following transfer function

$$\frac{O(s)}{E(s)} = G(s) = K_p\left(1 + \frac{1}{T_i s} + T_d s\right). \quad (6.3)$$

Alternatively, in terms of all gains, the right hand side can be rewritten as

$$\frac{O(s)}{E(s)} = G(s) = K_p + \frac{K_i}{s} + K_d s.$$

Transforming Equation 6.3 back into the time domain results in

$$O(t) = K_p\left[E(t) + \frac{1}{T_i}\int_{-\infty}^{t} E(\tau)d\tau + T_d \frac{dE(t)}{dt}\right],$$

where $O(t)$ is the output in the time domain, and $E(t)$ is the error signal in the time domain. These gains, or time constants, become the control parameters that are adjusted so that structural response meets design criteria. Note that transfer function definitions assume zero initial conditions. In the next section, we consider the design process and the selection of appropriate control actions leading to a desired response.

6.4.2 Control of Transient Response

In this section we discuss and demonstrate how the transient response of an oscillating system can be tailored using some of the above control actions.

6.4. AUTOMATIC CONTROL OF TRANSIENT RESPONSE

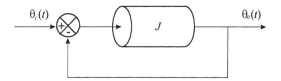

Figure 6.10: System diagram for rotational element.

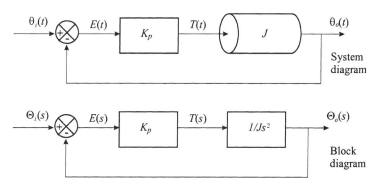

Figure 6.11: Block diagram for rotational element with proportional controller.

To do this consider the simple rotational element with inertia J that has been studied in Chapter 3. Figure 6.10 shows the block diagram representation, where $\theta_i(t)$ is the input or *reference signal* and $\theta_o(t)$ is the response or *output signal*.

If torque $T(t)$ is applied to the rotor, its response $\theta_o(t)$ is governed by $J\ddot{\theta}_o = T(t)$, or in the Laplace transformed domain,

$$Js^2 \Theta_o(s) = T(s),$$

with the resulting transfer function

$$\frac{\Theta_o(s)}{T(s)} = \frac{1}{Js^2}.$$

Suppose it is necessary to modify or control the response but the value of J is fixed. Then the applied torque $T(t)$ must be controlled as a function of response. Initially use a proportional controller with transfer function gain K_p, as shown in Figure 6.11.

Derive the closed–loop transfer function in the following way. Begin with the relation between $\theta_o(t)$ and $T(t)$: $J\ddot{\theta}_o = T(t)$, where $T(t) = K_p E(t)$

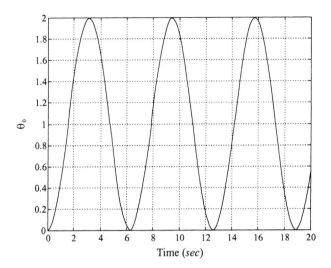

Figure 6.12: Step response with proportional controller.

from the system diagram. $E(t)$ is the *error* and is given by the difference between input and output signals. In the transformed domain,

$$T(s) = K_p E(s) = K_p(\Theta_i(s) - \Theta_o(s)).$$

Therefore,[4]

$$Js^2\Theta_o(s) = K_p(\Theta_i(s) - \Theta_o(s))$$
$$(Js^2 + K_p)\Theta_o(s) = K_p\Theta_i(s). \quad (6.4)$$

The transfer function is then

$$\frac{\Theta_o(s)}{\Theta_i(s)} = \frac{K_p}{Js^2 + K_p},$$

where the proportional gain acts as a stiffness element in the characteristic equation. The roots of the characteristic equation, $Js^2 + K_p = 0$, are imaginary and the response to a unit step loading will continue to oscillate without decay, as shown in Figure 6.12.

While this response does not grow, neither does it decay, as is necessary for most control applications. Examine the effect of adding a derivative controller to the proportional controller.

---

[4] In the time domain, Equation 6.4 is

$$J\ddot{\theta}_o + K_p\theta_o = K_p\theta_i.$$

6.4. AUTOMATIC CONTROL OF TRANSIENT RESPONSE

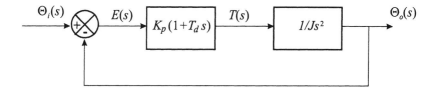

Figure 6.13: Block diagram for rotational element with PD controller.

The block diagram representation becomes that of Figure 6.13. Here,

$$\begin{aligned} T(s) &= K_p(1+T_d s)E(s) \\ &= K_p(1+T_d s)(\Theta_i(s) - \Theta_o(s)), \end{aligned}$$

and, in conjunction with the relation $T(s) = Js^2\Theta_o(s)$, we have[5]

$$(Js^2 + K_p T_d s + K_p)\Theta_o(s) = K_p(1+T_d s)\Theta_i(s), \qquad (6.5)$$

where the transfer function is given by

$$\frac{\Theta_o(s)}{\Theta_i(s)} = \frac{K_p(1+T_d s)}{Js^2 + K_p T_d s + K_p}, \qquad (6.6)$$

with characteristic equation $Js^2 + K_p T_d s + K_p = 0$. There are two roots with negative real parts because J, K_p, and T_d are positive quantities. Thus, derivative control action introduces the additional term in the characteristic equation, $K_p T_d$, which is a damping effect. The effective damping coefficient $c = K_p T_d$ and effective stiffness coefficient $k = K_p$ can now be selected according to given design criteria. The response curve is of the form of Figure 6.14.

In a similar way, if *PID* control is applied to the uncontrolled second order system with external torque $T_{ext}(t)$, governed by the equation

$$J\ddot{\theta}_o + c\dot{\theta}_o + k\theta_o = T_{ext}(t),$$

then the transfer function Equation 6.6 becomes

$$\frac{\Theta_o(s)}{\Theta_i(s)} = \frac{K_p(1+T_d s) + T_{ext}(s)}{Js^2 + (c + K_p T_d)s + (k + K_p)}. \qquad (6.7)$$

---

[5]In the time domain, Equation 6.5 becomes
$$J\ddot{\theta}_o + K_p T_d \dot{\theta}_o + K_p \theta_o = K_p \theta_i + K_p T_d \dot{\theta}_i.$$
This is the same equation as for the base excited structure of Chapter 3.

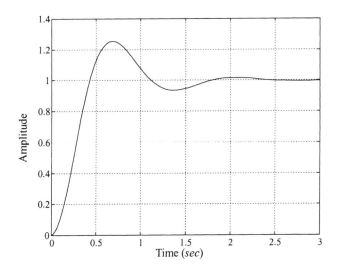

Figure 6.14: Step response with PD controller.

The block diagram representation is given in Figure 6.15. In the time domain, Equation 6.7 becomes

$$J\ddot{\theta}_o + (c + K_p T_d)\dot{\theta}_o + (k + K_p)\theta_o = K_p \theta_i + K_p T_d \dot{\theta}_i + T_{ext}(t),$$

where the forces on the right hand side are due to control actions and external torque. The designer needs to adjust K_p and T_d so that the response is appropriate to the application at hand.

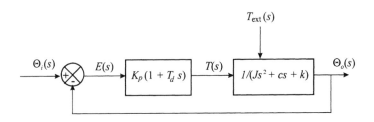

Figure 6.15: Block diagram for second-order system with PD control and external torque.

6.5 Sensitivity to Parameter Variations

We are aware from Chapters 4 and 5 that it is common and expected that there will be uncertainties in the models used to represent physical systems. One way to understand and model such uncertainties is to use the probabilistic approach of Chapter 5, as for random loadings. Another way is to examine the *sensitivity of the system response to parameter variations*. The question that needs to be answered is: *How does the structural response change as a result of a change in a particular parameter value?*[6] An understanding of the sensitivity of system response to parameter variation is useful to the control design process. Parameter sensitivity is also valuable in determining which additional tests must be carried out to better understand system characteristics.

For linear systems, the transfer function and the frequency response function, see Equation 3.23, define the system. As we know, the frequency response function can be obtained from the expression for the transfer function by replacing the complex variable s in the Laplace transform by $i\omega$.

The frequency response function, as well as the transfer function, are functions of the parameters that define them. For the structures we study, these parameters are the stiffness, damping, and mass properties, or in other terms, the natural frequency ω_n and the viscous damping factor ζ. The accuracy of the frequency response and transfer functions depend directly on the accuracy of these physical parameter values. The *nominal* values are those used in a calculation. How the frequency response function varies as a function of deviations of the parameter values from nominal is called the *sensitivity* of the system.

The frequency response can be defined by its magnitude and phase angle, as plotted against the frequency ratio ω/ω_n. Equations 3.24–3.26 are reproduced here;

$$x(t) = A|H(i\omega)|e^{i(\omega t - \phi)}$$

$$|H(i\omega)| = \frac{1/\omega_n^2}{\sqrt{(1 - \omega^2/\omega_n^2)^2 + (2\zeta\omega/\omega_n)^2}}$$

$$\phi = \tan^{-1}\frac{2\zeta\omega/\omega_n}{1 - \omega^2/\omega_n^2}.$$

The graphs that these functions represent are often determined experimentally, since the above parameters may not be known, but the experiment

---

[6] Should you need to find out much more about sensitivities, especially regarding how they relate to control, look up the very detailed book by P.M. Frank, **Introduction to System Sensitivity Theory**, Academic Press, 1978.

generally does not provide us with enough data for exact graphs. In principle, an infinite number of data values are required for an exact representation over all frequencies.

The accuracy of the model then depends on how closely the amplitude and phase graphs approximate the actual functions. The sensitivity of the system is in this case a measure of the amount by which its frequency response function differs from its nominal value when a parameter differs from its nominal value.

One measure of sensitivity is given by the ratio of the percent change in transfer function to percent change in parameter value. Suppose transfer function T changes in some sense by amount ΔT due to the change Δp in parameter p. The *sensitivity* is defined as

$$S = \frac{\Delta T/T}{\Delta p/p} = \frac{\Delta T}{\Delta p}\frac{p}{T}.$$

By definition, the *sensitivity function* is evaluated in the limit as $\Delta p \to 0$, or

$$\begin{aligned}S_p^T &= \lim_{\Delta p \to 0} \frac{\Delta T}{\Delta p}\frac{p}{T} \\ &= \frac{\partial T}{\partial p}\frac{p}{T},\end{aligned} \quad (6.8)$$

where T is a function of Laplace transform variable s and parameter p. It is usually easier to interpret the sensitivities of the frequency response function than of the transfer function.

Let the frequency response function be written as $H(p)$, where p may be any of the parameters ω_n, ω or ζ. There are three possible sensitivity functions, two if the ratio ω/ω_n is taken as a single parameter. The sensitivity function for the frequency response function is then

$$S_p^{H(p)} = \frac{\partial H(p)}{\partial p}\frac{p}{H(p)}. \quad (6.9)$$

Note that the sensitivity function can also be written as

$$S_p^{H(p)} = \frac{d[\ln H(p)]}{d[\ln p]}.$$

Following Equation 6.9, the sensitivity functions for the magnitude and phase angle of the frequency response functions are, respectively,

$$\begin{aligned}S_p^{|H(p)|} &= \frac{\partial |H(p)|}{\partial p}\frac{p}{|H(p)|} \\ S_p^{\phi(p)} &= \frac{\partial \phi(p)}{\partial p}\frac{p}{\phi(p)}.\end{aligned}$$

6.5. SENSITIVITY TO PARAMETER VARIATIONS

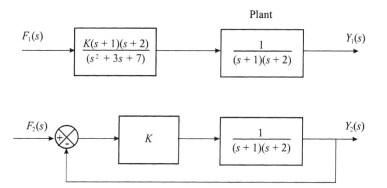

Figure 6.16: Two block diagrams.

These two sensitivity functions can be related by the sum

$$S_p^{H(p)} = S_p^{|H(p)|} + i\phi(p) S_p^{\phi(p)}, \tag{6.10}$$

where in general the terms on the right–hand side are complex numbers. If parameter p is real, as it is in our problems, then the terms in Equation 6.10 are real numbers.

Example 6.3 Sensitivity Analysis as Part of Control Design
To examine how sensitivity can play a role in control systems design, consider the two system block diagrams in Figure 6.16.

The respective transfer functions are

$$T_1(s) = \frac{Y_1(s)}{F_1(s)} = \frac{K}{s^2 + 3s + 7}$$

$$T_2(s) = \frac{Y_2(s)}{F_2(s)} = \frac{K}{(s+1)(s+2) + K}.$$

For gain value $K = 5$, $T_1(s) = T_2(s)$.

To determine the sensitivity of each transfer function, let us look first at a general transfer function that includes the above as special cases:

$$T(s) = \frac{C_1(s) + pC_2(s)}{C_3(s) + pC_4(s)},$$

where $C_i(s)$ are polynomials in s, and p is any parameter of interest, as before. Equation 6.8 is applied to the general transfer function, resulting in

$$S_p^T = \frac{p(C_2 C_3 - C_1 C_4)}{(C_3 + pC_4)(C_1 + pC_2)}.$$

Going back to the particular problem for $T_1(s)$, we have $p = K$, $C_1 = C_4 = 0$, $C_2 = 1$ and $C_3 = s^2 + 4s + 7$. Then, $S_K^{T_1} = 1$, a constant regardless of the value of gain K. For $T_2(s)$, $p = K$, $C_1 = 0$, $C_2 = C_4 = 1$ and $C_3 = s^2 + 3s + 2$. The sensitivity function for $T_2(s)$ is then

$$S_K^{T_2} = \frac{K(s^2 + 3s + 2)}{(s^2 + 3s + 2 + K)K}$$
$$= \frac{1}{1 + K/(s^2 + 3s + 2)}.$$

Here, for the closed loop system, the sensitivity is a function of gain K and the complex variable s. Thus, to reduce sensitivity to parameter variation, a designer would select an appropriate gain or maintain the frequencies of the input within an appropriate range. Since the sensitivity is a function of complex variable s, the operation frequency range of the application is important. ∎

6.6 State Variable Models

The *state* of a system is the minimum number of variables that define the current configuration and, with the governing dynamic equations, determine the future behavior of the system. The state of a system is described by the set of *first order differential equations* written in terms of the state variables. It is a framework for which there exists a theory and computational procedures. Therefore, whether the model consists of two first–order equations or one hundred first–order equations, the same matrix format of first–order equations is representative of the system. Such models are viewed as a unifying framework for numerical solution, and many numerical integration algorithms are based on solving systems of first order equations. The same concept is applied to multi degree of freedom structural systems in Chapters 8 and 9, where matrix methods apply regardless of the number of degrees of freedom.

In this section, the state space model is introduced by way of example and applied to single degree of freedom oscillators with control forces.

Example 6.4 State Variables for an Oscillator
The equation governing the oscillation of a single degree of freedom system is $m\ddot{y} + c\dot{y} + ky = F(t)$. To write this in state variable formalism, the state variables need to be defined. These are the *position* and *velocity* of the oscillator. A second–order equation can be written as two first order

6.6. STATE VARIABLE MODELS

equations. Define

$$\begin{aligned} x_1(t) &= y(t) \\ x_2(t) &= \frac{dy}{dt} = \dot{y}. \end{aligned}$$

With these definitions, the equation of motion becomes

$$\dot{x}_2 + \frac{c}{m}x_2 + \frac{k}{m}x_1 = \frac{F(t)}{m}.$$

The two first-order equations will be for $\dot{x}_1$ and $\dot{x}_2$,

$$\dot{x}_1 = \dot{y} = x_2 \tag{6.11}$$

$$\dot{x}_2 = \ddot{y} = -\frac{c}{m}x_2 - \frac{k}{m}x_1 + \frac{F(t)}{m}, \tag{6.12}$$

or in matrix form

$$\frac{d}{dt}\begin{Bmatrix} x_1 \\ x_2 \end{Bmatrix} = \begin{bmatrix} 0 & 1 \\ -k/m & -c/m \end{bmatrix}\begin{Bmatrix} x_1 \\ x_2 \end{Bmatrix} + \begin{Bmatrix} 0 \\ 1/m \end{Bmatrix}F(t). \tag{6.13}$$

The solution to this equation is then used to solve for $y(t)$,

$$y = \begin{bmatrix} 1 & 0 \end{bmatrix}\begin{Bmatrix} x_1 \\ x_2 \end{Bmatrix},$$

and for the structural velocity,

$$\dot{y} = \begin{bmatrix} 0 & 1 \end{bmatrix}\begin{Bmatrix} x_1 \\ x_2 \end{Bmatrix}.$$

Equation 6.13 is generalized below. ∎

Example 6.5 Base Excitation in State Space
A structure loaded at its base through stiffness and damping elements is governed by the equation

$$\ddot{y} + 2\zeta\omega_n\dot{y} + \omega_n^2 y = 2\zeta\omega_n\dot{F} + \omega_n^2 F,$$

where F is the base displacement and $\dot{F}$ is the base velocity. We are not able to follow the previous procedure to transform the second order equation into two first order equations because of the derivative term on the right hand side. To get around this, introduce the following change of variables,

$$x_1 = y$$
$$x_2 = \dot{x}_1 - 2\zeta\omega_n F.$$

This transformation works in the following way: substitute for y in the governing equation,

$$\ddot{x}_1 + 2\zeta\omega_n \dot{x}_1 + \omega_n^2 x_1 = 2\zeta\omega_n \dot{F} + \omega_n^2 F.$$

Next, replace $\dot{x}_1$ by $x_2 + 2\zeta\omega_n F$ and $\ddot{x}_1$ by $\dot{x}_2 + 2\zeta\omega_n \dot{F}$ to find

$$\dot{x}_2 + 2\zeta\omega_n \dot{F} + 2\zeta\omega_n(x_2 + 2\zeta\omega_n F) + \omega_n^2 x_1 = 2\zeta\omega_n \dot{F} + \omega_n^2 F,$$

where this change of variables leads to the $\dot{F}$ term canceling. Simplifying and solving for $\dot{x}_2$ results in

$$\dot{x}_2 = -\omega_n^2 x_1 - 2\zeta\omega_n x_2 + (\omega_n^2 - [2\zeta\omega_n]^2)F.$$

In matrix form, the two first–order state equations are written as,

$$\begin{Bmatrix} \dot{x}_1 \\ \dot{x}_2 \end{Bmatrix} = \begin{bmatrix} 0 & 1 \\ -\omega_n^2 & -2\zeta\omega_n \end{bmatrix} \begin{Bmatrix} x_1 \\ x_2 \end{Bmatrix} + \begin{Bmatrix} 2\zeta\omega_n \\ \omega_n^2 - [2\zeta\omega_n]^2 \end{Bmatrix} F(t)$$

with

$$y = \begin{bmatrix} 1 & 0 \end{bmatrix} \begin{Bmatrix} x_1 \\ x_2 \end{Bmatrix}.$$

State variable x_1 is the output signal that can be measured. The above procedure can be extended to cases where higher order derivatives appear on the right hand side. ∎

The notation of Equation 6.13 can be generalized for any number of first order equations,

$$\{\dot{x}\} = [A]\{x\} + \{B\}F(t),$$

with

$$y = \{C\}\{x\},$$

where $\{B\}$ is a column vector and $\{C\}$ is a row vector. The first question that comes to mind is how to solve this matrix set of first order equations. We have a hint by recalling the solution of the single first–order equation,

$$\dot{x} = ax + bF.$$

6.6. STATE VARIABLE MODELS

Take the Laplace transform of both sides to find
$$sX(s) - x(0) = aX(s) + bF(s).$$
The transform of the solution is then
$$X(s) = \frac{x(0)}{s-a} + \frac{b}{s-a}F(s),$$
and the inverse transform is obtained by using *Borel's theorem*, resulting in
$$x(t) = x(0)e^{at} + \int_0^t e^{a(t-\tau)}bF(\tau)d\tau. \tag{6.14}$$
For matrix first-order equations, we need the definition for the exponential of matrix $[A]$,
$$e^{[A]t} = [I] + [A]t + \frac{[A]^2 t^2}{2!} + \cdots + \frac{[A]^n t^n}{n!} + \cdots,$$
which converges for all finite t and any $[A]$. Therefore, the matrix equivalent to Equation 6.14 is
$$\{x(t)\} = \exp\left([A]t\right)\{x(0)\} + \int_0^t \exp\left([A](t-\tau)\right)\{B\}F(\tau)d\tau. \tag{6.15}$$
The matrix exponential function describes the unforced response of the system and is called the *state transition matrix*,
$$\Phi(t) = \exp\left([A]t\right),$$
and Equation 6.15 can be written in terms of $\Phi(t)$,
$$\{x(t)\} = \Phi(t)\{x(0)\} + \int_0^t \Phi(t-\tau)\{B\}F(\tau)d\tau,$$
where the unforced response is found by setting $F(\tau) = 0$,
$$\{x(t)\} = \Phi(t)\{x(0)\}.$$

For the second order oscillator defined by Equations 6.11 and 6.12, the response is given by
$$\left\{ \begin{array}{c} x_1(t) \\ x_2(t) \end{array} \right\} = \exp\left(\left[\begin{array}{cc} 0 & 1 \\ -k/m & -c/m \end{array} \right]t\right) \left\{ \begin{array}{c} x_1(0) \\ x_2(0) \end{array} \right\}$$
$$+ \int_0^t \exp\left(\left[\begin{array}{cc} 0 & 1 \\ -k/m & -c/m \end{array} \right](t-\tau)\right) \left\{ \begin{array}{c} 0 \\ 1/m \end{array} \right\} F(\tau)d\tau.$$

Equations in state space format can be solved numerically by commercially available programs such as *MATLAB*.

6.6.1 Matrix Derivatives and Integrals

In the previous section, we saw the integration of a matrix. The study of larger scale systems requires the differentiation and integration of matrices. This means the differentiation or integration of all the elements of a matrix. The *time derivative* of matrix $\mathbf{A}(t)$ is given by the matrix whose elements are the respective time derivatives of the original matrix,[7]

$$\frac{d}{dt}\mathbf{A}(t) = \left(\frac{d}{dt}a_{ij}(t)\right),$$

where each element is differentiated. Implicit is the assumption that all elements have derivatives.

Similarly, the *time integral* of matrix $\mathbf{A}(t)$ is given by the matrix whose elements are the respective time integrals of the original matrix,

$$\int \mathbf{A}(t)dt = \left(\int a_{ij}(t)dt\right).$$

Some rules of matrix differentiation follow from the scalar rules, but, as we know, multiplication of matrices (or of their derivatives) is not commutative,

$$\frac{d}{dt}(\mathbf{AB}) = \frac{d\mathbf{A}}{dt}\mathbf{B} + \mathbf{A}\frac{d\mathbf{B}}{dt}$$

$$\frac{d}{dt}(\mathbf{A} + \mathbf{B}) = \frac{d\mathbf{A}}{dt} + \frac{d\mathbf{B}}{dt}$$

$$\frac{d}{dt}(\mathbf{A}\alpha) = \frac{d\mathbf{A}}{dt}\alpha + \mathbf{A}\frac{d\alpha}{dt}.$$

A final rule of importance is the time derivative of the inverse of matrix $\mathbf{A}$. The rule is derived in the following steps:

$$\frac{d}{dt}(\mathbf{A}\mathbf{A}^{-1}) = \frac{d\mathbf{A}}{dt}\mathbf{A}^{-1} + \mathbf{A}\frac{d\mathbf{A}^{-1}}{dt} = \frac{d}{dt}\mathbf{I} = 0$$

$$\frac{d\mathbf{A}}{dt}\mathbf{A}^{-1} = -\mathbf{A}\frac{d\mathbf{A}^{-1}}{dt}.$$

Therefore,

$$\frac{d\mathbf{A}^{-1}}{dt} = -\mathbf{A}^{-1}\frac{d\mathbf{A}}{dt}\mathbf{A}^{-1},$$

which we notice is the matrix counterpart of the scalar derivative,

$$\frac{d}{dt}(x^{-1}) = -x^{-2}\frac{dx}{dt}.$$

---

[7] Matrices are shown in bold face in this section for clarity.

6.7 Concepts Summary

This brief chapter is meant to be an introduction to some of the concepts of structural control. As such, the reader has been introduced to the *connections* between vibration and its control. Some basic concepts regarding structural control have been discussed, the most important of these being the concepts of feedback, sensitivity, and stability, to varying degrees of detail.

All of the discussion is limited to the time domain, even though a large portion of control analysis and design is accomplished in the frequency domain. However, within the context of a vibration text, the sense about control that we would like to impart is best done in the time domain and focuses on structural response.

6.8 Problems

Problems for Section 6.1 – Introduction

1. In each of the following systems, is there a control component(s)? If so, is it engineered, natural or biological, or hybrid? Describe qualitatively how the control system affects system behavior: *(i)* airplane wing, *(ii)* pen, *(iii)* elevator, *(iv)* standard transmission in car, *(v)* automatic transmission in car, *(vi)* eye, *(vii)* marathon runner, *(viii)* hand, *(ix)* hand holding a pen and writing, *(x)* computer.

   Problems for Section 6.2 – Feedback Control

2. Derive the transfer functions for the following governing equations of motion:
   (i) $\ddot{x} + 2\dot{x} + 4x = \cos \omega t$
   (ii) $m\ddot{x} + c\dot{x} + kx = A \cos \omega t$
   (iii) $m\ddot{x} + c\dot{x} + kx = A_1 \cos \omega t + A_2 \sin \omega t$
   (iv) $m\ddot{x} + kx = u(t)$, $u(t) = 1$, $t \geq 0$
   (v) $m\ddot{x} + c\dot{x} + kx = e^{-\omega t} \cos \Omega t$.

3. For each of the equations in the previous problem, discuss how appropriate choices of m, c and k can be made to either maximize or minimize the effect of the transfer function. Can you think of applications where either maximum or minimum effects are desirable?

   Problems for Section 6.3 – Performance of Feedback Control Systems

4. For the step response $x_s(t)$ governed by the equation

$$\ddot{x}_s + 2\zeta\omega_n \dot{x}_s + \omega_n^2 x_s = u(t),$$

where $u(t)$ is the unit step function, evaluate rise time t_r, peak time t_p, overshoot, and settling time where $\delta = 0.02$ for the following cases:

(i) $\zeta = 0.01$, $\omega_n = 1$ rad/sec

(ii) $\zeta = 0.01$, $\omega_n = 2$ rad/sec

(iii) $\zeta = 0.01$, $\omega_n = 0.5$ rad/sec

(iv) $\zeta = 0.1$, $\omega_n = 1$ rad/sec

(v) $\zeta = 0.1$, $\omega_n = 2$ rad/sec

(vi) $\zeta = 0.1$, $\omega_n = 0.5$ rad/sec

(vii) $\zeta = 0.5$, $\omega_n = 1$ rad/sec

(viii) $\zeta = 0.5$, $\omega_n = 2$ rad/sec

(ix) $\zeta = 0.5$, $\omega_n = 0.5$ rad/sec.

5. For the equation of motion $m\ddot{x} + c\dot{x} + kx = F(t)$, with $m = 9$ kg, $c = 4$ Nsec/m, $k = 4$ N/m and $F(t)$ is the unit step load, find the response. If the response is oscillatory, determine the possible modifications that need to be implemented to make the system critically damped. Plot both original and modified step response.

6. What applications exist where swiftness of response is more important than accuracy of response? Are there applications where the opposite is true? For what applications are both of critical importance?

7. What is the maximum value of the step response as a function of ζ.

8. Find the poles and zeros for the following second order systems, and discuss system stability:

(i) $\ddot{y} + 2\dot{y} + y = A\cos 3t$,

(ii) $\ddot{y} + \dot{y} + 0.1y = A\cos 3t$,

(iii) $\ddot{y} + 300\dot{y} + 10y = A\cos 3t$,

(iv) $\ddot{y} + 2\dot{y} + y = A_1 x + A_2 \dot{x}$,

(v) $\ddot{y} + y = A\cos\omega t$,

(vi) $\ddot{y} + y = B[x(t) + \dot{y}]$.

6.8. PROBLEMS

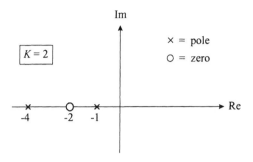

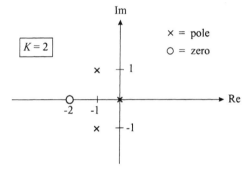

Figure 6.17: Pole–zero maps.

9. For the given transfer functions, find the governing differential equation of motion:
   (i) $T(s) = (2s + 2)/(s^2 + 2s + 5)$
   (ii) $T(s) = (s - 1)/(3s^2 + 4)$
   (iii) $T(s) = (s + 1)/(s^2 + s + 1)$.

10. For a gain factor $K = 2$, use the pole–zero maps in Figure 6.17 to find each transfer function.

11. For each of Figures 6.18(a)–(d), explain why the response appears stable or unstable.

    Problems for Section 6.4 – Automatic Control of Transient Response

12. For the transfer function
$$\frac{\Theta_o(s)}{\Theta_i(s)} = \frac{K_p}{Js^2 + K_p},$$

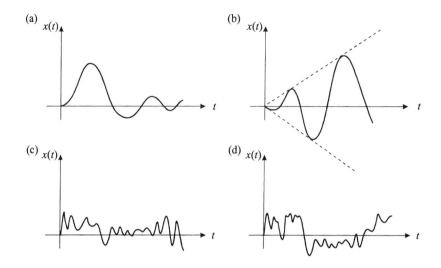

Figure 6.18: Four possible time-histories.

solve for $\theta_o(t)$ in terms of J, K_p and $\theta_i(t)$. How do variations in the values of parameters J and K_p affect the behavior of $\theta_o(t)$? Can a bounded $\theta_i(t)$ be selected to destabilize response $\theta_o(t)$? Discuss and show this.

13. For a rotating element with PD controller, the transfer function is given by Equation 6.6, repeated here,

$$\frac{\Theta_o(s)}{\Theta_i(s)} = \frac{K_p(1+T_d s)}{J s^2 + K_p T_d s + K_p}.$$

Select control parameters K_p and T_d to maximize the rate of the decay to zero of a step response. What is the effect of parameter J on the response?

Problems for Section 6.5 – Sensitivity to Parameter Variations

14. A system transfer function is given by

$$T(s) = \frac{Ks}{Ks^2 + 1}.$$

Derive the sensitivity function and determine the value(s) of K that minimize sensitivity. Plot $T(s)$ for $K = 1, 10, 100$ on one set of axes and draw conclusions from the comparison. Plot the sensitivity function for each of the K values, also on one set of axes. Which value of K reduces sensitivity?

15. Repeat *Problem* 15 for the transfer function

$$T(s) = \frac{Ks^2}{Ks^2 + s + 1}.$$

16. For the general transfer function

$$T(s) = \frac{C_1(s) + pC_2(s)}{C_3(s) + pC_4(s) + p^2 C_5(s)},$$

where $C_i(s)$ are polynomials in s, find the general sensitivity as a function of parameter p.

6.9 Mini–Projects

The following papers provide an opportunity for the student to examine and understand some of the current concerns of the control community. Control theory is, however, an extremely broad discipline, ranging from simple design considerations to the very mathematical and theoretical concerns. We have limited ourselves to the more engineering and applied topics. Additional control related papers are referred to in the *Mini–Project* sections of the subsequent chapters.

1. *New Concepts in Control Theory*, A.E. Bryson, Jr., *J. Guidance, Control and Dynamics*, Vol. 8, No. 4, July–Aug. 1985, 417–425.

   Control theory is a branch of applied mathematics that deals with the analysis and synthesis of logic for the control of man–made systems. It is applied to a wide range of fields from aerospace to robotics and economics. Its applications to aerospace can be divided into four areas: flight planning, navigation, guidance, and control. These four categories often overlap and are discussed here.

2. *Control of Wind–Induced Instabilities Through Application of Nutation Dampers: A Brief Overview*, V.J. Modi, F. Welt, M.L. Seto, *Engineering Structures*, Vol. 17, No. 9, 626–638, 1995.

   Developments in the area of vibration suppression through application of nutation dampers are briefly reviewed. The focus is on industrial aerodynamics problems involving vortex–induced and galloping–type instabilities.

3. *A Method of Vibration Control of a Magnetic Coupling Carrying a Rotor in a Fluid with Large System Variations*, K. Nagaya, T. Aiba, *J. Sound and Vibration* (1995) **183**(3), 435–450.

   This study is concerned with a magnetic coupling in which a shielding wall exists between coupling disks. Analyses are presented of this coupling and a method of control for accelerating a shaft with a magnetic coupling carrying a rotor in a fluid.

4. *Microgravity Isolation System Design: A Case Study*, R.D. Hampton, C.R. Knospe, C.M. Grodsinsky, *J. Spacecraft and Rockets*, Vol. 33, No. 1, Jan.–Feb. 1996, 120–125.

   Many acceleration–sensitive, microgravity science experiments will require active vibration isolation from manned orbiters on which they will be mounted. The isolation problem, especially in the case of a tethered payload, is a complex three–dimensional one that is best suited to modern control design methods.

5. *Parametric and Classical Resonance in Passive Satellite Aerostabilization*, R.R. Kumar, D.D. Mazanek, M.L. Heck, *J. Spacecraft and Rockets*, Vol. 33, No. 2, Mar.–Apr. 1996, 228–234.

   Purely passive aerostabilization of satellites has never been flight demonstrated. The Shuttle hitchhiker passive aerodynamically stabilized magnetically damped satellite experiment would be the first flight experiment of its kind that, in conjunction with results from a high–fidelity computer simulator, would corroborate attitude stability. Discussion with analysis is part of the paper.

6. *Acceleration Feedback Control of MDOF Structures*, S.J. Dyke, B.F. Spencer, Jr., P. Quast, M.K. Sain, D.C. Kaspari, Jr., T.T. Soong, *J. of Engineering Mechanics*, Vol. 122, No. 9, Sept. 1996, 907–918.

   Current practice in structural control is primarily based on displacement and velocity feedback. But accelerometers, which are reliable and inexpensive, may provide a better way to control using acceleration feedback. The present paper experimentally demonstrates that acceleration feedback control strategies are effective and robust, and they can achieve performance levels comparable to full–state (displacement and velocity) feedback controllers.

7. *Passive and Active Mass Damper Control of the Response of Tall Buildings to Wind Gustiness*, F. Ricciardelli, A.D. Pizzimenti, M. Mattei, *Engineering Structures*, 25 (2003) 1199-1209.

6.9. MINI-PROJECTS

The performance of passive, active and hybrid mass dampes for the reduction of the buffeting response of tal buildings is investigated. Simple and complex structures are considered.

8. *H_2 Active Vibration Control for Offshore Platform Subjected to Wave Loading*, H.J. Li, S-L.J. Hu, C. Jakubiak, *J. of Sound and Vibration* 263 (2003) 709–724.

   Offshore platforms are usually located in hostile environments. These platforms undergo excessive vibrations due to wave loads for both normal operating and extreme conditions. To ensure safety, the displacements of the platforms need to be limited, whereas for the comfort of people who work at the structures, accelerations also need to be restricted. This article is devoted to developing a proper procedure on applying H_2 control algorithms for controlling the lateral vibration of a jacket–type offshore drilling platforms by using an active mass damper.

9. *Nonlinear Control of Torsional and Bending Vibrations of Oilwell Drill strings*, S.S. Al–Hiddabi, B. Samanta, A. Seibi, *J. Sound and Vibration* 265 (2003) 401–415.

   Drillstring dynamics is highly nonlinear in nature and its model can only be described by a set of nonlinear differential equations. In this paper a nonlinear dynamic inversion control design method is used to suppress the lateral and the torsional vibrations of a nonlinear drillstring.

10. *Passive Control of the Seismic Rocking Response of Art Objects*, I. Caliò, M. Marletta, *Engineering Structures* 25 (2003) 1009–1018.

    The paper deals with the passive control of the vibrations of art objects subjected to base excitations. The art object is modelled as a rigid bloc simply supported on a pedestal which is connected to a viscoelastic device in order to obtain a passive control system.

Chapter 7

Variational Principles

"Energy is the basis."

7.1 Introduction

This chapter presents several of the most important concepts from analytical dynamics. The most important of these concepts to us is *Lagrange's equation* and how it can be used for the derivation of governing equations of motion. We have already been introduced to Lagrange's equation in Chapter 4, and derive it in this chapter. The equation is especially useful for the derivation of the equations of motion for systems, discrete or continuous, with more than one degree of freedom, where the Newtonian free body diagrams become more difficult to apply. We will also derive *Hamilton's principle*, an integral energy formulation, also applicable to both discrete and continuous systems, and see how it is related to Lagrange's equation.[1]

The basis of this chapter is the *principle of virtual work*. There are many advantages to the *analytical approach* of Lagrange and Hamilton over Newton's force analysis. This is especially true for systems of interacting bodies, where each exerts a force on the other and where constraints, such as boundaries, also exert forces on the system, limiting motion. Such *auxiliary conditions* can be more easily handled using the analytical approach.

---

[1] The interested reader should consider looking up two excellent books that are completely dedicated to analytical dynamics. They are important works and are highly recommended. The first and more recent is **Methods of Analytical Dynamics** by L. Meirovitch, McGraw-Hill, 1970. The second is **The Variational Principles of Mechanics** by C. Lanczos, Dover, 1986. This book is rigorous, yet the reader feels as though a private conversation is taking place with the author. Much emphasis is placed on physical understanding. It was originally published in 1949.

The analytical approaches are based on *variational principles*, which are the unifying basis of the equations that follow. The term variational is from the *calculus of variations*,[2] the foundation for such techniques. An important advantage of the analytical method is that the equations of motion are coordinate–independent. Newton's second law of motion is not.

To motivate and explain the procedure, consider the simple function: $y = f(x)$. The variable y can represent a displacement curve of a cable or beam. The variational approach is based on comparing the function $f(x)$ with a slightly modified function $f_\epsilon(x) = f(x) + \epsilon\phi(x)$, where ϵ can be as small as necessary, including zero, and ϕ must be continuous and differentiable. For any value of the independent variable x, the *variation*, or difference is defined as

$$\delta y \equiv f_\epsilon(x) - f(x) = \epsilon\phi(x).$$

There are two fundamental points to be emphasized here: *(i)* the variation is *arbitrary* or *virtual*, and *(ii)* it is an infinitesimal change since ϵ can be made arbitrarily small. Note that while both δy and dy represent infinitesimal changes in the function $f(x)$, dy refers to a change in $f(x)$ caused by an infinitesimal change of the independent variable dx, while δy is an infinitesimal change of y that results in a *new function* $y + \delta y$.

This process of variation is for each fixed value of x. Therefore, x is not varied, meaning that $\delta x = 0$, and the two end points of this function are prescribed and therefore also not varied. The variation is between definite limits. When we work with time as the independent variable, the beginning and ending times are prescribed and therefore not varied.

As we will discover below, in applying the variational procedures to a particular system, in addition to finding the governing the equation of motion, the necessary number of boundary conditions is also derived. *The stationary value conditions imposed by the variational principles result in both the differential equations and the boundary conditions.*

Before proceeding with the details of the analytical techniques, it is useful to summarize the key topics to be examined in this chapter:

- The *principle of virtual work* is introduced along with its relation to the equilibrium of a body.

- The principle of virtual work, in conjunction with *d'Alembert's* principle, is extended to include dynamic systems.

---

[2] The best introduction that I have read on the calculus of variations, especially for those interested in applications, is by F.B. Hildebrand in his **Methods of Applied Mathematics**, Second Edition, Prentice–Hall 1965. It is currently available from Dover Publications, New York. An excellent but more advanced introduction is given by I.M. Gelfand and S.V. Fomin, **Calculus of Variations**, Prentice–Hall, 1963.

7.2. VIRTUAL WORK

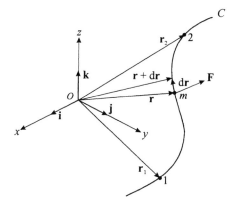

Figure 7.1: Path of mass m due to force $\mathbf{F}$.

- Lagrange's equation and Hamilton's variational principle are then derived from d'Alembert's principle, and will be applied to several problems of interest.

Many examples are presented in this chapter on the application of these principles. In subsequent chapters, further applications will be demonstrated for multi–degree of freedom systems in Chapters 8 and 9, and for continuous systems in Chapters 10 and 11. The notation used in this chapter is standard.

7.2 Virtual Work

The principle of virtual work is the basis for the remainder of this chapter and also forms the foundation for the variational principles of mechanics. Some of the most powerful computational models are based on a variational approach.

7.2.1 Work and Energy

The concepts of work and energy are reviewed before proceeding to *virtual work*. Consider a particle of mass m moving along a curve C under the action of a force $\mathbf{F}$ as shown in Figure 7.1. In this chapter, we follow the custom in dynamics of showing vectors as bold face variables.

The position of the particle with respect to an origin O is given by the vector $\mathbf{r}$, which is a function of time. The work necessary to move the mass a distance $d\mathbf{r}$ is $dW = \mathbf{F} \cdot d\mathbf{r}$. The *work done* to move the particle from position $\mathbf{r}_1$ to position $\mathbf{r}_2$ is

$$W_{12} = \int_{\mathbf{r}_1}^{\mathbf{r}_2} \mathbf{F} \cdot d\mathbf{r}.$$

Assuming the mass of the particle to be constant, Newton's second law of motion can be written as

$$\mathbf{F} = m\frac{d\dot{\mathbf{r}}}{dt} = m\frac{d}{dt}\left(\frac{d\mathbf{r}}{dt}\right).$$

This equation is in a slightly different form than used in Chapter 2 to derive the equation of motion for an oscillator. The goal here is to connect force, work, and energy. Using $d\mathbf{r} = \dot{\mathbf{r}}dt$, and the above equations,

$$\begin{aligned}
W_{12} &= \int_{t_1}^{t_2} m\frac{d\dot{\mathbf{r}}}{dt} \cdot \dot{\mathbf{r}}\,dt \\
&= \frac{1}{2}\int_{t_1}^{t_2} m\frac{d}{dt}\left(\dot{\mathbf{r}} \cdot \dot{\mathbf{r}}\right)dt \\
&= \frac{1}{2}m[(\dot{\mathbf{r}}_2 \cdot \dot{\mathbf{r}}_2) - (\dot{\mathbf{r}}_1 \cdot \dot{\mathbf{r}}_1)] \\
&= \frac{1}{2}m(\dot{r}_2^2 - \dot{r}_1^2) \\
&= T_2 - T_1,
\end{aligned}$$

where the limits of integration have been transformed from $\mathbf{r}$ to t, and T is the kinetic energy of the mass, $T = \frac{1}{2}m\dot{\mathbf{r}} \cdot \dot{\mathbf{r}}$. As expected, we started with a scalar, the work, and ended with a scalar, the change in kinetic energy. The kinetic energy of a body is defined as the total work that must be done on the body to bring it from a state of rest to a velocity $\dot{\mathbf{r}}$. Thus, for $v = \dot{\mathbf{r}}$,

$$\begin{aligned}
T &= \int_0^v mv\,dv \\
&= \frac{1}{2}mv^2.
\end{aligned}$$

Next, the work done by the force is related to the respective change in position of the mass. To do this, define a *conservative force field* as one where the work done depends only on the *initial* and the *final* positions of the particle and is *independent* of the path connecting these positions. An example of a conservative force field is gravity. Nonconservative forces, such as friction and external forces, are *energy–dissipating*, and for these the work done is path–dependent.

7.2. VIRTUAL WORK

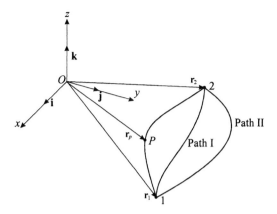

Figure 7.2: Path in a conservative force field.

From Figure 7.2, any path within the conservative force field which connects points 1 and 2 can be selected, and the work done bringing the particle from 1 to 2 will be the same, and is denoted by

$$W_{12c} = \underbrace{\int_{\mathbf{r}_1}^{\mathbf{r}_2} \mathbf{F} \cdot d\mathbf{r}}_{\text{Path I}} = \underbrace{\int_{\mathbf{r}_1}^{\mathbf{r}_2} \mathbf{F} \cdot d\mathbf{r}}_{\text{Path II}}.$$

The *potential energy* $V(\mathbf{r}_1)$ is associated with position $\mathbf{r}_1$ and is defined as the work done[3] by a conservative force moving a particle from position $\mathbf{r}_1$ to a *reference position* $\mathbf{r}_p$,

$$V(\mathbf{r}_1) = \int_{\mathbf{r}_1}^{\mathbf{r}_p} \mathbf{F} \cdot d\mathbf{r}.$$

Relate the work done moving a particle in a conservative force field to the potential energy of the particle. To do this, consider again W_{12c} but choose the arbitrary path through reference position $\mathbf{r}_p$, then

$$\begin{aligned} W_{12c} &= \int_{\mathbf{r}_1}^{\mathbf{r}_p} \mathbf{F} \cdot d\mathbf{r} + \int_{\mathbf{r}_p}^{\mathbf{r}_2} \mathbf{F} \cdot d\mathbf{r} \\ &= \int_{\mathbf{r}_1}^{\mathbf{r}_p} \mathbf{F} \cdot d\mathbf{r} - \int_{\mathbf{r}_2}^{\mathbf{r}_p} \mathbf{F} \cdot d\mathbf{r} \\ &= -[V(\mathbf{r}_2) - V(\mathbf{r}_1)] \\ &= -(V_2 - V_1). \end{aligned}$$

---

[3] Had the limits been interchanged, subsequent equations would be of opposite signs.

By this equation, *the work done in a conservative force field is the negative of the change in potential energy.* From vector calculus,[4] a conservative force equals the negative of the gradient of the potential energy function.

Finally, if we denote W_{12nc} as the nonconservative work, then

$$\begin{aligned} W_{12nc} &= W_{12} - W_{12c} \\ &= (T_2 - T_1) + (V_2 - V_1) \\ &= (T_2 + V_2) - (T_1 + V_1) \\ &= E_2 - E_1, \end{aligned}$$

where E_i denotes the total energy in state i. Therefore, W_{12nc} is a measure of the change in particle energy due to dissipation, and if $W_{12nc} = 0$ then $E_2 = E_1$. That is, the energy of the particle is constant and there is conservation of energy.

Gravitational potential energy is defined as the work (mgh) done against the gravitational field to elevate a body of mass m a distance h above an arbitrary reference plane (datum).

7.2.2 Virtual Work

The *Principle of Virtual Work* states that *the virtual work performed by the applied forces undergoing infinitesimal virtual displacements compatible with the system constraints is zero,*

$$\delta W = \sum_{i=1}^{N} \mathbf{F}_i \cdot \delta \mathbf{r}_i = 0. \qquad (7.1)$$

A *constraint* is a physical barrier to free motion, for example, a wall, a string connecting two bodies, or a magnetic field. Equation 7.1 applies to static systems, or quasi-static systems where inertia effects can be ignored. The dynamic version of the principle of virtual work, known as *d'Alembert's principle*, is developed in the next section. These principles form the basis for the variational principles that follow.

Our formulation is for a system of N particles moving in three dimensions. The results are applicable for discrete as well as continuous systems. *Virtual displacements* are defined in each of the three dimensions for each particle,

$$\delta x_i, \ \delta y_i, \ \delta z_i,$$

---

[4] For a very readable introduction to vector calculus, look up **Div, Grad, Curl and all that**, by H.M. Schey, Norton, 1973.

7.2. VIRTUAL WORK

where $1 \leq i \leq N$. *Virtual displacements may be interpreted as possible alternate configurations of the system of particles.* These alternate configurations must be consistent with the system constraints. We consider the system at its initial configuration and at its alternate configuration due to the virtual displacement. Time is not a variable here since we are only examining the system in two possible configurations. Time will be soon considered with d'Alembert's principle.

Example 7.1 Constraint Equation
Considering a mass on an inextensible string of length r, the position of the mass is given by the equation of the circle: $x^2 + y^2 = r^2$. This is a *geometrical constraint* relation between coordinates x, y and parameter r. It is not a dynamical relation. Time is not a parameter and therefore we do not know the relation between time and mass position in x, y space. ∎

Consider the initial configuration of the N masses along with the constraints on them via a *constraint equation*,

$$g(x_1, y_1, z_1, x_2, y_2, z_2, \ldots, x_N, y_N, z_N, t) = c, \tag{7.2}$$

and the alternate configuration resulting from a virtual displacement,

$$g(x_1 + \delta x_1, y_1 + \delta y_1, z_1 + \delta z_1, \ldots, z_N + \delta z_N, t) = c, \tag{7.3}$$

where parameter t is included to demonstrate that it is not varied. The system constraints are within the constant c

A goal is to examine in detail the rules that govern the variations in Equation 7.3. When completed, we will be able to relate the virtual work done by forces undergoing a virtual displacement. This is interpreted as a statement of static equilibrium.

Proceed by expanding Equation 7.3 about the *unvaried* path via a Taylor series representation. Only first order terms are retained,

$$g(x_1, y_1, z_1, \ldots, x_N, y_N, z_N, t) + \sum_{i=1}^{N} \left(\frac{\partial g}{\partial x_i} \delta x_i + \frac{\partial g}{\partial y_i} \delta y_i + \frac{\partial g}{\partial z_i} \delta z_i \right) = c. \tag{7.4}$$

We know from Equation 7.2 that $g() = c$, and upon substitution, Equation 7.4 yields the relations that must be satisfied so that the virtual displacements are compatible with system constraints,

$$\sum_{i=1}^{N} \left(\frac{\partial g}{\partial x_i} \delta x_i + \frac{\partial g}{\partial y_i} \delta y_i + \frac{\partial g}{\partial z_i} \delta z_i \right) = 0. \tag{7.5}$$

Each of the N masses can move in three possible coordinate directions. Therefore, in general, Equation 7.5 relates $3N$ unknowns, but with one equation. Since one variable may be written in terms of the remainder, there are $3N - 1$ variables.

Example 7.2 Two Particle Case
For two particles, Equation 7.5 becomes

$$\left(\frac{\partial g}{\partial x_1}\delta x_1 + \frac{\partial g}{\partial y_1}\delta y_1 + \frac{\partial g}{\partial z_1}\delta z_1\right) + \left(\frac{\partial g}{\partial x_2}\delta x_2 + \frac{\partial g}{\partial y_2}\delta y_2 + \frac{\partial g}{\partial z_2}\delta z_2\right) = 0,$$

where there are six unknowns and one equation. Any five of these unknowns are independent, with the sixth unknown being a function of those five. ∎

Next, assume that the N particles are subject to resultant force $\mathcal{F}_i = \mathbf{F}_i + \mathbf{f}_i$, where $\mathbf{F}_i$ is an applied force, and $\mathbf{f}_i$ is a constraint force. For the system to be in static equilibrium, every particle is at rest, and $\mathcal{F}_i = 0$ in any possible configuration. For the virtual displacement configuration, static equilibrium requires that

$$\mathcal{F}_i \cdot \delta \mathbf{r}_i = 0,$$

or

$$\mathcal{F}_i \cdot (\delta x_i \cdot \mathbf{i} + \delta y_i \cdot \mathbf{j} + \delta z_i \cdot \mathbf{k}) = 0,$$

where $\mathbf{i}, \mathbf{j},$ and $\mathbf{k}$ are the unit vectors in three–dimensional space.

Given virtual displacements, one can proceed to define *virtual work* as the product of a force and its corresponding virtual displacement. For the system in equilibrium, the virtual work for the entire system vanishes according to the relation

$$\begin{aligned}\delta W &= \sum_{i=1}^{N} \mathcal{F}_i \cdot \delta \mathbf{r}_i = 0 \\ &= \sum_{i=1}^{N} \mathbf{F}_i \cdot \delta \mathbf{r}_i + \sum_{i=1}^{N} \mathbf{f}_i \cdot \delta \mathbf{r}_i = 0.\end{aligned}$$

Before proceeding, consider the types of constraints to which a structure may be exposed. Likely examples include physical boundaries, in which case the boundary force is perpendicular to the motion of the body and there is no work performed. It is possible that contact friction will do work in resisting a motion. Dissipative forces such as friction will be introduced later in this chapter when dynamic motion is added. Therefore, $\sum_{i=1}^{N} \mathbf{f}_i \cdot \delta \mathbf{r}_i = 0$.

7.2. VIRTUAL WORK

The remaining equation is given the name *principle of virtual work* for a static system,
$$\delta W = \sum_{i=1}^{N} \mathbf{F}_i \cdot \delta \mathbf{r}_i = 0,$$
where $\mathbf{F}_i$ represents the external forces on the system.

For the special but useful case of a conservative system,
$$\begin{aligned}\delta W &= \sum_{i=1}^{N} \mathbf{F}_i \cdot \delta \mathbf{r}_i \\ &= -\delta V \\ &= -\sum_{i=1}^{N} \left(\frac{\partial V}{\partial x_i} \delta x_i + \frac{\partial V}{\partial y_i} \delta y_i + \frac{\partial V}{\partial z_i} \delta z_i \right) \\ &= 0,\end{aligned}$$
where V is the potential energy of the system. Since the variations are independent and arbitrary, the coefficients of the variations must equal zero,
$$\begin{aligned}F_{x_i} &= \frac{\partial V}{\partial x_i} = 0 \\ F_{y_i} &= \frac{\partial V}{\partial y_i} = 0 \\ F_{z_i} &= \frac{\partial V}{\partial z_i} = 0.\end{aligned}$$

These three equations can be used to define the static equilibrium configuration for the system. (See *Example 7.3.*) We proceed next with d'Alembert's principle, which extends the principle of virtual work to time–dependent problems.

7.2.3 d'Alembert's Principle

d'Alembert extended the applicability of the principle of virtual work to dynamic problems. Newton's law of motion can be rewritten as d'Alembert's principle in the following form for N particles,
$$\mathbf{F}_i + \mathbf{f}_i - m_i \ddot{\mathbf{r}}_i = 0, \quad i = 1, 2, \ldots, N. \tag{7.6}$$

The term $-m_i \ddot{\mathbf{r}}_i$ is considered an *inertia force*. Each force in Equation 7.6 may be a constant or a function of time. The virtual work performed by the ith particle is
$$(\mathbf{F}_i + \mathbf{f}_i - m_i \ddot{\mathbf{r}}_i) \cdot \delta \mathbf{r}_i = 0,$$

where the virtual displacements $\delta \mathbf{r}_i$ are compatible with the constraints. Assuming virtual work due to constraint forces equals zero, the virtual work for the system is

$$\sum_{i=1}^{N}(\mathbf{F}_i - m_i\ddot{\mathbf{r}}_i) \cdot \delta \mathbf{r}_i = 0, \tag{7.7}$$

where this is called the *generalized principle of d'Alembert*. $(\mathbf{F}_i - m_i\ddot{\mathbf{r}}_i)$ is sometimes called the *effective force*. d'Alembert's principle will be used in the next section to derive Lagrange's equation.

> "The importance of d'Alembert's principle lies in the fact that it is *more* than a reformulation of Newton's equation. It is the expression of a *principle*. We know that the vanishing of a force in Newtonian mechanics means equilibrium. Hence [Equation 7.6] says that the addition of the force of inertia to the other forces produces equilibrium. But this means that if we have any criterion for the equilibrium of a mechanical system, we can immediately extend that criterion to a system which is in motion. All we have to do is add the new 'force of inertia' to the previous forces. By this device *dynamics is reduced to statics.*"[5]

The linking of Newton's second law of motion with the principle of virtual work clarifies that the principle is equally applicable to masses at rest and to masses in motion. The virtual displacement involves a possible but purely mathematical experiment that can be applied at any specific time. At that instant, the actual motion of the body does not enter into account and the dynamic problem is reduced to a static one.

Example 7.3   d'Alembert's Principle for the Derivation of an Equation of Motion

Derive the equation of motion for the system in Figure 7.3, consisting of two masses connected by an inextensible string.

The free body forces are superimposed on the system schematic. In order to use d'Alembert's principle, the system must undergo a virtual displacement compatible with the constraints that guide the motion (string and platforms upon which masses sit). Assume mass m_2 is displaced a virtual displacement δr_2 as shown. The string constraint equation is its length: $l = |\mathbf{r}_1| + |\mathbf{r}_2| + p$, where p is the length of string on the pulley. Since l is constant, $\delta \mathbf{r}_1 + \delta \mathbf{r}_2 = 0$. From d'Alembert's principle, Equation 7.7, with $i = 1, 2$, we have

$$[(m_1\mathbf{g} + \mathbf{F}_1) - m_1\ddot{\mathbf{r}}_1] \cdot \delta \mathbf{r}_1 + [(m_2\mathbf{g} + \mathbf{F}_2) - m_2\ddot{\mathbf{r}}_2] \cdot \delta \mathbf{r}_2 = 0,$$

---

[5]Lanczos, p.89.

7.2. VIRTUAL WORK

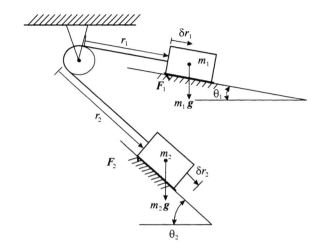

Figure 7.3: Two body single degree of freedom system.

and, after expanding the dot products, we find

$$m_1 g(\sin\theta_1 - \mu\cos\theta_1)\delta r_1 + m_2 g(\sin\theta_2 - \mu\cos\theta_2)\delta r_2$$
$$- (m_1 \ddot{r}_1 \,\delta r_1 + m_2 \ddot{r}_2 \,\delta r_2) = 0.$$

In the above equations, $\mathbf{F}_i$ are the friction forces and μ is the dynamic coefficient of friction between the block and the surface. We use the relations $\delta r_1 = -\delta r_2$ and $\ddot{r}_1 = -\ddot{r}_2$ to find

$$[-m_1 g(\sin\theta_1 - \mu\cos\theta_1) + m_2 g(\sin\theta_2 - \mu\cos\theta_2) - (m_1 + m_2)\ddot{r}_2]\delta r_2 = 0.$$

Since the variation δr_2 is arbitrary, the expression in the square brackets must be equal to zero,

$$(m_1 + m_2)\ddot{r}_2 = -m_1 g(\sin\theta_1 - \mu\cos\theta_1) + m_2 g(\sin\theta_2 - \mu\cos\theta_2). \quad (7.8)$$

This is the (nonlinear) equation of motion for the single degree of freedom system in terms of coordinate r_2. If $\ddot{r}_2 = 0$, we recover the equation for static equilibrium,

$$m_1(\sin\theta_1 - \mu\cos\theta_1) = m_2(\sin\theta_2 - \mu\cos\theta_2).$$

∎

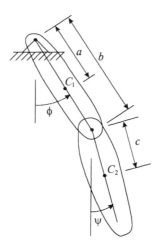

Figure 7.4: Double compound pendulum.

Example 7.4 Equations of Motion by Virtual Work and d'Alembert's Principle

To show the power of the principle of virtual work used in conjunction with the inertia forces of d'Alembert, the equations of motion of a double compound pendulum are derived. Consider Figure 7.4, draw the free body diagrams for each component, and derive the equations of motion.

Solution Since the two components of the pendulum are assumed to be perfectly rigid, two coordinates are sufficient to describe the motion of the system: ϕ for the upper mass m_1 and ψ for the lower mass m_2.

First sketch the free body force diagrams for each mass, as shown in Figure 7.5. Note that inertia terms are included with each free body diagram. This is the only free body diagram in which inertia terms are included, but we do so here to emphasize their physical meaning; they are part of the force balance as interpreted by d'Alembert. Let J_1 and J_2 be the polar moments of inertia of the respective bodies about their pivots. C_1 and C_2 are the respective centers of gravity.

Derive the virtual work expression for each coordinate and set it equal to zero to find the respective equation of motion. First consider the virtual work done by forces and moments corresponding to ψ when it is increased by a small virtual amount $\delta\psi$, while ϕ is held constant, that is $\delta\phi = 0$. We can do this since the two generalized coordinates are independent. Take one of the components of the virtual work to see how the formulation proceeds. For the lower pendulum, the force equals the product of mass m_2,

7.2. VIRTUAL WORK

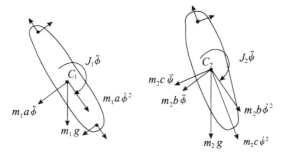

Figure 7.5: Free body force diagrams for each mass of the compound pendulum.

at mass center C_2, and the rectilinear acceleration at that point $c\ddot{\psi}$. The virtual displacement is $c\delta\psi$. Therefore, this component of the virtual work is the product $-c\ddot{\psi}\,c\delta\psi$, where the minus sign indicates a negative rotation according to the *right–hand rule*. In the next equation, the forces are shown in square brackets. The complete virtual work is then

$$\begin{aligned}\delta W_\psi &= -[m_2 c\ddot{\psi}]c\,\delta\psi - [m_2 b\ddot{\phi}]c\cos(\phi-\psi)\delta\psi - [m_2 g]c\sin\psi\,\delta\psi \\ &\quad -[J_2\ddot{\psi}]\delta\psi + [m_2 b\dot{\phi}^2]c\sin(\phi-\psi)\delta\psi \\ &= \Big[-(m_2 c^2 + J_2)\ddot{\psi} - m_2 bc(\ddot{\phi}\cos(\phi-\psi) \\ &\quad -\dot{\phi}^2\sin(\phi-\psi)) - m_2 gc\sin\psi\Big]\delta\psi,\end{aligned}$$

where $m_2 c\dot{\psi}^2$, being perpendicular to the motion $\delta\psi$, and torque $J_1\ddot{\phi}$ do no work.

Similarly, for the virtual rotation $\delta\phi$, with ψ held constant, $\delta\psi = 0$, the virtual work in the ϕ coordinate is

$$\begin{aligned}\delta W_\phi &= [-m_1 a^2 \ddot{\phi} - m_1 ga\sin\phi - J_1\ddot{\phi} - m_2 bc\ddot{\psi}\cos(\phi-\psi) \\ &\quad - m_2 b^2 \ddot{\phi} - m_2 gb\sin\phi - m_2 bc\dot{\psi}^2 \sin(\phi-\psi)]\delta\phi,\end{aligned}$$

where the following do no work, $m_1 a\dot{\phi}^2$, $m_2 b\dot{\phi}^2$, and $J_2 \ddot{\psi}$, and have been removed from the expression.

Since the virtual work done by all forces, including inertia forces, equals zero by d'Alembert's principle, we can immediately write down the equation of motion for each coordinate system by setting the generalized forces, that is the expression in each of the above square brackets, to zero. Suppose we go one step further by assuming that only small oscillations take place, meaning that ϕ and ψ are small and higher order terms such as $\dot{\phi}^2$ can be neglected. The linearized equations of motion are then

$$(m_2 c^2 + J_2)\ddot{\psi} + m_2 bc\ddot{\phi} + m_2 cg\psi = 0$$
$$m_2 bc\ddot{\psi} + (m_1 a^2 + J_1 + m_2 b^2)\ddot{\phi} + (m_1 a + m_2 b)g\phi = 0.$$

These two *coupled* governing equations must be solved *simultaneously* using techniques to be discovered in Chapter 8. ∎

d'Alembert

Jean Le Rond d'Alembert lived during 1717–1783 and was born and died in Paris, France. d'Alembert was a pioneer in the study of differential equations and pioneered their use of in physics. He was a friend of Voltaire. d'Alembert grew up in Paris. In 1741, he was admitted to the Paris Academy of Sciences, where he worked for the rest of his life. He helped to resolve the controversy in physics over the conservation of kinetic energy by improving Newton's definition of force in his *Traité de Dynamique* (1742), which articulates d'Alembert's principle of mechanics. In 1744 he applied the results to the equilibrium and motion of fluids. He did important work in the foundations of analysis and, in 1754, in an article entitled "Differentiel", in volume 4 of *Encyclopédie*, suggested that the theory of limits be put on a firm foundation. He was one of the first to understand the importance of functions and, in this article, he defined the derivative of a function as the limit of a quotient of increments. In fact, he wrote most of the mathematical articles in the 28 volume *Encyclopédie*. d'Alembert also studied hydrodynamics, the mechanics of rigid bodies, the three–body problem in astronomy and atmospheric circulation. d'Alembert turned down a number of offers in his life. He declined an offer from Frederick II to go to Prussia as President of the Berlin Academy. He also turned down an invitation from Catherine II to go to Russia as a tutor for her son.

7.3 Lagrange's Equation

Lagrange's equation is an energy–based expression that provides a general formulation for the equations of motion of a dynamical system.[6] The behavior of the system may be linear or nonlinear, and the advantage of the method becomes evident for multi degree of freedom systems. In addition, this approach is based on the energies of the system, the kinetic, potential,

---

[6] We go through the detail of its derivation because it is important to understand how such an equation is derived and the thought processes that are used. Perhaps there is another Lagrange out there who will be motivated by being exposed to the derivation!

7.3. LAGRANGE'S EQUATION

and strain energies. Therefore, it is not necessary to invoke the vectorial approach in applying Lagrange's equation as one must with Newton's second law of motion.

The equations derived below are written in terms of the *generalized coordinates*[7] q_k. The physical coordinates, $\mathbf{r}_i$, of an n degree of freedom system for N particles can be related to the generalized coordinates by an appropriate set of equations,

$$\mathbf{r}_i = \mathbf{r}_i(q_1, q_2, \ldots, q_n), \quad i = 1, 2, \ldots, N. \tag{7.9}$$

The purpose of these transformations from physical coordinates, which are vectorial, to generalized coordinates, which are not, is to recast the vectorial d'Alembert's principle into Lagrange's equation.

In the following derivations, d'Alembert's generalized principle is expanded and rewritten in terms of potential and kinetic energies, all functions of the generalized coordinates. First, derive the relations between physical and generalized coordinates.

The total derivative of Equation 7.9 is

$$\begin{aligned}
\dot{\mathbf{r}}_i &= \frac{\partial \mathbf{r}_i}{\partial q_1}\frac{dq_1}{dt} + \frac{\partial \mathbf{r}_i}{\partial q_2}\frac{dq_2}{dt} + \cdots + \frac{\partial \mathbf{r}_i}{\partial q_n}\frac{dq_n}{dt} \\
&= \frac{\partial \mathbf{r}_i}{\partial q_1}\dot{q}_1 + \frac{\partial \mathbf{r}_i}{\partial q_2}\dot{q}_2 + \cdots + \frac{\partial \mathbf{r}_i}{\partial q_n}\dot{q}_n \\
&= \sum_{k=1}^{n} \frac{\partial \mathbf{r}_i}{\partial q_k}\dot{q}_k, \quad i = 1, 2, \ldots, N.
\end{aligned}$$

Then differentiate these equations with respect to $\dot{q}_k$,

$$\frac{\partial \dot{\mathbf{r}}_i}{\partial \dot{q}_k} = \frac{\partial \mathbf{r}_i}{\partial q_k}, \quad i = 1, 2, \ldots, N, \quad k = 1, 2, \ldots, n.$$

Since variations $\delta \mathbf{r}_i$ follow the same rules as differentials $d\mathbf{r}_i$, the variations of $\mathbf{r}_i$ and $\dot{\mathbf{r}}_i$ are

$$\begin{aligned}
\delta \mathbf{r}_i &= \frac{\partial \mathbf{r}_i}{\partial q_1}\delta q_1 + \frac{\partial \mathbf{r}_i}{\partial q_2}\delta q_2 + \cdots + \frac{\partial \mathbf{r}_i}{\partial q_n}\delta q_n \\
&= \sum_{k=1}^{n} \frac{\partial \mathbf{r}_i}{\partial q_k}\delta q_k, \quad i = 1, 2, \ldots, N, \\
\delta \dot{\mathbf{r}}_i &= \frac{\partial \mathbf{r}_i}{\partial q_1}\delta \dot{q}_1 + \frac{\partial \mathbf{r}_i}{\partial q_2}\delta \dot{q}_2 + \cdots + \frac{\partial \mathbf{r}_i}{\partial q_n}\delta \dot{q}_n \\
&= \sum_{k=1}^{n} \frac{\partial \mathbf{r}_i}{\partial q_k}\delta \dot{q}_k, \quad i = 1, 2, \ldots, N.
\end{aligned}$$

---

[7] A generalized coordinate is a degree of freedom of the system being modeled.

Now consider the second term of d'Alembert's principle, Equation 7.7, written in terms of the generalized coordinates,

$$\sum_{i=1}^{N} m_i \ddot{\mathbf{r}}_i \cdot \delta \mathbf{r}_i = \sum_{i=1}^{N} \left(m_i \ddot{\mathbf{r}}_i \cdot \sum_{k=1}^{n} \frac{\partial \mathbf{r}_i}{\partial q_k} \delta q_k \right)$$

$$= \sum_{k=1}^{n} \left(\sum_{i=1}^{N} m_i \ddot{\mathbf{r}}_i \cdot \frac{\partial \mathbf{r}_i}{\partial q_k} \right) \delta q_k. \qquad (7.10)$$

Examine Equation 7.10 more closely in order to recast it in an energy form. For a particular term within the interior sum, we can perform the following algebra,

$$m_i \ddot{\mathbf{r}}_i \cdot \frac{\partial \mathbf{r}_i}{\partial q_k} = \frac{d}{dt}\left(m_i \dot{\mathbf{r}}_i \cdot \frac{\partial \mathbf{r}_i}{\partial q_k}\right) - m_i \dot{\mathbf{r}}_i \cdot \frac{d}{dt}\left(\frac{\partial \mathbf{r}_i}{\partial q_k}\right)$$

$$= \frac{d}{dt}\left(m_i \dot{\mathbf{r}}_i \cdot \frac{\partial \mathbf{r}_i}{\partial q_k}\right) - m_i \dot{\mathbf{r}}_i \cdot \frac{\partial \dot{\mathbf{r}}_i}{\partial q_k}$$

$$= \left[\frac{d}{dt}\left(\frac{\partial}{\partial \dot{q}_k}\right) - \frac{\partial}{\partial q_k}\right]\left(\frac{1}{2} m_i \dot{\mathbf{r}}_i \cdot \dot{\mathbf{r}}_i\right). \qquad (7.11)$$

Then Equation 7.10 becomes

$$\sum_{i=1}^{N} m_i \ddot{\mathbf{r}}_i \cdot \delta \mathbf{r}_i = \sum_{k=1}^{n} \left[\frac{d}{dt}\left(\frac{\partial T}{\partial \dot{q}_k}\right) - \frac{\partial T}{\partial q_k}\right] \delta q_k,$$

where the kinetic energy T is defined as

$$T = \frac{1}{2} \sum_{i=1}^{N} m_i \dot{\mathbf{r}}_i \cdot \dot{\mathbf{r}}_i = T(q_1, \ldots, q_n, \dot{q}_1, \ldots, \dot{q}_n).$$

The other term in d'Alembert's principle is the virtual work $\delta W = \sum_{i=1}^{N} \mathbf{F}_i \cdot \delta \mathbf{r}_i$, and so d'Alembert's equation can be written as

$$\sum_{k=1}^{n} \left[\frac{d}{dt}\left(\frac{\partial T}{\partial \dot{q}_k}\right) - \frac{\partial T}{\partial q_k}\right] \delta q_k = \delta W. \qquad (7.12)$$

To explicitly show the virtual work in terms of forces $\mathbf{F}_i$, write them in terms of the generalized coordinates. Consider the virtual work done by

7.3. LAGRANGE'S EQUATION

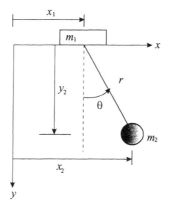

Figure 7.6: Pendulum suspended from a moving block.

these forces,

$$\begin{aligned}
\delta W &= \sum_{i=1}^{N} \mathbf{F}_i \cdot \delta \mathbf{r}_i \\
&= \sum_{i=1}^{N} \mathbf{F}_i \cdot \sum_{k=1}^{n} \frac{\partial \mathbf{r}_i}{\partial q_k} \delta q_k \\
&= \sum_{k=1}^{n} \left(\sum_{i=1}^{N} \mathbf{F}_i \cdot \frac{\partial \mathbf{r}_i}{\partial q_k} \right) \delta q_k \\
&= \sum_{k=1}^{n} Q_k \, \delta q_k, \quad\quad\quad (7.13)
\end{aligned}$$

where Q_k are called the *generalized forces*. The generalized force also may be a torque, and will likely be a complicated expression. Before further specializing the generalized force, look at an example that shows how the generalized force can be identified.

Example 7.5 The Generalized Force
In this example we demonstrate the procedure implied by Equations 7.12 and 7.13. Consider the pendulum suspended from an unattached block resting on a smooth surface, as drawn in Figure 7.6. Gravity acts in the direction y. Derive the equations of motion.
  Solution In this procedure, the kinetic energy of each mass is derived and appropriate derivatives are set equal to the respective generalized

force,
$$\left[\frac{d}{dt}\left(\frac{\partial T}{\partial \dot{q}_k}\right) - \frac{\partial T}{\partial q_k}\right] \equiv Q_k, \tag{7.14}$$
for each generalized coordinate k. To find the generalized forces, evaluate the virtual work done in each generalized coordinate.

The kinetic energy of the system of two masses is given by
$$T = \frac{1}{2}m_1\dot{x}_1^2 + \frac{1}{2}m_2(\dot{x}_2^2 + \dot{y}_2^2),$$
where the following substitutions can be made,
$$x_2 = x_1 + r\sin\theta \tag{7.15}$$
$$y_2 = r\cos\theta. \tag{7.16}$$

Equations 7.15 and 7.16 are constraint equations. We select x_1 and θ as the two generalized coordinates. Eliminate $\dot{x}_2$ and $\dot{y}_2$ from the expression for kinetic energy to find
$$T = \frac{1}{2}m_1\dot{x}_1^2 + \frac{1}{2}m_2(\dot{x}_1^2 + 2r\dot{x}_1\dot{\theta}\cos\theta + r^2\dot{\theta}^2).$$

The following derivatives need to be evaluated in order to apply Lagrange's equation for the equation of motion in generalized coordinate x_1,
$$\frac{\partial T}{\partial \dot{x}_1} = m_1\dot{x}_1 + m_2\dot{x}_1 + m_2 r\dot{\theta}\cos\theta$$
$$\frac{d}{dt}\left(\frac{\partial T}{\partial \dot{x}_1}\right) = m_1\ddot{x}_1 + m_2\ddot{x}_1 + m_2 r(\ddot{\theta}\cos\theta - \dot{\theta}^2\sin\theta)$$
$$\frac{\partial T}{\partial x_1} = 0.$$

Substitute these into Equation 7.14 and equate to the generalized force in the x_1 direction,
$$(m_1 + m_2)\ddot{x}_1 + m_2 r(\ddot{\theta}\cos\theta - \dot{\theta}^2\sin\theta) = F_{x_1}.$$

This is the equation of motion in the x_1 direction. The expression for F_{x_1} will be evaluated after the equation of motion in the θ direction is derived. To do this calculate the following derivatives,
$$\frac{\partial T}{\partial \dot{\theta}} = m_2 r\dot{x}_1\cos\theta + m_2 r^2\dot{\theta}$$
$$\frac{d}{dt}\left(\frac{\partial T}{\partial \dot{\theta}}\right) = m_2 r(\ddot{x}_1\cos\theta - \dot{x}_1\dot{\theta}\sin\theta) + m_2 r^2\ddot{\theta}$$
$$\frac{\partial T}{\partial \theta} = -m_2 r\dot{x}_1\dot{\theta}\sin\theta.$$

7.3. LAGRANGE'S EQUATION

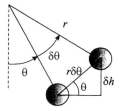

Figure 7.7: Virtual displacement $\delta\theta$.

As before, we can directly write the second equation equation of motion,

$$m_2 r \ddot{x}_1 \cos\theta + m_2 r^2 \ddot{\theta} = F_\theta.$$

Next, evaluate the virtual work performed by forces in the x_1 and θ directions. The general virtual work expression is

$$\delta W = F_{x_1} \delta x_1 + F_\theta \delta\theta,$$

where δx_1 and $\delta\theta$ are independent of each other. Since there is no friction between the block and the surface upon which it rests, and there are no other forces in the x_1 direction, $F_{x_1} = 0$. To evaluate F_θ it is helpful to sketch the virtual displacement $\delta\theta$. See Figure 7.7.

The virtual work is performed in raising the pendulum mass a vertical distance δh, which can be related to $\delta\theta$ by $\delta h = r \, \delta\theta \sin\theta$. Therefore, the virtual work done by gravity is $F_\theta \, \delta\theta$, or

$$-m_2 g r \sin\theta \, \delta\theta,$$

where the negative sign signifies that the work is performed by the mass against the gravitational force. Thus,

$$F_\theta = -m_2 g r \sin\theta.$$

Therefore, the equations of motion are

$$(m_1 + m_2)\ddot{x}_1 + m_2 r(\ddot{\theta}\cos\theta - \dot{\theta}^2 \sin\theta) = 0$$
$$m_2 r \ddot{x}_1 \cos\theta + m_2 r^2 \ddot{\theta} = -m_2 g r \sin\theta.$$

Simultaneous equations of this type will be linearized and solved in Chapter 8. Had there been other external forces, they would have been components in F_{x_1} or F_θ and have appeared on the right hand sides of the equations of motion.

Dissipation forces such as damping or friction would also be part of the virtual work expression and thereby also find their way to the right hand side of the equation of motion. This is discussed next. ∎

D'Alembert's equation can be further specialized on the way to Lagrange's equation by separately considering the *conservative* (derivable from potential energy V) and *non–conservative* forces acting on the system. From Equation 7.13,

$$\sum_{i=1}^{N} \mathbf{F}_i \cdot \delta \mathbf{r}_i = \delta W = \delta W_c + \delta W_{nc}.$$

Work in a conservative vector field equals the negative of the change in potential, $-\delta V$. The virtual work of non–conservative generalized forces $Q_{k_{nc}}$ undergoing virtual displacements δq_k is given by $\sum_{k=1}^{n} Q_{k_{nc}} \delta q_k$. Therefore,

$$\begin{aligned} \delta W &= -\left(\frac{\partial V}{\partial q_1} \delta q_1 + \ldots + \frac{\partial V}{\partial q_n} \delta q_n \right) + \sum_{k=1}^{n} Q_{k_{nc}} \delta q_k \\ &= -\sum_{k=1}^{n} \left(\frac{\partial V}{\partial q_k} - Q_{k_{nc}} \right) \delta q_k. \end{aligned}$$

D'Alembert's generalized principle, Equation 7.12, becomes

$$\sum_{k=1}^{n} \left[\frac{d}{dt} \left(\frac{\partial T}{\partial \dot{q}_k} \right) - \frac{\partial T}{\partial q_k} + \frac{\partial V}{\partial q_k} - Q_{k_{nc}} \right] \delta q_k = 0.$$

Since the virtual displacements δq_k are arbitrary, the expression in the square brackets must equal zero for each k. Therefore,

$$\frac{d}{dt} \left(\frac{\partial T}{\partial \dot{q}_k} \right) - \frac{\partial T}{\partial q_k} + \frac{\partial V}{\partial q_k} = Q_k, \quad k = 1, 2, \ldots, n, \qquad (7.17)$$

where $Q_k \equiv Q_{k_{nc}}$ includes dissipative forces such as damping and non–conservative external forces. These are *Lagrange's equations of motion*, one equation for each of the n degrees of freedom. It is customary to define the *Lagrangian* function as

$$L = T - V.$$

Since potential energy V is a function of position only, it cannot vary with velocity, and therefore,

$$\frac{\partial T}{\partial \dot{q}_k} = \frac{\partial (T-V)}{\partial \dot{q}_k} = \frac{\partial L}{\partial \dot{q}_k},$$

7.3. LAGRANGE'S EQUATION

and Equation 7.17 becomes

$$\frac{d}{dt}\left(\frac{\partial L}{\partial \dot{q}_k}\right) - \frac{\partial L}{\partial q_k} = Q_k, \quad k = 1, 2, \ldots, n. \tag{7.18}$$

There are several key advantages to Lagrange's equation:

- Lagrange's equation contains only scalar quantities, eliminating the force and acceleration vectors inherent in Newton's second law of motion.

- There is one Lagrange equation for each degree of freedom, whereas the use of free body diagrams in Newton's formulation leads to extraneous equations resulting from the internal forces between bodies that are attached to each other. Via Newton's approach, such internal forces have to be eliminated after the equations of motion are derived. Of course, in some applications, these internal forces are needed.

- Lagrange's equation is independent of the coordinate system since the energy functions T and V are scalar.

Example 7.6 Lagrange's the Two Body Single Degree of Freedom System
We now formulate the problem of *Example* 7.3 using Lagrange's equation. Figure 7.8 depicts the system along with the relevant coordinates. Since this is a one degree of freedom system, only one generalized coordinate is needed to define the state of the system. However, we will select the two coordinates

$$q_1 = l - r \qquad q_2 = r$$
$$\dot{q}_1 = -\dot{r} \qquad \dot{q}_2 = \dot{r}.$$

Other coordinates are possible, but these were selected because they show the link between the two coordinates and the string that constrains them. The equation of motion is derived *(i)* first ignoring friction forces and *(ii)* then including friction forces.

Solution *(i)* The kinetic energy is given by

$$T = \frac{1}{2}m_1\dot{q}_1^2 + \frac{1}{2}m_2\dot{q}_2^2 = \frac{1}{2}(m_1 + m_2)\dot{r}^2.$$

The potential energy is given by

$$\begin{aligned} V &= -m_1gh_1 - m_2gh_2 \\ &= -m_1g(l-r)\sin\theta_1 - m_2gr\sin\theta_2. \end{aligned}$$

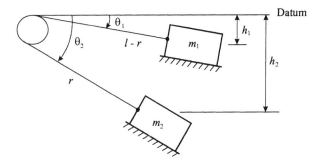

Figure 7.8: Lagrange's equation for the two body single degree of freedom system.

Thus the Lagrangian is

$$L = T - V$$
$$= \frac{1}{2}(m_1 + m_2)\dot{r}^2 + m_1 g(l - r)\sin\theta_1 + m_2 g r \sin\theta_2.$$

Now we can operate on the Lagrangian as necessary to derive the equation of motion

$$\frac{d}{dt}\left(\frac{\partial L}{\partial \dot{q}_2}\right) = \frac{d}{dt}\left(\frac{\partial L}{\partial \dot{r}}\right) = (m_1 + m_2)\ddot{r}$$

$$\frac{\partial L}{\partial q_2} = \frac{\partial L}{\partial r} = -m_1 g \sin\theta_1 + m_2 g \sin\theta_2.$$

Substituting these into Equation 7.18, we find the equation of motion for this single degree of freedom system to be

$$(m_1 + m_2)\ddot{r} = -m_1 g \sin\theta_1 + m_2 g \sin\theta_2.$$

This is a nonlinear equation valid for large θ_1 and θ_2. Note that the same equation is derived if Lagrange's equation is applied to q_1, although care must be given with the signs since $\dot{q}_1 = -\dot{r}$.

(ii) For the second part of the problem friction is included, and therefore the nonconservative forces Q_k from Lagrange's equation must be included. Since this is a two particle, one degree of freedom, system, there is only one dissipative force,

$$Q = -\mu(m_1 g \cos\theta_1 + m_2 g \cos\theta_2),$$

resulting in the corresponding governing equation

$$(m_1 + m_2)\ddot{r} = -m_1 g(\sin\theta_1 + \mu\cos\theta_1) + m_2 g(\sin\theta_2 - \mu\cos\theta_2),$$

7.3.1 Lagrange's Equation for Small Oscillations

We have learned that Lagrange's equation can be utilized to derive the fully nonlinear equations of motion for a dynamic system. But in many applications, vibration is essentially linear. Therefore, it is of interest to examine how Lagrange's equation simplifies for small amplitude oscillations about equilibrium.

Expand the expression for the potential energy $V(q_1, q_2, \ldots, q_n)$ in an n–variable Taylor series[8] about an arbitrary equilibrium reference position $V(0, 0, \ldots, 0)$,

$$V(q_1, q_2, \ldots, q_n) = \frac{1}{2}\left(\frac{\partial^2 V}{\partial q_1^2}q_1^2 + \frac{\partial^2 V}{\partial q_2^2}q_2^2 + \cdots + 2\frac{\partial^2 V}{\partial q_1 \partial q_2}q_1 q_2 + \cdots\right) + \cdots.$$

Use is made that $V(0, 0, \ldots, 0) = 0$ and $\partial V/\partial q_i = 0$ in the equilibrium position. For small amplitudes, q_i to powers two and higher can be ignored, leaving the approximation

$$V \approx \frac{1}{2}\sum_{i=1}^{n}\sum_{j=1}^{n}\frac{\partial^2 V}{\partial q_i \partial q_j}q_i q_j = \frac{1}{2}\sum_{i=1}^{n}\sum_{j=1}^{n}k_{ij}q_i q_j,$$

where k_{ij} are known as the *stiffness coefficients*. The kinetic energy is given by

$$T = \frac{1}{2}\sum_{i=1}^{n}\sum_{j=1}^{n}m_{ij}\dot{q}_i \dot{q}_j.$$

Substituting the above expressions into Lagrange's equation leads to the following n *coupled* equations of motion,

$$[m]\{\ddot{q}\} + [k]\{q\} = \{0\}. \tag{7.19}$$

This is a matrix equation of motion, to be derived again (and solved) in Chapter 8 via Newton's second law of motion.

---

[8]
$$f(x, y, z) = f(0, 0, 0) + f_x(0, 0, 0) \cdot (x - x(0)) + f_y(0, 0, 0) \cdot (y - y(0))$$
$$+ f_z(0, 0, 0) \cdot (z - z(0)) + \frac{1}{2}(f_{xx}(0, 0, 0) \cdot (x - x(0))^2 + \cdots$$
$$+ 2f_{xy}(0, 0, 0) \cdot (x - x(0))(y - y(0)) + \cdots) + \cdots.$$

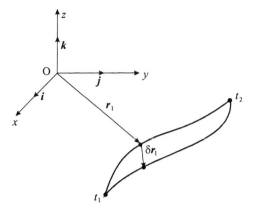

Figure 7.9: Path and varied path for Hamilton's principle.

7.4 Hamilton's Principle

We now offer an alternate approach to the derivation of Lagrange's equation. Along the way we derive *Hamilton's principle*, a very powerful integral variational statement. Begin with Equation 7.10, and continue along a new path by rewriting the left hand side of that equation as follows,

$$\begin{aligned}
\sum_{i=1}^{N} m_i \ddot{\mathbf{r}}_i \cdot \delta \mathbf{r}_i &= \sum_{i=1}^{N} m_i \frac{d}{dt}(\dot{\mathbf{r}}_i \cdot \delta \mathbf{r}_i) - \delta \sum_{i=1}^{N} \frac{1}{2} m_i (\dot{\mathbf{r}}_i \cdot \dot{\mathbf{r}}_i) \\
&= \sum_{i=1}^{N} m_i \frac{d}{dt}(\dot{\mathbf{r}}_i \cdot \delta \mathbf{r}_i) - \delta T.
\end{aligned} \qquad (7.20)$$

Use has been made of the relation

$$\frac{d}{dt}(\dot{\mathbf{r}}_i \cdot \delta \mathbf{r}_i) = \ddot{\mathbf{r}}_i \cdot \delta \mathbf{r}_i + \dot{\mathbf{r}}_i \cdot \delta \dot{\mathbf{r}}_i = \ddot{\mathbf{r}}_i \cdot \delta \mathbf{r}_i + \delta(\frac{1}{2}\dot{\mathbf{r}}_i \cdot \dot{\mathbf{r}}_i).$$

Substitute Equation 7.20, and $\sum_{i=1}^{N} \mathbf{F}_i \cdot \delta \mathbf{r}_i = \delta W$ into d'Alembert's principle, and find

$$\delta T + \delta W = \sum_{i=1}^{N} m_i \frac{d}{dt}(\dot{\mathbf{r}}_i \cdot \delta \mathbf{r}_i).$$

Consider a *varied path*, as shown in Figure 7.9, where the paths coincide

7.4. HAMILTON'S PRINCIPLE

at the initial and final times, and integrate between t_1 and t_2,

$$\begin{aligned}
\int_{t_1}^{t_2} (\delta T + \delta W) dt &= \int_{t_1}^{t_2} \sum_{i=1}^{N} m_i \frac{d}{dt} (\dot{\mathbf{r}}_i \cdot \delta \mathbf{r}_i) dt \\
&= \sum_{i=1}^{N} \int_{t_1}^{t_2} m_i \frac{d}{dt} (\dot{\mathbf{r}}_i \cdot \delta \mathbf{r}_i) dt \\
&= \sum_{i=1}^{N} m_i \dot{\mathbf{r}}_i \cdot \delta \mathbf{r}_i \Big|_{t_1}^{t_2} \\
&= 0,
\end{aligned}$$

where, based on previous discussion, $\delta r_i = 0$ at t_1 and t_2. Therefore,

$$\int_{t_1}^{t_2} (\delta T + \delta W) dt = 0. \tag{7.21}$$

This is called *the extended Hamilton's principle*. δW includes both conservative and non–conservative work. If the forces are only conservative, then $\delta W = -\delta V$, and

$$\delta \int_{t_1}^{t_2} (T - V) dt = 0,$$

where the *Lagrangian* $L = T - V$. This equation may be physically interpreted as nature trying to equalize the kinetic and potential energies of a system, in absence of dissipation.

Lagrange's equation can be derived from Hamilton's principle. To do this, in Equation 7.21, vary $T(q_i, \dot{q}_i)$ for each generalized coordinate,

$$\delta T = \sum_{i=1}^{N} \frac{\partial T}{\partial q_i} \delta q_i + \sum_{i=1}^{N} \frac{\partial T}{\partial \dot{q}_i} \delta \dot{q}_i. \tag{7.22}$$

Let

$$\delta \dot{q}_i = \frac{d(\delta q_i)}{dt},$$

and integrate by parts the ith component of the second term in Equation 7.22:

$$\int_{t_1}^{t_2} \frac{\partial T}{\partial \dot{q}_i} \frac{d(\delta q_i)}{dt} dt = \frac{\partial T}{\partial \dot{q}_i} \delta q_i \Big|_{t_1}^{t_2} - \int_{t_1}^{t_2} \delta q_i \frac{d}{dt} \left(\frac{\partial T}{\partial \dot{q}_i} \right) dt,$$

where

$$\frac{\partial T}{\partial \dot{q}_i} \delta q_i \Big|_{t_1}^{t_2} = 0$$

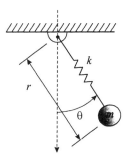

Figure 7.10: An elastic pendulum.

at the end times. Equation 7.21 then becomes

$$\int_{t_1}^{t_2} \sum_{i=1}^{N} \left[\left(\frac{\partial T}{\partial q_i} - \frac{d}{dt}\left(\frac{\partial T}{\partial \dot{q}_i}\right) \right) + Q_i \right] \delta q_i \, dt = 0, \qquad (7.23)$$

where $\delta W = -\sum_i (\partial V/\partial q_i - Q_{i_{nc}})\delta q_i \equiv \sum_i Q_i \delta q_i$.

Since all δq_i are arbitrary except at the end times, for each i in Equation 7.23 the expression within the parentheses equals zero. This is again Lagrange's equation, one for each generalized coordinate, as in Equation 7.17.

Example 7.7 Hamilton's Principle for the Derivation of the Equations of Motion of an "Elastic Pendulum"

To demonstrate the utility of Hamilton's principle for the derivation of the equations of motion of complex dynamical systems, consider the system depicted in Figure 7.10. This is a pendulum with an additional degree of freedom due to the spring. The system is assumed to have no dissipation.

The Lagrangian is

$$\begin{aligned} L &= T - V \\ &= \frac{1}{2}m[\dot{r}^2 + (r\dot{\theta})^2] - [-mgh + \frac{1}{2}k(r - r_0)^2], \end{aligned}$$

where $h = r\cos\theta$ and r_0 is the undeformed spring length. Substituting the

7.4. HAMILTON'S PRINCIPLE

variation of the Lagrangian into Hamilton's principle,

$$\begin{aligned}\int_{t_1}^{t_2} \delta L\, dt &= \int_{t_1}^{t_2} [m\dot{r}\,\delta\dot{r} + m(r\dot{\theta}^2\,\delta r + r^2\dot{\theta}\,\delta\dot{\theta}) \\ &\quad + mg\,\delta r \cos\theta - mgr\,\delta\theta \sin\theta - k(r-r_0)\delta r]dt \\ &= \int_{t_1}^{t_2} \Big\{ m\dot{r}\,\delta\dot{r} + mr^2\dot{\theta}\,\delta\dot{\theta} \\ &\quad + [m(r\dot{\theta}^2 + mg\cos\theta - k(r-r_0)]\delta r - mgr\sin\theta\,\delta\theta \Big\}\,dt.\end{aligned}$$
(7.24)

As is necessary in such developments, convert variations of time derivatives of parameters into variations of the parameters themselves. There are two such terms in Equation 7.24,

$$m\dot{r}\,\delta\dot{r}\,dt, \quad mr^2\dot{\theta}\,\delta\dot{\theta}\,dt,$$

that need to be converted. The essential approach is relatively straightforward and involves an "inverse" chain rule. We show the procedure here. For the first case above, transform a term such as $(\cdots)\delta\dot{r}\,dt$ to one such as $(\cdots)\delta r\,dt$. Begin with the time derivative

$$\frac{d}{dt}(m\dot{r}\,\delta r) = m\ddot{r}\,\delta r + m\dot{r}\,\delta\dot{r}.$$

Then,

$$m\dot{r}\,\delta\dot{r}\,dt = d(m\dot{r}\,\delta r) - m\ddot{r}\,\delta r\,dt.$$

For expression $mr^2\dot{\theta}\,\delta\dot{\theta}\,dt$, which is a product of three functions of time, begin with the time derivative

$$\frac{d}{dt}(mr^2\dot{\theta}\,\delta\theta) = m(2r\dot{r}\dot{\theta}\,\delta\theta) + m(r^2\ddot{\theta}\,\delta\theta) + mr^2\dot{\theta}\,\delta\dot{\theta}.$$

Therefore,

$$mr^2\dot{\theta}\,\delta\dot{\theta}\,dt = d(mr^2\dot{\theta}\,\delta\theta) - (mr^2\ddot{\theta} + 2mr\dot{r}\dot{\theta})\delta\theta\,dt.$$

Putting all this together in Equation 7.24 we have

$$\begin{aligned}-\int_{t_1}^{t_2} \delta L\,dt &= \int_{t_1}^{t_2} \Big[\big\{m\ddot{r} - mr\dot{\theta}^2 - mg\cos\theta + k(r-r_0)\big\}\delta r \\ &\quad + \big\{mr^2\ddot{\theta} + 2mr\dot{r}\dot{\theta} + mgr\sin\theta\big\}\delta\theta\Big]dt \\ &\quad - \int_{t_1}^{t_2} \Big[d(m\dot{r}\,\delta r) + d(mr^2\dot{\theta}\,\delta\theta)\Big] = 0. \end{aligned}$$
(7.25)

Since $\delta r = 0$ and $\delta\theta = 0$ at t_1 and t_2, the second integral on the right hand side of Equation 7.25 equals zero. Further, since δr is independent of $\delta\theta$ and both are arbitrary,

$$m\ddot{r} - mr\dot{\theta}^2 - mg\cos\theta + k(r - r_0) = 0$$
$$mr^2\ddot{\theta} + 2mr\dot{r}\dot{\theta} + mgr\sin\theta = 0$$

are the nonlinear equations of motion for the "elastic pendulum." It is assumed that the spring behaves linearly; but even if in addition we limit the second degree of freedom to small angles, $\cos\theta \approx 1$, $\sin\theta \approx \theta$, we cannot remove the nonlinearity $\dot{\theta}^2$ without additional assumptions. ∎

Hamilton

Sir William Rowan Hamilton lived during 1805–1865 and was born and died in Dublin, Ireland. By the age of five, Hamilton had already learned Latin, Greek, and Hebrew, and he soon mastered additional languages. At age 15 he started studying the works of Newton and Laplace. In 1822 Hamilton found an error in Laplace's *Méchanique Celeste* and as a result of this he came to the attention of the Astronomer Royal of Ireland. Hamilton entered University at the age of 18. While still an undergraduate at Trinity College, Dublin, he predicted theoretically conical refraction in biaxial crystals, which was later confirmed experimentally. He published his results in several papers, including the *Theory of Systems of Rays* in 1827. In the same year, Hamilton, although he had never graduated, was appointed Astronomer Royal at Dunsink Observatory and professor of astronomy at Trinity College. In 1833, he published a study of vectors as ordered pairs. He used algebra in treating dynamics in *On a General Method in Dynamics* in 1834. Hamilton's later life was unhappy and he became addicted to alcohol. He died from a severe attack of gout.

7.5 Lagrange's Equation with Damping

Prior to this section, damping was not formally considered in the variational formulation. Here, the inclusion of damping is examined. Rather than proceeding with a full derivation, as we have for Lagrange's equation and Hamilton's principle, it is preferable to state a final result, and refer the reader to an excellent source[9] that includes many examples.

As we briefly discussed in Chapter 2, there are many types of damping, and the particular application will determine which is most suitable. For example, when damping has been included, it has been exclusively viscous

---

[9]D.A. Wells, **Lagrangian Dynamics**, Schaum's Outline Series, McGraw–Hill, 1967.

damping, which is proportional to the first power of the speed and opposite in direction to its motion. This form of damping is adequate if the speed is "not too great." At higher speed, the damping may be proportional to the speed taken to a power greater than one.

For viscous damping, there is a special form for the generalized force,

$$Q_D = -\frac{\partial R}{\partial \dot{q}}$$

for each generalized coordinate, where R is known as the *Rayleigh dissipation function* and is given by

$$R = \frac{1}{2}\sum_k \sum_l c_{kl}\dot{q}_k\dot{q}_l,$$

where the c_{kl} are damping coefficients. For the kth generalized coordinate then,

$$Q_{D_k} = -\frac{\partial R}{\partial \dot{q}_k} = -\sum_l c_{kl}\dot{q}_l,$$

with the resulting Lagrange's equation,

$$\frac{d}{dt}\left(\frac{\partial T}{\partial \dot{q}_k}\right) - \frac{\partial T}{\partial q_k} + \frac{\partial V}{\partial q_k} + \frac{\partial R}{\partial \dot{q}_k} = Q_k \quad k = 1, 2, \ldots, n.$$

We will also apply Lagrange's equation and Hamilton's principle in Chapters 8–11 for structures modeled as multi degree of freedom and continuous systems, respectively.

7.6 Concepts Summary

We have introduced the fundamental concepts underlying the variational approaches to mechanics and vibration. First among these is the principle of virtual work. d'Alembert's principle extends virtual work to dynamic problems. From these, Lagrange's equation and Hamilton's principle can be derived. Their importance for us is that they provide tools for the derivation of governing equations of motion and their boundary conditions. Chapters 8 and 9 develop idealized multi degree of freedom models of vibration. Chapters 10 and 11 develop idealized continuous system models of vibration. In each of these chapters, the equations of motion are derived in a variety ways, by the application of Newton's second law of motion, by Lagrange's equation, and by Hamilton's principle.

While this has been a very technical chapter, with many mathematical manipulations, the interested reader benefits from an understanding of how Lagrange's equation and Hamilton's principle are derived and their theoretical underpinnings..

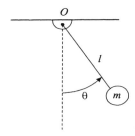

Figure 7.11: Simple pendulum.

7.7 Problems

The following problems are meant to give the reader a facility with the variational procedures for deriving equations of motion and accounting for boundary conditions. Most of the problems you are asked to formulate are based on Lagrange's equation, the most widely used approach. Also you will note how numerous are pendulum problems. It is quite an interesting fact that many dynamic problems can be interpreted as various types of pendula. In subsequent chapters, you will be shown further applications of these methods.

1. For the simple pendulum of Figure 7.11, derive the governing equation of motion *(i)* assuming that m is a point mass, and *(ii)* that the mass is a sphere with small but finite mass moment of inertia. Assuming $l = 10 \ cm$ and $m = 5 \ kg$, how significant is the effect of the mass inertia if $r = 0.5 \ cm$? Derive the governing equation of motion using *(a)* Newton's second law of motion, *(b)* Lagrange's equation, and *(c)* Hamilton's principle.

2. The light–weight bar in Figure 7.12 is released from rest when the spring is undeformed at $\theta = 0$. The two masses at A and B slide in frictionless guides. Formulate the equation of motion using *(i)* Newton's second law of motion, *(ii)* Lagrange's equation, and *(iii)* Hamilton's principle.

3. The spring supporting the rod–sphere system in Figure 7.13 is undeformed when the rod is horizontal. If the system is in this position when it is released from rest, derive the equation of motion using *(i)* the principle of virtual work along with d'Alembert's principle, and *(ii)* using Lagrange's equation. Assume rod mass is negligible. The

7.7. PROBLEMS

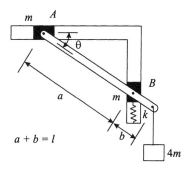

Figure 7.12: Oscillation of constrained bar.

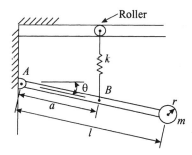

Figure 7.13: Oscillation of rod-sphere system.

roller from which the spring is suspended permits the spring to maintain a vertical configuration.

4. Verify Equation 7.11.

5. Derive Equation 7.19.

6. Formulate *Example* 7.5, but this time use Newton's second law to derive the equations of motion.

7. For the elastic pendulum discussed in *Example* 7.7 and shown in Figure 7.10, derive the governing equations of motion using Lagrange's equation, and compare with the equations derived using Hamilton's principle.

8. The simple pendulum is modified as per Figure 7.14 so that the length r is variable. Derive the equations of motion. Discuss the possibility and implications if $\omega = \dot{\theta}$.

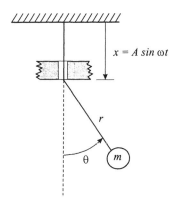

Figure 7.14: Variable length pendulum.

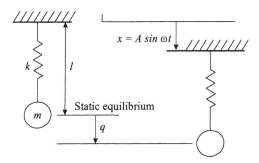

Figure 7.15: Elastically supported pendulum on a moving base.

9. *(i)* For the elastically supported pendulum on a moving base, derive the governing equation of motion first using d'Alembert's principle, and then directly using Lagrange's equation. Discuss the effect of base motion and that it is equivalent to a force on mass m equal to $kA\sin\omega t$. Refer to Figure 7.15. *(ii)* Repeat the above problem where a viscous damping exists due to the motion of the mass through the surrounding medium. Assume a damping constant denoted by c.

10. Consider the motion of a pendulum that is supported by springs that elastically restrain horizontal motion, as depicted in Figure 7.16. Assume that the springs are massless and remain horizontal, that θ is small, and r is a constant. Formulate the equation of motion using *(i)* Newton's second law of motion, *(ii)* Lagrange's equation, and *(iii)*

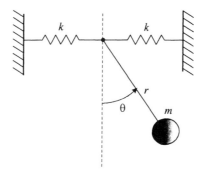

Figure 7.16: Pendulum supported by horizontal springs.

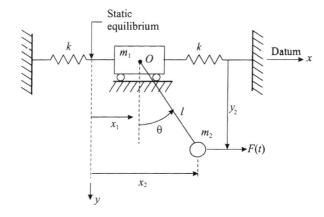

Figure 7.17: Pendulum suspended from a horizontally restrained mass.

Hamilton's principle.

11. For a pendulum supported from a horizontally restrained mass, as per Figure 7.17, derive the governing equations of motion using *(i)* Lagrange's equation and *(ii)* by Hamilton's principle. Identify the constraint equations. Show and discuss the effects of setting $m_1 = 0$. Finally, simplify the governing equations if small motions are assumed.

12. Suppose a pendulum is suspended from a torsionally restrained disk, as in Figure 7.18. Derive the governing equation of motion using *(i)* Lagrange's equation, and *(ii)* by Hamilton's principle. Compare the equation of motion to that of the simple pendulum.

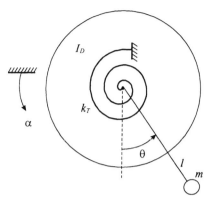

Figure 7.18: Pendulum supported from a torsionally restrained disk.

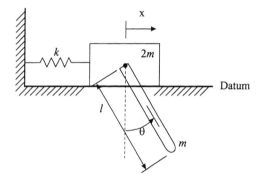

Figure 7.19: Compound pendulum suspended from a block.

13. A rigid beam acts as a compound pendulum, suspended from an elastically restrained block that can undergo horizontal motion, as drawn in Figure 7.19. Derive the full nonlinear equations of motion using Lagrange's equation. Consider the following cases: *(i)* where there is no friction between block and surface, and *(ii)* where viscous damping restrains the motion of the block.

14. The mechanism in Figure 7.20 oscillates about the equilibrium position. The coil springs are each undeformed when $\theta_1 = \theta_2 = 0$, and have units of torque/radian. If the bar masses are neglected, derive the equations of motion.

15. Pulleys are simple machines that are found in one form or another in

7.8. ADDITIONAL READINGS

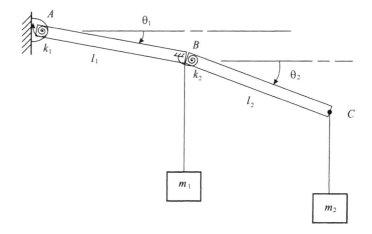

Figure 7.20: Torsionally restrained beams.

mechanical devices. Figure 7.21 shows a schematic of such a pulley. A torque $M(t)$ is applied to the pulley, resulting in oscillatory motion. Use Lagrange's equation to derive the equations of motion. Identify the generalized forces Q_θ and Q_x.

16. Power systems depend on gear systems to transmit energy. A simple gear–shaft system is sketched in Figure 7.22, depicting two gear systems coupled through connected gears. Assume gears 2 and 4 have gear ratio $n = r_2/r_4$, and that external torque $M(t)$ is applied to gear 1. Identify the constraint equation and the generalized coordinates, and derive the governing equations of motion. Discuss how the value of n will affect the resulting behavior.

7.8 Additional Readings

As possible mini–projects, selective readings from some of the advanced texts referenced in this chapter would be of great interest and benefit to the motivated student. Additional problems from the book by Wells could supplement the problems provided above. Additional but more advanced readings can be suggested from the following:

1. Junkins, J.L., Kim, Y., **Introduction to Dynamics and Control of Flexible Structures**, AIAA Education Series, 1993.

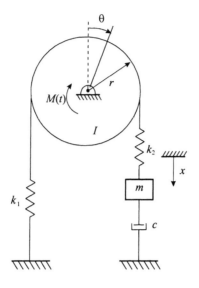

Figure 7.21: A pulley subjected to a torque.

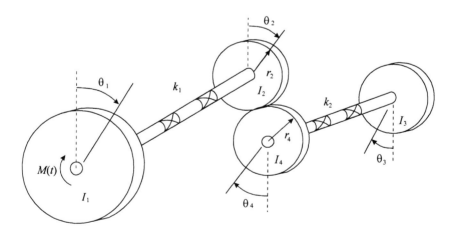

Figure 7.22: Gear-shaft system.

7.8. ADDITIONAL READINGS

2. Tabarrok, B., Rimrott, F.P.J., **Variational Methods and Complementary Formulations in Dynamics**, Kluwer Academic Publishers, 1994.

3. *Equations of Motion for an N-Link Planar System for Simulation and Optimization*, B.L. Biswell, J. Puig–Suari, *American Astronautical Society* Paper 97–720, 1997, 1843–1858.

Chapter 8

Multi Degree of Freedom Vibration

"And now we no longer operate in isolation since we are aware of others."

Engineered structures, such as turbomachinery, bridges, or aircraft, have thousands of degrees of freedom. Each degree of freedom has a frequency of vibration, much in the same way as did the single degree of freedom oscillator. However, once structures have two or more degrees of freedom, there is a *coupling* between each pair of degrees of freedom. Another new concept to be understood is the *mode* of vibration. A vibrating linear structure, while appearing to move in very complicated patterns, can be modeled as n uncoupled vibrating oscillators when the principal coordinates, are chosen. An exciting aspect of what we will learn here is the single degree of freedom oscillator is the basis for solving many of the problems encountered next.

8.1 Example Problems and Motivation

Two interesting applications motivate the study of multi–degree of freedom discrete models. Additional applications are introduced throughout the chapter.

8.1.1 Periodic Structures

Periodic structures have geometry and material properties that are repetitive and have a certain pattern. Such structures are very important in

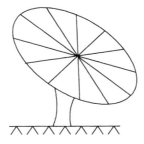

Figure 8.1: A periodic structure: an antenna dish.

applications such as the circularly periodic antenna dish of Figure 8.1, the truss lattice of a space station, or the repetitive pattern of stiffeners on the inside of the shell of an aircraft fuselage. These are well known applications.

An example of a less obvious application is a dynamic model of DNA,[1] a biological polymer that plays an essential role in the conservation and transportation of genetic information.

8.1.2 Inverse Problems

Inverse problems are so named because the output, such as frequency or displacement, is known, and the analysis leads to estimates of parameters such as the mass and stiffness of a structure. Such problems are sometimes also known as *identification problems*. For example, an engineer applies a known force to a vibrating structure with unknown mass and stiffness properties, resulting in measured vibrational data. An inverse analysis uses this frequency test data, along with the known input force, to estimate these structural parameters. We return to this problem after studying the basic tools of vibration analysis for multi–degree of freedom structures.

8.2 Stiffness and Flexibility

The modeling of multi degree of freedom structures requires one governing equation of motion for each degree of freedom. These *simultaneous* equations can be written in matrix form, leading to the matrix counterparts of mass, damping, and stiffness.

---

[1] *Nonlinear DNA dynamics: hierarchy of the models*, L.V. Yakushevich, *Physica D*, 79 (1994) 77–86. This is a very interesting paper demonstrating how five increasingly complex models of the DNA molecule can be used to study its dynamics.

8.2. STIFFNESS AND FLEXIBILITY

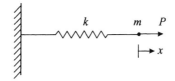

Figure 8.2: The concepts of flexibility and stiffness.

Several methods are utilized to derive the equations of motion and the property matrices. These include the application of Newton's second law of motion, Lagrange's equation, Hamilton's principle, and a way using influence methods. The method of influence coefficients provides additional insights to the concepts of flexibility and stiffness that arise naturally in the derivation of the equations of motion.

Regarding notation, *in this chapter we adopt braces { } for vectors and brackets [ ] for matrices.* This is different than the bold notation used in Chapter 7, but serves us well by making it easy to distinguish between vectors and matrices. It is also easier to follow the derivations using such explicit notation since the reader can compare general matrix equations of motion with, say, a two degree of freedom matrix equation of motion.

8.2.1 Influence Coefficients

In systems with many degrees of freedom a force acts on mass m_i resulting the in the motion of mass m_j. The concept of *flexibility* and *stiffness* leads to relations between a force and a displacement. The following is a static analysis, but d'Alembert's principle is used to extend these to dynamic problems.

Begin with the simple, single mass–stiffness system in Figure 8.2. Let the stiffness element k be subjected to static force P, resulting in displacement x. Hooke's law applies, $P = kx$, along with the inverse relation $x = k^{-1}P = fP$, where k = *stiffness influence coefficient* and f = *flexibility influence coefficient*. If one applies a *unit force*, $P = 1$, and measures the displacement x, then the flexibility coefficient can be evaluated directly since $f = x$. Similarly, if a *unit displacement* is applied, $x = 1$, then the stiffness coefficient is given by $k = P$. These ideas are generalized next for systems of more than one mass.

Consider now a system of three masses connected by two springs, with a third spring connecting these to a wall, as in Figure 8.3. There exist

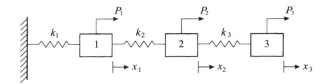

Figure 8.3: Three mass system.

displacement–force relations of the form

$$\begin{Bmatrix} x_1 \\ x_2 \\ x_3 \end{Bmatrix} = \begin{bmatrix} f_{11} & f_{12} & f_{13} \\ f_{21} & f_{22} & f_{23} \\ f_{31} & f_{32} & f_{33} \end{bmatrix} \begin{Bmatrix} P_1 \\ P_2 \\ P_3 \end{Bmatrix}.$$

Assuming that one can fix x_i and measure P_j, how are the f_{ij} evaluated? One possibility is to set $P_1 = 1$, and $P_2 = P_3 = 0$, and the first relation in the above matrix equation,

$$x_1 = f_{11}P_1 + f_{12}P_2 + f_{13}P_3,$$

becomes $x_1 = f_{11}$ providing the value of flexibility coefficient f_{11}. In a similar manner, the other influence coefficients can be determined. Thus, the *flexibility influence coefficient* f_{ij} is defined as the displacement at i due to a unit force at j, with all other forces set to 0.

One can also proceed from the reciprocal force–displacement relations:

$$\begin{Bmatrix} P_1 \\ P_2 \\ P_3 \end{Bmatrix} = \begin{bmatrix} k_{11} & k_{12} & k_{13} \\ k_{21} & k_{22} & k_{23} \\ k_{31} & k_{32} & k_{33} \end{bmatrix} \begin{Bmatrix} x_1 \\ x_2 \\ x_3 \end{Bmatrix}. \qquad (8.1)$$

One can fix x_i and measure P_j, then set $x_1 = 1$, and $x_2 = x_3 = 0$. The first equation,

$$P_1 = k_{11}x_1 + k_{12}x_2 + k_{13}x_3,$$

becomes $P_1 = k_{11}$, yielding the value of stiffness coefficient k_{11}. The *stiffness influence coefficient* k_{ij} is the force at i producing unit displacement at j, with all other displacements equal to 0. The stiffness matrix of Equation 8.1 also arises naturally when the equations of motion for the structure are derived, as we will see in the next section.

Thus, we see the beginnings of a procedure by which the flexibility and stiffness coefficients can be determined, as shown in the two examples that follow.

8.2. STIFFNESS AND FLEXIBILITY

Example 8.1 Flexibility Matrix
Consider again the three mass, three spring system. Determine the flexibility matrix. Apply force $P_1 = 1$, that is, a unit force on mass 1, and $P_2 = P_3 = 0$. Since k_2 and k_3 are unstretched, they displace as a rigid body with displacement

$$x_1 = P_1/k_1 = 1/k_1$$
$$= x_2 = x_3.$$

Thus,

$$\begin{Bmatrix} x_1 \\ x_2 \\ x_3 \end{Bmatrix} = \begin{bmatrix} 1/k_1 & 0 & 0 \\ 1/k_1 & 0 & 0 \\ 1/k_1 & 0 & 0 \end{bmatrix} \begin{Bmatrix} 1 \\ 0 \\ 0 \end{Bmatrix}.$$

Next, set $P_1 = P_3 = 0$, and $P_2 = 1$. We can see from the figure depicting the system that $x_2 = x_3$. The force displacement relation for mass 2 is $P_2 = (x_2 - x_1)k_2$, and for mass 1, $P_2 = x_1 k_1$. Combining these two equations, we have

$$x_2 = \left(\frac{1}{k_1} + \frac{1}{k_2}\right).$$

Therefore,

$$\begin{Bmatrix} x_1 \\ x_2 \\ x_3 \end{Bmatrix} = \begin{bmatrix} 0 & 1/k_1 & 0 \\ 0 & \left(\frac{1}{k_1} + \frac{1}{k_2}\right) & 0 \\ 0 & \left(\frac{1}{k_1} + \frac{1}{k_2}\right) & 0 \end{bmatrix} \begin{Bmatrix} 0 \\ 1 \\ 0 \end{Bmatrix}.$$

Finally, set $P_1 = P_2 = 0$, and $P_3 = 1$. Then displacement of mass 1 is governed by $x_1 = P_3/k_1$, of mass 2 by $x_2 = P_3(1/k_1 + 1/k_2)$, and of mass 3 by $x_3 = P_3(1/k_1 + 1/k_2 + 1/k_3)$, and in matrix form,

$$\begin{Bmatrix} x_1 \\ x_2 \\ x_3 \end{Bmatrix} = \begin{bmatrix} 0 & 0 & \frac{1}{k_1} \\ 0 & 0 & \left(\frac{1}{k_1} + \frac{1}{k_2}\right) \\ 0 & 0 & \left(\frac{1}{k_1} + \frac{1}{k_2} + \frac{1}{k_3}\right) \end{bmatrix} \begin{Bmatrix} 0 \\ 0 \\ 1 \end{Bmatrix}.$$

The complete flexibility matrix is then

$$\begin{Bmatrix} x_1 \\ x_2 \\ x_3 \end{Bmatrix} = \begin{bmatrix} \frac{1}{k_1} & \frac{1}{k_1} & \frac{1}{k_1} \\ \frac{1}{k_1} & \left(\frac{1}{k_1} + \frac{1}{k_2}\right) & \left(\frac{1}{k_1} + \frac{1}{k_2}\right) \\ \frac{1}{k_1} & \left(\frac{1}{k_1} + \frac{1}{k_2}\right) & \left(\frac{1}{k_1} + \frac{1}{k_2} + \frac{1}{k_3}\right) \end{bmatrix} \begin{Bmatrix} P_1 \\ P_2 \\ P_3 \end{Bmatrix},$$

which can be used for arbitrary forces, and, as we will see below, for dynamic forces where inertia becomes an important factor. ∎

Example 8.2 Stiffness Matrix
In this example, the stiffness influence coefficients are derived for the system of three masses linked by four springs to fixed boundaries. Note that by setting $k_4 = 0$, the system of the previous example is regained.

Proceeding according to the definition of stiffness influence coefficient, let $x_1 = 1$ with $x_2 = x_3 = 0$. The forces required at masses 1, 2, and 3 to enforce the required displacements are the following:

$$\begin{aligned} P_1 &= (k_1 + k_2)x_1 = k_1 + k_2 \equiv k_{11} \\ P_2 &= -k_2 x_1 = -k_2 \equiv k_{21} \\ P_3 &= 0 \equiv k_{31}. \end{aligned}$$

Repeat this procedure with $x_2 = 1$ and $x_1 = x_3 = 0$, and find the required forces to be

$$\begin{aligned} P_1 &= -k_2 x_2 = -k_2 \equiv k_{12} \\ P_2 &= (k_2 + k_3)x_2 \equiv k_{22} \\ P_3 &= -k_3 x_2 = -k_3 \equiv k_{32}. \end{aligned}$$

Finally, let $x_3 = 1$, with $x_1 = x_2 = 0$, and find

$$\begin{aligned} P_1 &= 0 \equiv k_{13} \\ P_2 &= -k_3 x_3 = -k_3 \equiv k_{23} \\ P_3 &= (k_3 + k_4)x_3 = k_3 + k_4 \equiv k_{33}. \end{aligned}$$

Combining all these force displacement relations into matrix form, we have

$$\left\{ \begin{array}{c} P_1 \\ P_2 \\ P_3 \end{array} \right\} = \left[\begin{array}{ccc} k_1 + k_2 & -k_2 & 0 \\ -k_2 & k_2 + k_3 & -k_3 \\ 0 & -k_3 & k_3 + k_4 \end{array} \right] \left\{ \begin{array}{c} x_1 \\ x_2 \\ x_3 \end{array} \right\}. \qquad (8.2)$$

The matrix of stiffness coefficients will be derived again using Newton's second law of motion and by using Lagrange's equation. By examining a variety of methods we gain a deeper understanding of these equations. ∎

It is important to note that the influence matrices are symmetrical with respect to the diagonal. Therefore, $f_{ij} = f_{ji}$ and $k_{ij} = k_{ji}$. Furthermore,

8.2. STIFFNESS AND FLEXIBILITY

$[f] = [k]^{-1}$, $[k] = [f]^{-1}$, and therefore $[f][k] = [I]$. This is known as the *Maxwell–Betti* reciprocity theorem, which is discussed in the next example.

Example 8.3 Reciprocity
Reciprocity means that load and deflection can be interchanged for a linear system in the following way. The deflection at location 2 due to a load at location 1 equals the deflection at 1 due to the same load at 2. This can be shown using work/energy considerations. Recall that $work = force \times displacement$, where the displacement is in the direction of the force. In terms of the influence coefficients, we need to integrate the product of force and displacement to find $W = kx^2/2 = fP^2/2$, where x is the maximum displacement, and P is the force at the maximum displacement. Consider the work done by the two loads P_i and P_j acting in this order at locations i and j, respectively. In terms of the flexibility coefficients, the work due to the first force P_i is $W_i = P_i^2 f_{ii}/2$, and due to the second force P_j is $W_j = P_j^2 f_{jj}/2$. But note that there is an additional displacement at location i due to the second force. Therefore, the additional work by P_j at location i is $W_{ij} = (f_{ij}P_j)P_i$. The total work is

$$W_{tot} = \frac{P_i^2 f_{ii}}{2} + \frac{P_j^2 f_{jj}}{2} + f_{ij}P_jP_i.$$

Doing the same for the case where load P_j is applied first results in a total work of

$$W_{tot} = \frac{P_j^2 f_{jj}}{2} + \frac{P_i^2 f_{ii}}{2} + f_{ji}P_iP_j.$$

For a linear system, the total work is a scalar and does not depend on path or order of loading. Therefore, these two work values must be equal, leading to the conclusion that $f_{ij} = f_{ji}$, confirming the previous observation of symmetry. ■

Example 8.4 Equations of Motion via Flexibility
Consider a slightly more intricate and useful example of the formulation of the equations of motion for a two degree of freedom oscillator.[2] Figure 8.4 is of an idealized model of a tower that oscillates transversely. In this idealized model, the mass of the tower is lumped at the top and bottom. The bottom lumped mass is incorporated into the foundation. For the purposes of this example, the tower is fixed at the bottom. The lumped mass at the top of the tower will translate and rotate as the oscillation progresses. We neglect

---
[2] This example is based on J.F. Wilson, **Dynamics of Offshore Structures**, Wiley–Interscience, 1984, p. 208.

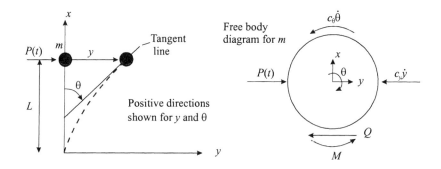

Figure 8.4: Idealization of a tower using an inverted pendulum.

the axial extension of the tower and assume that the lumped mass remains essentially at height L from the base. In reality, the lumped mass will go down as θ increases. However, for a linear oscillation, θ remains small and the position of m remains essentially along the horizontal line.

If the tower oscillates in a fluid such as water, then the *added mass* of the water that oscillates with the structure is similarly lumped at the ends. This concept has been discussed in Chapter 2. There are two coordinates describing this oscillation, a translation $y(t)$ and a rotation $\theta(t)$. From the free body diagram for the lumped mass in Figure 8.4, sum the forces acting on the mass to obtain the rectilinear equation of motion using Newton's second law of motion,

$$m\ddot{y} + c_y\dot{y} + Q(t) = P(t),$$

and sum the moments acting on the mass about the base to obtain the rotational equation of motion,

$$J\ddot{\theta} + c_\theta\dot{\theta} + M(t) = 0,$$

where c_y and c_θ are the viscous damping coefficients in the two directions, J is the polar moment of inertia for the lumped mass about the base, $Q(t)$ is the shear force between lumped mass and tower, $M(t)$ is the moment between lumped mass and tower, and $P(t)$ is the externally applied force on the structure. If there is an *external* moment acting on the structure, it would appear on the right hand side of the moment equation of motion.

The equations of motion for the tower system cannot yet be solved because force–displacement expressions relating Q and M to y and θ need to be derived. To do this, we need to derive the force–displacement relations as follows. There are two options for deriving these relations. Force–

8.2. STIFFNESS AND FLEXIBILITY

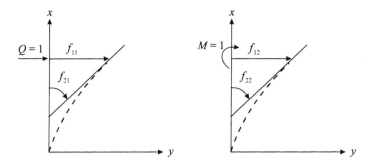

Figure 8.5: Physical meaning of flexibility coefficients.

displacement is related by a stiffness matrix,

$$\left\{ \begin{array}{c} Q \\ M \end{array} \right\} = \left[\begin{array}{cc} k_{11} & k_{12} \\ k_{21} & k_{22} \end{array} \right] \left\{ \begin{array}{c} y \\ \theta \end{array} \right\}. \qquad (8.3)$$

Alternatively, the displacement–force relationship, the inverse of Equation 8.3, is related by a flexibility matrix,

$$\left\{ \begin{array}{c} y \\ \theta \end{array} \right\} = \left[\begin{array}{cc} f_{11} & f_{12} \\ f_{21} & f_{22} \end{array} \right] \left\{ \begin{array}{c} Q \\ M \end{array} \right\}.$$

We see the relation between stiffness and flexibility as

$$\begin{aligned} \left[\begin{array}{cc} k_{11} & k_{12} \\ k_{21} & k_{22} \end{array} \right] &= \left[\begin{array}{cc} f_{11} & f_{12} \\ f_{21} & f_{22} \end{array} \right]^{-1} \\ &= \frac{1}{f_{11}f_{22} - f_{12}f_{21}} \left[\begin{array}{cc} f_{22} & -f_{12} \\ -f_{21} & f_{11} \end{array} \right]. \end{aligned} \qquad (8.4)$$

It turns out to be simpler, in general, to derive the flexibility matrix and then invert for the stiffness matrix.[3] It is the stiffness matrix that emerges in the matrix form of the equations of motion that are derived in the next section. For the flexibility equations, first let $Q = 1$ and $M = 0$. The first column of the flexibility matrix is the deflection of a cantilever beam due to a unit load at the end. See Figure 8.5.

---

[3] Experimentally, it is easier to apply a specified force and measure displacements than to apply a specified displacement and measure forces.

Let I be the moment of inertia about the axis perpendicular to bending, then $y = f_{11}Q = f_{11}$. From strength of materials, the deflection at the end of a cantilever beam due to a unit force at the free end is

$$f_{11} = \frac{L^3}{3EI}.$$

Similarly, the rotation at the end of a cantilever beam due to a unit force at the free end is $\theta = f_{21}Q = f_{21}$, where

$$f_{21} = \frac{L^2}{2EI}.$$

To determine the second column of the flexibility matrix, set $Q = 0$ and $M = 1$. Then, $y = f_{12}M = f_{12}$ and the deflection of the free end of a cantilever beam due to a unit moment applied at the free end is

$$f_{12} = \frac{L^2}{2EI}.$$

Similarly, $\theta = f_{22}M = f_{22}$ and the rotation of the free end of the cantilever beam due to unit moment applied at the free end is

$$f_{22} = \frac{L}{EI}.$$

As predicted by the Maxwell–Betti reciprocity theorem, $f_{12} = f_{21}$, and the flexibility matrix is

$$[f] = \frac{1}{EI} \begin{bmatrix} L^3/3 & L^2/2 \\ L^2/2 & L \end{bmatrix},$$

and from Equation 8.4

$$[k] = \frac{EI}{L} \begin{bmatrix} 12/L^2 & -6/L \\ -6/L & 4 \end{bmatrix}.$$

The matrix equation of motion is then

$$\begin{bmatrix} m & 0 \\ 0 & J \end{bmatrix} \begin{Bmatrix} \ddot{y} \\ \ddot{\theta} \end{Bmatrix} + \begin{bmatrix} c_y & 0 \\ 0 & c_\theta \end{bmatrix} \begin{Bmatrix} \dot{y} \\ \dot{\theta} \end{Bmatrix}$$
$$+ \frac{EI}{L} \begin{bmatrix} 12/L^2 & -6/L \\ -6/L & 4 \end{bmatrix} \begin{Bmatrix} y \\ \theta \end{Bmatrix} = \begin{Bmatrix} P(t) \\ 0 \end{Bmatrix}. \qquad (8.5)$$

8.3. DERIVATION OF EQUATIONS OF MOTION

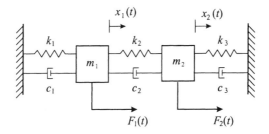

Figure 8.6: Two degree of freedom system.

We will learn how to solve such equations later in this chapter. ∎

The matrix framework is useful because it allows us to more straightforwardly examine key properties of multi–degree of freedom systems. The object of this chapter is twofold, first to learn how to derive matrix equations of motion such as Equation 8.5, and then to learn how to solve these equations. Several approaches to formulation and solution are examined. The example just completed showed a formulation based on Newton's second law of motion used in conjunction with the method of influence coefficients. Next we derive the equations of motion for a two degree of freedom system utilizing Newton's second law as applied directly to the free body diagram.

8.3 Derivation of Equations of Motion

Multi degree of freedom systems are natural extensions of two degree of freedom structures. For purposes of demonstration and discussion, the necessary concepts will be introduced by primarily working through the solution of two degree of freedom models. All the ideas transfer to larger systems, but with the two degree of freedom model key ideas can be demonstrated without the major algebraic and numerical demands made by the larger systems.

Consider the linear two degree of freedom system in Figure 8.6 that includes external forcing, damping and stiffness elements. The sketch shows the assumed positive directions of forces and displacements. To derive the governing equations of motion using Newton's second law of motion, a free body diagram is needed for each mass, as shown in Figure 8.7. This leads to one equation of motion for each mass.

Applying Newton's second law of motion to each free body diagram results in the following equations for the unknown displacement functions

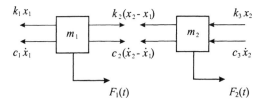

Figure 8.7: Two degree of freedom system: free body diagrams.

x_1 and x_2,

$$F_1(t) - c_1\dot{x}_1 - k_1 x_1 + c_2(\dot{x}_2 - \dot{x}_1) + k_2(x_2 - x_1) = m_1\ddot{x}_1$$
$$F_2(t) - c_3\dot{x}_2 - k_3 x_2 - c_2(\dot{x}_2 - \dot{x}_1) - k_2(x_2 - x_1) = m_2\ddot{x}_2.$$

These can be written in standard form as

$$m_1\ddot{x}_1 + (c_1 + c_2)\dot{x}_1 - c_2\dot{x}_2 + (k_1 + k_2)x_1 - k_2 x_2 = F_1(t)$$
$$m_2\ddot{x}_2 + (c_2 + c_3)\dot{x}_2 - c_2\dot{x}_1 + (k_2 + k_3)x_2 - k_2 x_1 = F_2(t).$$

Note that these two governing equations of motion are *coupled*, meaning that the independent variables appear in both equations. These equations must be solved *simultaneously*. The coupling terms are, respectively,

$$-c_2\dot{x}_2 - k_2 x_2$$
$$-c_2\dot{x}_1 - k_2 x_1.$$

This mathematical coupling represents the *physical* coupling between the masses, that is, the stiffness and damping elements c_2 and k_2 that connect the masses to each other.

It is customary and more convenient (especially for systems with more than two or three degrees of freedom) to recast the above equations in matrix form,

$$\begin{bmatrix} m_1 & 0 \\ 0 & m_2 \end{bmatrix} \begin{Bmatrix} \ddot{x}_1 \\ \ddot{x}_2 \end{Bmatrix} + \begin{bmatrix} c_1 + c_2 & -c_2 \\ -c_2 & c_2 + c_3 \end{bmatrix} \begin{Bmatrix} \dot{x}_1 \\ \dot{x}_2 \end{Bmatrix}$$
$$+ \begin{bmatrix} k_1 + k_2 & -k_2 \\ -k_2 & k_2 + k_3 \end{bmatrix} \begin{Bmatrix} x_1 \\ x_2 \end{Bmatrix} = \begin{Bmatrix} F_1(t) \\ F_2(t) \end{Bmatrix}. \qquad (8.6)$$

Of course, Equation 8.6, or any n degree of freedom equation, can be written as

$$[m]\{\ddot{x}\} + [c]\{\dot{x}\} + [k]\{x\} = \{F(t)\}, \qquad (8.7)$$

8.3. DERIVATION OF EQUATIONS OF MOTION

where the mass $[m]$, damping $[c]$, and stiffness $[k]$ matrices are symmetric:[4] $[m] = [m]^T$, $[c] = [c]^T$, $[k] = [k]^T$, and superscript T denotes transpose of a matrix. We can see that the coupling occurs in the off-diagonal terms of the property matrices. If coupling disappears, then the equations of motion could be solved independently as single degree of freedom oscillators. To numerically integrate the above equations, recast them in the state space format, as shown in Chapter 6, and use $MATLAB$, for example.

Also note that matrix equation of motion Equation 8.7 is identical to Lagrange's equation for *small oscillations*, derived in Section 7.3.1 for the undamped, unforced system.

Example 8.5 Derivation of Equation 8.6 by Lagrange's Equation and by Hamilton's Principle

The schematic of the system is provided in Figure 8.6. To use Lagrange's equation we need the kinetic, potential, and dissipation energies, respectively,

$$T = \frac{1}{2}m_1\dot{x}_1^2 + \frac{1}{2}m_2\dot{x}_2^2$$

$$V = \frac{1}{2}k_1x_1^2 + \frac{1}{2}k_2(x_2 - x_1)^2 + \frac{1}{2}k_3x_2^2$$

$$R = \frac{1}{2}c_1\dot{x}_1^2 + \frac{1}{2}c_2(\dot{x}_2 - \dot{x}_1)^2 + \frac{1}{2}c_3\dot{x}_2^2.$$

Lagrange's equation for generalized coordinate x_1 is

$$\frac{d}{dt}\left(\frac{\partial T}{\partial \dot{x}_1}\right) + \frac{\partial V}{\partial x_1} + \frac{\partial R}{\partial \dot{x}_1} = F_1(t).$$

Taking all the appropriate derivatives, the equation of motion for mass m_1 is

$$m_1\ddot{x}_1 + (c_1 + c_2)\dot{x}_1 - c_2\dot{x}_2 + (k_1 + k_2)x_1 - k_2x_2 = F_1(t).$$

Following the same procedure for mass m_2, as an exercise, show that the second governing equation is

$$m_2\ddot{x}_2 + (c_2 + c_3)\dot{x}_2 - c_2\dot{x}_1 + (k_2 + k_3)x_2 - k_2x_1 = F_2(t).$$

Hamilton's principle requires that the following variational equation be satisfied,

$$\int_{t_1}^{t_2} (\delta T + \delta W)dt = 0. \tag{8.8}$$

---

[4] This is not true for systems where electromagnetic forces or couples exist. Systems with symmetric matrices are known as *self-adjoint*. A good text on matrix methods will provide an introduction.

The virtual work can be written as the sum of conservative and nonconservative components. The conservative portion is equal to the negative of the potential energy. Non–conservative work W_{nc} is performed by dissipative forces such as friction, and by applied forces external to the system. Therefore, $\delta W = -\delta V + \delta W_{nc}$. Equation 8.8 can then be written as

$$\int_{t_1}^{t_2} (\delta L + \delta W_{nc}) dt = 0,$$

where δL is the variation of the Lagrangian, given by

$$\begin{aligned}
\delta L &= \delta(T - V) \\
&= \delta\left(\frac{1}{2}m_1\dot{x}_1^2 + \frac{1}{2}m_2\dot{x}_2^2 - \frac{1}{2}k_1 x_1^2 - \frac{1}{2}k_2(x_2 - x_1)^2 - \frac{1}{2}k_3 x_2^2\right) \\
&= m_1\dot{x}_1\delta\dot{x}_1 + m_2\dot{x}_2\delta\dot{x}_2 - k_1 x_1 \delta x_1 - k_2(x_2 - x_1)(\delta x_2 - \delta x_1) \\
&\quad - k_3 x_2 \delta x_2.
\end{aligned}$$

As you recall from the last chapter, we need to rewrite a term that is a variation of a derivative as a variation of the parameter itself. This can be accomplished in the following way, for example,

$$\frac{d}{dt}(m_1 \dot{x}_1 \delta x_1) = m_1 \ddot{x}_1 \delta x_1 + m_1 \dot{x}_1 \delta \dot{x}_1,$$

and therefore

$$m_1 \dot{x}_1 \delta \dot{x}_1 dt = d(m_1 \dot{x}_1 \delta x_1) - m_1 \ddot{x}_1 \delta x_1 dt.$$

First proceed for the case with no damping or external forces. Hamilton's principle becomes

$$\begin{aligned}
\int_{t_1}^{t_2} &\{[-m_1\ddot{x}_1 - k_1 x_1 + k_2(x_2 - x_1)]\delta x_1 \\
&+ [-m_2\ddot{x}_2 - k_3 x_2 - k_2(x_2 - x_1)]\delta x_2\} \, dt \\
&+ \int_{t_1}^{t_2} [d(m\dot{x}_1 \delta x_1) + d(m\dot{x}_2 \delta x_2)] = 0. \quad (8.9)
\end{aligned}$$

The second integral can be directly integrated, resulting in $|m\dot{x}_1 \delta x_1 + m\dot{x}_2 \delta x_2|_{t_1}^{t_2}$. Since variations δx_1 and δx_2 equal zero at t_1 and t_2, both these expressions equal zero as well. Furthermore, in the first integral above, variations δx_1 and δx_2 are independent of each other and are arbitrary; the equality can then only hold if the expressions in each set of square brackets add to zero, resulting in the equations of motion,

$$\begin{aligned}
m_1\ddot{x}_1 + (k_1 + k_2)x_1 - k_2 x_2 &= 0 \\
m_2\ddot{x}_2 + (k_2 + k_3)x_2 - k_2 x_1 &= 0.
\end{aligned}$$

8.3. DERIVATION OF EQUATIONS OF MOTION

Now add the effects of damping. The virtual work done by viscous damping equals $F_d \delta x = (-c\dot{x})\delta x$. For the system considered here,

$$\delta W_{nc} = -c_1 \dot{x}_1 \delta x_1 - c_2(\dot{x}_2 - \dot{x}_1)(\delta x_2 - \delta x_1) - c_3 \dot{x}_2 \delta x_2.$$

Insert these terms into Hamilton's principle, Equation 8.9, and group according to like variation, with the resulting damped equations of free motion,

$$m_1 \ddot{x}_1 + (c_1 + c_2)\dot{x}_1 + (k_1 + k_2)x_1 - c_2 \dot{x}_2 - k_2 x_2 = 0$$
$$m_2 \ddot{x}_2 + (c_2 + c_3)\dot{x}_2 + (k_2 + k_3)x_2 - c_2 \dot{x}_1 - k_2 x_1 = 0.$$

Finally, add the effects of the virtual work due to externally applied nonconservative forces,

$$\delta W_{nc} = F_1 \delta x_1 + F_2 \delta x_2.$$

These are substituted into Hamilton's principle, and we arrive at the same equations derived earlier by Lagrange's equation.

While Hamilton's principle appeared more involved than Lagrange's equation, there really was little difference between the two approaches. However, it is true that depending on the application, one method may have an advantage over the other. ∎

The major advantage of these energy methods is best observed with larger and more complex systems where Newton's second law of motion is cumbersome to apply. The next example is of a system with mass and stiffness coupling.

Example 8.6 Equations of Motion with Mass Coupling
Consider the two degree of freedom rigid model of Figure 8.8. Even though there is a single mass, two coordinates are required to specify its translation and rotation, as shown in the figure. If the force is applied at the center of mass C, then the beam only translates rather than both translate and rotate. The location of C depends on the distribution of mass on the beam. If the mass is uniformly distributed, then C is at the geometric center. Derive the governing equations of motion from a free body diagram, where the force is applied at an arbitrary point O.

Solution   From the free body diagram, consider an element of the rigid beam dm at a distance ε from O. Assume that the beam has a total mass m, and that it undergoes small rotations and translations. In this way we can reasonably say that the acceleration of the mass element is in the vertical direction and has a magnitude of

$$\frac{d^2}{dt^2}(x + \varepsilon \theta) = \ddot{x} + \varepsilon \ddot{\theta}.$$

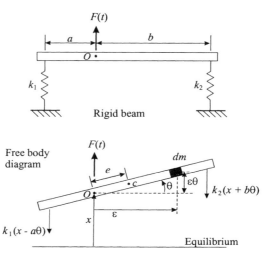

Figure 8.8: Two degree of freedom idealization with mass coupling.

By Newton's second law of motion, the sum of the forces in the vertical direction due to the elongation of the springs can be set equal to the mass times acceleration,

$$\begin{aligned}
F(t) - k_1(x - a\theta) - k_2(x + b\theta) &= \int_{beam} (\ddot{x} + \varepsilon\ddot{\theta}) dm \\
&= \ddot{x} \int_{beam} dm + \ddot{\theta} \int_{beam} \varepsilon dm \\
&= \ddot{x}m + \ddot{\theta}me,
\end{aligned}$$

where the location of the mass center is defined by $e = (\int \varepsilon dm)/m$.

Similarly, the moment equation about O becomes

$$\begin{aligned}
k_1(x - a\theta)a - k_2(x + b\theta)b &= \int_{beam} (\ddot{x} + \varepsilon\ddot{\theta})\varepsilon dm \\
&= \ddot{x} \int_{beam} \varepsilon dm + \ddot{\theta} \int_{beam} \varepsilon^2 dm \\
&= \ddot{x}me + \ddot{\theta}I_O,
\end{aligned}$$

where the mass moment of inertia of the body about O is $I_O = \int \varepsilon^2 dm$. The sign of the moment is taken as positive if it is in the positive θ direction.

8.3. DERIVATION OF EQUATIONS OF MOTION

These two equations of motion can be written in standard form as

$$m\ddot{x} + me\ddot{\theta} + (k_1 + k_2)x + (k_2b - k_1a)\theta = F(t)$$
$$me\ddot{x} + I_O\ddot{\theta} + (k_2b - k_1a)x + (k_1a^2 + k_2b^2)\theta = 0,$$

or in matrix form,

$$\begin{bmatrix} m & me \\ me & I_O \end{bmatrix} \begin{Bmatrix} \ddot{x} \\ \ddot{\theta} \end{Bmatrix} + \begin{bmatrix} k_1 + k_2 & k_2b - k_1a \\ k_2b - k_1a & k_1a^2 + k_2b^2 \end{bmatrix} \begin{Bmatrix} x \\ \theta \end{Bmatrix} = \begin{Bmatrix} F(t) \\ 0 \end{Bmatrix}.$$

Both mass and stiffness matrices are coupled. If C and O coincide, $e = 0$, and the mass but not stiffness decouples. The stiffness decouples if $k_2b = k_1a$. Later in this chapter we will discover how we can decouple both mass and stiffness matrices by transforming the problem to a new coordinate system using a procedure known as modal analysis. ∎

The mass and stiffness matrices are related to the energies of the system. This makes sense to us since the stiffness elements store energy and the inertia elements are related to the force by Newton's second law of motion. In the next section an understanding of energy properties will help us understand system behavior.

8.3.1 $[m]$ and $[k]$ Matrix Properties

The kinetic T and potential V energies of a system can be related to the stiffness (flexibility) and mass matrices, as shown here. (Recall that these energies have been used to derive the equation of motion via Lagrange's equation and Hamilton's principle.) For a single linear spring element with constant k, the area under the force–displacement curve is its potential energy: $V = \frac{1}{2}kx^2 = \frac{1}{2}Fx$, where Hooke's law of static linear displacement $F = kx$ is used. For an n degree of freedom vibrating system, the total potential energy can be written in two ways as well,

$$V = \frac{1}{2}\sum_{i=1}^{n}\sum_{j=1}^{n} k_{ij}x_i x_j = \frac{1}{2}\{x\}^T[k]\{x\}$$

$$= \frac{1}{2}\sum_{i=1}^{n}\sum_{j=1}^{n} f_{ij}F_i F_j = \frac{1}{2}\{F\}^T[f]\{F\}.$$

The kinetic energy for a single mass element is $T = \frac{1}{2}m\dot{x}^2$. For n masses it is

$$T = \frac{1}{2}\sum_{i=1}^{n} m_i \dot{x}_i^2 = \frac{1}{2}\{\dot{x}\}^T[m]\{\dot{x}\}.$$

Since kinetic energy is always positive, $T > 0$, $[m]$ is *positive definite*. The potential energy will always be positive for a restrained system undergoing oscillation. For the case of an unrestrained system, however, the potential energy may take a minimum value of $V = 0$. That is, there is no relative motion between masses, and the spring elements remain undeformed. In this case $[k]$ is called *positive semi–definite*. We will look more closely at such systems in Section 9.3.

Example 8.7 Potential Energy and the Stiffness Matrix
Consider the simple three degree of freedom system of Figure 8.9. The masses oscillate along the horizontal axis. It is straightforward to write the system potential energy as a function of the positions of the masses

$$\begin{aligned} V &= \frac{1}{2}[k_1 x_1^2 + k_2(x_2 - x_1)^2 + k_3(x_3 - x_2)^2 + k_4 x_3^2] \\ &= \frac{1}{2}[(k_1 + k_2)x_1^2 + (k_2 + k_3)x_2^2 + (k_3 + k_4)x_3^2 - 2k_2 x_1 x_2 - 2k_3 x_2 x_3]. \end{aligned}$$

In matrix form, $V = \frac{1}{2}\{x\}^T[k]\{x\}$, where

$$[k] = \begin{bmatrix} k_1 + k_2 & -k_2 & 0 \\ -k_2 & k_2 + k_3 & -k_3 \\ 0 & -k_3 & k_3 + k_4 \end{bmatrix},$$

which is the same stiffness matrix as in Equation 8.2. As an exercise, perform the triple product to verify the previous expression for V. If the system above is unrestrained, that is, $k_1 = k_4 = 0$, then

$$[k] = \begin{bmatrix} k_2 & -k_2 & 0 \\ -k_2 & k_2 + k_3 & -k_3 \\ 0 & -k_3 & k_3 \end{bmatrix},$$

which is a singular matrix. This means that the determinant of the stiffness matrix equals zero, $|k| = 0$, and its inverse does not exist. $[f]$ then does not exist, which is expected since the concept of flexibility is not applicable to an unrestrained system. Such systems are important in practice and will be discussed in Section 9.3. ∎

In the next section, we will solve the free vibration problem without damping, $[m]\{\ddot{x}\} + [k]\{x\} = \{0\}$. In solving the coupled equations, we will be introduced to the new concept of *mode of vibration*. Modes turn out to be a useful tool for understanding and mathematically modeling intricate vibration problems.

8.4. UNDAMPED VIBRATION

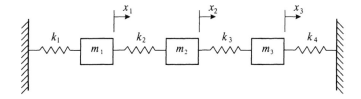

Figure 8.9: Three degree of freedom system.

8.4 Undamped Vibration

The undamped free vibration problem for an n degree of freedom structure or machine is governed by the matrix equation of motion

$$[m]\{\ddot{x}\} + [k]\{x\} = \{0\}, \quad (8.10)$$

where $n \times n$ matrices $[m]$ and $[k]$ are positive definite and the vector of displacements $\{x\}$ is of dimension $n \times 1$.

The approach here is the same as for single degree of freedom systems, look for harmonic solution. We will find that there are as many natural frequencies as there are degrees of freedom, and the complete response for each mass is a sum of all the harmonic components. Consider the following two degree of freedom system.

8.4.1 Two Degree of Freedom Motion: Solution by the Direct Method

Referring to Figures 8.6 and 8.7, with $c_1 = 0$, $c_2 = 0$ and $F_1(t) = 0$, $F_2(t) = 0$, the equations governing free vibration are

$$m_1 \ddot{x}_1 + (k_1 + k_2)x_1 - k_2 x_2 = 0 \quad (8.11)$$
$$m_2 \ddot{x}_2 + (k_2 + k_3)x_2 - k_2 x_1 = 0, \quad (8.12)$$

or in matrix form,

$$\begin{bmatrix} m_1 & 0 \\ 0 & m_2 \end{bmatrix} \begin{Bmatrix} \ddot{x}_1 \\ \ddot{x}_2 \end{Bmatrix} + \begin{bmatrix} k_1 + k_2 & -k_2 \\ -k_2 & k_2 + k_3 \end{bmatrix} \begin{Bmatrix} x_1 \\ x_2 \end{Bmatrix} = \begin{Bmatrix} 0 \\ 0 \end{Bmatrix}. \quad (8.13)$$

For simplicity and generality rewrite the stiffness matrix as

$$\begin{bmatrix} k_1 + k_2 & -k_2 \\ -k_2 & k_2 + k_3 \end{bmatrix} = \begin{bmatrix} k_{11} & k_{12} \\ k_{12} & k_{22} \end{bmatrix}.$$

Several points are important to understand about the solution of matrix Equation 8.13. The individual governing equations are homogeneous and therefore it is only possible to obtain a solution to within an arbitrary constant.[5] When the initial conditions are satisfied, the solution becomes specific to that problem and the constants are no longer arbitrary. While a harmonic solution can be assumed, we would like to show how this can be motivated by physical considerations in a general solution.

The governing equations for $x_1(t)$ and $x_2(t)$ are the same mathematically, implying a *synchronicity of motion*. The displacement coordinates $x_1(t)$ and $x_2(t)$ are then expected to vary in a constant proportion as a function of time, that is, the motions of the two masses are either *in phase*[6] or completely out of phase.

Then $x_1(t)$ and $x_2(t)$ are assumed to be respectively the product of constants u_1 and u_2 and a function of time $Y(t)$, all of which are determined in the following analysis. Because the equations of motion are identical, we expect the ratio of the amplitudes u_1 and u_2 to be independent of time. This is a result of harmonic time dependence. The ratios between the two displacements are called the *modes of vibration*.

Therefore, let

$$x_1(t) = u_1 Y(t) \qquad (8.14)$$
$$x_2(t) = u_2 Y(t), \qquad (8.15)$$

where it is noted that

$$\frac{x_1(t)}{x_2(t)} = \frac{u_1}{u_2}.$$

Since Equations 8.14 and 8.15 are solutions, they are differentiated and substituted into the governing Equations 8.11 and 8.12 they must satisfy. The result is

$$m_1 u_1 \ddot{Y} + (k_{11} u_1 + k_{12} u_2) Y = 0$$
$$m_2 u_2 \ddot{Y} + (k_{12} u_1 + k_{22} u_2) Y = 0.$$

The following ratios must all hold:

$$\frac{k_{11} u_1 + k_{12} u_2}{m_1 u_1} = \frac{k_{12} u_1 + k_{22} u_2}{m_2 u_2} = -\frac{\ddot{Y}}{Y} \equiv \lambda.$$

---

[5] If $x_1(t)$ and $x_2(t)$ are solutions that satisfy the governing differential equations, then so are $Ax_1(t)$ and $Bx_2(t)$, since the constants A and B can be canceled due to the zero right hand sides. This fact was also true for the single degree of freedom solution.

[6] We will see how damping results in a phase difference between the responses of the two masses.

8.4. UNDAMPED VIBRATION

The third equality results from the fact that the first two ratios are constant, and therefore must be equal to some constant λ, evaluated below. Therefore, the following three equations, one differential and two algebraic, must be satisfied:

$$\ddot{Y} + \lambda Y = 0 \tag{8.16}$$
$$(k_{11} - \lambda m_1)u_1 + k_{12}u_2 = 0 \tag{8.17}$$
$$k_{12}u_1 + (k_{22} - \lambda m_2)u_2 = 0. \tag{8.18}$$

Equation 8.16–8.18 have three unknowns, λ, u_1, and u_2. First consider Equation 8.16 in order to establish the meaning of λ. From Chapters 2 and 3, we know that its solution is $Y(t) = A \exp(rt)$ with characteristic roots of the form $r = \pm\sqrt{-\lambda}$. The complete solution is

$$Y(t) = A_1 e^{\sqrt{-\lambda}t} + A_2 e^{-\sqrt{-\lambda}t}. \tag{8.19}$$

If λ is negative, the first solution goes to infinity with time, contradicting the assumption of a stable motion. Therefore, λ must be positive, resulting in the solution for the harmonic oscillator. Let $\lambda = \omega^2$ and Equation 8.19 can be written as $Y(t) = A_1 \exp(i\omega_1 t) + A_2 \exp(-i\omega_2 t)$, or as

$$Y(t) = C_1 \cos(\omega_1 t - \phi_1) + C_2 \cos(\omega_2 t - \phi_2). \tag{8.20}$$

Equations 8.17 and 8.18 may be put in matrix form,

$$\begin{bmatrix} k_{11} - \lambda m_1 & k_{12} \\ k_{12} & k_{22} - \lambda m_2 \end{bmatrix} \begin{Bmatrix} u_1 \\ u_2 \end{Bmatrix} = \begin{Bmatrix} 0 \\ 0 \end{Bmatrix}. \tag{8.21}$$

The simultaneous solution of Equations 8.17 and 8.18 results in the values of parameter ω^2 for which nontrivial solutions exist. The ω values are called the *characteristic values* or *eigenvalues* of the problem. From linear algebra, it is known that Equations 8.17 and 8.18 have nontrivial solutions[7] only if the *determinant* of the coefficients of u_1 and u_2 equals zero.

This requirement is due to the way in which a matrix equation is solved. Let Equation 8.21 be written in short-hand as $[K]\{u\} = \{0\}$. A trivial solution arises when multiplying both sides by the inverse of $[K]$, that is, $[K]^{-1}[K]\{u\} = \{0\}$. Therefore, $\{u\} = \{0\}$. A non–trivial solution can be imposed if the inverse does not exist. The inverse is given by $[K]^{-1} = \text{adj}[K]/\det[K]$, which exists only if $\det[K] \neq 0$. Therefore, requiring that $\det[K] = 0$ implies that there is no inverse and, therefore, no trivial solution.

---
[7] The trivial solutions are $u_1 = u_2 = 0$, where there is no motion.

From Equation 8.21,

$$\det \begin{bmatrix} k_{11} - \omega^2 m_1 & k_{12} \\ k_{12} & k_{22} - \omega^2 m_2 \end{bmatrix} = 0. \tag{8.22}$$

Equation 8.22 is called the *characteristic determinant*. Expand the determinant to obtain the *characteristic* or *frequency equation*,

$$m_1 m_2 \omega^4 - (m_1 k_{22} + m_2 k_{11})\omega^2 + k_{11} k_{22} - k_{12}^2 = 0,$$

with roots given by the quadratic formula,

$$\omega_{1,2}^2 = \frac{m_1 k_{22} + m_2 k_{11}}{2 m_1 m_2} \mp \sqrt{\left(\frac{m_1 k_{22} + m_2 k_{11}}{2 m_1 m_2}\right)^2 - \frac{k_{11} k_{22} - k_{12}^2}{m_1 m_2}}. \tag{8.23}$$

The $\mp$ ordering of the roots follows the convention that the frequencies are numbered from lowest to highest, with ω_1 called the *fundamental frequency*.

The two degree of freedom system has two natural frequencies of oscillation. Once ω_1 and ω_2 have been evaluated, we go back to either Equation 8.17 or 8.18 in order to evaluate the *ratios* u_1/u_2 for each ω that solves the two equations. It is emphasized that the equations are homogeneous and, therefore only relative values of u_1 and u_2 can be found. There is one ratio for each ω, and we will use the notation $(u_1/u_2)_1$ or u_{11}/u_{21} for ω_1 and $(u_1/u_2)_2$ or u_{12}/u_{22} for ω_2, where the first subscript denotes mass number and the second denotes frequency number. For ω_1^2, Equations 8.17 and 8.18 become

$$(k_{11} - \omega_1^2 m_1) u_{11} + k_{12} u_{21} = 0 \tag{8.24}$$
$$k_{12} u_{11} + (k_{22} - \omega_1^2 m_2) u_{21} = 0. \tag{8.25}$$

Note that these equations can be written as

$$k_{11} u_{11} + k_{12} u_{21} = \omega_1^2 m_1 u_{11}$$
$$k_{12} u_{11} + k_{22} u_{21} = \omega_1^2 m_2 u_{21},$$

or in matrix form

$$[k]\{u\}_1 = \omega_1^2 [m]\{u\}_1, \tag{8.26}$$

where $\{u\}_1 = [u_{11}\ u_{21}]^T$. *Either* Equation 8.24 *or* 8.25 can be solved for the necessary ratios or *eigenvectors*,

$$\frac{u_{21}}{u_{11}} = -\frac{k_{11} - \omega_1^2 m_1}{k_{12}} = -\frac{k_{12}}{k_{22} - \omega_1^2 m_2}, \tag{8.27}$$

8.4. UNDAMPED VIBRATION

and similarly for ω_2^2: $[k]\{u\}_2 = \omega_2^2[m]\{u\}_2$, with the eigenvectors given by

$$\frac{u_{22}}{u_{12}} = -\frac{k_{11} - \omega_2^2 m_1}{k_{12}} = -\frac{k_{12}}{k_{22} - \omega_2^2 m_2}, \qquad (8.28)$$

where ω_1^2 and ω_2^2 are substituted from Equation 8.23. Equations 8.27 and 8.28 represent the relative magnitudes of the response of each mass for the frequencies ω_1 and ω_2, respectively, and are known as the *modal ratios* for the structure. Such an equation exists for each mode. The modal characteristics of a structure include frequency and mode shape, and is denoted by $[\omega_i, \{u\}_i]$, $i = 1, 2$. The absolute response magnitudes are evaluated by satisfying the initial conditions, as we see next.

The complete motion of the structure is equal to the linear superposition of the two solutions (modes), Equation 8.20, or

$$\left\{ \begin{array}{c} x_1(t) \\ x_2(t) \end{array} \right\} = C_1 \left\{ \begin{array}{c} u_{11} \\ u_{21} \end{array} \right\} \cos(\omega_1 t - \phi_1) + C_2 \left\{ \begin{array}{c} u_{12} \\ u_{22} \end{array} \right\} \cos(\omega_2 t - \phi_2), \quad (8.29)$$

with four *initial conditions*: $x_1(0), x_2(0), \dot{x}_1(0)$ and $\dot{x}_2(0)$. In order to substitute the modal ratios of Equations 8.27 and 8.28 in Equation 8.29, rewrite the latter in the following form,

$$\left\{ \begin{array}{c} x_1(t) \\ x_2(t) \end{array} \right\} = \bar{C}_1 \left\{ \begin{array}{c} 1 \\ u_{21}/u_{11} \end{array} \right\} \cos(\omega_1 t - \phi_1) + \bar{C}_2 \left\{ \begin{array}{c} 1 \\ u_{22}/u_{12} \end{array} \right\} \cos(\omega_2 t - \phi_2), \tag{8.30}$$

where $(C_1 u_{11}) = \bar{C}_1$ and $(C_2 u_{12}) = \bar{C}_2$ are the new constants of integration. Physically, Equation 8.30 demonstrates that the response of each mass is a sum of natural harmonics. The proportion contributed by each harmonic to the total response is given by the modal ratio. The numerical values of $\bar{C}_1, \bar{C}_2, \phi_1$ and ϕ_2 are determined by satisfying the four initial conditions, that is, by simultaneously solving the following four equations,

$$\left\{ \begin{array}{c} x_1(0) \\ x_2(0) \end{array} \right\} = \bar{C}_1 \left\{ \begin{array}{c} 1 \\ u_{21}/u_{11} \end{array} \right\} \cos(-\phi_1)$$

$$+ \bar{C}_2 \left\{ \begin{array}{c} 1 \\ u_{22}/u_{12} \end{array} \right\} \cos(-\phi_2) \qquad (8.31)$$

$$\left\{ \begin{array}{c} \dot{x}_1(0) \\ \dot{x}_2(0) \end{array} \right\} = -\bar{C}_1 \omega_1 \left\{ \begin{array}{c} 1 \\ u_{21}/u_{11} \end{array} \right\} \sin(-\phi_1)$$

$$- \bar{C}_2 \omega_2 \left\{ \begin{array}{c} 1 \\ u_{22}/u_{12} \end{array} \right\} \sin(-\phi_2). \qquad (8.32)$$

It is interesting to note that if for some particular initial conditions $\bar{C}_2 = 0$, then the two degree of freedom structure will oscillate only in its first mode at its first natural frequency. Similarly, if $\bar{C}_1 = 0$, the structure will oscillate solely in its second mode at its second natural frequency.

Modes generally have amplitudes of alternating signs, and therefore, the modal amplitude lines cross the zero $(n-1)$ times where n is the mode number. The crossing points are called *nodes*.

To emphasize, a normal mode of oscillation is one in which all points on the structure execute simple harmonic motion at the same frequency and phase. This is known as synchronous *motion*. The amplitude generally varies from point to point. Figure 8.10 shows how the two modes of a two degree of freedom system undergo synchronous motion.

Let us consider a representative example of the above procedure.

Example 8.8 Two Degrees of Freedom, No Damping
Begin with the model of Figure 8.6, and simplify for the undamped, unforced case as shown in Figure 8.11.

Let $m_1 = m$ kg, $m_2 = 2m$ kg, and $k_1 = k$ N/m, $k_2 = 2k$ N/m, $k_3 = 3k$ N/m, and obtain the natural frequencies and modes of vibration. The equations of motion become

$$m\ddot{x}_1 + 3kx_1 - 2kx_2 = 0$$
$$2m\ddot{x}_2 + 5kx_2 - 2kx_1 = 0.$$

In matrix form, it is easy to note the symmetry of the mass and stiffness matrices,

$$\begin{bmatrix} m & 0 \\ 0 & 2m \end{bmatrix} \begin{Bmatrix} \ddot{x}_1 \\ \ddot{x}_2 \end{Bmatrix} + \begin{bmatrix} 3k & -2k \\ -2k & 5k \end{bmatrix} \begin{Bmatrix} x_1 \\ x_2 \end{Bmatrix} = \begin{Bmatrix} 0 \\ 0 \end{Bmatrix}. \quad (8.33)$$

Assume harmonic responses of the form

$$x_1(t) = A_1 \exp(i\omega t)$$
$$x_2(t) = A_2 \exp(i\omega t).$$

Differentiate and substitute into governing Equation 8.33 to find

$$-\omega^2 \begin{bmatrix} m & 0 \\ 0 & 2m \end{bmatrix} \begin{Bmatrix} A_1 e^{i\omega t} \\ A_2 e^{i\omega t} \end{Bmatrix} + \begin{bmatrix} 3k & -2k \\ -2k & 5k \end{bmatrix} \begin{Bmatrix} A_1 e^{i\omega t} \\ A_2 e^{i\omega t} \end{Bmatrix} = \begin{Bmatrix} 0 \\ 0 \end{Bmatrix}.$$

Combine like terms to obtain

$$\begin{bmatrix} -\omega^2 m + 3k & -2k \\ -2k & -2\omega^2 m + 5k \end{bmatrix} \begin{Bmatrix} A_1 \\ A_2 \end{Bmatrix} = \begin{Bmatrix} 0 \\ 0 \end{Bmatrix}. \quad (8.34)$$

8.4. UNDAMPED VIBRATION

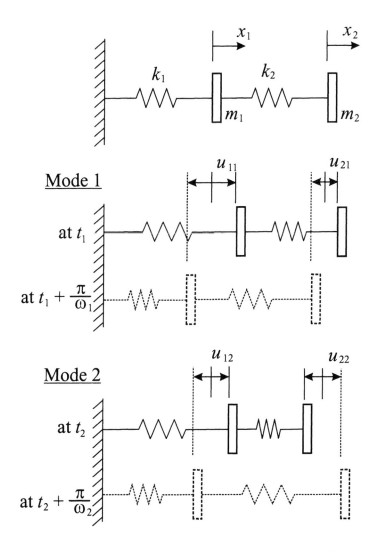

Figure 8.10: Schematic showing the synchronous motion of the two modes of a two degree of freedom system. The actual motion is a combination of the two modes of vibration.

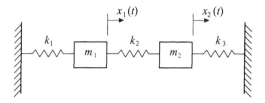

Figure 8.11: Two degrees of freedom, no damping.

Setting the determinant of the coefficient matrix to zero leads to the characteristic equation

$$2m^2\omega^4 - 11mk\omega^2 + 11k^2 = 0,$$

with roots

$$\omega_{1,2}^2 = \frac{11mk}{4m^2} \mp \sqrt{\frac{(11mk)^2}{16m^4} - 4\frac{22(mk)^2}{16m^4}} = [2.75 \mp 1.44]\frac{k}{m}.$$

The two natural frequencies are then

$$\omega_1 = 1.14\sqrt{\frac{k}{m}} \text{ rad/sec}, \quad \omega_2 = 2.05\sqrt{\frac{k}{m}} \text{ rad/sec}.$$

Once the natural frequencies are known, Equation 8.27 (or Equation 8.34) can be used directly to obtain the respective (constant) modes of vibration,

$$\frac{A_{21}}{A_{11}} = -\frac{k_{11} - \omega_1^2 m_1}{k_{12}} = -\frac{3k - 1.31(k/m)m}{-2k} = 0.845$$

$$\frac{A_{22}}{A_{12}} = -\frac{k_{11} - \omega_2^2 m_1}{k_{12}} = -\frac{3k - 4.19(k/m)m}{-2k} = -0.595.$$

The denominators can never equal zero since physically this would imply that one mass is motionless while the other is vibrating. This can only occur if the masses are not coupled. The modes are sketched in Figure 8.12, and are then used to write down the complete response as per Equation 8.30,

$$\left\{\begin{array}{c} x_1(t) \\ x_2(t) \end{array}\right\} = \bar{C}_1 \left\{\begin{array}{c} 1 \\ 0.845 \end{array}\right\} \cos\left(1.14\sqrt{\frac{k}{m}}t - \phi_1\right)$$

$$+ \bar{C}_2 \left\{\begin{array}{c} 1 \\ -0.595 \end{array}\right\} \cos\left(2.05\sqrt{\frac{k}{m}}t - \phi_2\right).$$

8.4. UNDAMPED VIBRATION

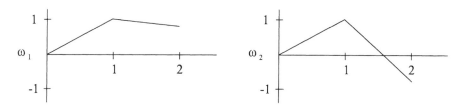

Figure 8.12: Two modes of vibration.

Mode two in the figure has one node. The response of each mass is a combination of two harmonics that are weighted by the respective modal ratios.

Given specific initial conditions, such as $x_1(0) = 1$ cm, $x_2(0) = -1$ cm and $\dot{x}_1(0) = 3$ cm/sec, $\dot{x}_2(0) = -2$ cm/sec, the constants of integration $\bar{C}_1, \bar{C}_2, \phi_1$ and ϕ_2 can be evaluated by solving the four simultaneous algebraic Equations 8.31 and 8.32. (Be sure to use consistent units.) ∎

Example 8.9 Elevator Cable System[8]

This example considers various models of an elevator-cable system for the purpose of estimating its natural frequency, or natural frequencies. The models considered are a one–mass, a one–mass with spring correction, and a three–mass system. The system parameters are given:

weight of car	$W_{car} = 1000$ lb
length of cable	$l = 750$ ft
cable $\frac{\text{weight}}{\text{length}}$	$\gamma = 1\frac{\text{lb}}{\text{ft}}$
cable cross–section area	$A = 1$ in^2
cable modulus of elasticity	$E = 4.5 \times 10^7$ psi.

For the one mass model, Figure 8.13, the spring stiffness is

$$k_{tot} = \frac{EA}{l} = \frac{4.5 \times 10^7 \text{ psi} \cdot 1 \text{ in}^2}{750 \text{ ft}} = 6.0 \times 10^4 \frac{\text{lb}}{\text{ft}}.$$

The total weight of the cable is

$$W_{cable} = \gamma l = 1\frac{\text{lb}}{\text{ft}} \cdot 750 \text{ ft} = 750 \text{ lb}.$$

---

[8] Professor Mark Nagurka is gratefully acknowledged for this problem.

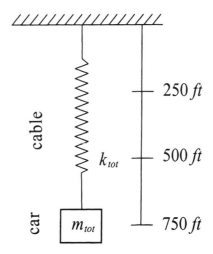

Figure 8.13: One mass model of elevator.

For the one–mass model, the cable weight is added to the car weight in the calculation of the natural frequency of oscillation,

$$W_{total} = W_{car} + W_{cable}$$
$$= 1000 \text{ lb} + 750 \text{ lb} = 1750 \text{ lb}.$$

The natural frequency is then

$$\omega_n = \sqrt{\frac{k_{tot}}{m_{tot}}} = \sqrt{\frac{kg}{W_{total}}}$$
$$= \sqrt{\frac{(6.0 \times 10^4 \frac{\text{lb}}{\text{ft}})(32.2 \frac{\text{ft}}{\text{sec}^2})}{(1750 \text{ lb})}}$$
$$= 33.3 \text{ rad/sec}$$
$$= 33.3 \frac{\text{rad}}{\text{sec}} \cdot \frac{1 \text{ cycle}}{2\pi \text{ rad}}$$
$$= 5.25 \text{ Hz}.$$

Next, it is of interest to improve the model by accounting for the inertia of the spring by using an equivalent weight for the mass. See Figure 8.14. As per *Example 2.6*,

$$W_{equiv} = W_{car} + \frac{1}{3}W_{cable}$$
$$= 1000 \text{ lb} + \frac{1}{3} \cdot 750 \text{ lb} = 1250 \text{ lb}.$$

8.4. UNDAMPED VIBRATION

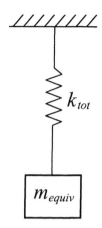

Figure 8.14: One mass model with spring correction.

In this case, the natural frequency increases,

$$\omega_n = \sqrt{\frac{k_{tot}\, g}{W_{total}}} = \sqrt{\frac{6.0 \times 10^4 \frac{\text{lb}}{\text{ft}} \cdot 32.2 \frac{\text{ft}}{\text{sec}^2}}{1250 \text{ lb}}}$$
$$= 39.3 \text{ rad/sec} = 6.26 \text{ Hz}.$$

This is a more accurate representation of the system since only a part of the cable mass can be viewed to be acting at the car location.

Finally, it is of interest to distribute the cable weight more realistically, at two discrete locations, at 250 ft and 500 ft from the top. See Figure 8.15. It is necessary to recalculate the spring stiffness for the shorter lengths,

$$k = \frac{EA}{l} = \frac{4.5 \times 10^7 \text{ psi} \cdot 1 \text{ in}^2}{250 \text{ ft}} = 1.8 \times 10^5 \frac{\text{lb}}{\text{ft}}.$$

The mass of the cable is distributed in thirds, one third to each mass and the other third to the mass of the car, as before,

$$m_1 = m_2 = \frac{1}{3} \frac{W_{cable}}{g} = \frac{250}{g} \text{ lb}$$
$$m_3 = \frac{1250 \text{ lb}}{g}.$$

To calculate the three natural frequencies, derive the equations of motion,

$$m_1 \ddot{x}_1 + 2kx_1 - x_2 = 0$$
$$m_2 \ddot{x}_2 - kx_1 + 2kx_2 - kx_3 = 0$$
$$m_3 \ddot{x}_3 - kx_2 + kx_3 = 0.$$

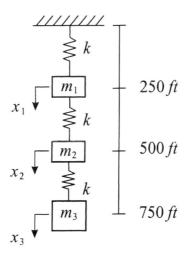

Figure 8.15: Three mass model of elevator.

The equations of motion can be put into matrix form. Assume harmonic responses of the form $A_i \exp(i\omega t)$, differentiate and substitute into the matrix equation of motion to find,

$$-\omega^2 \begin{bmatrix} m_1 & 0 & 0 \\ 0 & m_2 & 0 \\ 0 & 0 & m_3 \end{bmatrix} \begin{Bmatrix} A_1 \exp(i\omega t) \\ A_2 \exp(i\omega t) \\ A_3 \exp(i\omega t) \end{Bmatrix}$$
$$+ \begin{bmatrix} 2k & -k & 0 \\ -k & 2k & -k \\ 0 & -k & k \end{bmatrix} \begin{Bmatrix} A_1 \exp(i\omega t) \\ A_2 \exp(i\omega t) \\ A_3 \exp(i\omega t) \end{Bmatrix} = \begin{Bmatrix} 0 \\ 0 \\ 0 \end{Bmatrix}.$$

Use the fact that $m_1 = m_2$, combine like terms, and multiply all terms by $\exp(-i\omega t)$, to find,

$$\begin{bmatrix} -\omega^2 m_1 + 2k & -k & 0 \\ -k & -\omega^2 m_2 + 2k & -k \\ 0 & -k & -\omega^2 m_3 + k \end{bmatrix} \begin{Bmatrix} A_1 \\ A_2 \\ A_3 \end{Bmatrix} = \begin{Bmatrix} 0 \\ 0 \\ 0 \end{Bmatrix}.$$

Set the determinant of the coefficient matrix equal to zero to obtain the characteristic equation,

$$\omega^6 \left(-m_1^2 m_3\right) + \omega^4 \left[km_1 \left(m_1 + 4m_3\right)\right] + \omega^2 \left[-3k^2 \left(m_1 + m_3\right)\right] + k^3 = 0.$$

8.4. UNDAMPED VIBRATION

Substitute the parameter values and find the frequencies of oscillation,

$$\omega_{1,2,3} = 46.3, 165, 266 \text{ rad/sec}$$
$$= 7.37, 26.26, 42.34 \text{ Hz}.$$

The conclusion from this example is that a variety of models can be used to represent a complex system, and the numerical results can be only as good as the initial assumptions. The next level of realism in this model could be to use a string or a beam for the cable and therefore represent its physical properties via a continuum. ∎

8.4.2 Forced Vibration by the Direct Method

The considerations of the last section are extended here to the forced undamped structure. Begin with the generic two degree of freedom matrix equation of motion, Equation 8.13,

$$\begin{bmatrix} m_1 & 0 \\ 0 & m_2 \end{bmatrix} \begin{Bmatrix} \ddot{x}_1 \\ \ddot{x}_2 \end{Bmatrix} + \begin{bmatrix} k_{11} & k_{12} \\ k_{21} & k_{22} \end{bmatrix} \begin{Bmatrix} x_1 \\ x_2 \end{Bmatrix} = \begin{Bmatrix} F_1(t) \\ F_2(t) \end{Bmatrix}, \quad (8.35)$$

where forcing has been added and $k_{12} = k_{21}$. Since our purpose here is just to introduce the procedure, assume the very simplest of dynamic loading,

$$F_1(t) = f_1 \cos \omega t$$
$$F_2(t) = f_2 \cos \omega t,$$

where the forcing frequency ω is identical for each force.[9] Since there is no damping, the response will be either in phase or 180° out of phase with the forcing.

Assume the forced or steady–state responses to be given by

$$x_1(t) = A \cos \omega t$$
$$x_2(t) = B \cos \omega t.$$

Substituting these solutions into the matrix equation of motion and canceling the common cosine factor results in

$$-\omega^2 \begin{bmatrix} m_1 & 0 \\ 0 & m_2 \end{bmatrix} \begin{Bmatrix} A \\ B \end{Bmatrix} + \begin{bmatrix} k_{11} & k_{12} \\ k_{21} & k_{22} \end{bmatrix} \begin{Bmatrix} A \\ B \end{Bmatrix} = \begin{Bmatrix} f_1 \\ f_2 \end{Bmatrix}.$$

---
[9] If each force is at a different frequency, then we can solve for the response due to each force separately, and then add the two for the complete response.

Combine terms and find an expression that looks very similar to the eigenvalue problem, except that the algebraic equations are non–homogeneous,

$$\begin{bmatrix} k_{11} - m_1\omega^2 & k_{12} \\ k_{21} & k_{22} - m_2\omega^2 \end{bmatrix} \begin{Bmatrix} A \\ B \end{Bmatrix} = \begin{Bmatrix} f_1 \\ f_2 \end{Bmatrix}.$$

Define the square matrix, which is the system characteristic matrix, by $[S]$. To solve for A and B, there must be an inverse of $[S]$,

$$\begin{Bmatrix} A \\ B \end{Bmatrix} = [S]^{-1} \begin{Bmatrix} f_1 \\ f_2 \end{Bmatrix}.$$

From matrix algebra,

$$[S]^{-1} = \frac{\text{adj}[S]}{\det[S]},$$

and for the system considered here,

$$\begin{Bmatrix} A \\ B \end{Bmatrix} = \frac{1}{\det[S]} \begin{bmatrix} k_{22} - m_2\omega^2 & -k_{12} \\ -k_{21} & k_{11} - m_1\omega^2 \end{bmatrix} \begin{Bmatrix} f_1 \\ f_2 \end{Bmatrix}.$$

Expanding the matrix equation, A and B are given by

$$A = \frac{(k_{22} - m_2\omega^2)f_1 - k_{12}f_2}{\det[S]}$$

$$B = \frac{-k_{12}f_1 + (k_{11} - m_1\omega^2)f_2}{\det[S]},$$

where $k_{11} \equiv k_1 + k_2$, $k_{12} \equiv -k_2$ and $k_{22} \equiv k_2 + k_3$. The expression for $\det[S]$ is known and is the characteristic polynomial with the natural frequencies as roots. Therefore, at the natural frequencies, the equations for A and B will have a zero in the denominator, resulting in an unbounded response. This we already know; *if the forcing frequency equals the natural frequency, the response is unstable and grows without bound.* Look back at Equation 2.31. For a multi degree of freedom system, such an instability will occur for each natural frequency.

Consider the plot of the response amplitudes A and B as functions of driving frequency ω, the key characteristic of these curves is that they go to *infinity* about each natural frequency. Figures 8.16(a) and (b) show an example of such a set of curves for parameter values $m_1 = m_2 = 1$ kg, and $k_1 = k_2 = k_3 = 1$ N/m. The natural frequencies are 0.87 and 1.00 rad/sec. Compare these to the corresponding resonance diagram for the single degree of freedom system, Figure 2.15(b).

8.4. UNDAMPED VIBRATION

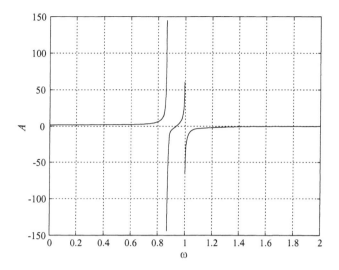

Figure 8.16: (a) Response magnitude A as a function of driving frequency.

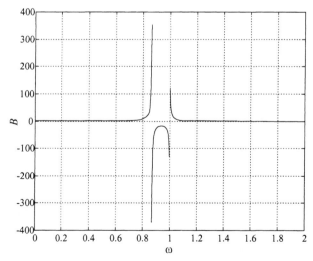

Figure 8.16: (b) Response magnitude B as a function of driving frequency.

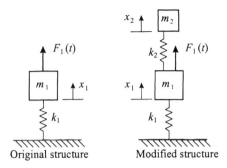

Figure 8.17: Primary structure without and with absorber.

Undamped Vibration Absorber

A simple application of the above theory to a very important class of problems is the *vibration absorber*.[10] The simplest of such devices is the undamped vibration absorber. Figure 8.17 shows the original structure subjected to a harmonic force $F_1(t)$, and the modified structure, where a second mass is attached to it via a spring. The equations of motion of the modified two degree of freedom structure are

$$m_1\ddot{x}_1 + (k_1 + k_2)x_1 - k_2 x_2 = f_1 \sin \omega t \qquad (8.36)$$
$$m_2\ddot{x}_2 - k_2 x_1 + k_2 x_2 = 0. \qquad (8.37)$$

An alternative modification would be to subject the absorber mass m_2 to a force $F_2(t)$ that can be *controlled*. In this case, the control parameters that can be specified by the designer are m_2, k_2 and $F_2(t)$. The more parameters that can be modified, the finer the tuning that is possible. We proceed with the simpler case where $F_2(t) = 0$.

Although it is not necessary to do so, put each equation of motion into standard form,

$$\ddot{x}_1 + \frac{(k_1 + k_2)}{m_1}x_1 - \frac{k_2}{m_1}x_2 = \frac{f_1}{m_1}\sin \omega t$$
$$\ddot{x}_2 - \frac{k_2}{m_2}x_1 + \frac{k_2}{m_2}x_2 = 0,$$

---

[10] There are numerous ways by which to modify the character of structural oscillations, including structural control. Vibration absorbers are passive or active control devices that are attached to a primary structure with the intent to change its behavior. Such absorbers may be passive, active or hybrid devices. This is a field of study in its own right that is based on vibration theory. One reference is by B.G. Korenev and L.M. Reznikov, **Dynamic Vibration Absorbers**, John Wiley & Sons, 1993.

8.4. UNDAMPED VIBRATION

and define the following dimensionless parameters,

$$m_r = \frac{m_2}{m_1}, \quad \omega_n^2 = \frac{k_1}{m_1}, \quad \omega_a^2 = \frac{k_2}{m_2},$$

where m_r is the mass ratio, ω_n is the natural frequency of the primary structure, and ω_a is the natural frequency of the absorber structure. Substituting these, along with the derived relation $k_2/m_1 = m_r \omega_a^2$, into the equations of motion results in

$$\ddot{x}_1 + (\omega_n^2 + m_r \omega_a^2) x_1 - m_r \omega_a^2 x_2 = \frac{f_1}{m_1} \sin \omega t$$

$$\ddot{x}_2 - \omega_a^2 x_1 + \omega_a^2 x_2 = 0.$$

The direct method of the previous section can be used here, where it is assumed that the harmonic forcing on mass m_1 results in the harmonic response

$$x_1(t) = A(\omega) \sin \omega t, \quad x_2(t) = B(\omega) \sin \omega t.$$

The anticipated dependence of the response amplitudes on driving frequency ω is shown explicitly. Differentiating the assumed responses twice and substituting the resulting expressions into the equations of motion, simplifying and using matrix form, leads to the matrix equation

$$\begin{bmatrix} \omega_n^2 + m_r \omega_a^2 - \omega^2 & -m_r \omega_a^2 \\ -\omega_a^2 & \omega_a^2 - \omega^2 \end{bmatrix} \begin{Bmatrix} A(\omega) \\ B(\omega) \end{Bmatrix} = \begin{Bmatrix} f_1/m_1 \\ 0 \end{Bmatrix}.$$

The response magnitudes can be solved by taking the inverse of the matrix in brackets, the characteristic matrix $[S]$,

$$\begin{Bmatrix} A(\omega) \\ B(\omega) \end{Bmatrix} = \frac{1}{\det[S]} \begin{bmatrix} \omega_a^2 - \omega^2 & m_r \omega_a^2 \\ \omega_a^2 & \omega_n^2 + m_r \omega_a^2 - \omega^2 \end{bmatrix} \begin{Bmatrix} f_1/m_1 \\ 0 \end{Bmatrix}. \quad (8.38)$$

Then,

$$A(\omega) = \frac{(\omega_a^2 - \omega^2) f_1/m_1}{\det[S]} \quad (8.39)$$

$$B(\omega) = \frac{\omega_a^2 f_1/m_1}{\det[S]}. \quad (8.40)$$

Note that if another harmonic force at the same frequency ω is applied to the added mass m_2, as mentioned before as a possibility, its effect would appear as a second term in the numerator of each expression. The force vector in Equation 8.38 would be $\{f_1/m_1 \quad f_2/m_2\}^T$. If there are two different frequencies, then solve for each forcing separately and add the results.

Equations 8.39 and 8.40 can be simplified and written in a more useful form where there are dimensionless frequency ratios, as follows. Replace f_1/m_1 by its equivalent, $f_1\omega_n^2/k_1$, and divide top and bottom by $1/\omega_n^2\omega_a^2$. Note that $f_1/k_1 = x_{static}$, the static displacement of mass m_1. Performing these operations on both equations results in

$$A(\omega) = \frac{[1-(\omega/\omega_a)^2]x_{static}}{[1+m_r(\omega/\omega_a)^2-(\omega/\omega_n)^2][1-(\omega/\omega_a)^2]-m_r(\omega_a/\omega_n)^2}$$

$$B(\omega) = \frac{x_{static}}{[1+m_r(\omega/\omega_a)^2-(\omega/\omega_n)^2][1-(\omega/\omega_a)^2]-m_r(\omega_a/\omega_n)^2}.$$

Note that when the driving frequency ω equals ω_a, amplitude $A(\omega_a) = 0$, that is, the primary mass does not move. Let us see if this makes physical sense. For $\omega = \omega_a$, secondary amplitude $B(\omega_a) = -f_1/k_2$, and the response of the absorber structure is

$$x_2(t) = -\frac{f_1}{k_2}\sin\omega t,$$

or $k_2 x_2 = -f_1 \sin \omega t$. The left hand side of this equation is the force exerted by m_2 on m_1. This is equal and opposite to the force applied to m_1. Since the force due to the absorber structure is exactly opposite to the applied force, the primary mass does not move at all.

Plot $A(\omega)/x_{static}$ versus the ratio ω/ω_a, as in Figure 8.18, and see that a slight shift away from $\omega/\omega_a = 1$ results in a large increase in amplitude. For this figure, $m_r = 1$, $\omega_a = \omega_n$, and the natural frequencies are $\omega_{1,2} = 0.618$, 1.618, seen in the figure where amplitudes go to infinity. Therefore, even though this is a good conceptual example, it may have serious practical limitations. The interested reader is urged to pursue the literature on vibration dampers and vibration isolation for viable ideas.

8.4.3 Coupled Pendula – Beating

An interesting and important phenomenon experienced by coupled systems with *close* natural frequencies is called *beating*. Because of this closeness, the system undergoes a high frequency oscillation that is modulated (enveloped) by a lower frequency oscillation. We will examine this behavior by first recalling the simple pendulum of Figure 8.19 and its free body diagram.

The equation of motion can be derived in either of two ways, the first via sum of the forces,

$$\sum F_\theta = ma_\theta = -mg\sin\theta$$
$$\implies mL\ddot{\theta} + mg\sin\theta = 0,$$

8.4. UNDAMPED VIBRATION

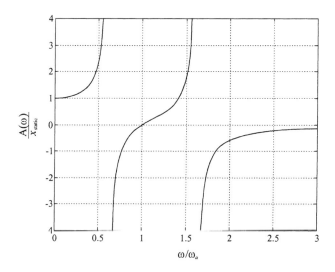

Figure 8.18: Nondimensional response of primary mass versus ratio ω/ω_a, where $\omega_n = \omega_a$ and $\omega_{1,2}$ are natural frequencies.

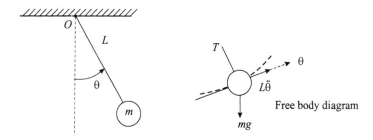

Figure 8.19: Simple pendulum.

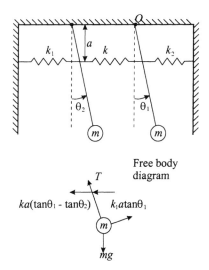

Figure 8.20: Simple coupled pendula.

where the angular acceleration a_θ is $L\ddot{\theta}$. For small motions, $\sin\theta \approx \theta$ and

$$\ddot{\theta} + \omega_n^2 \theta = 0, \quad \omega_n = \sqrt{\frac{g}{L}}. \tag{8.41}$$

The second way, summing moments, is

$$\sum M_o = I_o \alpha, \quad I_o = m_o + mL^2 \approx mL^2$$
$$\implies -mgL\sin\theta = mL^2\ddot{\theta},$$

which results in the same equation as before, of course. The simplification of the expression for I_o is valid since the physical size of the pendulum mass is so small that the moment of inertia about its centroid, m_o, is much smaller than the transfer distance contribution mL^2 required by the parallel axis theorem.

Next consider the coupled pendula of Figure 8.20 that includes the free body of the mass with position defined by θ_1. Taking the sum of the moments about o, one proceeds through the following sequence of equations leading to the equation of motion,

$$\sum M_o = I_o \alpha$$
$$mL^2 \ddot{\theta}_1 = -mgL\sin\theta_1$$
$$- [\{ka(\tan\theta_1 - \tan\theta_2) + k_1 a\tan\theta_1\}\cos\theta_1]a(1+\tan^2\theta_1)^{1/2}.$$

8.4. UNDAMPED VIBRATION

Make the assumption of small θ_1, leading to the simplifications $\cos\theta \approx 1$, $\tan\theta \approx \sin\theta \approx \theta$, and $\tan^2\theta_1 \ll 1$, to find

$$mL^2\ddot{\theta}_1 + mgL\theta_1 + (k+k_1)a^2\theta_1 - ka^2\theta_2 = 0.$$

Similarly for the θ_2 coordinate,

$$mL^2\ddot{\theta}_2 + mgL\theta_2 + (k+k_1)a^2\theta_2 - ka^2\theta_1 = 0.$$

In matrix form,

$$\begin{bmatrix} mL^2 & 0 \\ 0 & mL^2 \end{bmatrix} \begin{Bmatrix} \ddot{\theta}_1 \\ \ddot{\theta}_2 \end{Bmatrix} + \begin{bmatrix} mgL+(k+k_1)a^2 & -ka^2 \\ -ka^2 & mgL+(k+k_1)a^2 \end{bmatrix} \begin{Bmatrix} \theta_1 \\ \theta_2 \end{Bmatrix} = \begin{Bmatrix} 0 \\ 0 \end{Bmatrix}. \quad (8.42)$$

As expected, if there is no coupling between the pendula, $k = 0$ and Equation 8.41 is recovered for each uncoupled pendulum. To solve coupled Equations 8.42, assume a harmonic solution[11] for each degree of freedom,

$$\theta_1(t) = \Theta_1 e^{i\omega t}$$
$$\theta_2(t) = \Theta_2 e^{i\omega t},$$

appropriately differentiate each and substitute into Equation 8.42. Cancel the exponential from all terms to obtain

$$-\omega^2 \begin{bmatrix} mL^2 & 0 \\ 0 & mL^2 \end{bmatrix} \begin{Bmatrix} \Theta_1 \\ \Theta_2 \end{Bmatrix} + \begin{bmatrix} mgL+(k+k_1)a^2 & -ka^2 \\ -ka^2 & mgL+(k+k_1)a^2 \end{bmatrix} \begin{Bmatrix} \Theta_1 \\ \Theta_2 \end{Bmatrix} = \begin{Bmatrix} 0 \\ 0 \end{Bmatrix}.$$

Combine terms as before to obtain the characteristic matrix,

$$\begin{bmatrix} \begin{pmatrix} -\omega^2 mL^2 + mgL \\ +(k+k_1)a^2 \end{pmatrix} & -ka^2 \\ -ka^2 & \begin{pmatrix} -\omega^2 mL^2 + mgL \\ +(k+k_1)a^2 \end{pmatrix} \end{bmatrix} \begin{Bmatrix} \Theta_1 \\ \Theta_2 \end{Bmatrix} = \begin{Bmatrix} 0 \\ 0 \end{Bmatrix}. \quad (8.43)$$

---

[11] Recall that we could have just as easily assumed solutions of the form $\theta(t) = \Theta(\cos\omega t - \phi)$.

Setting the determinant of the matrix to zero, we find the characteristic equation for the system,
$$(-\omega^2 mL^2 + mgL + (k+k_1)a^2)^2 - (ka^2)^2 = 0,$$
or
$$-\omega^2 mL^2 + mgL + (k+k_1)a^2 = \pm ka^2,$$
which has two solutions, as required for a two degree of freedom system. For $+ka^2$,
$$\omega_1 = \sqrt{\frac{g}{L} + \frac{k_1}{m}\frac{a^2}{L^2}},$$
and for $-ka^2$,
$$\omega_2 = \sqrt{\frac{g}{L} + \frac{2k+k_1}{m}\frac{a^2}{L^2}}.$$
To evaluate the natural mode for each natural frequency, substitute each ω_i into Equation 8.43,
$$\begin{bmatrix} \left(\begin{array}{c} -\omega_i^2 mL^2 + mgL \\ +(k+k_1)a^2 \end{array}\right) & -ka^2 \\ -ka^2 & \left(\begin{array}{c} -\omega_i^2 mL^2 + mgL \\ +(k+k_1)a^2 \end{array}\right) \end{bmatrix} \left\{\begin{array}{c} \Theta_1 \\ \Theta_2 \end{array}\right\}_i = \left\{\begin{array}{c} 0 \\ 0 \end{array}\right\}.$$

Solving for the modal ratios, we find
$$\frac{\Theta_{21}}{\Theta_{11}} = 1, \quad \frac{\Theta_{22}}{\Theta_{12}} = -1.$$

These modes are sketched in Figure 8.21. The coupling parameter k affects ω_2 but not ω_1. The reason is that in the mode associated with ω_1, the masses move in tandem while for the second mode the masses move in opposite directions.

The general motion of the two coupled pendula is given by the superposition of the two natural modes of vibration,
$$\left\{\begin{array}{c} \theta_1(t) \\ \theta_2(t) \end{array}\right\} = \bar{C}_1 \left\{\begin{array}{c} \Theta_1 \\ \Theta_2 \end{array}\right\}_1 \cos(\omega_1 t - \phi_1) + \bar{C}_2 \left\{\begin{array}{c} \Theta_1 \\ \Theta_2 \end{array}\right\}_2 \cos(\omega_2 t - \phi_2)$$
$$= C_1 \left\{\begin{array}{c} 1 \\ \Theta_{21}/\Theta_{11} \end{array}\right\} \cos(\omega_1 t - \phi_1) + C_2 \left\{\begin{array}{c} 1 \\ \Theta_{22}/\Theta_{12} \end{array}\right\} \cos(\omega_2 t - \phi_2),$$
where the factors Θ_{11} and Θ_{12} have been incorporated into coefficients C_1 and C_2, respectively. The modification of the integration constants does

8.4. UNDAMPED VIBRATION

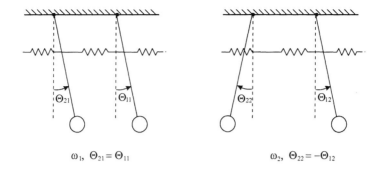

Figure 8.21: Modes for a coupled pendulum.

not change the final result since these constants are determined by the satisfaction of the initial conditions. Substituting the modal ratios results in

$$\left\{ \begin{array}{c} \theta_1(t) \\ \theta_2(t) \end{array} \right\} = C_1 \left\{ \begin{array}{c} 1 \\ 1 \end{array} \right\} \cos(\omega_1 t - \phi_1) + C_2 \left\{ \begin{array}{c} 1 \\ -1 \end{array} \right\} \cos(\omega_2 t - \phi_2). \quad (8.44)$$

All that remains is the evaluation of C_1, C_2, ϕ_1 and ϕ_2.

In order to demonstrate the beating phenomenon, it is easiest to assume the following initial conditions:[12]

$$\theta_1(0) = \theta_0 \quad \theta_2(0) = 0$$
$$\dot{\theta}_1(0) = 0 \quad \dot{\theta}_2(0) = 0.$$

Using Equation 8.44 and its derivative to satisfy the initial conditions results in the values

$$C_1 = C_2 = \frac{1}{2}\theta_0, \quad \phi_1 = \phi_2 = 0.$$

Before proceeding with the analysis, let us try to anticipate the behavior of the coupled pendula, which depends entirely on the coupling forces in the spring. If the coupling is "strong" (high stiffness), then we expect the two masses to oscillate almost in tandem, since the strong coupling will not allow one mass to lag far behind the motion of the other. If the coupling is "weak," defined precisely below, then we may expect that the two masses will oscillate, with one sometimes reinforcing the motion of the other (larger

---

[12] The initial conditions mean that one pendulum is pulled to one aside θ_0 rad while holding pendulum two fixed. Assuming zero initial velocities does not change the character of the beating except to provide a zero slope at the initial time on the displacement time history.

amplitudes) and other times canceling out the motion of the other (smaller amplitudes). The oscillations do not look like that of the separate pendula because energy is being transferred continuously between the two masses.

In order to better observe this behavior, modify Equation 8.44. Define

$$\alpha = \frac{(\omega_2 - \omega_1)t}{2}, \quad \beta = \frac{(\omega_2 + \omega_1)t}{2},$$

and using a trigonometric identity,[13] Equation 8.44 becomes

$$\theta_1(t) = \theta_0 \cos\frac{(\omega_2 - \omega_1)t}{2} \cos\frac{(\omega_2 + \omega_1)t}{2} \tag{8.45}$$

$$\theta_2(t) = \theta_0 \sin\frac{(\omega_2 - \omega_1)t}{2} \sin\frac{(\omega_2 + \omega_1)t}{2}. \tag{8.46}$$

The difference and sum of natural frequencies can be written as

$$\frac{(\omega_2 \mp \omega_1)}{2} = \frac{1}{2}\left(\sqrt{\frac{g}{L} + \frac{2k + k_1}{m}\frac{a^2}{L^2}} \mp \sqrt{\frac{g}{L} + \frac{k_1}{m}\frac{a^2}{L^2}}\right).$$

Furthermore, using the binomial series representation[14] for the square root, the following approximation for small $(2k + k_1)a^2/mgL$ can be used,

$$\sqrt{\frac{g}{L} + \frac{2k + k_1}{m}\frac{a^2}{L^2}} = \sqrt{\frac{g}{L}}\sqrt{1 + \frac{2k + k_1}{mg}\frac{a^2}{L}} \approx \sqrt{\frac{g}{L}}\left[1 + \frac{1}{2}\frac{(2k + k_1)a^2}{mgL}\right].$$

Weak coupling here implies $(2k + k_1)a^2 \ll mgL$, and therefore

$$\frac{(\omega_2 - \omega_1)}{2} \approx \frac{1}{2}\frac{k}{m}\frac{a^2}{\sqrt{gL^3}} \equiv \frac{\omega_b}{2} \tag{8.47}$$

$$\frac{(\omega_2 + \omega_1)}{2} \approx \sqrt{\frac{g}{L}} + \frac{1}{2}\frac{k + k_1}{m}\frac{a^2}{\sqrt{gL^3}} \equiv \omega_{avg}.$$

For the weak coupling case, Equations 8.45 and 8.46 can then be written as

$$\theta_1(t) \approx \theta_0 \cos\frac{\omega_b t}{2} \cos\frac{\omega_{avg} t}{2}$$

$$\theta_2(t) \approx \theta_0 \sin\frac{\omega_b t}{2} \sin\frac{\omega_{avg} t}{2}.$$

---

[13] $\cos(\alpha \pm \beta) = \cos\alpha\cos\beta \mp \sin\alpha\sin\beta$
[14] $(1+x)^{\pm\frac{1}{2}} = 1 \pm \frac{1}{2}x \mp (1+x)x^2 \pm \cdots$ for $|x| < 1$. The smaller the value of x, the more accurate is the two term approximation $1 \pm \frac{1}{2}x$.

8.4. UNDAMPED VIBRATION

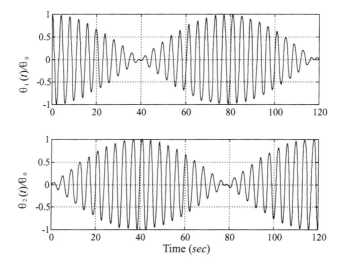

Figure 8.22: Beating behavior for weak coupling.

This behavior can be called an *amplitude modulated harmonic function*. The beat frequency is $\omega_b = ka^2/m\sqrt{gL^3}$, from Equation 8.47. The maximum time between two peaks of the beat is $T/2 = 2\pi/\omega_b$, where the period of the amplitude modulated envelope is T.

For parameter values $k = 10$ N/m, $k_1 = 10$ N/m, $a = 0.5$ m, $m = 10$ kg, $g = 9.8$ m/sec^2 and $L = 1$ m, the condition $(2k + k_1)a^2 << mgL$ is $7.5 << 98.0$ and is satisfied. The two characteristic frequencies are

$$\omega_b = 0.080 \text{ rad/sec}$$
$$\omega_{avg} = 3.210 \text{ rad/sec}.$$

Figure 8.22 shows the time histories for $\theta_1(t)/\theta_0$ and $\theta_2(t)/\theta_0$, which are given by

$$\frac{\theta_1(t)}{\theta_0} \approx \cos\frac{0.080t}{2}\cos 3.210t$$
$$\frac{\theta_2(t)}{\theta_0} \approx \sin\frac{0.080t}{2}\sin 3.210t.$$

We proceed next to the damped, free vibration problem.

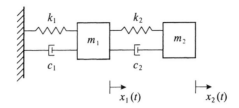

Figure 8.23: Two degree of freedom structure.

8.5 Direct Method: Free Vibration with Damping

The equations of free motion with viscous damping,

$$[m]\{\ddot{x}\} + [c]\{\dot{x}\} + [k]\{x\} = 0,$$

can still be solved in a direct manner by assuming a harmonic form for the response. Due to the presence of damping, the characteristic equation will be a polynomial that has complex conjugate roots. Given complex conjugate eigenvalues, there are complex conjugate eigenvectors. We explore this approach by way of a specific system.[15]

For the system sketched in Figure 8.23, the equations of motion are

$$m_1\ddot{x}_1 + (c_1 + c_2)\dot{x}_1 + (k_1 + k_2)x_1 - c_2\dot{x}_2 - k_2 x_2 = 0$$
$$m_2\ddot{x}_2 + c_2\dot{x}_2 + k_2 x_2 - c_2\dot{x}_1 - k_2 x_1 = 0.$$

Assume a solution of the form

$$x_1(t) = A e^{\lambda t}$$
$$x_2(t) = B e^{\lambda t},$$

and substitute these into the governing equations. Eliminating the common factor $\exp\{\lambda t\}$ leads to the two simultaneous algebraic equations

$$[m_1\lambda^2 + (c_1 + c_2)\lambda + (k_1 + k_2)]A - (c_2\lambda + k_2)B = 0 \quad (8.48)$$
$$-(c_2\lambda + k_2)A + (m_2\lambda^2 + c_2\lambda + k_2)B = 0. \quad (8.49)$$

For a non-trivial solution to exist, the characteristic determinant of the coefficients A and B must equal zero,

$$\begin{vmatrix} m_1\lambda^2 + (c_1 + c_2)\lambda + (k_1 + k_2) & -(c_2\lambda + k_2) \\ -(c_2\lambda + k_2) & m_2\lambda^2 + c_2\lambda + k_2 \end{vmatrix} = 0,$$

---

[15] This development follows that of W. Weaver, S.P. Timoshenko, D.H. Young, in **Vibration Problems in Engineering**, Fifth Edition, Wiley–Interscience, 1990, p.260.

8.5. DIRECT METHOD: FREE VIBRATION WITH DAMPING

with the resulting characteristic equation

$$[m_1\lambda^2 + (c_1 + c_2)\lambda + (k_1 + k_2)](m_2\lambda^2 + c_2\lambda + k_2) - (c_2\lambda + k_2)^2 = 0,$$

which can be easily expanded to

$$m_1 m_2 \lambda^4 + (m_1 c_2 + [c_1 + c_2]m_2)\lambda^3$$
$$+(m_1 k_2 + c_1 c_2 + [k_1 + k_2]m_2)\lambda^2 + (c_1 k_2 + k_1 c_2)\lambda + k_1 k_2 = 0.$$

The characteristic equation generally has four distinct roots, λ_1, λ_2, λ_3, and λ_4, that must be evaluated numerically.

From the theory of algebraic equations, since all coefficients are positive, the non–zero roots can be *neither* real and positive *nor* complex with positive real roots. The roots must be either real and negative, or complex with negative real parts.[16]

Physically, we know that most engineering structures are lightly damped and will vibrate. From our studies of single degree of freedom systems, we also know that complex roots of algebraic equations always occur in conjugate pairs.[17] For the one mass system, there is one complex conjugate pair. For this two degree of freedom system, there are two complex conjugate pairs of the form

$$\lambda_{11} = -a_1 + i\omega_{d1}$$
$$\lambda_{12} = -a_1 - i\omega_{d1}$$
$$\lambda_{21} = -a_2 + i\omega_{d2}$$
$$\lambda_{22} = -a_2 - i\omega_{d2},$$

where the damped natural frequencies are $\omega_{di} = \omega_{ni}\sqrt{1 - \zeta_i^2}$, and a_i are positive numbers (for a stable system) that define the rate of decay of the exponential envelope. These expressions correspond to similar ones for single degree of freedom structures.

Substitute each root in turn into Equations 8.48 and 8.49 to derive the amplitude ratios, or modes, A_{ij}/B_{ij} for each frequency,

$$\begin{bmatrix} \begin{pmatrix} m_1\lambda_{ij}^2 + (c_1 + c_2)\lambda_{ij} \\ +(k_1 + k_2) \end{pmatrix} & -c_2\lambda_{ij} - k_2 \\ -c_2\lambda_{ij} - k_2 & \begin{pmatrix} m_2\lambda_{ij}^2 + c_2\lambda_{ij} \\ +k_2 \end{pmatrix} \end{bmatrix} \begin{Bmatrix} A_{ij} \\ B_{ij} \end{Bmatrix} = \begin{Bmatrix} 0 \\ 0 \end{Bmatrix}. \quad (8.50)$$

---

[16] A positive real root or real part of a complex number results in a solution with a growing exponential factor; this is an unstable system.

[17] Recall that the real part appears as the coefficient of time in the exponential envelope, and the imaginary part becomes the damped frequency of oscillation.

By substituting each root λ_{ij} into Equation 8.50, four possible ratios, or modes, are obtained of the form

$$\frac{A_{ij}}{B_{ij}} = r_{ij} = \frac{c_2 \lambda_{ij} + k_2}{m_1 \lambda_{ij}^2 + (c_1 + c_2)\lambda_{ij} + (k_1 + k_2)} = \frac{m_2 \lambda_{ij}^2 + c_2 \lambda_{ij} + k_2}{c_2 \lambda_{ij} + k_2},$$

where $i = 1, 2$ and $j = 1, 2$. The resulting ratios r_{11}, r_{12}, r_{21} and r_{22} are complex conjugate pairs that relate the modal amplitudes. These are the modes of vibration, and are generally *complex modes*. Using the relations $A_{ij} = r_{ij} B_{ij}$, the complete solution may be written as

$$x_1(t) = r_{11} B_{11} e^{\lambda_{11} t} + r_{12} B_{12} e^{\lambda_{12} t} + r_{21} B_{21} e^{\lambda_{21} t} + r_{22} B_{22} e^{\lambda_{22} t} \quad (8.51)$$
$$x_2(t) = B_{11} e^{\lambda_{11} t} + B_{12} e^{\lambda_{12} t} + B_{21} e^{\lambda_{21} t} + B_{22} e^{\lambda_{22} t}, \quad (8.52)$$

where (B_{11}, B_{12}) and (B_{21}, B_{22}) are complex conjugate pairs that are determined using the initial conditions.

Proceed as for a single mass system by converting Equations 8.51 and 8.52 into their equivalent trigonometric expressions by combining complex conjugate pairs. From the equation for $x_2(t)$, let

$$B_{11} e^{\lambda_{11} t} + B_{12} e^{\lambda_{12} t} = e^{-a_1 t}(C_1 \cos \omega_{d1} t + C_2 \sin \omega_{d1} t),$$

where use is made of Euler's equation for the complex exponential, and where $C_1 = B_{11} + B_{12}$ and $C_2 = i(B_{11} - B_{12})$. Note that C_1 and C_2 are real numbers.

A similar procedure can be followed for the complex conjugate pairs in the solution for $x_1(t)$. The ratios r_{11} and r_{12} are a complex conjugate pair of the form

$$r_{11} = \alpha_1 + i\beta_1, \quad r_{12} = \alpha_1 - i\beta_1. \quad (8.53)$$

Recall that a complex number represents a magnitude and a phase. Then

$$r_{11} B_{11} e^{\lambda_{11} t} + r_{12} B_{12} e^{\lambda_{12} t}$$
$$= e^{-a_1 t}[(C_1 \alpha_1 + C_2 \beta_1) \cos \omega_{d1} t + (-C_1 \beta_1 + C_2 \alpha_1) \sin \omega_{d1} t].$$

The second pair of complex conjugates can be converted using the real constants $C_3 = B_{21} + B_{22}$ and $C_4 = i(B_{21} - B_{22})$, and introducing the notation

$$r_{21} = \alpha_2 + i\beta_2, \quad r_{22} = \alpha_2 - i\beta_2. \quad (8.54)$$

The response equations then become

$$x_1(t) = e^{-a_1 t}(r_1 C_1 \cos \omega_{d1} t + r_1' C_2 \sin \omega_{d1} t)$$
$$\qquad + e^{-a_2 t}(r_2 C_3 \cos \omega_{d2} t + r_2' C_4 \sin \omega_{d2} t) \quad (8.55)$$
$$x_2(t) = e^{-a_1 t}(C_1 \cos \omega_{d1} t + C_2 \sin \omega_{d1} t)$$
$$\qquad + e^{-a_2 t}(C_3 \cos \omega_{d2} t + C_4 \sin \omega_{d2} t), \quad (8.56)$$

8.5. DIRECT METHOD: FREE VIBRATION WITH DAMPING

where

$$r_1 = \frac{C_1\alpha_1 + C_2\beta_1}{C_1}, \quad r_1' = \frac{-C_1\beta_1 + C_2\alpha_1}{C_2}$$

$$r_2 = \frac{C_3\alpha_2 + C_4\beta_2}{C_3}, \quad r_2' = \frac{-C_3\beta_2 + C_4\alpha_2}{C_4},$$

are real numbers. Equations 8.55 and 8.56 can be compared to their counterparts for undamped vibration in Equation 8.30. However, they differ in the same way that a damped oscillator differs from the undamped one, *a phasing term is introduced.*

We also have four amplitude ratios that characterize the damped case, whereas only two ratios were needed in Equation 8.30 for the undamped modes of vibration. The extra ratios permit us to evaluate the *phasing difference* between respective complex conjugate pairs in the solutions for $x_1(t)$ and $x_2(t)$. The first complex conjugate pair in the solution for $x_1(t)$ is not in phase with the respective complex conjugate pair in the solution for $x_2(t)$. This can be observed easily by using the phase angle form for Equations 8.55 and 8.56:

$$x_1(t) = B_1' e^{-a_1 t} \cos(\omega_{d1} t - \phi_{d1}') + B_2' e^{-a_2 t} \cos(\omega_{d2} t - \phi_{d2}') \quad (8.57)$$

$$x_2(t) = B_1 e^{-a_1 t} \cos(\omega_{d1} t - \phi_{d1}) + B_2 e^{-a_2 t} \cos(\omega_{d2} t - \phi_{d2}), \quad (8.58)$$

where

$$B_1 = \sqrt{C_1^2 + C_2^2}, \quad B_2 = \sqrt{C_3^2 + C_4^2}$$

$$B_1' = B_1\sqrt{\alpha_1^2 + \beta_1^2}, \quad B_2' = B_2\sqrt{\alpha_2^2 + \beta_2^2}$$

$$\phi_{d1} = \arctan\left(\frac{C_2}{C_1}\right), \quad \phi_{d2} = \arctan\left(\frac{C_4}{C_3}\right)$$

$$\phi_{d1}' = \arctan\left(\frac{r_1' C_2}{r_1 C_1}\right), \quad \phi_{d2}' = \arctan\left(\frac{r_2' C_4}{r_2 C_3}\right).$$

Thus, principal modes that are in phase do not generally exist for a damped system of the type considered here. The natural modes that do exist have a phase relationship, that is, *they are out of phase,* as shown above.

Some simplifications are possible if certain reasonable assumptions can be made regarding the damping. Suppose that the specific structure under consideration has *very small viscous damping.* Then the characteristic equation will be very close to that for the undamped case, and the following approximations may be made,

$$\omega_{d1} \approx \omega_1, \quad r_1' \approx r_1$$

$$\omega_{d2} \approx \omega_2, \quad r_2' \approx r_2,$$

with the responses approximated by

$$\begin{aligned}x_1(t) &\approx r_1 e^{-a_1 t}(C_1 \cos \omega_1 t + C_2 \sin \omega_1 t) \\ &\quad + r_2 e^{-a_2 t}(C_3 \cos \omega_2 t + C_4 \sin \omega_2 t) \\ x_2(t) &\approx e^{-a_1 t}(C_1 \cos \omega_1 t + C_2 \sin \omega_1 t) \\ &\quad + e^{-a_2 t}(C_3 \cos \omega_2 t + C_4 \sin \omega_2 t),\end{aligned}$$

where C_1 through C_4 are constants depending on initial conditions. In this case, there will be negligible phase difference between the modes.

Where there is *very high damping*, there are two possibilities. The first is that all the roots, $p_1, \ldots, p_4$, of the characteristic equation are real and negative. As in the single degree of freedom problem, the response is *not oscillatory* but decaying exponentially,

$$\begin{aligned}x_1(t) &\approx r_1 D_1 e^{-p_1 t} + r_2 D_2 e^{-p_2 t} + r_3 D_3 e^{-p_3 t} + r_4 D_4 e^{-p_4 t} \\ x_2(t) &\approx D_1 e^{-p_1 t} + D_2 e^{-p_2 t} + D_3 e^{-p_3 t} + D_4 e^{-p_4 t},\end{aligned}$$

where $D_1, \ldots, D_4$ and $r_1, \ldots, r_4$ are real. The second possibility is that two of the four roots are real and negative, and the other two roots are complex conjugates with negative real parts. This solution has the form

$$\begin{aligned}x_1(t) &\approx e^{-at}(r_1 C_1 \cos \omega_d t + r_1' C_2 \sin \omega_d t) \\ &\quad + r_3 C_3 e^{-p_3 t} + r_4 C_4 e^{-p_4 t} \\ x_2(t) &\approx e^{-at}(C_1 \cos \omega_d t + C_2 \sin \omega_d t) \\ &\quad + C_3 e^{-p_3 t} + C_4 e^{-p_4 t}.\end{aligned}$$

These solutions show an exponentially decaying oscillation superposed with a purely exponentially decaying response. Two numerical examples are studied next to better understand the procedure, the meaning of phasing, and how it comes into a response.

Example 8.10 Free Vibration with Damping by the Direct Method: Case I

For the system of Figure 8.23, the system parameters are given as $m_1 = m_2 = 1$ kg, $c_1 = c_2 = 1$ N sec/m, and $k_1 = k_2 = 1$ N/m. The governing matrix equation of motion is then

$$\begin{bmatrix} 1 & 0 \\ 0 & 1 \end{bmatrix} \begin{Bmatrix} \ddot{x}_1 \\ \ddot{x}_2 \end{Bmatrix} + \begin{bmatrix} 2 & -1 \\ -1 & 1 \end{bmatrix} \begin{Bmatrix} \dot{x}_1 \\ \dot{x}_2 \end{Bmatrix} + \begin{bmatrix} 2 & -1 \\ -1 & 1 \end{bmatrix} \begin{Bmatrix} x_1 \\ x_2 \end{Bmatrix} = \begin{Bmatrix} 0 \\ 0 \end{Bmatrix}.$$

Assuming a harmonic response results in the characteristic matrix

$$\begin{bmatrix} \lambda^2 + 2\lambda + 2 & -(\lambda + 1) \\ -(\lambda + 1) & \lambda^2 + \lambda + 1 \end{bmatrix} \begin{Bmatrix} A \\ B \end{Bmatrix} = \begin{Bmatrix} 0 \\ 0 \end{Bmatrix}. \qquad (8.59)$$

8.5. DIRECT METHOD: FREE VIBRATION WITH DAMPING

The determinant of the characteristic matrix leads to the characteristic equation

$$\lambda^4 + 3\lambda^3 + 4\lambda^2 + 2\lambda + 1 = 0,$$

that must be solved numerically. Using a program such as MATLAB makes it easy to find the roots in the general form $\lambda_{jk} = -a_j \pm i\omega_{d_j}$, or specifically to this problem,

$$\lambda_{11} = -1.309 + 0.951i$$
$$\lambda_{12} = -1.309 - 0.951i$$
$$\lambda_{21} = -0.191 + 0.588i$$
$$\lambda_{22} = -0.191 - 0.588i.$$

Let us go through the steps used in proceeding from the value of λ_{11} to that of r_{11}. Of course, the same needs to be done for each root.

Substitute λ_{11} into the first equation of matrix Equation 8.59 to find

$$(\lambda_{11}^2 + 2\lambda_{11} + 2)A_{11} - (\lambda_{11} + 1)B_{11} = 0,$$

or

$$\frac{A_{11}}{B_{11}} = r_{11} = \frac{\lambda_{11} + 1}{\lambda_{11}^2 + 2\lambda_{11} + 2}$$
$$= \frac{-0.309 + 0.951i}{0.191 - 0.588i}. \tag{8.60}$$

By multiplying top and bottom of Equation 8.60 by the complex conjugate of its denominator results in

$$r_{11} = -1.618,$$

where a small imaginary part may remain because of round off error. The eigenvector in this case (not needed in this solution) can be written as $\{-1.618 \ \ 1\}^T$. Similarly, we have $r_{12} = -1.618$, $r_{21} = 0.618$, and $r_{22} = 0.618$. There are two distinct eigenvalues. Had these ratios included imaginary parts, then r_{11} and r_{12} would have been complex conjugates, as would have been r_{21} and r_{22}. In general, the ratios r_{ij} are complex numbers. See Equations 8.53 and 8.54. The next example will show this case.

What does it mean physically that the imaginary parts drop out completely? It means that there is no phasing in the modes, and the system behaves as though there is no damping, or, since we do have damping, the damping matrix is linearly related to the $[m]$ and $[k]$ matrices. This is a special type of damping, known as proportional damping, that we will learn more about in Section 8.6.3, and when we study modal analysis in the next section.

Using standard notation,

$$a_1 = 1.309, \quad \omega_{d1} = 0.951; \quad a_2 = 0.191, \quad \omega_{d2} = 0.588,$$

and

$$\alpha_1 = -1.618, \quad \beta_1 = 0; \quad \alpha_2 = 0.618, \quad \beta_2 = 0.$$

These parameters are substituted into the general Equations 8.55 and 8.56. In order that the procedure is clear, β_1 and β_2 will be retained even though they are equal to zero,

$$\begin{aligned} x_1(t) &= e^{-1.309t}\left(\frac{C_1(-1.618)+C_2(\beta_1)}{C_1}C_1\cos 0.951t\right. \\ &\quad \left. + \frac{-C_1(\beta_1)+C_2(-1.618)}{C_2}C_2\sin 0.951t\right) \\ &\quad + e^{-0.191t}\left(\frac{C_3(0.618)+C_4(\beta_2)}{C_3}C_3\cos 0.588t\right. \\ &\quad \left. + \frac{-C_3(\beta_2)+C_4(0.618)}{C_4}C_4\sin 0.588t\right) \\ x_1(t) &= e^{-1.309t}\left(-1.618C_1\cos 0.951t - 1.618C_2\sin 0.951t\right) \\ &\quad + e^{-0.191t}\left(0.618C_3\cos 0.588t + 0.618C_4\sin 0.588t\right) \quad (8.61) \\ x_2(t) &= e^{-1.309t}\left(C_1\cos 0.951t + C_2\sin 0.951t\right) \\ &\quad + e^{-0.191t}\left(C_3\cos 0.588t + C_4\sin 0.588t\right). \quad (8.62) \end{aligned}$$

The constants of integration, C_i, are obtained by satisfying the initial conditions. Find C_i for initial conditions $x_1(0) = 1$ m, $x_2(0) = 0$, $\dot{x}_1(0) = 0$ and $\dot{x}_2(0) = 0$. Using the initial displacements, we find $C_1 = -0.447$ and $C_2 = 0.447$. The initial velocities can be used to find the other two constants. The response Equations 8.61 and 8.62 are plotted in Figure 8.24 for these initial conditions.

In the next example, a more general damping is considered where ratios r_{ij} are fully complex conjugate. ∎

8.5. DIRECT METHOD: FREE VIBRATION WITH DAMPING

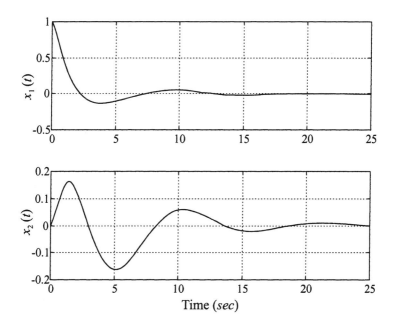

Figure 8.24: (a) Case I: Response time histories.

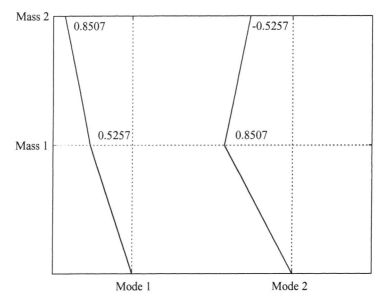

Figure 8.24: (b) Case I: Modes.

Example 8.11 Free Vibration with Damping by the Direct Method: Case II

We now re-examine the two degree of freedom system of *Example* 8.10, except that new property matrices $[m], [c]$ and $[k]$ are used. The matrix equation of motion is

$$\begin{bmatrix} 2 & 0 \\ 0 & 4 \end{bmatrix} \begin{Bmatrix} \ddot{x}_1 \\ \ddot{x}_2 \end{Bmatrix} + \begin{bmatrix} 4 & -3 \\ -3 & 3 \end{bmatrix} \begin{Bmatrix} \dot{x}_1 \\ \dot{x}_2 \end{Bmatrix} + \begin{bmatrix} 7 & -3 \\ -3 & 3 \end{bmatrix} \begin{Bmatrix} x_1 \\ x_2 \end{Bmatrix} = \begin{Bmatrix} 0 \\ 0 \end{Bmatrix}.$$

Following the usual procedures, the characteristic equation is

$$8\lambda^4 + 22\lambda^3 + 37\lambda^2 + 15\lambda + 12 = 0,$$

which is solved numerically, leading to the following four roots,

$$\begin{aligned} \lambda_{11} &= -1.260 + 1.430i \\ \lambda_{12} &= -1.260 - 1.430i \\ \lambda_{21} &= -0.115 + 0.632i \\ \lambda_{22} &= -0.115 - 0.632i. \end{aligned}$$

8.5. DIRECT METHOD: FREE VIBRATION WITH DAMPING

Following the procedure outlined earlier, we find the complex modes to be

$$\begin{aligned} r_{11} &= -2.178 + 1.004i \\ r_{12} &= -2.178 - 1.004i \\ r_{21} &= 0.511 + 0.131i \\ r_{22} &= 0.511 - 0.131i, \end{aligned}$$

where the complex conjugate nature of the modes are observed. There are two pairs of complex conjugate eigenvectors. Given the λ_{ij} and r_{ij} values, the following parameters can be calculated,

$$a_1 = -1.260, \quad \omega_{d1} = 1.430; \quad a_2 = -0.115, \quad \omega_{d2} = 0.0632,$$

and

$$\alpha_1 = -2.178, \quad \beta_1 = 1.004; \quad \alpha_2 = 0.511, \quad \beta_2 = 0.131.$$

Assume a simple set of initial conditions; $x_1(0) = 1$ cm and all other initial conditions equal zero. With these, the solutions can be derived,

$$\begin{aligned} \left\{ \begin{array}{c} x_1(t) \\ x_2(t) \end{array} \right\} &= e^{-1.260t} \left[\left\{ \begin{array}{c} 0.820 \\ -0.460 \end{array} \right\} \cos 1.430t + \left\{ \begin{array}{c} 0.859 \\ -0.182 \end{array} \right\} \sin 1.430t \right] \\ &+ e^{-0.115t} \left[\left\{ \begin{array}{c} 0.180 \\ 0.460 \end{array} \right\} \cos 0.632t + \left\{ \begin{array}{c} -0.276 \\ -0.422 \end{array} \right\} \sin 0.632t \right]. \end{aligned}$$

(8.63)

The time histories, the complex modes, r_{ij}, plotted in the complex plane, are shown in Figure 8.25(a) and (b).

The physical interpretation of the complex mode is that it represents a relative amplitude and a phase. The introduction of a phase implies that there is relative motion between the masses. Figure 8.25(b) shows the complex modes as vectors in the complex plane. If the modes are real, as in the last example, they appear as vectors along the real axis only. The angle the complex vector makes with the real axis is representative of the phase. ∎

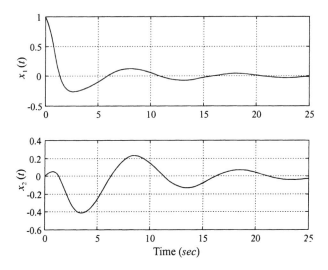

Figure 8.25: (a) Case II: Response time histories.

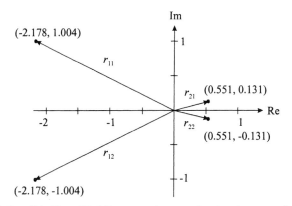

Figure 8.25: (b) Case II: The complex modes in the complex plane.

8.6 Modal Analysis

Modal analysis is a procedure by which orthogonality properties of the modes of vibration are utilized to transform the equations of motion from the physical coordinate system to the principal coordinate system and thereby decouple the equations of motion. Each of the decoupled equations can then be solved independently as a single oscillator rather than solving n simultaneous ordinary differential equations. This procedure is valid for undamped

8.6. MODAL ANALYSIS

systems as well as certain damped systems known as *proportionally damped*. First, we examine these important orthogonality properties.

8.6.1 Modal Orthogonality

The orthogonality of the modes of vibration is the single most important property they possess. Orthogonality is a property that indicates that two vectors are perpendicular to each other. The type of orthogonality the modes possess is one with respect to the mass and stiffness matrices. Rather than proving the orthogonality properties in general, it is more instructive to demonstrate these properties for the two degree of freedom system of Section 8.4.1 using modal ratio Equations 8.27 and 8.28. Recall the modal vectors,

$$\{u\}_1 = \begin{Bmatrix} u_{11} \\ u_{21} \end{Bmatrix} = u_{11} \begin{Bmatrix} 1 \\ u_{21}/u_{11} \end{Bmatrix} = u_{11} \begin{Bmatrix} 1 \\ -\frac{k_{11}-\omega_1^2 m_1}{k_{12}} \end{Bmatrix}$$

$$\{u\}_2 = \begin{Bmatrix} u_{12} \\ u_{22} \end{Bmatrix} = u_{12} \begin{Bmatrix} 1 \\ u_{22}/u_{12} \end{Bmatrix} = u_{12} \begin{Bmatrix} 1 \\ -\frac{k_{11}-\omega_2^2 m_1}{k_{12}} \end{Bmatrix},$$

where ω_1 and ω_2 are the natural frequencies, and only elastic (stiffness) coupling is included to simplify the demonstration. Form the matrix product

$$\{u\}_2^T [m]\{u\}_1$$

$$= u_{11}u_{12} \begin{Bmatrix} 1 \\ -\frac{k_{11}-\omega_2^2 m_1}{k_{12}} \end{Bmatrix}^T \begin{bmatrix} m_1 & 0 \\ 0 & m_2 \end{bmatrix} \begin{Bmatrix} 1 \\ -\frac{k_{11}-\omega_1^2 m_1}{k_{12}} \end{Bmatrix}$$

$$= u_{11}u_{12}[m_1 + \frac{m_2}{k_{12}^2}(k_{11}-\omega_1^2 m_1)(k_{11}-\omega_2^2 m_1)],$$

into which Equations 8.23 for ω_1 and ω_2 are substituted, respectively. A few pages of algebra[18] leads to

$$\{u\}_2^T [m]\{u\}_1 = u_{11}u_{12}[0] = 0,$$

which shows the *orthogonality of the modes of vibration with respect to the mass matrix*. Using Equation 8.26, $\omega_1^2[m]\{u\}_1 = [k]\{u\}_1$, and pre–multiplying by $\{u\}_2^T$, we can show that

$$\{u\}_2^T [k]\{u\}_1 = 0, \qquad (8.64)$$

---

[18] Try this once as an exercise. It is very instructive to proceed through the matrix manipulations of the type encountered in a modal analysis. Also, the end result will be the same for a fully populated mass matrix.

that is, *the modes are orthogonal with respect to the stiffness matrix.*

The following matrix product, $\{u\}_1^T[m]\{u\}_1$, equals a constant. It is customary to define a normalized mode $\{\hat{u}\}$ so that the matrix product equals one. To do this, multiply the mode by a constant, say a, such that $\{\hat{u}\} = a\{u\}$. Then, for mode i,

$$a_i^2\{u\}_i^T[m]\{u\}_i = 1,$$

from which the value of a_i can be found. Therefore,

$$\{\hat{u}\}_i^T[m]\{\hat{u}\}_i = 1, \; i = 1, 2, \tag{8.65}$$

and then using Equation 8.26 we obtain

$$\{\hat{u}\}_i^T[k]\{\hat{u}\}_i = \omega_i^2, \; i = 1, 2. \tag{8.66}$$

These modal orthogonality and normalization properties are needed to decouple the equations of motion as part of a modal analysis; the physical coordinate system, $x_1(t)$ and $x_2(t)$, is transformed into uncoupled *modal coordinates* $q_1(t)$ and $q_2(t)$. The notation $q_1(t)$ and $q_2(t)$ is customary for the modal coordinates, and also matches those for the generalized coordinates of Chapter 7. Sometimes, these coordinates are also known as *natural* or *principal* coordinates. These are discussed in a physical sense subsequently.

Consider a simple problem. Begin with the general matrix equation of free, undamped motion,

$$[m]\{\ddot{x}\} + [k]\{x\} = \{0\}, \tag{8.67}$$

but consider only two degrees of freedom. Expand the response matrix $\{x(t)\}$ in terms of the orthogonal modes,

$$\{x(t)\} = \{\hat{u}\}_1 q_1(t) + \{\hat{u}\}_2 q_2(t) = \sum_{i=1}^{2}\{\hat{u}\}_i q_i(t), \tag{8.68}$$

where the modal coordinates $q_1(t)$ and $q_2(t)$ are unknowns that are found in the modal analysis solution. The frequencies and modes have already been evaluated in an eigenvalue analysis using the property matrices $[m]$ and $[k]$ in the characteristic matrix equation $[k - \omega^2 m]\{u\} = \{0\}$.

Equation 8.68, which relates the physical coordinate $x(t)$ to the "normal," decoupled coordinate $q_i(t)$, is a reduced version of the general expression known as the *expansion theorem*,

$$\{x(t)\} = \sum_{i=1}^{n}\{\hat{u}\}_i q_i(t),$$

8.6. MODAL ANALYSIS

which, from matrix theory, is seen to be the expansion of an arbitrary function in terms of orthogonal functions. Equation 8.68 may be written in matrix form to show the generality of the procedure for n degrees of freedom,

$$\{x(t)\} = [P]\{q(t)\}, \tag{8.69}$$

where $[P]$ is called the normalized *modal matrix*. For a two degree of freedom structure, $[P]$ is the 2×2 *modal matrix*

$$[P] = [\{\hat{u}\}_1 \ \{\hat{u}\}_2]. \tag{8.70}$$

The braces are included to emphasize the respective modal vectors. It is emphasized that the eigenvalue problem has already been solved to obtain the necessary modes for Equation 8.68. Differentiate and substitute Equation 8.68 into governing Equation 8.67,

$$[m](\{\hat{u}\}_1\ddot{q}_1 + \{\hat{u}\}_2\ddot{q}_2) + [k](\{\hat{u}\}_1 q_1 + \{\hat{u}\}_2 q_2) = \{0\}. \tag{8.71}$$

Multiply on the left by the transpose of the first modal vector $\{\hat{u}\}_1^T$. This operation creates the matrix triple products that possess the orthogonality properties needed to decouple the physical equations of motion,

$$\{\hat{u}\}_1^T [m](\{\hat{u}\}_1\ddot{q}_1 + \{\hat{u}\}_2\ddot{q}_2) + \{\hat{u}\}_1^T [k](\{\hat{u}\}_1 q_1 + \{\hat{u}\}_2 q_2) = 0.$$

Orthogonality of the modes with respect to mass and stiffness matrices reduces this to

$$\{\hat{u}\}_1^T [m]\{\hat{u}\}_1 \ddot{q}_1 + \{\hat{u}\}_1^T [k]\{\hat{u}\}_1 q_1 = 0. \tag{8.72}$$

The normalization relations $\{\hat{u}\}_r^T [m]\{\hat{u}\}_r = 1$ and $\{\hat{u}\}_r^T [k]\{\hat{u}\}_r = \omega_r^2$ further reduce Equation 8.72 to the equation for the first modal coordinate

$$\ddot{q}_1 + \omega_1^2 q_1 = 0. \tag{8.73}$$

Similarly, multiplying Equation 8.71 by the transpose of the second modal vector $\{\hat{u}\}_2^T$ leads to the decoupled equation of motion in the second modal coordinate $q_2(t)$,

$$\ddot{q}_2 + \omega_2^2 q_2 = 0. \tag{8.74}$$

This is how the orthogonality properties of the modes transform the coupled equations of motion into uncoupled equations.

Equations 8.73 and 8.74 are solved for $q_1(t)$ and $q_2(t)$, respectively, to obtain

$$q_i(t) = C_i \cos(\omega_i t - \phi_i), \quad i = 1, 2. \tag{8.75}$$

Then insert Equations 8.75 into either of Equations 8.68 or 8.69 to recover the physical response,

$$\begin{aligned}
\{x(t)\} &= \sum_{i=1}^{2}\{\hat{u}\}_i q_i(t) = [P]\{q(t)\} = [\{\hat{u}\}_1 \ \{\hat{u}\}_2]\begin{Bmatrix} q_1(t) \\ q_2(t) \end{Bmatrix} \\
&= \sum_{i=1}^{2}\{\hat{u}\}_i C_i \cos(\omega_i t - \phi_i) \\
&= \sum_{i=1}^{2} C_i\{\hat{u}\}_i[\cos\omega_i t \cos\phi_i + \sin\omega_i t \sin\phi_i].
\end{aligned} \quad (8.76)$$

Given the initial conditions $\{x(0)\}$ and $\{\dot{x}(0)\}$, Equation 8.76 can be used to evaluate the arbitrary constants C_i and ϕ_i, as follows,

$$\{x(0)\} = \sum_{i=1}^{2} C_i\{\hat{u}\}_i \cos\phi_i \quad (8.77)$$

$$\{\dot{x}(0)\} = \sum_{i=1}^{2} C_i\omega_i\{\hat{u}\}_i \sin\phi_i. \quad (8.78)$$

Pre-multiply Equations 8.77 and 8.78 by $\{\hat{u}\}_j^T[m]$, $j = 1, 2$,

$$\{\hat{u}\}_j^T[m]\{x(0)\} = \sum_{i=1}^{2} C_i\{\hat{u}\}_j^T[m]\{\hat{u}\}_i \cos\phi_i$$

$$\{\hat{u}\}_j^T[m]\{\dot{x}(0)\} = \sum_{i=1}^{2} C_i\omega_i\{\hat{u}\}_j^T[m]\{\hat{u}\}_i \sin\phi_i,$$

from which, by orthogonality, the only non–zero terms are for $j = i$. The two equations required to evaluate C and ϕ for each degree of freedom,

$$\{\hat{u}\}_i^T[m]\{x(0)\} = C_i \cos\phi_i, \quad i = 1, 2,$$

$$\frac{\{\hat{u}\}_i^T[m]\{\dot{x}(0)\}}{\omega_i} = C_i \sin\phi_i, \quad i = 1, 2.$$

Substitute these last two expressions into Equation 8.76 to obtain the complete response,

$$\{x(t)\} = \sum_{i=1}^{2}\{\hat{u}\}_i\left(\{\hat{u}\}_i^T[m]\{x(0)\}\cos\omega_i t + \frac{1}{\omega_i}\{\hat{u}\}_i^T[m]\{\dot{x}(0)\}\sin\omega_i t\right). \quad (8.79)$$

8.6. MODAL ANALYSIS

This is the free vibration response using a modal analysis approach.[19] It is important to remember that the modes $\{\hat{u}\}_i$ in this solution are normalized modes.

It is informative to consider the solution Equation 8.79 with the following initial conditions:

$$\{\dot{x}(0)\} = 0$$
$$\{x(0)\} = x_0\{u\}_s,$$

where the initial displacement is related linearly to the sth normal mode. Then,

$$\{x(t)\} = \sum_{i=1}^{2} x_0\{\hat{u}\}_i [\{\hat{u}\}_i^T [m]\{u\}_s] \cos\omega_i t$$
$$= \sum_{i=1}^{2} x_0\{\hat{u}\}_i [\delta_{is}] \cos\omega_i t$$
$$= x_0\{\hat{u}\}_s \cos\omega_s t,$$

where the Kroenecker delta $\delta_{is} = 1$ for $i = s$. This represents synchronous harmonic oscillation at the natural frequency ω_s; the system configuration resembles the sth mode at all times. Thus, any of the normal modes can be excited independently in a linear system, which is not the case for nonlinear systems.

Why does modal analysis work? It works because *the mass and stiffness matrices* can be *simultaneously diagonalized*. Modal analysis transforms the need to solve n simultaneous equations of motion to a procedure that only involves a series of matrix multiplications and additions, as we see in Equation 8.79. Consider the following example problem.

Example 8.12 Free Vibration via Modal Analysis
Try a solution of *Example* 8.8 using modal analysis. The matrix equation of motion is

$$\begin{bmatrix} m & 0 \\ 0 & 2m \end{bmatrix} \begin{Bmatrix} \ddot{x}_1 \\ \ddot{x}_2 \end{Bmatrix} + \begin{bmatrix} 3k & -2k \\ -2k & 5k \end{bmatrix} \begin{Bmatrix} x_1 \\ x_2 \end{Bmatrix} = \begin{Bmatrix} 0 \\ 0 \end{Bmatrix}$$

with natural frequencies

$$\omega_1 = 1.14\sqrt{\frac{k}{m}} \text{ rad/sec}, \quad \omega_2 = 2.05\sqrt{\frac{k}{m}} \text{ rad/sec}$$

---

[19] Recall from linear algebra that vector–matrix multiplication is not associative and therefore in Equation 8.79 we cannot move the factor $\{\hat{u}\}_i$ arbitrarily inside the parenthesis within the matrix products to obtain triple products such as $\{\hat{u}\}_i^T [m]\{\hat{u}\}_i$.

and respective modal ratios

$$\frac{u_{21}}{u_{11}} = 0.845$$
$$\frac{u_{22}}{u_{12}} = -0.595.$$

Use the initial conditions: $x_1(0) = 1$ cm, $x_2(0) = -1$ cm, $\dot{x}_1(0) = 3$ cm/sec, and $\dot{x}_2(0) = -2$ cm/sec. We retain three places after the decimal.

Solution First normalize the modes with respect to the mass matrix as follows. Let the normalized modes be

$$\{\hat{u}\}_1 = a_1 \left\{ \begin{array}{c} 1.000 \\ 0.845 \end{array} \right\}, \quad \{\hat{u}\}_2 = a_2 \left\{ \begin{array}{c} 1.000 \\ -0.595 \end{array} \right\},$$

where a_1 and a_2 are the factors needed to normalize the modes. Then

$$\{\hat{u}\}_1^T [m] \{\hat{u}\}_1 = \frac{1}{a_1^2},$$

or, substituting the appropriate vectors and matrix,

$$\{1.000 \quad 0.845\} \begin{bmatrix} m & 0 \\ 0 & 2m \end{bmatrix} \left\{ \begin{array}{c} 1.000 \\ 0.845 \end{array} \right\} = \frac{1}{a_1^2}.$$

Solve for a_1,

$$a_1 = \frac{0.642}{\sqrt{m}}.$$

Doing the same computation for the second modal vector,

$$\{1.000 \quad -0.595\} \begin{bmatrix} m & 0 \\ 0 & 2m \end{bmatrix} \left\{ \begin{array}{c} 1.000 \\ -0.595 \end{array} \right\} = \frac{1}{a_2^2},$$

with the result

$$a_2 = \frac{0.765}{\sqrt{m}}.$$

The normalized modes are then

$$\{\hat{u}\}_1 = \frac{1}{\sqrt{m}} \left\{ \begin{array}{c} 0.642 \\ 0.542 \end{array} \right\}, \quad \{\hat{u}\}_2 = \frac{1}{\sqrt{m}} \left\{ \begin{array}{c} 0.765 \\ -0.455 \end{array} \right\}.$$

8.6. MODAL ANALYSIS

Now we have what we need to solve for the response, using Equation 8.79:

$$\left\{ \begin{array}{c} x_1(t) \\ x_2(t) \end{array} \right\} =$$

$$\frac{1}{\sqrt{m}} \left\{ \begin{array}{c} 0.642 \\ 0.542 \end{array} \right\} \left(\frac{1}{\sqrt{m}} \{0.642\ 0.542\} \left[\begin{array}{cc} m & 0 \\ 0 & 2m \end{array} \right] \left\{ \begin{array}{c} 1 \\ -1 \end{array} \right\} \cos \omega_1 t \right.$$

$$+ \left. \frac{1}{\omega_1\sqrt{m}} \{0.642\ 0.542\} \left[\begin{array}{cc} m & 0 \\ 0 & 2m \end{array} \right] \left\{ \begin{array}{c} 3 \\ -2 \end{array} \right\} \sin \omega_1 t \right)$$

$$+ \frac{1}{\sqrt{m}} \left\{ \begin{array}{c} 0.765 \\ -0.455 \end{array} \right\} \left(\frac{1}{\sqrt{m}} \{0.765\ -0.455\} \left[\begin{array}{cc} m & 0 \\ 0 & 2m \end{array} \right] \left\{ \begin{array}{c} 1 \\ -1 \end{array} \right\} \cos \omega_2 t \right.$$

$$+ \left. \frac{1}{\omega_2\sqrt{m}} \{0.765\ -0.455\} \left[\begin{array}{cc} m & 0 \\ 0 & 2m \end{array} \right] \left\{ \begin{array}{c} 3 \\ -2 \end{array} \right\} \sin \omega_2 t \right) \text{ cm.}$$

In order to complete this problem, we need the values of k and m. Assume that $k = 1$ N/m and $m = 1$ kg. Then, $\omega_1 = 1.14$ rad/sec and $\omega_2 = 2.05$ rad/sec. Substituting all numbers into the general solution, we find

$$x_1(t) = -0.284 \cos 1.14t - 0.136 \sin 1.14t + 1.281 \cos 2.05t + 1.536 \sin 2.05t$$
$$x_2(t) = -0.240 \cos 1.14t - 0.115 \sin 1.14t - 0.762 \cos 2.05t - 0.914 \sin 2.05t.$$

These are plotted in Figure 8.26. A good check that all the matrix products and other calculations were done correctly is to satisfy the initial conditions using these final equations. (Try this for yourself, and you will see the round off errors introduced by retaining only three terms after the decimal.) ∎

We need to emphasize that the above example has been carried out to show a procedure, not a realistic application with all its inherent difficulties. For free, undamped vibration, there is no difference between the direct method and a modal analysis. Modal analysis begins to be advantageous for a forced system. A full discussion of the advantages of the modal approach follows.

8.6.2 Modal Analysis with Forcing

Begin with the general matrix equation of forced, undamped motion,

$$[m]\{\ddot{x}\} + [k]\{x\} = \{F(t)\}, \tag{8.80}$$

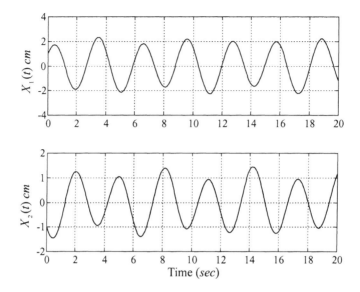

Figure 8.26: Response of two coupled masses in time.

but again consider only a two degree of freedom system so that

$$\{x(t)\} = \{\hat{u}\}_1 q_1(t) + \{\hat{u}\}_2 q_2(t) = \sum_{i=1}^{2} \{\hat{u}\}_i q_i(t) \quad (8.81)$$
$$= [P]\{q(t)\}.$$

Substituting Equation 8.81 and its second derivative into 8.80 results in

$$[m](\{\hat{u}\}_1 \ddot{q}_1 + \{\hat{u}\}_2 \ddot{q}_2) + [k](\{\hat{u}\}_1 q_1 + \{\hat{u}\}_2 q_2) = \{F(t)\}. \quad (8.82)$$

Multiply on the left of each term by $\{\hat{u}\}_1^T$ to obtain the matrix triple products that possess the orthogonality properties needed,

$$\{\hat{u}\}_1^T [m](\{\hat{u}\}_1 \ddot{q}_1 + \{\hat{u}\}_2 \ddot{q}_2) + \{\hat{u}\}_1^T [k](\{\hat{u}\}_1 q_1 + \{\hat{u}\}_2 q_2) = \{\hat{u}\}_1^T \{F(t)\}.$$

Orthogonality of the modes with respect to mass and stiffness matrices reduces this to

$$\{\hat{u}\}_1^T [m]\{\hat{u}\}_1 \ddot{q}_1 + \{\hat{u}\}_1^T [k]\{\hat{u}\}_1 q_1 = \{\hat{u}\}_1^T \{F(t)\},$$

and normalization Equations 8.65 and 8.66 reduces this further to the equation governing the first modal coordinate,

$$\ddot{q}_1 + \omega_1^2 q_1 = N_1(t), \quad (8.83)$$

8.6. MODAL ANALYSIS

where $N_1(t) = \{\hat{u}\}_1^T \{F(t)\}$. Similarly, multiplying Equation 8.82 by $\{\hat{u}\}_2^T$, leads to the decoupled equation for the second modal coordinate,

$$\ddot{q}_2 + \omega_2^2 q_2 = N_2(t). \tag{8.84}$$

Equations 8.83 and 8.84 are solved for $q_1(t)$ and $q_2(t)$, respectively,

$$\begin{aligned} q_i(t) &= \frac{1}{\omega_i} \int_0^t N_i(\tau) \sin \omega_i(t-\tau) d\tau \\ &+ q_i(0) \cos \omega_i t + \frac{\dot{q}_i(0)}{\omega_i} \sin \omega_i t, \quad i = 1, 2. \end{aligned} \tag{8.85}$$

The first term in the solution is the Duhamel integral, and the second and third terms are the free responses due to initial displacement and velocity. The initial conditions in modal coordinates can be related to the physical initial conditions using Equation 8.81, and setting $t = 0$,

$$\{x(0)\} = [P]\{q(0)\} = \sum_{i=1}^{2} \{\hat{u}\}_i q_i(0) \tag{8.86}$$

$$\{\dot{x}(0)\} = [P]\{\dot{q}(0)\} = \sum_{i=1}^{2} \{\hat{u}\}_i \dot{q}_i(0).$$

Therefore, the relation between initial displacements is

$$\left\{ \begin{array}{c} x_1(0) \\ x_2(0) \end{array} \right\} = \left\{ \begin{array}{c} \hat{u}_1 \\ \hat{u}_2 \end{array} \right\}_1 q_1(0) + \left\{ \begin{array}{c} \hat{u}_1 \\ \hat{u}_2 \end{array} \right\}_2 q_2(0).$$

Similarly for the initial velocities,

$$\left\{ \begin{array}{c} \dot{x}_1(0) \\ \dot{x}_2(0) \end{array} \right\} = \left\{ \begin{array}{c} \hat{u}_1 \\ \hat{u}_2 \end{array} \right\}_1 \dot{q}_1(0) + \left\{ \begin{array}{c} \hat{u}_1 \\ \hat{u}_2 \end{array} \right\}_2 \dot{q}_2(0).$$

We now have four equations to solve for four unknowns. In matrix form, begin with Equation 8.86, and pre-multiply by $\{\hat{u}\}_j^T [m]$

$$\{\hat{u}\}_j^T [m] \{x(0)\} = \{\hat{u}\}_j^T [m] \sum_{i=1}^{2} \{\hat{u}\}_i q_i(0), \quad j = 1, 2.$$

By orthogonality, only one term on the right hand side will remain ($j = i$):

$$\{\hat{u}\}_i^T [m] \{x(0)\} = q_i(0),$$

and similarly for the velocity relations,

$$\{\hat{u}\}_i^T[m]\{\dot{x}(0)\} = \dot{q}_i(0).$$

Now Equation 8.85 can be solved for each modal degree of freedom, and then substituted into Equation 8.81 to obtain the complete response in the physical coordinates: $\{x(t)\} = \{\hat{u}\}_1 q_1(t) + \{\hat{u}\}_2 q_2(t)$,

$$\begin{aligned}
\{x(t)\} &= \sum_i^2 \{\hat{u}\}_i \left(\frac{1}{\omega_i}\int_0^t N_i(\tau)\sin\omega_i(t-\tau)d\tau \right.\\
&+ \left. q_i(0)\cos\omega_i t + \frac{\dot{q}_i(0)}{\omega_i}\sin\omega_i t\right) \\
&= \sum_i^2 \{\hat{u}\}_i \left(\frac{1}{\omega_i}\int_0^t N_i(\tau)\sin\omega_i(t-\tau)d\tau + \{\hat{u}\}_i^T[m]\{x(0)\}\cos\omega_i t\right.\\
&+ \left.\{\hat{u}\}_i^T[m]\{\dot{x}(0)\}\frac{1}{\omega_i}\sin\omega_i t\right) \qquad (8.87)
\end{aligned}$$

$$N_i(t) = \sum_{j=1}^2 \hat{u}_{ji} F_j(t), \quad i=1,2.$$

In Equation 8.87, the first term is the steady state response given by a Duhamel integral, and the next two terms are the responses due to initial displacements and initial velocities, respectively. Since there is no damping in this model, there are no decaying terms.

Modal Participation Factor

Before proceeding to an example, additional insight can be gained by examining the forcing term $N_1(t)$ in the modal Equation 8.83,

$$\ddot{q}_1 + \omega_1^2 q_1 = N_1(t),$$

where $N_1(t) = \{\hat{u}\}_1^T\{F(t)\}$. Let us explicitly show the normalized denominator as follows,

$$N_1(t) = \frac{\{\hat{u}\}_1^T\{F(t)\}}{\{\hat{u}\}_1^T[m]\{\hat{u}\}_1}.$$

The ratio

$$\frac{\{\hat{u}\}_1^T}{\{\hat{u}\}_1^T[m]\{\hat{u}\}_1}$$

is given the name *mode participation factor* in order to signify the effect, in this case, of the first mode on the response at the first natural frequency.

8.6. MODAL ANALYSIS

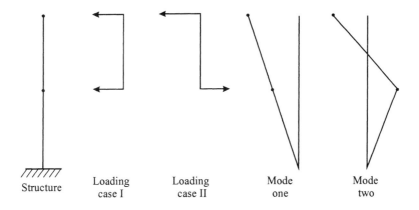

Figure 8.27: Modal participation: two loading cases and the two modes.

Of course, the denominator equals one, and therefore the factor is actually the numerator $\{\hat{u}\}_1^T$. One can use the non–normal modes as well. Various such factors are defined in applications as aides to the design of efficient and accurate computations of very large degree of freedom systems.

Consider Figure 8.27 of a two degree of freedom structure, which shows schematics of two possible load combinations and the two modes of the structure. For the first load, the two forces are acting in the same direction and would therefore primarily excite the first mode of vibration. The second load, where the two forces act in opposite directions, will excite the second mode more than the first mode.

The practical importance of understanding this arises in the attempt to decide how many modes are needed to accurately model structural response to a particular load. The response to loading case one would be adequately modeled using only mode one. Similarly, for loading case two, the response would be primarily in the second mode. Extending our thoughts to a structure with hundreds of degrees of freedom, we can see that a structure will generally respond in the modes that resemble the loading. Such understanding can facilitate the development of realistic but simplified models for particular loading cases. This concept is also applicable to the continuous systems to be studied in Chapters 10 and 11.

Example 8.13 Forced Vibration via Modal Analysis
Consider the example of a two degree of freedom system without damping, from formulation to modal analysis and solution. The system under consideration is shown in Figure 8.28.

The free body diagrams, in conjunction with Newton's second law of

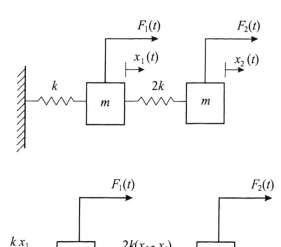

Figure 8.28: Two degree of freedom undamped system.

motion, lead to the two governing equations of motion,

$$F_1(t) - kx_1 + 2k(x_2 - x_1) = m\ddot{x}_1$$
$$F_2(t) - 2k(x_2 - x_1) = m\ddot{x}_2,$$

or in the more useful matrix form,

$$\begin{bmatrix} m & 0 \\ 0 & m \end{bmatrix} \begin{Bmatrix} \ddot{x}_1 \\ \ddot{x}_2 \end{Bmatrix} + \begin{bmatrix} 3k & -2k \\ -2k & 2k \end{bmatrix} \begin{Bmatrix} x_1 \\ x_2 \end{Bmatrix} = \begin{Bmatrix} F_1(t) \\ F_2(t) \end{Bmatrix}.$$

The procedure is to first solve the free vibration problem for the natural frequencies and modes. For free vibration, assume harmonic behavior,

$$x_1(t) = X_1 e^{-i\omega t} \qquad x_2(t) = X_2 e^{-i\omega t}$$
$$\ddot{x}_1(t) = -X_1 \omega^2 e^{-i\omega t} \qquad \ddot{x}_2(t) = -X_2 \omega^2 e^{-i\omega t},$$

which are substituted into the matrix equation of motion. Combining and canceling the exponential which appears in all terms leads to

$$\begin{bmatrix} 3k - m\omega^2 & -2k \\ -2k & 2k - m\omega^2 \end{bmatrix} \begin{Bmatrix} X_1 \\ X_2 \end{Bmatrix} = \begin{Bmatrix} 0 \\ 0 \end{Bmatrix}. \qquad (8.88)$$

8.6. MODAL ANALYSIS

The determinant of the characteristic matrix set equal to 0, leading to the characteristic equation

$$\omega^4 - 5\frac{k}{m}\omega^2 + 2\left(\frac{k}{m}\right)^2 = 0.$$

This is a quadratic in ω^2 with solutions

$$\omega_1^2 = 0.44\frac{k}{m} \text{ (rad/sec)}^2, \quad \omega_2^2 = 4.56\frac{k}{m} \text{ (rad/sec)}^2.$$

To obtain the modes, substitute

$$\omega_1 = 0.66\sqrt{k/m}\,\text{rad/sec}$$
$$\text{and } \omega_2 = 2.14\sqrt{k/m}\,\text{rad/sec},$$

respectively, into one of Equations 8.88 and solve for the modal ratios as follows,

$$\left(3k - m \cdot 0.44\frac{k}{m}\right)X_1 - 2kX_2 = 0$$

$$\left(\frac{X_2}{X_1}\right)_1 = 1.28$$

and

$$\left(3k - m \cdot 4.56\frac{k}{m}\right)X_1 - 2kX_2 = 0$$

$$\left(\frac{X_2}{X_1}\right)_2 = -0.78.$$

The two modes, in standard notation, are

$$\{u\}_1 = \left\{\begin{array}{c} 1.00 \\ 1.28 \end{array}\right\}, \quad \{u\}_2 = \left\{\begin{array}{c} 1.00 \\ -0.78 \end{array}\right\}.$$

In order to proceed with a modal analysis and decouple the equations of motion, it is necessary to normalize the modes. Introduce constant multiples a_1 and a_2 of the modal vectors,

$$\{\hat{u}\}_1 = a_1 \left\{\begin{array}{c} 1.00 \\ 1.28 \end{array}\right\}, \quad \{\hat{u}\}_2 = a_2 \left\{\begin{array}{c} 1.00 \\ -0.78 \end{array}\right\}.$$

Normalizing the first mode requires that

$$\{\hat{u}\}_1^T [m]\{\hat{u}\}_1 = 1$$

$$a_1^2 \begin{Bmatrix} 1.00 \\ 1.28 \end{Bmatrix}^T \begin{bmatrix} m & 0 \\ 0 & m \end{bmatrix} \begin{Bmatrix} 1.00 \\ 1.28 \end{Bmatrix} = 1$$

$$2.64 a_1^2 m = 1$$

$$a_1 \simeq \frac{0.62}{\sqrt{m}}$$

$$\{\hat{u}\}_1 = \frac{1}{\sqrt{m}} \begin{Bmatrix} 0.62 \\ 0.79 \end{Bmatrix},$$

and

$$\{\hat{u}\}_2^T [m]\{\hat{u}\}_2 = 1$$

$$a_2^2 \begin{Bmatrix} 1.00 \\ -0.78 \end{Bmatrix}^T \begin{bmatrix} m & 0 \\ 0 & m \end{bmatrix} \begin{Bmatrix} 1.00 \\ -0.78 \end{Bmatrix} = 1$$

$$1.61 a_2^2 m = 1$$

$$a_2 \simeq \frac{0.79}{\sqrt{m}}$$

$$\{\hat{u}\}_2 = \frac{1}{\sqrt{m}} \begin{Bmatrix} 0.79 \\ -0.62 \end{Bmatrix}.$$

Next in the solution consider external forcing of the form

$$F_1(t) = A_1 u(t)$$
$$F_2(t) = A_2 u(t),$$

where $u(t)$ is the unit step function and A_1 and A_2 are known constants. To decouple the matrix equations of motion, use the modal transformation Equation 8.69 $\{x(t)\} = [P]\{q(t)\}$, where

$$[P] = \frac{1}{\sqrt{m}} \begin{bmatrix} 0.62 & 0.79 \\ 0.79 & -0.62 \end{bmatrix}$$

is the modal matrix, and the symmetry is coincidental. Substituting the modal transformation into the matrix equation of motion and pre–multiplying by $[P]^T$ leads to

$$[P]^T [m][P]\{\ddot{q}\} + [P]^T [k][P]\{q\} = [P]^T \{F(t)\},$$

8.6. MODAL ANALYSIS

which are now decoupled. By using the modal matrix $[P]$, all the equations are simultaneously decoupled rather than one equation at a time. The right hand side of the above equation can be expanded,

$$\{N(t)\} \equiv [P]^T\{F(t)\} = [P]^T \left\{ \begin{array}{c} A_1 u(t) \\ A_2 u(t) \end{array} \right\}$$

$$= \frac{1}{\sqrt{m}} \left[\begin{array}{cc} 0.62 & 0.79 \\ 0.79 & -0.62 \end{array} \right] \left\{ \begin{array}{c} A_1 u(t) \\ A_2 u(t) \end{array} \right\}$$

$$\left\{ \begin{array}{c} N_1(t) \\ N_2(t) \end{array} \right\} = \frac{1}{\sqrt{m}} \left\{ \begin{array}{c} 0.62 A_1 + 0.79 A_2 \\ 0.79 A_1 - 0.62 A_2 \end{array} \right\} u(t).$$

The uncoupled modal equations of motion are then

$$\ddot{q}_1 + \omega_1^2 q_1 = N_1(t)$$
$$\ddot{q}_2 + \omega_2^2 q_2 = N_2(t).$$

The solution of each, for zero initial conditions, is the undamped Duhamel integral,

$$q_i(t) = \frac{1}{\omega_i} \int_0^t N_i(\tau) \sin \omega_i (t - \tau) d\tau, \quad i = 1, 2.$$

Therefore,

$$q_1(t) = \frac{1}{\omega_1} \frac{1}{\sqrt{m}} (0.62 A_1 + 0.79 A_2) \int_0^t u(\tau) \sin \omega_1 (t - \tau) d\tau$$

$$= \frac{(0.62 A_1 + 0.79 A_2)}{\omega_1 \sqrt{m}} \frac{(1 - \cos \omega_1 t)}{\omega_1}, \quad (8.89)$$

$$q_2(t) = \frac{1}{\omega_2} \frac{1}{\sqrt{m}} (0.79 A_1 - 0.62 A_2) \int_0^t u(\tau) \sin \omega_2 (t - \tau) d\tau$$

$$= \frac{(0.79 A_1 - 0.62 A_2)}{\omega_2 \sqrt{m}} \frac{(1 - \cos \omega_2 t)}{\omega_2}. \quad (8.90)$$

To complete the solution, we need to transform the above modal solution back to physical space using $\{x(t)\} = [P]\{q(t)\}$,

$$\left\{ \begin{array}{c} x_1(t) \\ x_2(t) \end{array} \right\} = \frac{1}{\sqrt{m}} \left[\begin{array}{cc} 0.62 & 0.79 \\ 0.79 & -0.62 \end{array} \right] \left\{ \begin{array}{c} q_1(t) \\ q_2(t) \end{array} \right\}.$$

Expanded, the forced response for each mass is

$$x_1(t) = \frac{1}{\sqrt{m}} (0.62 q_1(t) + 0.79 q_2(t)) \quad (8.91)$$

$$x_2(t) = \frac{1}{\sqrt{m}} (0.79 q_1(t) - 0.62 q_2(t)), \quad (8.92)$$

where $q_1(t)$ and $q_2(t)$ are given by Equations 8.89 and 8.90. We can clearly see the contribution of each mode to the total motion of each mass. If the initial conditions above are not zero, then Equations 8.91 and 8.92 would be added to the free vibration response as in Equation 8.85, resulting in the complete response. ∎

We next proceed to examine the possibility of utilizing a modal analysis for a system with damping.

8.6.3 Modal Analysis with Proportional Damping

Damping can be directly incorporated into modal analysis if it is *proportional damping*, $[c] \sim [m] + [k]$, that is, if the damping matrix is proportional to a linear combination of mass and stiffness matrices. In this approach, modal analysis of a damped system leads to a decoupled set of governing equations. Otherwise, the first derivative damping related terms prevent such decoupling and the direct method of Section 8.5 or other procedures become necessary. The physical meaning of such a damping model is discussed at the end of this section.

Begin with the general matrix equation of forced, damped motion,

$$[m]\{\ddot{x}\} + [c]\{\dot{x}\} + [k]\{x\} = \{F(t)\}, \tag{8.93}$$

and consider a two degree of freedom system with a two term solution,

$$\{x(t)\} = [P]\{q(t)\} = \sum_{i=1}^{2}\{\hat{u}\}_i q_i(t). \tag{8.94}$$

Differentiate and substitute Equation 8.94 into governing Equation 8.93 to find

$$[m](\{\hat{u}\}_1\ddot{q}_1 + \{\hat{u}\}_2\ddot{q}_2) + [c](\{\hat{u}\}_1\dot{q}_1 + \{\hat{u}\}_2\dot{q}_2) + [k](\{\hat{u}\}_1 q_1 + \{\hat{u}\}_2 q_2) = \{F(t)\}. \tag{8.95}$$

In terms of the modal matrix $[P] = [\{\hat{u}\}_1 \ \{\hat{u}\}_2]$, Equation 8.95 can be written as

$$[m][P]\{\ddot{q}\} + [c][P]\{\dot{q}\} + [k][P]\{q\} = \{F(t)\}.$$

Multiply on the left by $[P]^T$ in order to obtain the matrix triple products that possesses the needed orthogonality properties,[20]

$$[P]^T[m][P]\{\ddot{q}\} + [P]^T[c][P]\{\dot{q}\} + [P]^T[k][P]\{q\} = [P]^T\{F(t)\}.$$

---
[20] We could have pre-multiplied by $\{u\}_1^T$ and $\{u\}_2^T$ in sequence as in Section 8.6.1, but by showing this procedure we see how to decouple the equations of motion in one step. This will be shown in more detail in Section 9.2.1.

8.6. MODAL ANALYSIS

We know that orthogonality of the modes with respect to mass and stiffness matrices permits the substitutions $[P]^T[m][P] = [I]$ and $[P]^T[k][P] = [diag\ \omega^2] \equiv [\Omega]$, but what about the triple product $[P]^T[c][P]$? In general, there is no decoupling and it is not possible to *simultaneously* diagonalize the mass, stiffness *and* damping matrices. However, suppose the structural viscous damping can be represented as a numerical proportion of mass and stiffness, in the following way,

$$[c] = C_m[m] + C_k[k],$$

where C_m and C_k are constants. Then,

$$\begin{aligned}[P]^T[c][P]\{\dot{q}\} &= [P]^T(C_m[m] + C_k[k])[P]\{\dot{q}\} \\ &= C_m[P]^T[m][P]\{\dot{q}\} + C_k[P]^T[k][P]\{\dot{q}\} \\ &= (C_m[I] + C_k[\Omega])\{\dot{q}\} \\ &= (C_m + C_k\omega_i^2)\dot{q}_i.\end{aligned}$$

Such damping is known as proportional[21] or *Rayleigh damping*.

The equations can now be decoupled into a governing equation for each degree of freedom,

$$\ddot{q}_i + (C_m + C_k\omega_i^2)\dot{q}_i + \omega_i^2 q_i = N_i(t), \qquad (8.96)$$

where it is customary to define

$$2\zeta_i\omega_i \equiv C_m + C_k\omega_i^2,$$

or

$$\zeta_i = \frac{1}{2}\left(\frac{C_m}{\omega_i} + C_k\omega_i\right), \qquad (8.97)$$

and $N_i(t) = \sum_{j=1}^n u_{ji}F_j(t)$. Constant parameters C_m and C_k have units to make both sides of Equation 8.97 dimensionless. C_m has dimensions sec^{-1} and C_k has dimensions sec.

Equation 8.96 is solved by the Duhamel integral

$$\begin{aligned}q_i(t) &= \frac{1}{\omega_{d_i}}\int_0^t N_i(\tau)e^{-\zeta_i\omega_{d_i}(t-\tau)}\sin\omega_{d_i}(t-\tau)d\tau \\ &+ e^{-\zeta_i\omega_i t}\left[\frac{q_i(0)}{(1-\zeta_i^2)^{1/2}}\cos(\omega_{d_i}t - \phi_i) + \frac{\dot{q}_i(0)}{\omega_{d_i}}\sin\omega_{d_i}t\right],\end{aligned}$$

---

[21] *Example* 8.10 can be re-examined to see that the damping given was indeed proportional and therefore that problem can be solved by a modal analysis. *Example* 8.11, on the other hand, cannot be solved by a modal analysis.

where
$$\omega_{d_i} = \omega_i\sqrt{1-\zeta_i^2}, \quad \phi_i = \arctan\frac{\zeta_i}{\sqrt{1-\zeta_i^2}},$$

and
$$q_i(0) = \{\hat{u}\}_i^T[m]\{x(0)\}, \quad \dot{q}_i(0) = \{\hat{u}\}_i^T[m]\{\dot{x}(0)\}.$$

Once each $q_i(t)$ is evaluated, the response $\{x(t)\}$ is found by performing the matrix–vector multiplication $[P]\{q(t)\}$ of Equation 8.94, as we have done in earlier examples.

Rayleigh

John William Strutt, Lord Rayleigh was born on 12 November 1842 in Langford Grove (near Maldon), Essex, England and died on 30 June 1919 in Terling Place, Witham, Essex, England. John Strutt suffered from poor health and his schooling at Eton and Harrow was disrupted and for four years he had a private tutor. He entered Trinity College, Cambridge in 1861, graduating in 1864. His first paper in 1865 was on Maxwell's electromagnetic theory. He worked on propagation of sound and, while on an excursion to Egypt taken for health reasons, Strutt wrote his *Treatise on Sound* (1870-1871). In 1879 he wrote a paper on travelling waves, this theory has now developed into the theory of solitons. His theory of scattering (1871) was the first correct explanation of why the sky is blue.

In 1873 he succeeded to the title of Baron Rayleigh. From 1879 to 1884 he was the second Cavendish professor of experimental physics at Cambridge, succeeding Maxwell. Then in 1884 he became secretary of the Royal Society. Rayleigh discovered the inert gas argon in 1895, work which earned him a Nobel Prize, in 1904.

He was awarded the De Morgan Medal of the London Mathematical Society in 1890 and was president of the Royal Society between 1905 and 1908. He became chancellor of Cambridge University in 1908.

Example 8.14 Forced Vibration with Proportional Damping via Modal Analysis

Here we continue *Example* 8.13, except that damping is added to the equations of motion. Solve using a modal analysis procedure.

Solution Experimental data leads to proportional damping of the form

$$[c] = 2[m] + 3[k]$$
$$= \begin{bmatrix} 2m & 0 \\ 0 & 2m \end{bmatrix} + \begin{bmatrix} 9k & -6k \\ -6k & 6k \end{bmatrix} \text{ N sec/m}.$$

8.6. MODAL ANALYSIS

This leads to the decoupling of the equations of motion. In modal coordinates, the equations are

$$\ddot{q}_1 + 2\zeta_1\omega_1\dot{q}_1 + \omega_1^2 q_1 = N_1(t)$$
$$\ddot{q}_2 + 2\zeta_2\omega_2\dot{q}_2 + \omega_2^2 q_2 = N_2(t),$$

where $N_1(t)$ and $N_2(t)$ are the same as in *Example* 8.12, and

$$\zeta_1 = \frac{1}{2}\left(\frac{2}{\omega_1} + 3\omega_1\right), \quad \zeta_2 = \frac{1}{2}\left(\frac{2}{\omega_2} + 3\omega_2\right).$$

The remainder of the solution is straightforward. ∎

Damping in Multi–Degree of Freedom Systems

The normal mode method, or modal analysis, applies only to undamped systems or systems where the damping matrix can be made mathematically equivalent to the mass and/or stiffness matrices. Sometimes damping can be ignored in the forced response of a vibratory system. For example, if the force acts for a relatively short time period, a small amount of damping will not significantly affect the response, especially during the time immediately after the force stops.

Also, damping plays an insignificant role in the steady state response of a periodically forced structure if the forcing frequency is not near a resonance. Recall that in the magnification factor curves of Figure 3.12, except for the region about resonance $\omega/\omega_n = 1$, the curves are quite close to each other for all ζ values. Of course, if the periodic forcing frequency is very near or at a resonant frequency, damping plays a crucial role in the response amplitude.

Since the analyst generally does not know in advance the importance of damping in a particular problem, it must be included in any vibration analysis until it can be shown to be insignificant to the response.

For special systems where the damping matrix is linearly related to the mass and stiffness matrices, that is the proportional damping case, the simultaneous diagonalization of the stiffness and mass matrices can be accomplished along with that of the damping matrix. This results in the equation we have found above, $\zeta_i = (C_m/\omega_i + C_k\omega_i)/2$. For such systems, two cases are of particular interest. If C_m is very small or $C_m = 0$, then the damping matrix is proportional only to the stiffness matrix, that is $\zeta_i = C_k\omega_i/2$, which means that the damping ratio for each normal mode is proportional to the undamped resonant frequency of that mode. Therefore, the response of the higher modes of such a system will be damped out more rapidly than those of the lower modes.

The other extreme case is when C_k is very small or $C_k = 0$. Now the damping matrix is proportional to the mass matrix, and $\zeta_i = C_m/2\omega_i$, meaning that the damping ratio in each mode is inversely proportional to the undamped resonant frequency. The lower modes for this type of system will damp quicker than the higher modes. In this way, for systems that include both C_m and C_k components, the analyst can develop a physical sense of how the system model will behave, given the relative values of C_m and C_k. In the same vein, an analyst working with damping data can create a proportional damping model for a structure using characteristic vibration data by choosing appropriate values of C_m and C_k so that the damping of the modes resembles that found in the data.

Sometimes, if damping is light, the assumption is made that the damping matrix satisfies the same modal orthogonality properties as the mass and stiffness matrices,

$$\{u\}_i^T [c] \{u\}_j = 0, \quad i \neq j,$$

which is valid exactly only if the off–diagonal terms are zero. The expression is approximately valid when the off–diagonal terms are very small in comparison to those of the diagonal. Light damping is a qualitative description, but such damping is common to many engineering structures and machines where $0 < \zeta_i \leq 0.20$ for all modes.

As we have already seen, when the damping cannot be considered proportional, the eigenvectors as well as the eigenvalues are complex. Physically, this is because the components of each mode (eigenvector) have different phase relationships and therefore do not reach their maximum amplitudes at the same time. Another way of visualizing this is that the *nodes* for each mode are not fixed but travel along the vibrating structure. A complex number with real and imaginary parts provides information on amplitude and phase for each component. For structures with proportional or no damping, the components of each mode are either in phase or 180° out of phase. If both components have the same sign they are in phase, if they are of opposite sign they are completely out of phase. There is no in between, except for nonproportional damping.

Real and Complex Modes

We summarize[22] the key aspects of real normal modes and complex normal modes, and show some of these in Figure 8.29. Real normal modes have the following characteristics:

---

[22] Based on *Modal Space*, P. Avitable, **Experimental Techniques,** May/June 2002, p. 17–18.

8.6. MODAL ANALYSIS

1. The mode shape is described by a standing wave (stationary wave). Nodes (where the zero axis is crossed) of standing waves are fixed.
2. All points pass through their maxima and minima at the same instant of time.
3. All points pass through the zero at the same instant of time.
4. The mode shape can be described by signed real numbers.
5. All points are either totally in phase or out of phase with any other point on the structure.
6. The mode shapes from the undamped case are the same as the mode shapes from the proportionally damped case. These modes uncouple the property matrices: $[m]$, $[c]$, $[k]$.

The complex modes have the following characteristics:

1. The mode shape is described by a traveling wave and appears to have moving node points on the structure.
2. All points do not pass through their maxima at the same instant of time. Some points lag.
3. All points do not pass through zero at the same instant of time.
4. The mode shapes are complex valued.
5. The different degrees of freedom will have phase differences between them.
6. The mode shapes from the undamped case will not uncouple the damping matrix.

8.6.4 Modal Analysis Compared to the Direct Method

It is appropriate to discuss and compare the direct method with modal analysis. The direct method has the advantage that a solution for the response does not require a special form for the damping, as does modal analysis. However, in general, the solution for the roots of the characteristic determinant becomes computationally intensive for systems with more than a few degrees of freedom. In fact, there are no exact solutions for the roots of a characteristic polynomial of order greater than four. In these instances, numerical techniques for determining the eigenvalues are used.

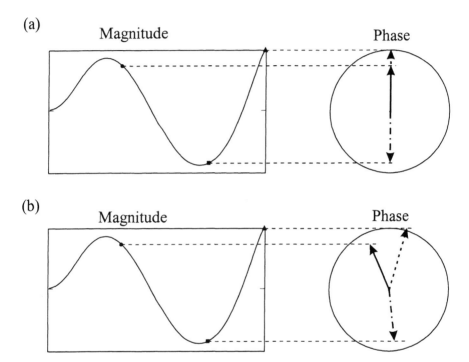

Figure 8.29: Mode Schematic: *(a)* Proportional (real normal) mode, *(b)* Non–proportional (complex) mode. This set of figures shows how real normal modes are either in phase or 180° out of phase, and that complex modes have the wave–like quality that different points can be out of phase generally.

8.6. MODAL ANALYSIS

Various techniques have been developed for solving eigenvalue problems because of their importance in all the technical and scientific disciplines. The most suitable method for a specific eigenvalue problem depends to some extent on the size of the matrix (number of vibration degrees of freedom) and on how many eigenvalues and eigenvectors are needed to model the response. Matrix iteration techniques are the basis for many eigenproblem solution techniques used in practice.[23] It is because the numerical methods for estimating frequencies and vectors are iterative, that is the estimates are sequential, that the modal approach permits the analyst to retain only as many degrees of freedom as needed for an accurate result. Some techniques[24] for the solution of eigenvalue problems include the *power method*, *Hotelling's deflation method*, *Jacobi's method*, and generally the most efficient for large systems (20 and more degrees of freedom), *Householder's method* used in conjunction with the QL algorithm.

In modal analysis, the symmetry of the stiffness and mass matrices are key to the orthogonality of the natural modes. The orthogonality relations are also useful as a check on the accuracy of a numerical computation of the modes. The physical meaning of the normal coordinates are discussed in Section 9.4. They are the principal coordinates, in the same sense that the term is used in solid mechanics. In single degree of freedom systems, by considering the vibration to be in only one direction, we imply that direction to be a normal or principal coordinate.

Modal analysis is especially useful where the structure is forced. A forced response by the direct method requires the *simultaneous solution* of as many equations as there are degrees of freedom. The modal analysis approach proceeds with decoupled oscillator equations that are solved independently in modal space. Modal analysis, even with forcing, transforms the solution of n simultaneous equations of motion to a procedure that only involves a series of matrix multiplications and additions.

Modal analysis is also useful as an approximate method for systems with a large number of degrees of freedom, for example, hundreds to thousands of degrees of freedom. If the forcing has only lower frequency components, modal analysis with only a small number of lower modes can be effectively used. It is numerically possible to selectively pick particular eigenvalue ranges within which the response is needed. Generally, this is due to loading that has a particular frequency band. In this way, the critical modes are

---

[23] R.W. Clough and J. Penzien, **Dynamics of Structures**, Second Edition, McGraw–Hill, 1993.

[24] See **Vibration of Mechanical and Structural Systems**, Second Edition, by M.L. James, G.M. Smith, J.C. Wolford, and P.W. Whaley, Harper Collins, 1994. A very nice introductory discussion is provided by C.F. Gerald, and P.O. Wheatley in **Applied Numerical Analysis**, Fifth Edition, Addison Wesley, 1994.

solved for without wasting computational time on those modes that have little or no contribution.

For example, suppose that only the first *three* modes of a *one thousand* degree of freedom model are the primary contributors to the response of the structure. The response can then be approximated by

$$\underbrace{\{x(t)\}}_{1000 \times 1} \approx \underbrace{[P]}_{1000 \times 3} \underbrace{\{q(t)\}}_{3 \times 1},$$

where the modal matrix $[P]$ has been previously determined.

The question of how many and which modes to retain generally depends on the problem. But if the forcing spectrum is limited to a frequency band of $\omega_i \to \omega_j$, then the analyst would retain modes corresponding to frequencies as high as two to four times the highest frequency ω_j. Of course, this is only a rule of thumb and exceptions exist. Our previous discussion of the modal participation factor is relevant here.

In the next Chapter, some of the procedures developed for two degree of freedom systems are generalized for n degree of freedom systems.

8.7 Concepts Summary

In this chapter we studied the vibration of structures idealized as discrete degrees of freedom. The concept of modes is introduced, the direct method for solving a set of coupled ordinary differential equations of motion is developed, and techniques for decoupling these equations using the modal transformation is demonstrated and explained. Such a transformation is possible for undamped systems as well as systems that have the property of proportional damping, wherein the property matrices are diagonalized simultaneously.

8.8 Problems

Problems for Section 8.2 – Preliminary Concepts of Stiffness and Flexibility

1. The two degree of freedom structure in Figure 8.30 undergoes rectilinear motion. *(i)* Derive the flexibility influence coefficients. *(ii)* Derive the stiffness influence coefficients. *(iii)* Find the inverse of the flexibility matrix and show that this equals the stiffness matrix. *(iv)* Write the matrix equation of motion.

8.8. PROBLEMS

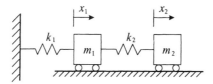

Figure 8.30: Two degree of freedom system.

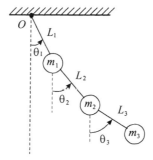

Figure 8.31: Triple pendulum.

2. For the triple pendulum of Figure 8.31, determine the flexibility influence coefficients. Find the inverse of the matrix of flexibility coefficients, and then write the matrix equations of motion for this system.

Problems for Section 8.3 – Derivation of Equations of Motion

3. For the double pendulum of Figure 8.32, derive the equations of motion in two ways: *(i)* using Newton's second law and *(ii)* using Lagrange's equations of motion.

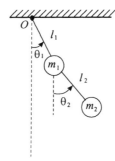

Figure 8.32: Double pendulum.

4. A simple lumped parameter model of a building which can be used for preliminary earthquake dynamics is provided in Figure 8.33. Derive the equations of motion for this two degree of freedom system in two ways: *(i)* Use Newton's Second Law, *(ii)* Use Lagrange's Equation.

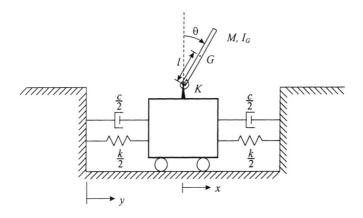

Figure 8.33: Simple model of building under base excitation, an earthquake loading model.

5. Derive the equation of motion for the elastically restrained rigid beam shown in Figure 8.34 in two ways: *(i)* using Newton's second law and *(ii)* using Lagrange's equations of motion.

6. Use Hamilton's principle to the derive the equations of motion for the structures shown in Figures *(i)* 8.32, *(ii)* 8.33, *(iii)* 8.34.

8.8. PROBLEMS

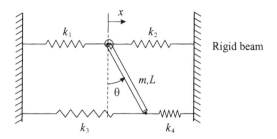

Figure 8.34: Elastically restrained beam.

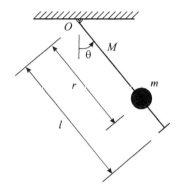

Figure 8.35: Sliding pendulum.

7. Derive the equations of motion for the problem of *Example* 8.6 using *(i)* Lagrange's equation, *(ii)* Hamilton's principle, and then solve for the response.

   Problems for Section 8.4 – Undamped Vibration

8. Refer to *Example* 8.8. Evaluate $\bar{C}_1$, $\bar{C}_2$, ϕ_1 and ϕ_2 using the given initial conditions.

9. Consider the problem of a "sliding pendulum" drawn in Figure 8.35, where a mass slides along a rod which is assumed to be uniform and thin. Use Lagrange's equation to derive the equations of motion. Can the resulting governing equations be linearized? Under what conditions? What happens when $r = l$?

10. Linearize and solve the equations of motion derived in the last problem. State all assumptions necessary for the linearization and explain

them physically.

11. For the system of *Example* 8.8, solve for the response where $k_1 = k_2 = k_3 = k$ N/cm, $m_1 = m_2 = m$ kg, and where the initial conditions are $x_1(0) = 1$ cm, $x_2(0) = 1$ cm with zero initial velocities. Plot the modes, identifying the nodes. Plot $\omega_{1,2}$ vs. m and k.

12. Solve the matrix governing equation of *Problem* 1.

13. Solve the governing equation of *Problem* 2.

14. Linearize the governing equations of *Problem* 3 and solve. State all assumptions in the linearization process.

15. In *Problem* 4, assume a harmonic base motion and find the undamped response.

16. Solve the free vibration problem of *Problem* 5.

17. Suppose system Equation 8.35 is forced by the vector $\{F_1(t) \; F_2(t)\}^T = \{f_1 \cos\omega_1 t \quad f_2 \cos\omega_2 t\}^T$. Solve for the general response using the direct method.

18. For the undamped vibration absorber problem with Equations 8.36 and 8.37, assume that the absorber mass is also forced by $f_2 \sin\omega t$. Solve for the responses $x_1(t)$ and $x_2(t)$. Plot $A(\omega)$ and $B(\omega)$ as a function of ω/ω_a.

19. The vibration characteristics of an airplane are very complex due to the intricate system of structures and substructures required. Suppose we consider the much simplified model of Figure 8.36, which is a three degree of freedom representation of the fuselage and wings. Derive the equations of motion using Lagrange's equation. Initially derive assuming arbitrary θ_1 and θ_2. Then consider two simplifications: *(i)* Assume symmetrical vibration of the wing–body combination, that is, $\theta_1 = \theta_2 \equiv \theta$, $m_1 = m_2 = m$, and $k_1 = k_2 = k$, and, finally, *(ii)* assume θ is small. Solve for the free vibration response after simplifying and linearizing the equations of motion. Discuss assumptions.

8.8. PROBLEMS

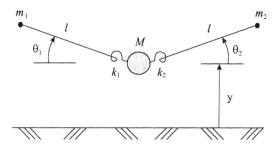

Figure 8.36: A simple airplane model.

20. What will change in our formulation and solution if the symmetry assumption in the previous problem is relaxed or removed? Discuss fully.

21. Complete the solution of Equation 8.44 if the initial conditions are given by $\theta_1(0) = \theta_a$, $\theta_2(0) = \theta_b$, $\dot{\theta}_1(0) = \theta_{av}$, $\dot{\theta}_2(0) = \theta_{bv}$.

Problems for Section 8.5 – Direct Method: Free Vibration with Damping

22. Derive Equations 8.55 and 8.56.

23. Derive Equations 8.57 and 8.58.

24. Use the direct method to solve the equations of motion for a system with the following property matrices,
$$[m] = \begin{bmatrix} 1 & 0 \\ 0 & 1 \end{bmatrix}, \quad [c] = \begin{bmatrix} 2 & -1 \\ -1 & 1 \end{bmatrix}, \quad [k] = \begin{bmatrix} 2 & -1 \\ -1 & 1 \end{bmatrix}.$$
Sketch this system.

25. Use the direct method to solve the equations of motion for a system with the following property matrices,
$$[m] = \begin{bmatrix} 1 & 0 \\ 0 & 1 \end{bmatrix}, \quad [c] = \begin{bmatrix} 5 & -2 \\ -2 & 5 \end{bmatrix}, \quad [k] = \begin{bmatrix} 2 & -1 \\ -1 & 1 \end{bmatrix}.$$
Sketch this system.

26. Use the direct method to solve the equations of motion for a system with the following property matrices,
$$[m] = \begin{bmatrix} 1 & 0 \\ 0 & 1 \end{bmatrix}, \quad [c] = \begin{bmatrix} 2 & -1 \\ -1 & 1 \end{bmatrix}, \quad [k] = \begin{bmatrix} 3 & -2 \\ -2 & 4 \end{bmatrix}.$$

Sketch this system.

27. Use the direct method to solve the equations of motion for a system with the following property matrices,

$$[m] = \begin{bmatrix} 1 & 0 \\ 0 & 1 \end{bmatrix}, \quad [c] = \begin{bmatrix} 5 & -2 \\ 2 & 3 \end{bmatrix}, \quad [k] = \begin{bmatrix} 2 & -1 \\ -1 & 1 \end{bmatrix}.$$

Sketch this system.

28. Derive Equation 8.63 showing all steps.

Problems for Section 8.6 – Modal Analysis

29. Derive Equation 8.64.

30. Show that the orthogonality relations hold where there is coupling in both mass and stiffness matrices. Doing this requires a derivation of the characteristic matrix and equation, the eigenvalues and the eigenvectors. The algebra will be extensive and is, therefore, a good candidate for a symbolic code such as MAPLE.

31. For the two degree of freedom system in Figure 8.37 undergoing longitudinal motion, derive the equations of motion utilizing each of the following approaches: *(i)* flexibility coefficients, *(ii)* Newton's second law of motion, *(iii)* Lagrange's equation, *(iv)* Hamilton's principle. Assume the springs behave linearly. Then, with the equations of motion in matrix form, derive the natural frequencies and mode shapes en route to solving for the responses. Do this for the case $m_1 = m_2 = 1$ kg and all stiffnesses have the value 1 N/m via *(a)* the direct method and *(b)* modal analysis.

Find the response of each mass if the initial velocities equal zero and the initial displacements are in the ratio of the first mode. For a design application, it is necessary that $|x_2 - x_1| \geq x_{cr}$, where x_{cr} is some critical separation distance between the masses. Describe how you would check that this condition is met, and what options exist for a redesign of the structure or a feedback control if the condition is not met.

8.8. PROBLEMS

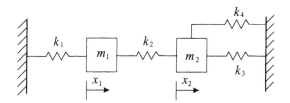

Figure 8.37: Two degree of freedom structure.

32. In the previous problem, the system is forced with $F_1(t) = A_1 \cos \omega_1 t$ and $F_2(t) = A_2 \sin \omega_2 t$. Using modal analysis, solve for the forced response assuming initial conditions $x_1(0) = 0$, $\dot{x}_1(0) = v_1$, $x_2(0) = d_2$ and $\dot{x}_2(0) = 0$.

33. If the previous problem is modified to include some damping between the two masses as shown in Figure 8.38, solve for the case with no forcing using *(i)* modal analysis, *(ii)* the direct method.

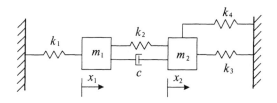

Figure 8.38: Two degree of freedom structure with damping.

34. The two degree of freedom damped and forced system of Figure 8.39 moves in the plane of the paper. Use Lagrange's equation to formulate the problem, and then solve the equations of motion for $F(t) = A \cos \omega t$. Check whether proportional damping is a valid model here. If not, then how can the structure be modified in order that such a model is valid? Do this modification, and then solve via modal analysis.

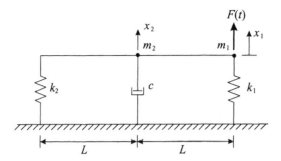

Figure 8.39: Two degree of freedom general system.

35. Solve *Example* 8.13 where the equation of motion is

$$\begin{bmatrix} m & 0 \\ 0 & m \end{bmatrix} \begin{Bmatrix} \ddot{x}_1 \\ \ddot{x}_2 \end{Bmatrix} + \begin{bmatrix} 3k & -2k \\ -2k & 2k \end{bmatrix} \begin{Bmatrix} x_1 \\ x_2 \end{Bmatrix} = \begin{Bmatrix} F_1(t) \\ F_2(t) \end{Bmatrix},$$

with $F_1(t) = \cos(0.66\sqrt{k/mt})$ and $F_2(t) = 0$. Plot the results.

36. In the previous problem, solve the same equation of motion, except with $F_1(t) = \cos(2\sqrt{k/mt})$ and $F_2(t) = 0$. Plot the results.

37. In the previous problem, solve for the response where the system damping is *(i)* $[c] = 2[m]$, *(ii)* $[c] = 3[k]$, *(iii)* $[c] = 2[m] + 3[k]$. In each of these discuss the results, especially regarding how the damping for each mode depends on the proportional damping model.

38. A vibrating mass is found to be oscillating with an amplitude that is too large. To reduce this amplitude, an auxiliary system is added as shown in Figure 8.40. This problem generally occurs when the forcing frequency is too close to the natural frequency of the primary system.

The auxiliary system acts as a damped vibration absorber. Derive the equation of motion of the system plus absorber, and then solve to determine what values of absorber mass m, stiffness k_2, and damping c must be selected in order to minimize the vibration amplitude of primary mass M. Assume that $m = 0.1M$ and $F(t) = F_0 \cos \omega t$.

Suppose that we can accept a design where k_2/m is selected so that no natural frequency of the combined system, ω_1 and ω_2, is closer than 5% to the driving frequency ω. Devise the procedure that achieves this criterion.

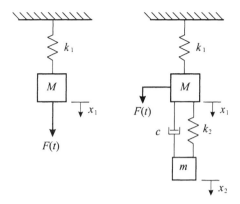

Figure 8.40: Primary system M and auxiliary mass m.

8.9 Mini–Projects

The following papers from the recent published literature use multi–degree of freedom models to study problems of interest to the vibration community. They may be used as bases for mini–projects for interested students that would appreciate the opportunity to explore more advanced applications of the material they learned in this chapter.

1. *Classical Normal Modes in Damped Linear Dynamic Systems*, T.K. Caughey, M.E.J. O'Kelly, *J. Applied Mechanics*, Vol. 27, Sept. 1965, 583–588.

   The purpose of this paper is to determine necessary and sufficient conditions under which both discrete and continuous damped linear systems possess classical normal modes.

2. *A Study of the Modal Behavior of the Human Hand–Arm System*, C. Thomas, S. Rakheja, R.B. Bhat, I. Stiharu, *J. Sound and Vibration* (1996) **191**(1), 171–176.

   The severe health risks posed by prolonged occupational exposure to the vibration of hand–held power tools have prompted a strong desire to enhance an understanding of the dynamic characteristics of the human hand–arm. An in–plane five degree of freedom biomechanical model has been developed to derive the vibration transmission characteristics of the hand–arm.

3. *Torsional Vibrations in Electrical Induction Motor Drives During Start–Up*, L. Ran, R. Yacamini, K.S. Smith, *J. of Vibration and Acoustics*, Vol. 118, April 1996, 242–251.

When electrical induction motors are started direct on–line they generate a considerable pulsating torque. This start up torque can create problems when the motor is connected to mechanical loads such as fans or pumps and there are reported cases of the interconnecting shafts having been sheared on start–up. This interrelationship between the electric motor and the mechanical system is effectively modeled as a multi–mass system.

4. *Nonlinear DNA Dynamics: Hierarchy of the Models*, L.V. Yakushevich, *Physica D*, 79 (1994) 77–86.

   The main types of the nonlinear models of the internal DNA dynamics are considered. It is shown that (i) the models can be arranged in the order of increasing complexity and they can form the hierarchy, (ii) the hierarchy of the nonlinear models is a particular case of the hierarchy of the DNA dynamical models, and (iii) the latter correlates with the hierarchy of the structural DNA models.

5. *Extending the Logarithmic Decrement Method to Analyze Two Degree of Freedom Transient Responses*, J.E. Cooper, *Mechanical Systems and Signal Processing*, (1996) **10**(4), 497–500.

   The classical logarithmic decrement method is extended to estimate modal parameters from the transient response of a lightly damped system with two close modes. This is a graphical method.

6. *Response Errors of Non–Proportionally Lightly Damped Structures*, W. Gawronski, J.T. Sawicki, *J. Sound and Vibration* (1997) **200**(4), 543–550.

   This paper studies the case of lightly but non–proportionately damped systems. Examples include steel structures, rotating equipment, space structures, and large antennas.

7. *The Role of Eigenvectors in Aeroelastic Analysis*, R.M.V. Pidaparti, D. Afolabi, *J. Sound and Vibration* (1996) **193**(4), 934–940.

   Many engineering structures experience dynamic instability known as *flutter* in aeroelastic analysis when a coupling of two or more modes takes place. Examples include slightly mistuned bladed disk assemblies, aeroelastic behavior of plates and shells, and in helicopter and rotor dynamics. This study focuses on coalescence flutter where the energy required to drive the instability is extracted from one of the stable modes, and damping is not a driver as in single mode flutter, where positive damping becomes negative.

8. *Performance of Tuned Mass Dampers Under Wind Loads*, K.C.S. Kwok, B. Samali, *Engineering Structures*, Vol. 17, No. 9, 655–667, 1995.

The performance of both passive and active tuned mass damper systems can be readily assessed by parametric studies have been the subject of much research. Here, such theories are verified experimentally.

9. *Effectiveness of Tuned Liquid Column Dampers for Vibration Control of Towers*, T. Balendra, C.M. Wang, H.F. Cheong, *Engineering Structures*, Vol. 17, No. 9, 668–675, 1995.

    The effectiveness of tuned liquid column dampers in controlling the wind–induced vibration of towers is studied. The nonlinear governing equation is linearized to obtain the stochastic response of the towers due to along–wind turbulence.

10. *The Relationship Between the Real and Imaginary Parts of Complex Modes*, S.D. Garvey, J.E.T. Penny, M.I. Friswell, *J. Sound and Vibration* (1998) **212**(1), 75–83.

    It is shown that a simple relationship exists between the real and imaginary parts of complex modes of all systems which can be represented by real and symmetric mass, stiffness and damping matrices.

11. *Mode Shapes During Asynchronous Motion and Non–Proportionality Indices*, A. Bhaskar, *J. Sound and Vibration* (1999) 224(1), 1–16.

    When synchronous motion does not exist, it is not possible to draw the classical mode shapes. In this paper, a representative shape of motion during free vibration of a non–classically damped system is sought.

12. *Understanding the Physics of Electrodynamic Shaker Performance*, G.F. Lang, D. Snyder, *Sound and Vibration*, October 2001, 24–33.

    This article discusses the basic electromechanical shaker model, how to determine maximum drive performance for sinusoidal testing, the merits of pneumatic load–leveling suspensions and the often overlooked side effects of shaker isolation.

13. *Modal Analysis of Living Spruce Using a Combined Prony and DFT Multichannel Method for Detection of Internal Decay*, J. Axmon, M. Hansson, L. Sörnmo, *Mechanical Systems and Signal Processing* (2002) **16**(4), 561–584.

    In this paper, a partial modal analysis is used to examine whether internal decay in living trees can be detected by studying the resonance frequencies of the trees.

14. *Vibrational Modes of Trumpet Bells*, T.R. Moore, J.D. Kaplon, G.D. McDowall and K.A. Martin, *J. Sound and Vibration* (2002) **254**(4), 777–786.

    This paper reports on an investigation of the normal modes of vibration of the bells of several modern trumpets.

15. *Forced Nonlinear Vibrations of a Symmetrical Two–Mass System*, L. Cveticanin, *J. Sound and Vibration*, 265 (2003) 451–458.

The aim of this paper is to analyze the forced vibrations of a symmetric two–mass system connected to fixed supports with linear springs. The connecting spring between the masses has strong nonlinear elastic properties.

Chapter 9

Multi Degree of Freedom Vibration: Advanced Topics

"Generalizations and complexities abound."

9.1 Overview

Continuing with the introduction to the vibration of multi degree of freedom systems of Chapter 8, we explore systems of a more general nature, and systems with some unique and complicating factor. Imperfections and uncertainties are now included in some of the models. A deeper understanding of the eigenvalue problem is gained by considering its meaning geometrically. The vibration problem is considered from the "inverse" perspective: given the response and the forcing, find the system properties of mass and stiffness. These additional studies helps the development of an appreciation for the breadth of vibration studies.

9.2 Generalization to n Degrees of Freedom

We proceed to briefly generalize the results of the previous section. The harmonic solution for an n degree of freedom system is separable in time and amplitude,

$$x_i(t) = u_i Y(t), \quad i = 1, 2, \ldots, n.$$

Solve for the function $Y(t)$ and for the amplitudes of vibration u_i. Note that for the above solution, the following holds,

$$\frac{x_i(t)}{x_{i+1}(t)} = \frac{u_i}{u_{i+1}} = \text{constant}$$

$$\implies x_{i+1} = \frac{u_{i+1}}{u_i} x_i$$

$$x_{i+p} = \frac{u_{i+p}}{u_{i+p-1}} \cdots \frac{u_{i+2}}{u_{i+1}} \frac{u_{i+1}}{u_i} x_i$$

$$= \frac{u_{i+p}}{u_i} x_i.$$

Substitute $x_i(t)$ into the matrix equation of motion for free vibration, Equation 8.10, to find

$$[m]\{u\}\ddot{Y} + [k]\{u\}Y = \{0\},$$

which, following Section 8.4.1, can be put in the form

$$\frac{\ddot{Y}}{Y} = -\frac{\sum_{j=1}^n k_{ij} u_j}{\sum_{j=1}^n m_{ij} u_j} = -\omega^2, \quad i = 1, 2, \ldots, n. \tag{9.1}$$

Equation 9.1 implies that the $[m]$ and $[k]$ property matrices can be fully populated. Since the first ratio is independent of amplitude u, and the second ratio is independent of time, both sides must equal a constant, say $-\omega^2$. The negative sign is to assure that the resulting differential equation for $Y(t)$ has a harmonic solution. The system is now governed by the following set of equations,

$$\ddot{Y} + \omega^2 Y = 0 \tag{9.2}$$

$$\sum_{j=1}^n \left(k_{ij} - \omega^2 m_{ij}\right) u_j = 0, \quad i = 1, 2, \ldots, n. \tag{9.3}$$

The solution to Equation 9.2 is, of course, a harmonic function $Y(t) = A\cos(\omega t - \phi)$, where A and ϕ are constants determined by the initial conditions. For every degree of freedom, there is a frequency and phase, as we will see below. To evaluate the *allowable* frequencies ω_i that the system can support, solve Equation 9.3, the *characteristic value* or *eigenvalue* problem. In matrix form, Equation 9.3 is

$$[k]\{u\} = \omega^2 [m]\{u\}. \tag{9.4}$$

From the theory of matrices, a nontrivial set of solutions exist if the *determinant* of the coefficients vanishes,

$$\left|[k] - \omega^2 [m]\right| = 0.$$

9.2. GENERALIZATION TO N DEGREES OF FREEDOM

This is known as the *characteristic determinant*, with the expansion known as the *characteristic equation*. Since $[k]$ and $[m]$ are symmetric and positive definite, the roots of the characteristic equation are *real* and *positive*. Physically, we expect this since the roots are the natural frequencies of oscillation.

The solution of the characteristic equation results in *characteristic values* ω_r, $r = 1, 2, \ldots, n$; once these are found, the *characteristic vectors* or *eigenvectors* can be evaluated. Physically, these vectors are the (relative) amplitudes u_i of the masses as they undergo harmonic motion. To determine these vectors, substitute each frequency, ω_r, into Equation 9.4

$$[k]\{u\}_r = \omega_r^2[m]\{u\}_r, \quad r = 1, 2, \ldots, n, \tag{9.5}$$

and solve for $\{u\}_r$. Note that there is one *modal vector* for each natural frequency. Also, due to the nature of Equation 9.5 with $\{u\}_r$ on both sides of the equation, the vectors are unique only to a constant. This means that if $\{u\}_r$ is a solution, then so is $a\{u\}_r$ since the constant a appears on both sides of the equation. Therefore, it is customary to fix the vectors by normalizing them according to the rule

$$\{\hat{u}\}_r^T[m]\{\hat{u}\}_r = 1, \tag{9.6}$$

where $\{\hat{u}\}_r$ is a normalized eigenvector. This normalization procedure does nothing more than adjust the constants of integration in the solution. From Equations 9.5 and 9.6, we have

$$\{\hat{u}\}_r^T[k]\{\hat{u}\}_r = \omega_r^2\{\hat{u}\}_r^T[m]\{\hat{u}\}_r = \omega_r^2, \quad r = 1, 2, \ldots, n.$$

These are the orthogonality relations for the mass and stiffness matrices for an n degree of freedom structure. The remaining procedure is the same as for the direct method.

9.2.1 Modal Matrix $[P]$

When the n normal modes for an n degree of freedom vibrating system are assembled into a square matrix, with each normal mode represented by a column, this forms the *modal matrix*, as introduced for two degrees of freedom in Equations 8.69 and 8.70. For a three degree of freedom system, the modal matrix may be written as

$$[P] = [\{\hat{u}\}_1 \; \{\hat{u}\}_2 \; \{\hat{u}\}_3] = \left[\left\{ \begin{array}{c} \hat{u}_1 \\ \hat{u}_2 \\ \hat{u}_3 \end{array} \right\}_1 \left\{ \begin{array}{c} \hat{u}_1 \\ \hat{u}_2 \\ \hat{u}_3 \end{array} \right\}_2 \left\{ \begin{array}{c} \hat{u}_1 \\ \hat{u}_2 \\ \hat{u}_3 \end{array} \right\}_3 \right]$$

$$= \begin{bmatrix} \hat{u}_{11} & \hat{u}_{12} & \hat{u}_{13} \\ \hat{u}_{21} & \hat{u}_{22} & \hat{u}_{23} \\ \hat{u}_{31} & \hat{u}_{32} & \hat{u}_{33} \end{bmatrix}.$$

Consider, then, the following matrix products that arise as part of a modal analysis,

$$[P]^T[m][P] = [\{\hat{u}\}_1 \; \{\hat{u}\}_2 \; \{\hat{u}\}_3]^T[m][\{\hat{u}\}_1 \; \{\hat{u}\}_2 \; \{\hat{u}\}_3]$$

$$= \begin{bmatrix} \{\hat{u}\}_1^T[m]\{\hat{u}\}_1 & \{\hat{u}\}_1^T[m]\{\hat{u}\}_2 & \{\hat{u}\}_1^T[m]\{\hat{u}\}_3 \\ \{\hat{u}\}_2^T[m]\{\hat{u}\}_1 & \{\hat{u}\}_2^T[m]\{\hat{u}\}_2 & \{\hat{u}\}_2^T[m]\{\hat{u}\}_3 \\ \{\hat{u}\}_3^T[m]\{\hat{u}\}_1 & \{\hat{u}\}_3^T[m]\{\hat{u}\}_2 & \{\hat{u}\}_3^T[m]\{\hat{u}\}_3 \end{bmatrix}$$

$$= \begin{bmatrix} 1 & 0 & 0 \\ 0 & 1 & 0 \\ 0 & 0 & 1 \end{bmatrix},$$

thus observing how the modal decoupling occurs simultaneously for larger degree of freedom systems. Similarly, for the matrix triple product involving the stiffness matrix,

$$[P]^T[k][P] = [\{\hat{u}\}_1 \; \{\hat{u}\}_2 \; \{\hat{u}\}_3]^T[k][\{\hat{u}\}_1 \; \{\hat{u}\}_2 \; \{\hat{u}\}_3]$$

$$= \begin{bmatrix} \{\hat{u}\}_1^T[k]\{\hat{u}\}_1 & \{\hat{u}\}_1^T[k]\{\hat{u}\}_2 & \{\hat{u}\}_1^T[k]\{\hat{u}\}_3 \\ \{\hat{u}\}_2^T[k]\{\hat{u}\}_1 & \{\hat{u}\}_2^T[k]\{\hat{u}\}_2 & \{\hat{u}\}_2^T[k]\{\hat{u}\}_3 \\ \{\hat{u}\}_3^T[k]\{\hat{u}\}_1 & \{\hat{u}\}_3^T[k]\{\hat{u}\}_2 & \{\hat{u}\}_3^T[k]\{\hat{u}\}_3 \end{bmatrix}$$

$$= \begin{bmatrix} \omega_1^2 & 0 & 0 \\ 0 & \omega_2^2 & 0 \\ 0 & 0 & \omega_3^2 \end{bmatrix}.$$

The method of *Example* 8.12 can be extended to any number of degrees of freedom, but numerical methods must be used to solve any but the most modest system.

We next consider a special class of structures, those that are unrestrained from translation and rotation.

9.3. UNRESTRAINED SYSTEMS

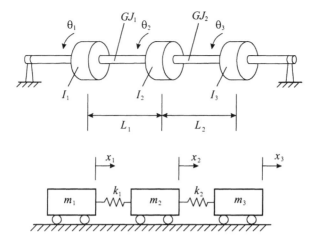

Figure 9.1: Examples of unrestrained translating and rotating systems.

9.3 Unrestrained Systems

So far we have only considered structures that are fixed to one location. In fixed structures, any vibration is constrained and oscillations are about a stationary equilibrium. But there are numerous important instances where the vibration is about a moving equilibrium. That is, the body is not constrained and can move; *vibration is then superimposed on the larger dynamic behavior*, which is known as *rigid body motion*. Examples of such systems that come to mind immediately are airplanes, rockets, trains, and autos. Also of significant importance are rotating machinery. All of these structures vibrate while undergoing large dynamic motion.

Two schematics are shown in Figure 9.1. We are interested in examining such systems in order to determine how the techniques of vibration analysis are altered by the fact that the system is unrestrained.

Physically, an unrestrained system is one which can translate or rotate without bound. The airplane is an unrestrained system, and therefore can move arbitrary distances relative to any fixed object. A turbine can rotate millions of revolutions without any restraint except friction. What is the common element in all such systems? It is that *rectilinear and rotational motion have unlimited magnitudes, and that the first structural mode is at zero frequency. There is no relative motion between any two points on the structure for this mode. This is known as a rigid body mode.*

Consider the three mass translating system of Figure 9.1. The strain energy V stored in the springs is a function of the squares of the differences in relative mass displacements,

$$V = \frac{1}{2}[k_1(x_2 - x_1)^2 + k_2(x_3 - x_2)^2].$$

The kinetic energy T is given by

$$T = \frac{1}{2}(m_1 \dot{x}_1^2 + m_2 \dot{x}_2^2 + m_3 \dot{x}_3^2).$$

The kinetic energy is always positive except when the system is static, and then $T = 0$. The potential energy is positive except when all the displacements are equal, $x_1 = x_2 = x_3$, then $V = 0$. In this instance, the three masses can translate as one, as a rigid body, and there are no oscillations, that is, *the rigid body mode has a zero natural frequency*.

For the rigid body mode, $\omega_0 = 0$ and the mode shape is $\{u\}_0 = \{1\ 1\ 1\}^T$, as we will show in a forthcoming example. Since the rigid body mode is a solution of the eigenvalue problem, it must be orthogonal to any other mode, say, $\{u\}_0^T[m]\{u\}_1 = 0$, and

$$\{1\ 1\ 1\} \begin{bmatrix} m_1 & 0 & 0 \\ 0 & m_2 & 0 \\ 0 & 0 & m_3 \end{bmatrix} \begin{Bmatrix} u_1 \\ u_2 \\ u_3 \end{Bmatrix}_1 = 0.$$

This matrix product can be written as $m_1 u_1 + m_2 u_2 + m_3 u_3 = 0$. We can replace the modal coordinate by the physical coordinate using $x_i = u_i \exp(i\omega_0 t)$, where, for $\omega_0 = 0$, $x_i = u_i$. Then,

$$\{x_0\ x_0\ x_0\} \begin{bmatrix} m_1 & 0 & 0 \\ 0 & m_2 & 0 \\ 0 & 0 & m_3 \end{bmatrix} \begin{Bmatrix} x_1 \\ x_2 \\ x_3 \end{Bmatrix} = 0,$$

which can be expanded and simplified as follows,

$$\begin{aligned} x_0(m_1 x_1 + m_2 x_2 + m_3 x_3) &= 0 \\ m_1 x_1 + m_2 x_2 + m_3 x_3 &= 0. \end{aligned} \quad (9.7)$$

Note that by differentiating Equation 9.7 with respect to time, the resulting equation is a statement that the linear momentum is a constant for the rigid body mode. Equation 9.7 can also be used to reduce the order of the system by one, since one of the displacements can be expressed as a function of the other two, for example,

$$x_3 = -\frac{m_1}{m_3} x_1 - \frac{m_2}{m_3} x_2.$$

9.3. UNRESTRAINED SYSTEMS

This equation is really a *constraint* on the system. Therefore, the system behavior can be written as follows,

$$\begin{Bmatrix} x_1 \\ x_2 \\ x_3 \end{Bmatrix} = \begin{bmatrix} 1 & 0 & 0 \\ 0 & 1 & 0 \\ -m_1/m_3 & -m_2/m_3 & 0 \end{bmatrix} \begin{Bmatrix} x_1 \\ x_2 \\ x_3 \end{Bmatrix}$$

$$= \begin{bmatrix} 1 & 0 \\ 0 & 1 \\ -m_1/m_3 & -m_2/m_3 \end{bmatrix} \begin{Bmatrix} x_1 \\ x_2 \end{Bmatrix}_\chi.$$

The subscript χ denotes a constrained system. In matrix shorthand,

$$\begin{aligned} \{x\} &= [\chi]\{x\}_\chi \\ \{\dot{x}\} &= [\chi]\{\dot{x}\}_\chi, \end{aligned}$$

where $[\chi]$ is the constraint matrix,

$$[\chi] = \begin{bmatrix} 1 & 0 \\ 0 & 1 \\ -m_1/m_3 & -m_2/m_3 \end{bmatrix},$$

that transforms between the original system of coordinates and the reduced system. By interpreting the three degree of freedom unrestrained system as a constrained system, the order has been reduced by one and the system has a new stiffness matrix that is positive definite. Thus,

$$\begin{aligned} T &= \frac{1}{2}\{\dot{x}\}^T[m]\{\dot{x}\} \\ &= \frac{1}{2}\{\dot{x}\}_\chi^T[\chi]^T[m][\chi]\{\dot{x}\}_\chi = \frac{1}{2}\{\dot{x}\}_\chi^T[M]\{\dot{x}\}_\chi \quad (9.8) \\ V &= \frac{1}{2}\{x\}^T[k]\{x\} \\ &= \frac{1}{2}\{x\}_\chi^T[\chi]^T[k][\chi]\{x\}_\chi = \frac{1}{2}\{x\}_\chi^T[K]\{x\}_\chi. \quad (9.9) \end{aligned}$$

$[M] = [\chi]^T[m][\chi]$ and $[K] = [\chi]^T[k][\chi]$ are matrices associated with the constrained system.

At this point we would work with the new constrained system as usual, finding the eigenvalues and eigenvectors of the reduced order model: ω_1, ω_2 and $\{\hat{u}\}_{1\chi}, \{\hat{u}\}_{2\chi}$. Then, to recover the complete description of the modes we need to perform the following operations,

$$\begin{aligned} \{\hat{u}\}_1 &= [\chi]\{\hat{u}\}_{1\chi} \\ \{\hat{u}\}_2 &= [\chi]\{\hat{u}\}_{2\chi}, \end{aligned}$$

where $\{\hat{u}\}_1$ and $\{\hat{u}\}_2$ are now 3×1 modal vectors for the original three degree of freedom system.

The question is then how to combine the above responses, those due to the rigid body motion and the rest due to the modal vibration. To do this we have to recognize that the rigid body mode has a different behavior in time. It is necessary to go back to either Equation 8.16 or 9.2, where the time function $Y(t)$ is solved. Equation 9.2 is repeated here,

$$\ddot{Y} + \omega^2 Y = 0.$$

To solve this equation for the rigid body mode, set $\omega = 0$. Therefore, the equation governing the time behavior of the rigid body mode is given by $\ddot{Y} = 0$, which can be integrated directly twice, resulting in

$$Y(t) = C_1 t + C_2,$$

where C_1 and C_2 are related to the initial displacement and velocity.

The complete solution is then

$$\{x(t)\} = (C_1 t + C_2)\{\hat{u}\}_0 + \sum_{i=1}^{2} A_i \{\hat{u}\}_i \cos(\omega_i t - \phi_i).$$

We clearly see here how modes $\{\hat{u}\}_1$ and $\{\hat{u}\}_2$ are superimposed upon the rigid body dynamics of the first mode.

Consider two examples, one which follows the procedure just discussed, and the other which proceeds by a direct analysis.

Example 9.1 Unrestrained Rotation
Let us work out some of the details of the problem of unrestrained rotation, which has applications to rotating machinery, in particular. A three degree of freedom problem provides a good sense of how such an analysis would proceed. See Figure 9.1.

From a free body diagram of each disk, derive Newton's three equations of motion for the system. Alternatively, using the approach discussed in Section 8.3.1, the inertia and stiffness matrices can be derived via the expressions of kinetic and potential energies, respectively. These are

$$T = \frac{1}{2}(I_1 \dot{\theta}_1^2 + I_2 \dot{\theta}_2^2 + I_3 \dot{\theta}_3^2) = \frac{1}{2}\{\dot{\theta}\}^T [I]\{\dot{\theta}\}$$

$$V = \frac{1}{2}\left[\frac{GJ_1}{L_1}(\theta_2 - \theta_1)^2 + \frac{GJ_2}{L_2}(\theta_3 - \theta_2)^2\right] = \frac{1}{2}\{\theta\}^T [k]\{\theta\},$$

9.3. UNRESTRAINED SYSTEMS

where the property matrices $[I]$ and $[k]$ are given by

$$[I] = \begin{bmatrix} I_1 & 0 & 0 \\ 0 & I_2 & 0 \\ 0 & 0 & I_3 \end{bmatrix}$$

$$[k] = \begin{bmatrix} GJ_1/L_1 & -GJ_1/L_1 & 0 \\ -GJ_1/L_1 & GJ_1/L_1 + GJ_2/L_2 & -GJ_2/L_2 \\ 0 & -GJ_2/L_2 & GJ_2/L_2 \end{bmatrix}.$$

Using standard procedures, assume that the rotation can be written as $\theta_i(t) = \Theta_i \exp\{i\omega t\}$, that, when substituted into the equations of motion, leads to the eigenvalue problem $\omega^2[I]\{\Theta\} = [k]\{\Theta\}$. We immediately note that stiffness matrix $[k]$ is singular, $|k| = 0$. The conclusion is that the first mode is a rigid body mode with frequency $\omega_0 = 0$ and mode shape $\{\Theta\}_0 = \Theta_0\{1\ 1\ 1\}^T$.

In order to generate the constraint equation, we make use of the fact that the eigenvectors are orthogonal to each other. Therefore,

$$\{\Theta\}_0[I]\{\Theta\} = \Theta_0(I_1\Theta_1 + I_2\Theta_2 + I_3\Theta_3) = 0,$$

and since $\Theta_0 \neq 0$,

$$I_1\Theta_1 + I_2\Theta_2 + I_3\Theta_3 = 0. \tag{9.10}$$

Using the relation between $\theta(t)$ and Θ, Equation 9.10 becomes

$$I_1\theta_1(t) + I_2\theta_2(t) + I_3\theta_3(t) = 0,$$

which shows us that the three degree of freedom system can be specified in terms of two coordinates since

$$\theta_3(t) = -\frac{I_1}{I_3}\theta_1(t) - \frac{I_2}{I_3}\theta_2(t).$$

The relation between the constrained motion and the arbitrary motion becomes

$$\begin{Bmatrix} \theta_1(t) \\ \theta_2(t) \\ \theta_3(t) \end{Bmatrix} = \begin{bmatrix} 1 & 0 & 0 \\ 0 & 1 & 0 \\ -I_1/I_3 & -I_2/I_3 & 0 \end{bmatrix} \begin{Bmatrix} \theta_1(t) \\ \theta_2(t) \\ \theta_3(t) \end{Bmatrix}$$

$$= \begin{bmatrix} 1 & 0 \\ 0 & 1 \\ -I_1/I_3 & -I_2/I_3 \end{bmatrix} \begin{Bmatrix} \theta_1(t) \\ \theta_2(t) \end{Bmatrix}_x.$$

In this problem, the constraint matrix is

$$[\chi] = \begin{bmatrix} 1 & 0 \\ 0 & 1 \\ -I_1/I_3 & -I_2/I_3 \end{bmatrix}.$$

Using Equations 9.8 and 9.9, the new inertia matrix is

$$[M] = [\chi]^T[I][\chi] = \frac{1}{I_3}\begin{bmatrix} I_1(I_1 + I_3) & I_1 I_2 \\ I_1 I_2 & I_2(I_2 + I_3) \end{bmatrix}$$

and the new stiffness matrix

$$[K] = [\chi]^T[k][\chi]$$
$$= \frac{1}{I_3^2}\begin{bmatrix} k_1 I_3^2 + k_2 I_1^2 & -k_1 I_3^2 + k_2 I_1(I_2 + I_3) \\ -k_1 I_3^2 + k_2 I_1(I_2 + I_3) & (k_1 + k_2)I_3^2 + k_2 I_2(2I_3 + I_2) \end{bmatrix},$$

where $k_i = GJ_i/L_i$, $i = 1, 2, 3$. We now work with these matrices in the associated eigenvalue problem to find the two elastic frequencies and modes. These are then added to the rigid body mode for the complete solution.

Assume that the torsional stiffnesses are $k_1 = k$ Nm/rad, $k_2 = 2k$ Nm/rad with inertia $I_1 = I$ kg m²/rad, $I_2 = 2I$ kg m²/rad, $I_3 = 3I$ kg m²/rad. Then

$$[M] = \frac{2I}{3}\begin{bmatrix} 2 & 1 \\ 1 & 5 \end{bmatrix},$$

and

$$[K] = \frac{k}{9}\begin{bmatrix} 11 & 1 \\ 1 & 59 \end{bmatrix}.$$

Solving the eigenvalue problem for this two degree of freedom system, the characteristic equation is $\omega^4 - 3.17k\omega^2/I + 2.00(k/I)^2 = 0$, and

$$\omega_1 = 0.93\sqrt{\frac{k}{I}} \text{ rad/sec}, \quad \{\Theta\}_{1\chi} = \begin{Bmatrix} 1.0 \\ 0.13 \end{Bmatrix}$$

$$\omega_2 = 1.52\sqrt{\frac{k}{I}} \text{ rad/sec}, \quad \{\Theta\}_{2\chi} = \begin{Bmatrix} 0.77 \\ -1.0 \end{Bmatrix}.$$

Using the constraint matrix,

$$[\chi] = \begin{bmatrix} 1 & 0 \\ 0 & 1 \\ -1/3 & -2/3 \end{bmatrix},$$

9.3. UNRESTRAINED SYSTEMS

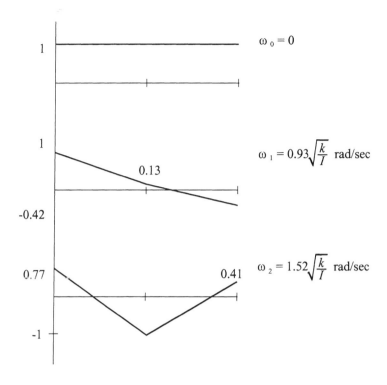

Figure 9.2: Three modes for an unrestrained system.

we can recover the elastic modes by multiplying modes $\{\Theta\}_{1\chi}$ and $\{\Theta\}_{2\chi}$ by $[\chi]$

$$\{\Theta\}_1 = \begin{bmatrix} 1 & 0 \\ 0 & 1 \\ -1/3 & -2/3 \end{bmatrix} \begin{Bmatrix} 1.0 \\ 0.13 \end{Bmatrix} = \begin{Bmatrix} 1.0 \\ 0.13 \\ -0.42 \end{Bmatrix}$$

$$\{\Theta\}_2 = \begin{bmatrix} 1 & 0 \\ 0 & 1 \\ -1/3 & -2/3 \end{bmatrix} \begin{Bmatrix} 0.77 \\ -1.0 \end{Bmatrix} = \begin{Bmatrix} 0.77 \\ -1.0 \\ 0.41 \end{Bmatrix}.$$

These two modes are in addition to the rigid body modes, as shown in Figure 9.2.

Note that the elastic modes and frequencies are the same as those for the respective two degree of freedom system.

The complete solution is

$$\{\theta(t)\} = (C_1 t + C_2)\{\Theta\}_0 + \sum_{i=1}^{2} A_i \{\Theta\}_i \cos(\omega_i t - \phi_i).$$

For this three degree of freedom system, three initial displacements and three initial velocities are required to solve for the six constants: C_1, C_2, A_1, A_2, ϕ_1, and ϕ_2. ∎

Example 9.2 Unrestrained Translation
The unrestrained translation problem can be approached in a direct manner without using a constraint equation. To demonstrate this solve the three degree of freedom unrestrained system shown in Figure 9.1, with parameters k and m, governed by

$$\begin{bmatrix} m & 0 & 0 \\ 0 & m & 0 \\ 0 & 0 & m \end{bmatrix} \begin{Bmatrix} \ddot{x}_1 \\ \ddot{x}_2 \\ \ddot{x}_3 \end{Bmatrix} + \begin{bmatrix} k & -k & 0 \\ -k & 2k & -k \\ 0 & -k & k \end{bmatrix} \begin{Bmatrix} x_1 \\ x_2 \\ x_3 \end{Bmatrix} = \begin{Bmatrix} 0 \\ 0 \\ 0 \end{Bmatrix}.$$

The characteristic equation for this system is

$$\lambda \left(\lambda - 3\frac{k}{m}\right)\left(\lambda - \frac{k}{m}\right) = 0,$$

with the roots $\lambda_{0,1,2} = 0$, k/m and $3k/m$, respectively. The case $\lambda_0 = \omega_0^2 = 0$ represents the rigid body motion. Next find the eigenvector corresponding to $\omega_0 = 0$,

$$\begin{bmatrix} k/m & -k/m & 0 \\ -k/m & 2k/m & -k/m \\ 0 & -k/m & k/m \end{bmatrix} \begin{Bmatrix} u_{10} \\ u_{20} \\ u_{30} \end{Bmatrix} = \begin{Bmatrix} 0 \\ 0 \\ 0 \end{Bmatrix},$$

from which we easily find

$$u_{10} = u_{20}$$
$$u_{20} = u_{30}.$$

The rigid body mode is $\{u\}_0 = \{1\ 1\ 1\}^T$, as we know. By the same procedure, the other two modes are $\{u\}_1 = \{1\ 0\ -1\}^T$ and $\{u\}_2 = \{1\ -2\ 1\}^T$. For the rigid body mode, integrate $\ddot{q}_0(t) = 0$ twice to find $q_0(t) = (C_1 t + C_2)$.

9.4. THE GEOMETRY OF THE EIGENVALUE PROBLEM

The remainder of the solution proceeds as in any modal analysis, with the complete solution given by

$$\{x(t)\} = (C_1 t + C_2)\{u\}_0 + \sum_{i=1}^{2} A_i \{u\}_i \cos(\omega_i t - \phi_i).$$

For this three degree of freedom system, three initial displacements and three initial velocities are required to solve for the six constants: C_1, C_2, A_1, A_2, ϕ_1, and ϕ_2. ∎

9.3.1 Repeated Frequencies

The question of what happens to the analyses if two of the structural natural frequencies have the identical numerical value is briefly discussed here. Recall the modal Equations 8.27 and 8.28,

$$\frac{u_{21}}{u_{11}} = -\frac{k_{11} - \omega_1^2 m_1}{k_{12}}$$

$$\frac{u_{22}}{u_{12}} = -\frac{k_{11} - \omega_2^2 m_1}{k_{12}}.$$

What happens to the mode shapes if $\omega_1 = \omega_2$? The eigenvectors are *not uniquely defined* and the problem has become *indeterminate*. Since both eigenvectors have the same natural frequency, any initial displacement of the structure will oscillate at ω_1. If an n degree of freedom system has $n-1$ eigenvectors because two of the frequencies are identical, then the nth independent eigenvector must be constructed. The details will not be discussed here, but it is noted that numerically determined eigenvalues that are more closely spaced than the round off errors of the computer will be viewed as repeated values. The physical meaning of repeated eigenvalues is further discussed from a geometrical perspective in the next section.

9.4 The Geometry of the Eigenvalue Problem

It is of interest and important to add physical insights to the matrix manipulations that led to the eigenvalue problem and the normal modes. In particular, we will explore some of the geometric underpinnings[1] of the eigenvalue problem.

---

[1] Our discussion is based on the book by C. Lanczos, **Applied Analysis**, Dover Publications, 1988, pp. 81–95.

Operations with matrices of the kind that were used in the previous sections to derive the eigenvalues and eigenvectors of a vibratory system are closely linked to the analytical geometry of *second order surfaces* such as ellipsoids and hyperboloids. Such second order surfaces can be visually depicted in three dimensions, but the analogies which are drawn are valid for higher order dimensional spaces, and the interpretations will hold for any n degree of freedom system and its eigenproblem. Thus, *"the entire theory of linear operators – whether they appear as systems of linear algebraic equations, or as linear ordinary or partial differential equations, or linear integral equations – can be formulated as a* geometrical *problem, associated with a certain second order surface."*[2]

Consider the equation of an ellipse (two dimensions),

$$\lambda_1 x_1^2 + \lambda_2 x_2^2 = 1, \tag{9.11}$$

or that of an ellipsoid (three dimensions),

$$\lambda_1 x_1^2 + \lambda_2 x_2^2 + \lambda_3 x_3^3 = 1, \tag{9.12}$$

or that of an n dimensional generalized ellipsoid,

$$\lambda_1 x_1^2 + \lambda_2 x_2^2 + \ldots + \lambda_n x_n^n = 1. \tag{9.13}$$

These forms of the surface equations are such that the major and minor axes coincide with the axes of our frame of reference. However, it is generally the case that the frame of reference of a vibration problem is determined by the physical problem at hand, and the ellipsoid will be in a slanted position with respect to all the axes of the n dimensional space, as in Figure 9.3. We can anticipate that the orientation is directly related to the physical parameters of the particular problem. Each different problem has a different orientation.

In this arbitrary orientation, the equation for the surface will include products of the coordinate axes. The above equations become, respectively, for the ellipse,

$$(a_{11}x_1 + a_{12}x_2)x_1 + (a_{21}x_1 + a_{22}x_2)x_2 = 1,$$

for the ellipsoid,

$$(a_{11}x_1 + a_{12}x_2 + a_{13}x_3)x_1 + (a_{21}x_1 + a_{22}x_2 + a_{23}x_3)x_2$$
$$+ (a_{31}x_1 + a_{32}x_2 + a_{33}x_3)x_3 = 1,$$

---

[2]Lanczos, p. 82

9.4. THE GEOMETRY OF THE EIGENVALUE PROBLEM

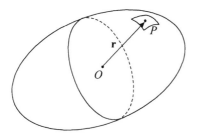

O is the origin
$\mathbf{r}$ is the vector from the origin to point P on the surface of the ellipsoid.

Figure 9.3: Arbitrarily oriented ellipsoid.

and in n arbitrary dimensions,

$$
\begin{aligned}
& (a_{11}x_1 + a_{12}x_2 + \ldots + a_{1n}x_n)x_1 \\
+\ & (a_{21}x_1 + a_{22}x_2 + \ldots + a_{2n}x_n)x_2 \\
+\ & \ldots \\
+\ & (a_{n1}x_1 + a_{n2}x_2 + \ldots + a_{nn}x_n)x_n = 1.
\end{aligned} \qquad (9.14)
$$

Note that terms such as $x_i x_j$ can be combined with $x_j x_i$ by adding coefficients $a_{ij} + a_{ji}$. Since this sum is symmetric with respect to an exchange of indices i, j, then $a_{ij} = a_{ji}$. Such a *symmetric* matrix is invariant under a transposition of rows and columns: $[A]^T = [A]$.

In matrix notation, general Equation 9.14 becomes

$$\{x\}^T [A]\{x\} = 1. \qquad (9.15)$$

The vector $\{x\}$ has the significance of a *radius vector*, connecting an arbitrary point on the surface P with the origin O. Our goal is to find the coordinate system in which the surface equation has no *cross terms* and has the form of Equations 9.11–9.13. We expect these axes to have special properties.

We need to explore what characterizes the principal axes of a quadratic surface. Figure 9.4 depicts the vector normal $\mathbf{n}$ to the tangent plane at point P. Such a normal can be constructed at every point of the surface. The radius vector $\mathbf{r}$ is generally not parallel to the normal vector. Only in exceptional cases is $\mathbf{r} \parallel \mathbf{n}$. For this particular set of directions the name *principal axes* is given.

In the equations above, the matrix–vector product $[A]\{x\}$ can be interpreted as a transformation of the vector $\{x\}$ into some new vector $\{y\}$, that is, $[A]\{x\} = \{y\}$. Thus, the original direction of vector $\{x\}$ is transformed to a new magnitude and direction $\{y\}$. In the case where the radius vector

438   CHAPTER 9. MDOF VIBRATION: ADVANCED TOPICS

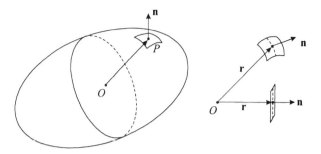

Figure 9.4: Normal at the surface: general case and principal case.

is parallel to the normal at the surface, the *direction is not changed, only the magnitude is changed*. Therefore, following our interpretation of the above matrix–vector product, we must have in this instance

$$[A]\{x\} = \lambda\{x\}. \tag{9.16}$$

This equation is quite familiar to us as the eigenvalue problem considered in Chapter 8. Here, this equation arises as a result of the search for the principal axes of a quadratic surface.

Consider now Equations 9.15 and 9.16. Matrix $[A]$ can be eliminated, leaving $\lambda\{x\}^T\{x\} = 1$, or $\{x\}^T\{x\} = 1/\lambda$. Thus λ_i is the reciprocal of the square of the distance from a point on the surface to the center.

Now that we understand the geometric significance of the principal axes, we also see that the frame of reference of the associated matrix can be changed to one where the matrix is in its principal form and diagonal. The counterpart in vibration is the simultaneous diagonalization of the mass and stiffness matrices in a modal analysis.

9.4.1 Repeated Frequencies

Again following Lanczos (p. 94) we consider the physical interpretation of repeated roots or frequencies. We have not studied this problem analytically because it is a topic that is too specialized for an introductory text. But perhaps some physical insight can be imparted by the discussion that follows.

As two or more distinct roots approach each other, leading to repeated or multiple frequencies, the ellipse of the last section gradually becomes a circle, but the principal axes do not disappear. Rather, they become two mutually perpendicular *diameters* on the circle. That same circle can

9.5. PERIODIC STRUCTURES

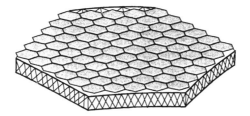

Figure 9.5: A structure with periodic properties.

be the limiting position of any ellipse and therefore any two perpendicular diameters of the circle can serve as principal axes. Similarly for the sphere, where any three perpendicular diameters can serve as the principal axes. This argument can be extended to any n dimensional case.

The existence of multiple roots does not invalidate the existence of mutually perpendicular axes, only that some of these axes are no longer uniquely determined and therefore can be replaced by other equally valid axes. The collapse of certain frequency values into one is not connected with a corresponding collapse of the associated axes since the mutual perpendicularity of the principal axes prevents them from ever collapsing into one. However, multiplicity of frequencies, or closely spaced frequencies, can cause numerical difficulties. Much research effort has been expended on the development of computational approaches to such problems.

In the remaining sections of this chapter, we study topics that apply what we have already learned to problems of greater complexity, requiring more sophistication. The first of these is the periodic structure.

9.5 Periodic Structures

Periodic structures are those with a repetitive pattern, where each *bay* is designed to be identical and joined to the next bay in the same manner. Examples of such structures are sections of aircraft fuselage that have repetitive stiffeners on a shell, turbine blades that have a circular periodicity, and antenna dishes. A schematic of a periodic structure is provided in Figure 9.5.

It is generally assumed that each bay of a repetitive structure is identical to the next one. This assumption tends to considerably simplify the analysis; only one bay plus the boundary conditions needs to be analyzed if the loading is also symmetric. However, in actual structures, the periodic nature can never be exact and there are at least very small differences in material properties and geometry when moving from one bay to the next.

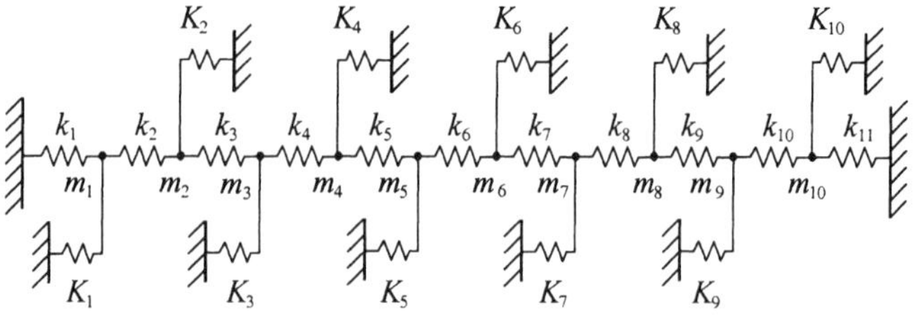

Figure 9.6: Ten bay structure in longitudinal motion.

Recent research has shown that even small imperfections can result in significant changes in structural response for such *near–periodic* structures. We will first examine the behavior of an exactly periodic structure, and then study how an imperfection in the periodicity affects structural response.

9.5.1 Perfect Lattice Models

Perfectly periodic discrete structures are sometimes called *lattice* models because, historically, such spring–mass systems looked like lattices to the physicists who used them to model the interactions of atoms in a solid. Here, a 10 degree of freedom structure undergoing longitudinal motion (along the axis of the structure) is formulated, and some numerical results are presented and discussed. Except for the larger number of degrees of freedom, the techniques of this chapter can be utilized for the analysis of a periodic structure. In this section, the structure is assumed to be perfectly periodic, and in the next section an imperfection is introduced so that we can examine its effects.

Consider the ten mass structure of Figure 9.6. Each mass represents the inertial properties of a *substructure* or *bay*. A mass is attached to a neighboring mass by a coupling spring k_i that represents the coupling stiffness between substructures. To represent the stiffness of a substructure, the mass is also attached to a spring that is fixed to some immovable point. This is meant to be a conceptual model of certain classes of structures that are weakly coupled internally, but are attached to a much stiffer base structure. Additional examples include space frame structures such as the space station, solar arrays attached to a satellite by highly stiff supports, and rotating machinery or other circular symmetric systems, where flexible blades are attached to a very stiff shaft. In Figure 9.6, if k_1 and k_{11} are made to be one and the same spring, then this model can be used for circular symmetric structures as well.

9.5. PERIODIC STRUCTURES

The matrix equation of motion for the periodic structure is

$$[m]\{\ddot{x}\} + [k]\{x\} = \{0\},$$

where

$$[m] = \begin{bmatrix} m_1 & 0 & \cdots & 0 \\ 0 & m_2 & \cdots & 0 \\ \cdots & & \ddots & 0 \\ 0 & \cdots & & m_{10} \end{bmatrix}, \quad \{x\} = \begin{Bmatrix} x_1 \\ x_2 \\ \vdots \\ x_{10} \end{Bmatrix}$$

and

$$[k] = \begin{bmatrix} k_1 + K_1 + k_2 & -k_2 & \cdots & & 0 \\ -k_2 & k_2 + K_2 + k_3 & & & 0 \\ \vdots & & 0 & \ddots & 0 \\ 0 & & & & -k_{10} \\ 0 & & \cdots & -k_{10} & k_{10} + K_{10} + k_{11} \end{bmatrix}.$$

The stiffness matrix is *tridiagonal*, meaning that nonzero elements appear only on the main and the two adjacent diagonals. The main diagonal is of the form $k_i + K_i + k_{i+1}$. All of the procedures we have learned for free vibration response apply here for the evaluation of natural frequencies, modes, and response. Our purpose is to examine the time history response for each mass where, in this discussion, $m_i = 10$ kg and $K_i = 100$ N/m for all i. A parameter found to be important in the behavior of such systems is the *coupling stiffness ratio*, defined as

$$\text{CSR} = \frac{k_i}{K_i}.$$

Once the CSR is prescribed, k_i is determined since K_i is already known. For example, a *weakly-coupled* structure may have a coupling stiffness ratio of CSR = 0.01 or 1%, and therefore, $k_i = 1$ N/m. The degree of coupling between bays, k_i, affects how fast energy can propagate from one bay to the next. This can be physically understood by recognizing that energy propagates due to the compression and elongation of the spring during oscillation. For larger k_i, energy from one mass is transferred faster to the next mass. This coincides with our studies of coupled pendula.

Figure 9.7 shows the response of each of the ten masses due to a unit initial velocity applied at mass m_1. The time history is 600 sec long, and we can see how the wave travels from position one to ten and then reflects back from the right end. Since the periodic system is perfectly periodic

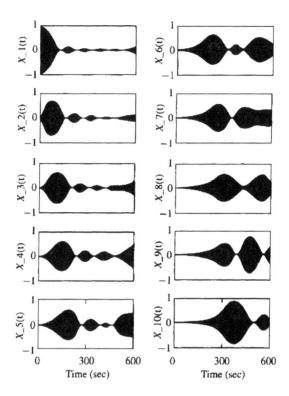

Figure 9.7: Response of ideal structure to a unit initial velocity at mass position one. (Impulse response.)

with no discontinuities or imperfections, there are no locations where a *mismatch* between the properties of adjacent cells or bays would result in some reflected energy.

The next section discusses the effects of an imperfection on the character of the response.

9.5.2 Effects of Imperfection

The effects of imperfection can be studied by introducing a parameter that is a measure of the physical differences between adjacent bays.[3] Assume

---
[3] See Mester, S. and H. Benaroya, "A Parameter Study of Localization," *Shock and Vibration*, Vol. 3, No. 1, pp. 1–10 (1996).

that imperfection is introduced due to differences between bay stiffnesses K_i. This *stiffness imperfection ratio* is defined as

$$\text{SIR} = \frac{K_d - K}{K},$$

where K_d is the introduced disordered bay stiffness and K is the ideal bay stiffness. For example, if SIR $= +10\%$ for an imperfectly periodic structure with ideal bay stiffness $K = 100$ N/m, then this implies that

$$\begin{aligned} K_d &= K(SIR+1) \\ &= 100(0.10+1) = 110 \text{ N/m.} \end{aligned}$$

If SIR $= -10\%$, then $K_d = 90$ N/m. Performing a standard free vibration analysis for the ten bay structure with 10% stiffness imperfection in the fifth bay has the effect of *localizing* vibrational energy about the fifth mass. The CSR $= 1\%$. A unit initial velocity at the first mass is used to initiate a free vibration of the system. The resulting responses are shown in Figure 9.8.

As an example of how the modes become distorted due to imperfections, see Figure 9.9. We see the change from a smooth mode curve to an irregular or distorted mode with the addition of imperfections.

9.6 Inverse Vibration: Estimate Mass and Stiffness

Inverse vibration problems come in many forms. Such problems are called inverse because what is known and what is evaluated are reversed. For example, in previous problems the system mass, damping, and stiffness were known quantities, and the known input force was used to solve for the response. In an inverse problem, the force and response are known and are used to evaluate the system mass, damping, and stiffness. Such problems are more difficult to solve than the usual *forward* problems because there may be more than one solution. That is, there may be more than one combination of system properties that satisfies the force–response relation.

In this section we consider the use of eigenvalue data to calculate the properties of a linear dynamic system. Suppose a set of experiments is run to estimate the natural frequencies of a structure. For example, a multi degree of freedom system can be driven by a variable–frequency load. At each resonance, there is a peak response and the phase angle is $\pi/2$ rad. (See Chapter 3.) Can this frequency data be used to evaluate the mass and stiffness properties of the structure? If not, then what additional information is necessary? This is a very simple statement of the inverse vibration problem.

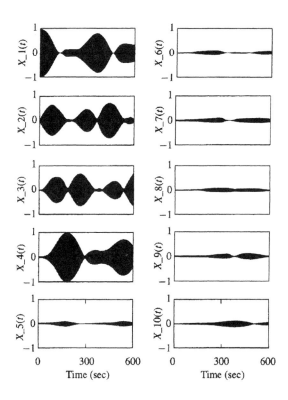

Figure 9.8: Response of a structure with SIR=10% located at the fifth bay, CSR=1%, with first bay loaded by unit initial velocity. (Impulse response.)

9.6. INVERSE VIBRATION: ESTIMATE MASS AND STIFFNESS 445

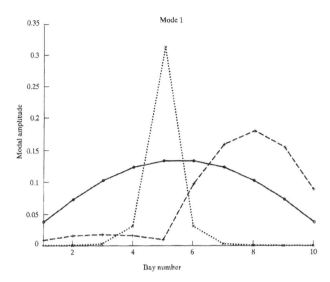

Figure 9.9: Distorted mode due to imperfection; [– × –] for negative disorder, [– ○ –] for no imperfection, [– + –] for positive disorder.

We will examine this fascinating problem using two approaches. The first is based on the work of Gladwell.[4] This is a deterministic approach that assumes all frequency data is exact. Such a study provides us with a new way of thinking about the relationships between structural properties and their respective free vibration characteristics.

In the second approach, it is more realistically assumed that the data has some small errors, regardless of the sophistication of the experimental setup. We are interested in finding out the mass and stiffness sensitivities to uncertainties in the frequency data.

A possible application of this type of work includes a method for nondestructive testing and evaluating structural integrity. Measurements at regular time intervals would be able to detect shifts in the spectral (frequency) properties of a given structure over time. In particular, such techniques could be utilized to estimate and locate changes in structural stiffness due to structural aging.

---

[4] G.M.L. Gladwell, **Inverse Problems in Vibration**, Martinus Nijhoff Publishers, 1986, and G.M.L. Gladwell, "Inverse Problems in Vibration," *Applied Mechanics Reviews*, Vol.39, No.7, July 1986, pages 1013–1018.

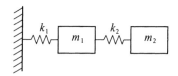

Figure 9.10: A two degree of freedom system.

9.6.1 Deterministic Inverse Vibration Problem

In a typical vibration problem, the physical parameters of the system are known at least approximately. These parameters are the masses and spring constants for a discrete system, or, density, modulus of elasticity, and physical dimensions for a continuous system. From an analysis of these parameters, the natural frequencies or the response to a particular excitation can be determined.

In an *inverse* vibration problem, the physical parameters of a system are determined from the spectral data, that is, frequencies and mode shapes, or eigenvalues and eigenvectors. Consider the simple spring-mass system shown in Figure 9.10. From vibration theory, it is known that this system has two distinct positive eigenvalues, λ_1 and λ_2, that are the roots of the characteristic equation,

$$\lambda^2 - \left[\frac{k_1 + k_2}{m_1} + \frac{k_2}{m_2}\right]\lambda + \frac{k_1 k_2}{m_1 m_2} = 0.$$

The respective natural frequencies are equal to $\sqrt{\lambda_1}$ and $\sqrt{\lambda_2}$. This equation yields the following relations between the eigenvalues:

$$\lambda_1 + \lambda_2 = \frac{k_1 + k_2}{m_1} + \frac{k_2}{m_2}, \quad \lambda_1 \lambda_2 = \frac{k_1 k_2}{m_1 m_2}. \tag{9.17}$$

In inverse vibration problems, the goal is to use the eigenvalue data to reconstruct the physical system properties. For this example, there are two equations for the four unknown values k_1, k_2, m_1, and m_2. This implies that there are an infinite number of two degree of freedom models which have the eigenvalues λ_1 and λ_2. It is, therefore, necessary to introduce two more equations so that the system can be completely specified.

In order to obtain more equations, consider the system shown in Figure 9.11. This system is identical to the previous one, except that the right end has been fixed, restricting it to a single degree of freedom. This constrained system has a single known eigenvalue, λ_3, given by

$$\lambda_3 = \frac{k_1 + k_2}{m_1}. \tag{9.18}$$

9.6. INVERSE VIBRATION: ESTIMATE MASS AND STIFFNESS

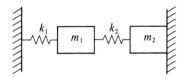

Figure 9.11: Constrained two degree of freedom system.

Constraining an end is one way of obtaining an additional equation.

By algebraically manipulating Equations 9.17 and 9.18, the ratios between the system properties can be obtained as

$$R_1 \equiv \frac{k_2}{m_2} = \lambda_1 + \lambda_2 - \lambda_3 \quad (9.19)$$

$$R_2 \equiv \frac{k_1}{m_1} = \frac{\lambda_1 \lambda_2}{\lambda_1 + \lambda_2 - \lambda_3} \quad (9.20)$$

$$R_3 \equiv \frac{k_2}{m_1} = \frac{(\lambda_3 - \lambda_1)(\lambda_2 - \lambda_3)}{\lambda_1 + \lambda_2 - \lambda_3}. \quad (9.21)$$

Ratios R_1, R_2, and R_3 must all be positive if there is to be a corresponding physical system, since all masses and stiffnesses are positive quantities. The ratios are physically the squares of frequencies. This requires that the eigenvalues satisfy

$$0 < \lambda_1 < \lambda_3 < \lambda_2, \quad (9.22)$$

which is predicted for this type of system by the *inclusion principle*.[5]

These ratios obviously reveal a great deal about the dynamic properties of the system, but they do not uniquely identify it. In order to do this, some further information is needed. For example, if the total mass of the system, $m = m_1 + m_2$, is known, then the parameters can be uniquely solved using

---

[5] The inclusion principle, sometimes called the *Sturmian separation theorem*, is a statement of how the natural frequencies of a system decrease as the number of degrees of freedom increase. For example, assume that there are two mathematical models of the same structure. One model is three degrees of freedom, and the other is two degrees of freedom. The first will have eigenvalues $\lambda_1 \leq \lambda_2 \leq \lambda_3$ while the other has eigenvalues $\Lambda_1 \leq \Lambda_2$. The inclusion principle can be used to show that $\lambda_1 \leq \Lambda_1 \leq \lambda_2 \leq \Lambda_2 \leq \lambda_3$.

This makes physical sense since the structure becomes less stiff, or more flexible, with more degrees of freedom, resulting in lower frequencies of oscillation. Such an understanding has implications on how we mathematically model and approximate a physical system.

Equations 9.19–9.21, resulting in

$$k_1 = \frac{R_1 R_2}{R_1 + R_3} m \qquad (9.23)$$

$$k_2 = \frac{R_1 R_3}{R_1 + R_3} m \qquad (9.24)$$

$$m_1 = \frac{R_1}{R_1 + R_3} m \qquad (9.25)$$

$$m_2 = \frac{R_3}{R_1 + R_3} m. \qquad (9.26)$$

The system's total mass serves only to scale the results, so in many cases it may be sufficient to assume a value if it is not known explicitly. Note that, from Equations 9.20 and 9.21, this approach will work if the eigenvalues are distinct and $\lambda_1 \neq \lambda_3 - \lambda_2$. Such degenerate cases require other techniques.

Example 9.3 A Two Degree of Freedom Inverse Vibration Problem
In the above development, assume that the following data were obtained from two experiments,

$$\omega_1 = \sqrt{\lambda_1} = 2 \text{ Hz}$$
$$\omega_2 = \sqrt{\lambda_2} = 22 \text{ Hz}$$
$$\omega_3 = \sqrt{\lambda_3} = 5 \text{ Hz},$$

and that $m_1 + m_2 = 11$ kg. The first experiment provided ω_1 and ω_2, while the second experiment provided ω_3. Solve for k_1, k_2, m_1 and m_2.
Solution   Note that ω_3 is the natural frequency of the structure when the second mass is fixed. See Equation 9.22. Using Equations 9.19–9.21, we find

$$R_1 = 463, \quad R_2 = 4.2, \quad R_3 = 20.8,$$

and using Equations 9.23–9.26,

$$k_1 = 44.2 \text{ N/m}, \quad k_2 = 218.9 \text{ N/m}$$
$$m_1 = 10.5 \text{ kg}, \quad m_2 = 0.47 \text{ kg}.$$

These results make physical sense since we expect a large disparity between mass and/or stiffness properties if there is a significant difference between the natural frequencies. ∎

While it has been shown that it is possible to derive closed–form solutions to the inverse vibration problem associated with a two degree of freedom

9.6. INVERSE VIBRATION: ESTIMATE MASS AND STIFFNESS 449

discrete system, for larger systems more intricate numerical approaches are necessary. Gladwell's book is a good start to examine such problems. Here, we restrict ourselves to problems that we can demonstrate analytically.

9.6.2 Effect of Uncertain Data

Let us now introduce some uncertainty into the experimentally determined parameters. Quantities involved in the design and analysis of engineering systems generally exhibit some degree of randomness. This can be attributed to several sources. One such source is the uncertainty involved in measurements. A measurement may be made, in some cases, as precisely as is needed for a particular application. In other situations, a measurement can be made only as precisely as the measuring system technology will allow. In either case, there is some uncertainty in the resulting values.

Even if quantities could be exactly measured, the inherent statistical nature of material properties and production techniques suggest a need for probabilistic methods. Two seemingly identical components will, in general, exhibit slight characteristic differences that may affect their respective performances. Assemblies of such components are even more likely to differ from one to another. Finally, the modeling of engineering systems usually requires some approximation. Reasons for this include a lack of understanding of the particular system, or a need to simplify a particularly complex equation. Such assumptions may introduce some form of uncertainty into the solution. This last form of randomness obviously depends on the particular system. The analysis of this type of uncertainty can be quite difficult.

Here, we consider only randomness of the first two types. The system can be mathematically modeled using deterministic equations, and randomness is introduced in the variables. It is important to note that a probabilistic analysis of the type shown here not only provides a better model of the system, but also provides the analyst with a tool for quantifying statistical confidence in the analytical results. We will draw on our discussion in Chapter 5 for probability concepts.

From the discussion of the previous section, Equations such as 9.19–9.21 must be solved. The λ_i's are random variables and, therefore, we need to be able to work with a function of random variables. In order to work analytically, it is necessary to approximate ratios such as R_1, R_2 and R_3 using the venerable Taylor series representation.

Consider[6] a function R of random variables λ_i, $i = 1, 2, \ldots, n$. Each of

---

[6] See Moss, D. and H. Benaroya, "A Discrete Inverse Vibration Problem with Parameter Uncertainties", *Applied Mathematics and Computation*, Vol. 69, 313–333 (1995).

these variables can be written as

$$\lambda_i = \mu_{\lambda_i} + \epsilon_i,$$

where μ_{λ_i} is the mean value of λ_i and ϵ_i is a (small) random parameter signifying some uncertainty about the actual (mean) value of the frequency. Therefore, $E[\epsilon_i] = 0$ and $E[\epsilon_i^2] = \sigma_{\lambda_i}^2$ since

$$\begin{aligned}\sigma_{\lambda_i}^2 &= E\left\{(\lambda_i - \mu_{\lambda_i})^2\right\} \\ &= E\left\{(\epsilon_i)^2\right\}.\end{aligned}$$

Before proceeding with the general expansion, consider the case of a two degree of freedom structure, with two distinct roots (frequency squared), each having uncertainties,

$$\begin{aligned}\lambda_1 &= \mu_{\lambda_1} + \epsilon_1 \\ \lambda_2 &= \mu_{\lambda_2} + \epsilon_2.\end{aligned}$$

For a general general nonlinear function R_1 of both λ_1 and λ_2, the Taylor series expansion about the mean values λ_1 and λ_2 is

$$\begin{aligned}R_1(\lambda_1, \lambda_2) =& \\ R_1(\mu_{\lambda_1}, \mu_{\lambda_2}) &+ \frac{\partial R_1(\mu_{\lambda_1}, \mu_{\lambda_2})}{\partial \lambda_1}(\lambda_1 - \mu_{\lambda_1}) + \frac{\partial R_1(\mu_{\lambda_1}, \mu_{\lambda_2})}{\partial \lambda_2}(\lambda_2 - \mu_{\lambda_2}) \\ +& \frac{1}{2}\left[\frac{\partial^2 R_1(\mu_{\lambda_1}, \mu_{\lambda_2})}{\partial \lambda_1^2}(\lambda_1 - \mu_{\lambda_1})^2 + \frac{\partial^2 R_1(\mu_{\lambda_1}, \mu_{\lambda_2})}{\partial \lambda_1 \partial \lambda_2}(\lambda_1 - \mu_{\lambda_1})(\lambda_2 - \mu_{\lambda_2})\right. \\ +& \left.\frac{\partial^2 R_1(\mu_{\lambda_1}, \mu_{\lambda_2})}{\partial \lambda_2^2}(\lambda_2 - \mu_{\lambda_2})^2\right] + \cdots,\end{aligned}$$

where $(\lambda_1 - \mu_{\lambda_1}) = \epsilon_1$ and $(\lambda_2 - \mu_{\lambda_2}) = \epsilon_2$. A similar expression can be derived for $R_2(\lambda_1, \lambda_2)$. It is important to observe that all terms in these expressions are evaluated at the respective mean values of λ_1 and λ_2, known quantities.

The function R_k for an n degree of freedom structure is a function of all λs, and can be expanded into a Taylor series as

$$R_k(\lambda_1, \lambda_2, \ldots, \lambda_n) =$$
$$R_k(\mu_{\lambda_1}, \mu_{\lambda_2}, \ldots, \mu_{\lambda_n}) + \sum_{i=1}^n \frac{\partial R_k}{\partial \lambda_i}\epsilon_i + \frac{1}{2}\sum_{i=1}^n \sum_{j=1}^n \frac{\partial^2 R_k}{\partial \lambda_i \partial \lambda_j}\epsilon_i \epsilon_j + \cdots,$$

for $k = 1, 2, \ldots, n$. If the frequencies exhibit only a small degree of randomness, that is, if $\lambda_i - \mu_{\lambda_i} = \epsilon_i \ll 1$, the expansion can be truncated after only a few terms with little error due to small terms such as ϵ_i^2 and $\epsilon_i \epsilon_j$.

9.6. INVERSE VIBRATION: ESTIMATE MASS AND STIFFNESS

We only take the first two terms of the Taylor series to demonstrate this procedure. This is a linear approximation for the actual value of R,

$$R_k(\lambda_1, \lambda_2, \ldots, \lambda_n) \approx R_k(\mu_{\lambda_1}, \mu_{\lambda_2}, \ldots, \mu_{\lambda_n}) + \sum_{i=1}^{n} \frac{\partial R_k}{\partial \lambda_i} \epsilon_i, \quad k = 1, 2, \ldots, n.$$

Since $E\{\epsilon_i\} = 0$, taking the expected value of R_k leads to the approximate result

$$E\{R_k\} \approx R_k(\mu_{\lambda_1}, \mu_{\lambda_2}, \ldots, \mu_{\lambda_n}), \quad k = 1, 2, \ldots, n. \quad (9.27)$$

Thus, the linear or first order approximation of the mean value of a complicated function can be obtained by substituting the mean values of all random variables in the function. To obtain an estimate of the standard deviation of R_k, assume the variables to be statistically independent,[7] so that $E[\epsilon_i \epsilon_j] = E[\epsilon_i]E[\epsilon_j] = 0$ for $i \neq j$. The standard deviation of R_k is estimated by

$$\sigma_{R_k}^2 = E\{R_k^2\} - E^2\{R_k\} = \sum_{i=1}^{n} \left(\frac{\partial R_k}{\partial \lambda_i}\right)^2 \sigma_{\lambda_i}^2, \quad k = 1, 2, \ldots, n, \quad (9.28)$$

where the partial derivatives on the right hand side are evaluated at the respective mean values. It should be noted that Equation 9.28 depends only on the mean and standard deviation of the random variables. It is independent of the particular distribution of these variables, with the only assumption being that they are independent. The method can therefore prove useful in cases where little is known about the probabilistic nature of the random variables, but that estimates of their means and variances can be obtained. *Example* 9.17 below demonstrates the procedure just developed.

A more accurate prediction for the statistics of random variable R_k is obtained by retaining the second-order term of the Taylor series,

$$R_k(\lambda_1, \lambda_2, \ldots, \lambda_n) \approx$$

$$R_k(\mu_{\lambda_1}, \mu_{\lambda_2}, \ldots, \mu_{\lambda_n}) + \sum_{i=1}^{n} \frac{\partial R_k}{\partial \lambda_i} \epsilon_i + \frac{1}{2} \sum_{i=1}^{n} \sum_{j=1}^{n} \frac{\partial^2 R_k}{\partial \lambda_i \partial \lambda_j} \epsilon_i \epsilon_j,$$

for $k = 1, 2, \ldots, n$. However, the convergence of approximate series such as these cannot be taken for granted.

It is necessary to verify solutions obtained using these approximate expansions since accuracy depends on the smallness of ϵ. To verify the accuracy of the truncated Taylor series, we have two options. The first is

---

[7] Such assumptions are generally made as a first approximation to the actual situation. For cases where this is not a valid assumption, it is necessary to somehow estimate, usually with experiments, what is the correlation so that $E\{\epsilon_i \epsilon_j\}$ can be evaluated.

452   CHAPTER 9. MDOF VIBRATION: ADVANCED TOPICS

to build an experiment that duplicates the vibrating structure and test it under various conditions. The other option is a powerful *numerical experiment* known as the Monte Carlo simulation technique, which is introduced in Section 9.11. The key point to be made here is that there must be an effort at verification of any and every approximation made in an analysis.

Example 9.4 An Uncertain Two Degree of Freedom System
Use the previous two term Taylor series approximation and the following data to demonstrate this procedure. The eigenvalues

$$\lambda_1 = 0.382 \text{ Hz}^2, \quad \lambda_2 = 2.618 \text{ Hz}^2$$

are for the original system shown in Figure 9.10, and

$$\lambda_3 = 1.000 \text{ Hz}^2$$

for the system with fixed end shown in Figure 9.11.

Use the following information: $m_1 + m_2 = 20$ kg exactly (zero standard deviation), where the mean values of m_1 and m_2 are each 10 kg, the stiffnesses are $k_1 = k_2 = 10$ kg/cm, and all λ values are approximate and assumed random with coefficients of variation $\delta = \sigma/\mu = 0.01$, a 1% variation. We are interested in estimating the mean values and variances of k_i and m_i, given the mean values and variances of the masses and eigenvalues.

Solution Utilize Equation 9.27 for the mean value calculations and Equation 9.28 for the standard deviation calculations. The procedure is in two parts:

1. Given the mean values and standard deviations of λ_i, and using Equations 9.19–9.21, derive the estimated mean values and variances of each ratio R_i.

2. With these results, Equations 9.23–9.26 are used to derive the estimated mean values and variances of each stiffness and mass.

The procedure will be demonstrated only for some of the variables since the algebra becomes very long. Begin with ratio $R_1 = \lambda_1 + \lambda_2 - \lambda_3$. The mean value of R_1 is estimated by

$$\mu_{R_1} = E\{R_1\} = R_1(\mu_{\lambda_1}, \mu_{\lambda_2}, \mu_{\lambda_3})$$
$$= \mu_{\lambda_1} + \mu_{\lambda_2} - \mu_{\lambda_3}.$$

The variance is estimated by

$$\sigma_{R_1}^2 = \left(\frac{\partial R_1}{\partial \lambda_1}\right)^2 \sigma_{\lambda_1}^2 + \left(\frac{\partial R_1}{\partial \lambda_2}\right)^2 \sigma_{\lambda_2}^2 + \left(\frac{\partial R_1}{\partial \lambda_3}\right)^2 \sigma_{\lambda_3}^2$$
$$= (1)^2 \sigma_{\lambda_1}^2 + (1)^2 \sigma_{\lambda_2}^2 + (-1)^2 \sigma_{\lambda_3}^2.$$

9.6. INVERSE VIBRATION: ESTIMATE MASS AND STIFFNESS

Next, follow the same procedure for ratio $R_2 = \lambda_1 \lambda_2 / (\lambda_1 + \lambda_2 - \lambda_3)$. The mean is estimated as

$$\mu_{R_2} = E\{R_2\} = \frac{\mu_{\lambda_1} \mu_{\lambda_2}}{\mu_{\lambda_1} + \mu_{\lambda_2} - \mu_{\lambda_3}},$$

and the variance as

$$\begin{aligned}
\sigma_{R_2}^2 &= \left(\frac{\partial R_2}{\partial \lambda_1}\right)^2 \sigma_{\lambda_1}^2 + \left(\frac{\partial R_2}{\partial \lambda_2}\right)^2 \sigma_{\lambda_2}^2 + \left(\frac{\partial R_2}{\partial \lambda_3}\right)^2 \sigma_{\lambda_3}^2 \\
&= \left(\frac{(\mu_{\lambda_1} + \mu_{\lambda_2} - \mu_{\lambda_3})\mu_{\lambda_2} - \mu_{\lambda_1} \mu_{\lambda_2}(1)}{(\mu_{\lambda_1} + \mu_{\lambda_2} - \mu_{\lambda_3})^2}\right)^2 \sigma_{\lambda_1}^2 \\
&+ \left(\frac{(\mu_{\lambda_1} + \mu_{\lambda_2} - \mu_{\lambda_3})\mu_{\lambda_1} - \mu_{\lambda_1} \mu_{\lambda_2}(1)}{(\mu_{\lambda_1} + \mu_{\lambda_2} - \mu_{\lambda_3})^2}\right)^2 \sigma_{\lambda_2}^2 \\
&+ \left(\frac{(\mu_{\lambda_1} + \mu_{\lambda_2} - \mu_{\lambda_3})(0) - \mu_{\lambda_1} \mu_{\lambda_2}(-1)}{(\mu_{\lambda_1} + \mu_{\lambda_2} - \mu_{\lambda_3})^2}\right)^2 \sigma_{\lambda_3}^2,
\end{aligned}$$

which can be algebraically simplified. The same procedure can be used to estimate the mean and variance of ratio R_3.

Now that the statistics of each ratio R_i has been estimated, proceed with step two to use these in the estimation of the statistics of real interest here, those of k_i and m_i. Beginning with the relation $k_2 = R_1 R_3 m / (R_1 + R_3)$, estimate the mean value of k_2 by

$$E\{k_2\} = \frac{\mu_{R_1} \mu_{R_3}}{\mu_{R_1} + \mu_{R_3}} m,$$

where the total mass m is assumed to be an exact value with no variance. For the estimated value of the variance of k_1, we have

$$\begin{aligned}
\sigma_{k_2}^2 &= \left(\frac{\partial k_2}{\partial R_1}\right)^2 \sigma_{R_1}^2 + \left(\frac{\partial k_2}{\partial R_2}\right)^2 \sigma_{R_2}^2 + \left(\frac{\partial k_2}{\partial R_3}\right)^2 \sigma_{R_3}^2 \\
&= \left(\frac{(\mu_{R_1} + \mu_{R_3})\mu_{R_3} m - \mu_{R_1} \mu_{R_3} m(1)}{(\mu_{R_1} + \mu_{R_3})^2}\right)^2 \sigma_{R_1}^2 \\
&+ \left(\frac{(\mu_{R_1} + \mu_{R_3})(0) - \mu_{R_1} \mu_{R_3} m(0)}{(\mu_{R_1} + \mu_{R_3})^2}\right)^2 \sigma_{R_2}^2 \\
&+ \left(\frac{(\mu_{R_1} + \mu_{R_3})\mu_{R_1} m - \mu_{R_1} \mu_{R_3} m(1)}{(\mu_{R_1} + \mu_{R_3})^2}\right)^2 \sigma_{R_3}^2,
\end{aligned}$$

where it is noted that the second expression on the right hand side equals zero since R_2 is not in the equation for k_2. The same procedure can then be

Variable	μ	σ	$\delta = \sigma/\mu$	(MC μ $\Delta\%$)	(MC σ $\Delta\%$)
m_1	10.00	0.235	0.024	0.07	0.54
m_2	10.00	0.235	0.024	0.07	0.54
k_1	10.00	0.212	0.021	0.21	2.45
k_2	10.00	0.158	0.016	0.22	3.05

Table 9.1: Taylor Expansion Results Compared to Monte Carlo Simulation. The values under the columns labeled (MC $*$ $\Delta\%$) show the percent differences between the perturbation results and the Monte Carlo results for μ and σ, respectively.

used to estimate the mean values and variances of the remaining parameters, k_1, m_1 and m_2. Substituting the mean values and variances given to us at the beginning of the problem statement, we find the results of the first order expansion, as listed in Table 9.1. Comparisons are made between the expansion values and a Monte Carlo (MC) simulation that is considered to be essentially exact.

We see larger errors for the approximate stiffness values than for the mass values when comparing with the Monte Carlo results. This is due to the greater complexity of the stiffness expressions, as seen in Equations 9.19–9.21. ∎

For many engineering applications, procedures such as those presented in this section can suitably model uncertainties.

9.7 Sloshing of Fluids in Containers

The vibration of structures *containing* fluids is of importance for a broad range of applications. Studies of liquid behavior in containers such as the one shown in Figure 9.12 have been used to evaluate the response of reservoirs and tanks to various kinds of excitation, such as earthquakes. Liquid motion is a concern in moving vehicles such as aircraft, spacecraft, ships, railroad cars and trucks transporting liquid cargo. The general goal in understanding fluid–structure interaction is to limit sloshing motions and thus prevent structural damage or interference with normal operations. One vital concern is the possibility that some of the frequencies of fluid oscillation will overlap with structural natural frequencies, causing resonances and vibration magnification. In critical situations, sloshing can result in structural

9.7. SLOSHING OF FLUIDS IN CONTAINERS

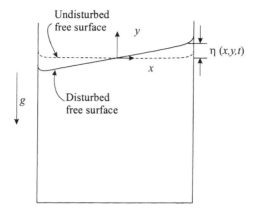

Figure 9.12: Container with moving liquid.

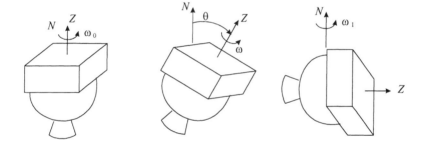

Figure 9.13: Spinning satellite.

failure or dynamic instability.

Two specific critical applications are the sloshing of fuel in aircraft and in spacecraft. There, sloshing of liquid fuel occurs during the launch phase at full gravity under high acceleration, and also during maneuvers in space under low gravity. See Figure 9.13.

In low gravity, fluid *capillary forces* dominate and the fluid equilibrium is a curved surface rather than the flat surface we are familiar with at full gravity. Analysis of sloshing can provide information about fluid motion during vehicle maneuvers, about spacecraft stability, and for the design of propellant management devices for reducing sloshing. These are difficult but important areas of ongoing research.[8]

---

[8] A comprehensive review of this area of engineering research can be found in *The Dynamic Behavior of Liquids in Moving Containers*, H.N. Abramson, Editor, NASA

There are several technical approaches to the analysis of the interaction between sloshing fluid and vibrating structure. Earlier work attempted to use *equivalent mechanical models* to represent the fluid motion and the forces the fluid exerts on the structure. Such efforts, which continue to this day, replace the interior fluid by an equivalent single or multi–degree of freedom system of masses, dampers, and springs, where the values of m, c and k are determined experimentally from scale models. Generally these equations of motion are nonlinear. As computational power improved, analytical mechanical models were partially replaced by computational models such as those based on the finite element method. However, even to this day there is much that is not understood about sloshing behavior, especially in severe environments such as impact or earthquake loading, and in the low gravity environment.

Two linear equivalent mechanical models are shown in Figures 9.14 and 9.15, although current state–of–the–art models are nonlinear. The models discussed here are a direct application of the multi degree of freedom vibratory motion studied in this chapter. Such mechanical models can provide significant general insights into the fluid–structure interaction.

Figures 9.14 and 9.15 show schematic representations of equivalent mechanical fluid representations. These can be very simple models with one sloshing mass m_1 and one mass m_0 representing that part of the fluid that does not slosh, as in Figure 9.14. Otherwise, the model may be more complex with many sloshing masses, as in Figure 9.15. Pendula are also popular models for a sloshing liquid.

The equivalent mechanical model of Figure 9.14 would be a two degree of freedom system, one equation of motion for each mass, m_0 and m_1. The container along with the fluid that does not appreciably move, represented by m_0, would have an equation of motion that includes any external forces acting on the container. This is coupled to the mechanical model of the sloshing fluid, represented by m_1. The degree of sophistication of the mechanical models is based on the available data. In order to capture the more complex behavior characteristics of the fluid, more degrees of freedom and nonlinear characteristics for the fluid are necessary, as in Figure 9.15.

Special Publication SP–106, U.S. Government Printing Office, 1966. Even though the volume is not at the state–of–the–art, it provides an excellent discussion of the key problem areas as well as some preliminary analyses. It is an excellent starting point for the study of liquid sloshing in containers.

9.7. SLOSHING OF FLUIDS IN CONTAINERS

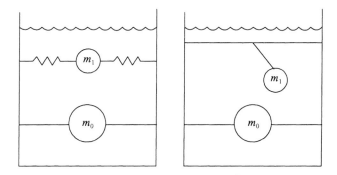

Figure 9.14: Simple mechanical models of linear, lateral sloshing.

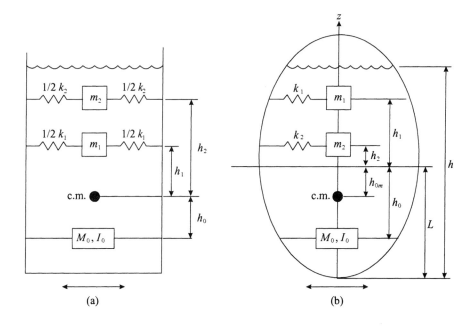

Figure 9.15: Multi degree of freedom mechanical models of linear, lateral sloshing in (a) a rectangular tank and (b) an ellipsoidal tank. Only two sloshing masses are shown, but any number of masses can be used.

9.8 Stability of Motion

Notions on stability and equilibrium have been mentioned in passing throughout this book. In our earlier studies of control in Chapter 6, it was noted that a primary goal was to stabilize the system. Stability can be understood as a characteristic that signifies whether a motion can continue without eventual failure. Here we would like to explore some aspects of the stability of motion by way of an example. In Chapter 12, an introduction is provided to nonlinear oscillations and stability.

When we study the vibration of a structure or machine, we are inherently assuming the oscillation is stable, that is, the oscillation is about a stable equilibrium. Otherwise, the instability dominates the structural behavior and is, therefore, of greater concern than any mode of vibration. In the following example, we show how a system which looks very similar to others studied earlier can have a very different governing equation and may become unstable in its motion for certain combinations of parameter values.

Example 9.5 Pendulum with Base Motion

Figure 9.16 is a sketch of a pendulum rotating about a vertically translating point. This model may be of an equivalent mechanical representation of a fluid in a container undergoing vertical motion. Examples include liquid fuel in a rocket, or stores of liquid in a tank on the ground. At first sight, this problem looks no different than many earlier problems. However, while we can derive its equation of motion as before, the resulting equation has a unique feature that demonstrates a behavior that is new to us.

The governing equation is derived using Lagrange's equation for each generalized coordinate, y and θ. Assume that the container is rigid and only vertical motion is significant.

The kinetic energy is given by (see Figure 9.17)

$$\begin{aligned} T &= \frac{1}{2}m[(L\dot{\theta}\cos\theta)^2 + (\dot{y} + L\dot{\theta}\sin\theta)^2] \\ &= \frac{1}{2}m[L^2\dot{\theta}^2 + 2L\dot{y}\dot{\theta}\sin\theta + \dot{y}^2]. \end{aligned}$$

The potential energy is the sum of the potential due to vertical displacement y and rotation θ,

$$V = mg[L(1 - \cos\theta) + y],$$

where V is measured positive from position $\theta = 0$.

The base motion results in a transmitted force, $F(t)$, on the pendulum. Since the only force on the system is $F(t)$, the virtual work simplifies as

9.8. STABILITY OF MOTION

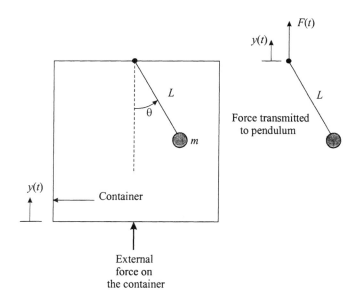

Figure 9.16: Pendulum with base motion.

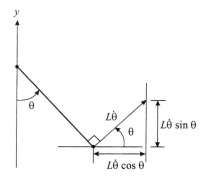

Figure 9.17: Pendulum with base motion: velocity components.

shown,
$$\delta W = F_\theta \delta\theta + F_y \delta y$$
$$= 0 \cdot \delta\theta + F\delta y.$$

We are now in a position to evaluate Lagrange's equation for each generalized coordinate. For the θ direction,

$$\frac{\partial T}{\partial \dot\theta} = \frac{1}{2}m[2L^2\dot\theta + 2L\dot y \sin\theta]$$
$$\frac{d}{dt}\left(\frac{\partial T}{\partial \dot\theta}\right) = \frac{1}{2}m[2L^2\ddot\theta + 2L\ddot y \sin\theta + 2L\dot y \dot\theta \cos\theta]$$
$$\frac{\partial T}{\partial \theta} = \frac{1}{2}m[2L\dot\theta \dot y \cos\theta]$$
$$\frac{\partial V}{\partial \theta} = mgL\sin\theta.$$

Substituting these into Lagrange's equation leads to
$$mL^2\ddot\theta + mL\ddot y \sin\theta + mgL\sin\theta = 0. \quad (9.29)$$

Following the same procedure for coordinate y,
$$\frac{\partial T}{\partial \dot y} = m[\dot y + L\dot\theta \sin\theta]$$
$$\frac{d}{dt}\left(\frac{\partial T}{\partial \dot y}\right) = m[\ddot y + L\ddot\theta \sin\theta + L\dot\theta^2 \cos\theta]$$
$$\frac{\partial T}{\partial y} = 0$$
$$\frac{\partial V}{\partial y} = mg.$$

The resulting equation of motion is
$$\ddot y + L\ddot\theta \sin\theta + L\dot\theta^2 \cos\theta + g = \frac{F}{m}. \quad (9.30)$$

Assume now that we are looking for the stability of motion about the equilibrium position $\theta = 0$. In the region about small θ, Equation 9.29 becomes
$$\ddot\theta + \frac{1}{L}(\ddot y + g)\theta = 0, \quad (9.31)$$

which becomes linearized but is still coupled to the motion in the y direction. Equation 9.30, on the other hand, becomes
$$\ddot y + L\ddot\theta\theta + L\dot\theta^2 + g = \frac{F}{m}, \quad (9.32)$$

9.9. STOCHASTIC RESPONSE OF A LINEAR MDOF SYSTEM

which is simplified but still nonlinear. If θ and $\dot{\theta}$ are small, we can neglect "higher order terms" such as $\dot{\theta}^2$ and $\ddot{\theta}\theta$. In this case, Equation 9.32 is linearized to

$$\ddot{y} + g = \frac{F}{m}. \tag{9.33}$$

This equation is decoupled from the θ motion. Assume that the support motion is harmonic, $y = A\cos\omega t$, suggesting base motion of a fluid filled container. From Equation 9.33, the force required is $F(t) = -mA\omega^2\cos\omega t + mg$. Equation 9.31 becomes

$$\ddot{\theta} + \frac{1}{L}\left(g - A\omega^2\cos\omega t\right)\theta = 0, \tag{9.34}$$

which is an equation with a time–varying coefficient. We immediately note that $(g - A\omega^2\cos\omega t)$ is a harmonic function that can have negative or positive values depending on the numerical values of A and ω. If $g > A\omega^2$ then $g > A\omega^2\cos\omega t$ and the coefficient of θ is always positive and response $\theta(t)$ is always stable. Otherwise, for $g < A\omega^2$, we obtain solutions with coefficients of the form $\exp(+\omega T)$ that grow without bound. For a sloshing fluid on a rocket, there would likely be a more complex forcing function. ∎

Such problems of dynamic stability are extremely important because they potentially affect all vibration studies. Equations such as 9.34, where the loading appears within a coefficient on the left hand side of the governing equation rather than the right hand side, are known as *parametrically excited* systems. See Chapter 12 for more discussion on such problems as well as nonlinear oscillations.

9.9 Stochastic Response of a Linear MDOF System

The response of a multi degree of freedom discrete system loaded by random forces is generally a very complicated problem to formulate and solve. Here, to provide the reader with one possible approach, the modal analysis of such a structure is worked through some of the intricacies.[9]

We start the analysis at the point where the modal equations of motion have been formulated, and the assumption of proportional damping has been made. Recalling that capital letters are used to signify random parameters,

---

[9] The approach of this section is standard, but some additional details are available, for example, in Chapter 23 of **Dynamics of Structures**, by R.W. Clough, and J. Penzien, McGraw–Hill, Second Edition, 1993.

the modal equations in indicial notation are

$$\ddot{Q}_i + 2\zeta_i\omega_i\dot{Q}_i + \omega_i^2 Q_i = F_i(t), \quad i = 1, 2, \ldots, n,$$

where the parameters are familiar. We further assume that the modal forces $F_i(t)$ are *ergodic* random excitations. The transformation between physical and modal spaces is

$$\{X(t)\} = \sum_{i=1}^{n}\{\hat{u}\}_i Q_i(t),$$

or

$$X_j(t) = \sum_{i=1}^{2} \hat{u}_{ji} Q_i(t) = \hat{u}_{j1} Q_1(t) + \hat{u}_{j2} Q_2(t), \tag{9.35}$$

for each degree of freedom j of a two degree of freedom structure. The goal in this analysis is to evaluate the statistics of the response, that is, to find autocorrelation, $R_{X_j X_j}(\tau)$, and its Fourier transform, the power spectrum, $S_{X_j X_j}(\omega)$. We will draw on the probability concepts of Chapter 5.

Begin with the definition of the autocorrelation, and substitute Equation 9.35 for $X_j(t)$,

$$\begin{aligned} R_{X_j X_j}(\tau) &= E\{X_j(t) X_j(t+\tau)\} \\ &= E\left\{\sum_{l=1}^{n}\sum_{m=1}^{n} \hat{u}_{jl}\hat{u}_{jm} Q_l(t) Q_m(t+\tau)\right\}, \end{aligned} \tag{9.36}$$

where

$$Q_i(t) = \int_0^t F_i(\tau) g_i(t-\tau) d\tau \tag{9.37}$$

$$g_i(t) = \frac{1}{\omega_{d_i}} \exp(-\zeta_i\omega_i t) \sin\omega_{d_i} t \tag{9.38}$$

$$\omega_{d_i} = \omega_i(1-\zeta_i^2)^{1/2}. \tag{9.39}$$

Note that since the impulse response function $g(t)$ is zero for $t < 0$, the lower limit on the integral defining $Q(t)$ can be made $-\infty$ without changing the value of the integral.

Substitute Equations 9.37–9.39 into Equation 9.36 and move the expectation operator to the stochastic terms,

$$\begin{aligned} R_{X_j X_j}(\tau) = \sum_l \sum_m \int_{-\infty}^{t+\tau}\int_{-\infty}^{t} u_{jl} u_{jm} E\{F_l(\theta_1) F_m(\theta_2)\} \cdot \\ \cdot g_l(t-\theta_1) g_m(t+\tau-\theta_2) d\theta_1 d\theta_2, \end{aligned} \tag{9.40}$$

9.9. STOCHASTIC RESPONSE OF A LINEAR MDOF SYSTEM

where θ_1 and θ_2 are dummy time variables, and the forcing cross–correlations are

$$R_{F_l F_m}(\theta_2 - \theta_1) = E\{F_l(\theta_1)F_m(\theta_2)\}$$

due to the assumed ergodicity (and thus stationarity) of the forcing. If the system is lightly damped and has well separated modal frequencies, as is the case in many engineering structures, it is of interest to note that the response due to $F_l(t)$ is almost statistically independent of the response due to $F_m(t)$. The cross correlation terms that arise in Equation 9.40 are then almost zero, with the only nonzero terms arising for $m = l$,

$$R_{F_l F_m}(\theta_2 - \theta_1) \approx E\{F_l(\theta_1)F_l(\theta_2)\}. \tag{9.41}$$

We have the correlation function for the response in terms of the correlation function for the random forcing. Next proceed to evaluate the response spectral density, from which probabilities of occurrence can be evaluated. To do this, the following transformation of variables[10] is necessary,

$$u_1 \equiv t - \theta_1 \qquad u_2 \equiv t + \tau - \theta_2$$
$$du_1 = -d\theta_1 \qquad du_2 = -d\theta_2,$$

resulting in the response correlation

$$R_{X_j X_j}(\tau) = \sum_l \sum_m \int_0^\infty \int_0^\infty u_{jl} u_{jm} R_{F_l F_m}(u_1 - u_2 + \tau) g_l(u_1) g_m(u_2) du_1 du_2.$$

The power spectral density for response $X(t)$ is equal to the Fourier transform of this correlation function,

$$S_{X_j X_j}(\omega) = \int_{-\infty}^{\infty} R_{X_j X_j}(\tau) e^{-i\omega\tau} d\tau.$$

Recall and utilize the assumption that the processes are ergodic, and, therefore, by averaging in time,[11]

$$S_{X_j X_j}(\omega)$$
$$= \sum_l \sum_m u_{jl} u_{jm} \left\{ \lim_{T \to \infty} \frac{1}{2T} \int_{-T}^{T} g_l(u_1) du_1 \cdot \lim_{T \to \infty} \frac{1}{2T} \int_{-T}^{T} g_m(u_2) du_2 \right.$$
$$\left. \cdot \lim_{T \to \infty} \frac{1}{2T} \int_{-T}^{T} R_{F_l F_m}(u_1 - u_2 + \tau) e^{-i\omega\tau} d\tau \right\},$$

---
[10] Do not forget to transform the integration limits when you transform the variables.
[11] See Equations 5.29 and 5.30.

where the lower limits on the integrals are set to $-T$ since $g(t)$ is zero for $t < 0$, and thus the change in lower limit does not affect the values of the integrals. Using the change of variables

$$\gamma \equiv u_1 - u_2 + \tau, \quad d\gamma = d\tau,$$

we obtain

$$\begin{aligned} S_{X_j X_j}(\omega) &= \sum_l \sum_m u_{jl} u_{jm} \left\{ \lim_{T \to \infty} \frac{1}{2T} \int_{-T}^{T} g_l(u_1) e^{i\omega u_1} du_1 \right. \\ &\quad \cdot \lim_{T \to \infty} \frac{1}{2T} \int_{-T}^{T} g_m(u_2) e^{-i\omega u_2} du_2 \\ &\quad \left. \cdot \lim_{T \to \infty} \frac{1}{2T} \int_{-T-u_2+u_1}^{T-u_2+u_1} R_{F_l F_m}(\gamma) e^{-i\omega \gamma} d\gamma \right\}. \end{aligned} \quad (9.42)$$

In the last integral, we make the physical argument that $R_{F_l F_m}(\gamma) \to 0$ as $|\gamma|$ increases, and, therefore, the limits can be replaced by $-T$ and T, respectively.[12] Then,

$$H_l(-i\omega) = \lim_{T \to \infty} \frac{1}{2T} \int_{-T}^{T} g_l(u_1) e^{i\omega u_1} du_1 \quad (9.43)$$

$$H_m(i\omega) = \lim_{T \to \infty} \frac{1}{2T} \int_{-T}^{T} g_m(u_2) e^{-i\omega u_2} du_2 \quad (9.44)$$

$$S_{F_l F_m}(\omega) = \lim_{T \to \infty} \frac{1}{2T} \int_{-T}^{T} R_{F_l F_m}(\gamma) e^{-i\omega \gamma} d\gamma, \quad (9.45)$$

with the resulting response spectral density

$$S_{X_j X_j}(\omega) = \sum_l \sum_m u_{jl} u_{jm} H_l(-i\omega) H_m(i\omega) S_{F_l F_m}(\omega),$$

where $H(i\omega)$ is given by Equation 3.23.

For lightly damped systems with well spaced modal frequencies, the cross terms in the double summation, those where $l \neq m$, contribute very little to the mean square response given by $(1/2\pi) \int_{-\infty}^{\infty} S_{X_j X_j}(\omega) d\omega$. In this case, we can use the approximation

$$S_{X_j X_j}(\omega) \cong \sum_l u_{jl}^2 |H_l(i\omega)|^2 S_{F_l F_l}(\omega),$$

---

[12] Physically, this is a statement that as time difference γ increases, there will be an exponentially decaying correlation. This is borne out by experiments on physical systems.

9.10 Rayleigh's Quotient

where $|H(i\omega)|$ is given by Equation 3.25. More details are available in specialized texts.[13]

Generally, the vibration response of engineering structures cannot be solved exactly for a number of reasons. The number of degrees of freedom may be too large for available computers. The material may be nonhomogeneous and the variability cannot be analytically described. The structure may be geometrically too complex for an explicit mathematical description. The forcing may be too intricate for a simple formulation. Finally, there may be uncertainties in the system or forcing, precluding an exact deterministic or probabilistic formulation or solution.

In such cases, alternate approaches are necessary. We wish to emphasize, however, that all of our previous work can be brought to bear in the development of such approaches. We build on the analytical models and generalize them for the computer. Material variability and geometrical complexity are formulated as computational models that are good representations of the physics of the problem. The most prominent and well developed group of such modeling techniques goes by the name *finite element methods*. The reader is urged to learn about such methods from the specialized texts on the subject.[14] We will not discuss these in this book.

An intermediate step between analytical models and computational models is the development of *approximate models*. Such approximate techniques permit the analyst to go beyond the simplest exact models and study more applicable systems, but without needing to resort to large scale computation. More modest computation is generally sufficient, generally utilizing commercially-based tools. Such approximate methods can still be understood physically, but are in a computational form.

In this section, *Rayleigh's quotient* for estimating the lower frequencies of a vibrating body is introduced. In Chapter 11 we will continue the study of approximation tools by applying these and other techniques to continuous systems.

Begin with the eigenvalue problem in the form

$$[k]\{u\}_r = \omega_r^2 [m]\{u\}_r, \quad r = 1, \ldots, n,$$

---

[13] See also P. Wirsching, T. Paez, and H. Ortiz, **Random Vibration: Theory and Practice**, Wiley–Interscience, 1995.

[14] An excellent introductory book on computational modeling, that is also theoretically comprehensive, is by W.D. Pilkey and W. Wunderlich, **Mechanics of Structures: Variational and Computational Methods**, CRC Press, 1994.

for an n degree of freedom system. Premultiply each side by $\{u\}_r^T$ and solve for ω_r^2 to find

$$\omega_r^2 = \frac{\{u\}_r^T [k] \{u\}_r}{\{u\}_r^T [m] \{u\}_r}, \quad r = 1, \ldots, n. \tag{9.46}$$

This equation is known as *Rayleigh's quotient*, which physically is the *ratio of the maxima of the potential energy to the kinetic energy*. (See the following example.) Its value as an estimator becomes apparent when a "reasonable" eigenvector is substituted for $\{u\}_r$, resulting in an estimate of the eigenvalue ω_r. A reasonable eigenvector at least satisfies boundary conditions such as slope and deflection. While r may represent any eigenvalue, it is generally difficult, if not impossible, to provide a good guess of the eigenvector for modes beyond the first several. Therefore, this approach is useful if we are interested in estimates of the lowest few modes of a system. Note also that in Equation 9.46 normal modes are not used because they are not known in this approximate method, they are *assumed*.

It can be shown that Rayleigh's quotient has a *minimum* in the neighborhood of a *fundamental* mode, and if the assumed vector differs from the eigenvector $\{u\}_r$ by a small quantity of *first order*, say ε, then the ratio differs from the eigenvalue ω^2 by a quantity of the *second order*, ε^2. As a practical matter, this means that even if our guess of the mode shape is of average accuracy, the calculated frequency can still be a good one. The example below will demonstrate this property. Two final important characteristics are that

- Rayleigh's quotient for the lowest frequency is never lower than the fundamental frequency, and

- Rayleigh's quotient for the highest frequency is never higher than the highest frequency.

This is another statement of the inclusion principle, cited earlier. For intermediate frequency values, the respective Rayleigh's quotient can be above or below the actual value. These properties have practical implications that are discussed in the following examples.

Example 9.6 Rayleigh's Quotient for a Two Degree of Freedom System

Apply Rayleigh's quotient to the simple two degree of freedom oscillator shown in Figure 9.18. The usefulness of Rayleigh's quotient depends on the ability of the analyst to guess the mode shape for the frequency of interest. Success depends on experience, especially for the higher mode shapes.

Solution Equate the maximum kinetic energy to the maximum potential energy (as we once did for a single degree of freedom oscillator) to

9.10. RAYLEIGH'S QUOTIENT

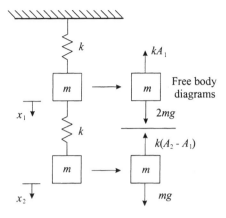

Figure 9.18: Two degree of freedom oscillator.

obtain Rayleigh's quotient. The maximum kinetic energy for a harmonic oscillator without damping is given when the harmonic function is at its maximum. Suppose

$$x_1(t) = A_1 \cos \omega t, \quad x_2(t) = A_2 \cos \omega t$$
$$\dot{x}_1(t) = -\omega A_1 \sin \omega t, \quad \dot{x}_2(t) = -\omega A_2 \sin \omega t.$$

In these expressions, we are considering *a single mode for both masses* and, therefore, do not distinguish their frequencies, since in each mode the motion is synchronous. Then,

$$T = \frac{1}{2}m\dot{x}_1^2 + \frac{1}{2}m\dot{x}_2^2$$
$$T_{max} = \frac{1}{2}m(A_1\omega)^2 + \frac{1}{2}m(A_2\omega)^2.$$

The corresponding potential energy expressions are given by

$$V = \frac{1}{2}kx_1^2 + \frac{1}{2}k(x_2 - x_1)^2$$
$$V_{max} = \frac{1}{2}kA_1^2 + \frac{1}{2}k(A_2 - A_1)^2.$$

Equate T_{max} to V_{max} to find

$$\omega^2 = \frac{k}{m} \frac{A_1^2 + (A_2 - A_1)^2}{(A_1^2 + A_2^2)} \; (\text{rad/sec})^2.$$

This equation is valid for any frequency and mode shape. Note that the mode shape is given by the ratio A_2/A_1. One has "only" to guess this ratio to obtain an estimate of ω.

For the first mode, use the *static deflection* of the two masses as the guess. From the free body diagrams, the static deflections are

$$A_1 = \frac{2mg}{k}, \quad A_2 = \frac{3mg}{k},$$

and then

$$\begin{aligned}\omega_1^2 &= \frac{k}{m}\frac{(2mg/k)^2 + (mg/k)^2}{(2mg/k)^2 + (3mg/k)^2} \\ &= \frac{5k}{13m} = 0.385\frac{k}{m} \; (\text{rad/sec})^2.\end{aligned}$$

The exact solution is $0.36k/m$ $(\text{rad/sec})^2$, showing that the estimate is quite accurate.

For the second mode, we know that the masses will be displaced in opposite directions. Therefore, let $A_1 = 2mg/k$ and $A_2 = -A_1$. Proceeding as before, we obtain

$$\omega_2^2 = 2.5\frac{k}{m} \; (\text{rad/sec})^2,$$

where the exact result is $2.618k/m$ $(\text{rad/sec})^2$. Again an accurate estimate is obtained.

Higher modes are more difficult to evaluate by this method because it is more difficult to guess the higher modal ratios. In the example we just completed, what would the results look like if a poorer guess was made for the assumed mode? Try a more naive first mode: $A_1 = A_2 = mg/k$. A simple computation results in $\omega_1^2 = 0.5k/m$ $(\text{rad/sec})^2$. This result is not as good as the one that resulted from the better physical understanding of the relative displacement of the masses. But both eigenvalue estimates of ω_1^2 are above the exact one, as we expect. One practical conclusion is that a variety of mode shapes can be guessed, and the smallest resulting eigenvalue is the best estimate.

The highest eigenvalue (here the second frequency ω_2) is below the exact value, as expected. ∎

Example 9.7 Rayleigh's Quotient for a Three Degree of Freedom System

For the three degree of freedom set of oscillators of Figure 9.19, use Rayleigh's quotient to estimate the three natural frequencies of vibration.

9.10. RAYLEIGH'S QUOTIENT

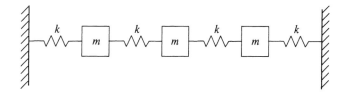

Figure 9.19: Three degree of freedom oscillator.

Solution Approach this problem in the same manner as the last one. The first mode is estimated using the ratio of static displacements. The mass and stiffness matrices are

$$[m] = m \begin{bmatrix} 1 & 0 & 0 \\ 0 & 1 & 0 \\ 0 & 0 & 1 \end{bmatrix} \text{ kg}, \quad [k] = k \begin{bmatrix} 2 & -1 & 0 \\ -1 & 2 & -1 \\ 0 & -1 & 2 \end{bmatrix} \text{ N/m}.$$

To obtain the ratio of static displacements, assume the force vector $\{F\} = \{1\ 1\ 1\}^T$. Then, for a static system,

$$\{u\}_1 = [k]^{-1}\{F\}$$

$$= \frac{1}{k} \begin{bmatrix} 0.75 & 0.50 & 0.25 \\ 0.50 & 1.00 & 0.50 \\ 0.25 & 0.50 & 0.75 \end{bmatrix} \begin{Bmatrix} 1 \\ 1 \\ 1 \end{Bmatrix} = \frac{1}{k} \begin{Bmatrix} 1.5 \\ 2.0 \\ 1.5 \end{Bmatrix},$$

where we can ignore the factor $1/k$ since we only need the ratios of the displacements in order to use Rayleigh's quotient. Use Equation 9.46 to find

$$\omega_1^2 = \frac{\{u\}_1^T[k]\{u\}_1}{\{u\}_1^T[m]\{u\}_1} = \frac{10k}{17m} = 0.5882\frac{k}{m} \text{ (rad/sec)}^2.$$

Then $\omega_1 = 0.7670\sqrt{k/m}$ rad/sec is an excellent estimate for the exact value of $\omega_1 = 0.7669\sqrt{k/m}$ rad/sec. (Note that the approximate value is higher than the exact value.)

To estimate the second natural frequency, use the knowledge that the second mode possesses one node at the center and assume $\{u\}_2 = \{1\ 0\ -1\}^T$. Proceeding as above, we find $\omega_2 = \sqrt{2k/m}$ rad/sec, which coincides with the exact value.

To estimate the third natural frequency, whose mode possesses two nodes, assume $\{u\}_3 = \{1\ -1\ 1\}^T$, and find $\omega_3 = 1.8257\sqrt{k/m}$ rad/sec as compared to the exact value of $1.8477\sqrt{k/m}$ rad/sec. It appears that we can do a good job of estimating the natural frequencies of a system with few degrees of freedom. (Note that the approximate value is lower than the exact value.) ∎

We will resume our study of approximate techniques in Chapter 11 for continuous systems. In the next section, the powerful Monte Carlo statistical numerical technique is introduced.

9.11 Monte Carlo Simulation

Monte Carlo simulation is a general label attached to probabilistic computational methods that are based on the repetitive sampling of the distributions of random variables. The term Monte Carlo was first introduced by von Neumann and Ulam[15] during World War II. The method was named after the resort city on the Riviera where gambling is a major industry.

Monte Carlo methods were initially used in the development of the atomic bomb, where they were applied to problems involving random neutron diffusion in fissile material. Monte Carlo simulation generates estimates of the statistics describing various functions, utilizing statistical descriptions of the involved variables. The simulation process consists of generating many values of the function of interest by computer calculations. Statistical tests are usually applied to the variable to be modeled to indicate the most likely probability distribution that will produce similar synthesized values. These values are normally obtained directly from already developed programs. Such *random numbers* are called *pseudorandom* because they are attained via deterministic equations. Sets of random values for the simulated variable are required in the Monte Carlo process. Each generated number produces a corresponding result, until a family of results exists for the set of random input variables. Finally, all the results are averaged to obtain a mean value and higher moments, from which confidence bounds can be obtained. It should be mentioned that since Monte Carlo results arise from observational data consisting of random numbers, it is necessary that the family of results be large to ensure convergence to a statistically significant result. The flow chart in Figure 9.20 illustrates the procedure.

The main drawback to Monte Carlo simulation is that it can often be a computationally time–consuming technique. Despite this problem, and considering the pace at which digital computers are evolving, the Monte Carlo method is now the most powerful and commonly used technique for

---

[15] Look up the book by J.M. Hammersley and D.C. Handscomb, **Monte Carlo Methods**, John Wiley and Sons, 1964. It is one of the most readable introductions to Monte Carlo methods.

9.11. MONTE CARLO SIMULATION

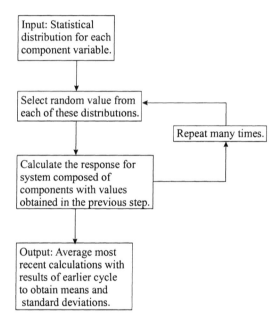

Figure 9.20: Flow chart of Monte Carlo simulation method.

analyzing complex probabilistic problems. However, the complexity and computational effort required has been increasing, since realism has demanded greater intricacy and more extensive descriptions.[16]

9.11.1 Random Number Generation

The most frequently used present–day method for generating pseudorandom numbers is the *linear congruential generator*. This generator is based on recursive calculations of the residues of modulus m of a linear transformation. Such a recursive relation may be expressed as

$$X_{i+1} = (aX_i + c)(mod\ m),$$

where a, c, and m are nonnegative integers and mod is the *modulus* which is defined such that X_{i+1} is the remainder of the division of m into $aX_i + c$. If k_i is the integer part of the ratio $(aX_i + c)/m$, that is,

$$k_i = \text{Int}\left(\frac{aX_i + c}{m}\right),$$

then the corresponding residue of modulus m is

$$X_{i+1} = aX_i + c - mk_i. \tag{9.47}$$

Normalizing the values obtained from Equation 9.47 by the modulus m, we obtain

$$u_{i+1} = \frac{X_{i+1}}{m},$$

which constitute a set of random values on the unit interval (0,1) with the standard uniform probability distribution (mean = 0.5 and standard deviation = 0.289).

Such a sequence of pseudorandom numbers is cyclic, and will repeat itself in at most m steps. To ensure randomness, the period of the cycle should be as long as possible, and therefore, in practical applications a large value of m should be assigned in the generation of u_i. The selection of values for m, a, and c is the most important step for creating a generator of this sort. Table 9.2 lists some choices for these constants that will yield a large period. These values have been tested statistically and shown to give satisfactory results. The choice of which sets of constants to select from Table 9.2, as well as the size of the randomizing shuffle, is essentially arbitrary. However, the larger the shuffling array, the less likely it is that sequential correlation will occur. Ultimately this choice must be balanced with computational time and required storage space for the array.

---

[16] A much more advanced text on this subject is by R.Y. Rubinstein, **Simulation and the Monte Carlo Method**, John Wiley and Sons, 1981.

9.11. MONTE CARLO SIMULATION

m	a	c
7875	421	1663
11979	859	2531
21870	1291	4621
81000	421	17117
86436	1093	18257
117128	1277	24749
121500	4081	25673
134456	8121	28411
243000	4561	51349
259200	7141	54773

Table 9.2: Constants for Random Number Generators.

9.11.2 Generation of Random Variates

The next question that needs addressing is how to use the uniformly distributed pseudorandom numbers to generate values according to a particular probability density.

This generation of random variates can be accomplished systematically from the uniform distribution on the interval $(0, 1)$ which was determined in the preceding section. This is done through one of several methods, such as the *inverse transform* method, *composition* method, and *acceptance-rejection* method. Two of the most important continuous distributions are the exponential and normal distributions, the former because it is commonly used to model dynamic loading and applied stress, and the latter because it provides a good representation of many variables, and its variates can be used to generate several other common distributions.

A nonstandard uniform distribution on any arbitrary interval (a, b) can be obtained using the inverse transform method. By this method, we solve for the random variable by inverting the cumulative distribution function. The cumulative distribution function is

$$F_x(x) = \begin{cases} 0, & x < a \\ (x-a)/(b-a), & a \leq x \leq b \\ 1, & x > b, \end{cases}$$

PDF to Simulate	Probability Density Function	Procedure to Obtain Random Value y
Exp	$f(y) = \lambda \exp[-\lambda(y-\mu)]$ $\mu \leq y < \infty$	$y = -\frac{1}{\lambda}\ln(1-R_U)$ $+\mu$
Lognormal	$f(y) = \frac{1}{\sqrt{2\pi}\sigma y}\exp\left[-\left(\frac{\ln y - \mu}{2\sigma}\right)^2\right]$ $0 \leq y < \infty$	$y = \exp(\mu + \sigma R_N)$
Normal	$f(y) = \frac{1}{\sigma\sqrt{2\pi}}\exp\left[-\frac{(y-\mu)^2}{2\sigma^2}\right]$ $-\infty < y < \infty$	$y = \mu + \sigma R_N$

Table 9.3: Recursive Relations to Obtain Common Random Variates

and the inverse function in this case is

$$x = F_x^{-1}(u) = a + (b-a)u.$$

To apply the inverse transform, the distribution function must exist in a form for which the corresponding inverse transform can be found analytically. Other distribution functions that can be inverted are the Weibull, logistic, and Cauchy distributions.

Other techniques must be employed to generate the random variates in a normal distribution, which has the probability density function

$$f_x(x) = \frac{1}{\sigma\sqrt{2\pi}}\exp\left[-\frac{(x-\mu)^2}{2\sigma^2}\right], \quad -\infty < x < \infty, \qquad (9.48)$$

denoted by $R_N(\mu, \sigma)$; μ is the mean and σ is the standard deviation.

It is known that if U_1 and U_2 are two statistically independent standard uniform variates, then the following functions constitute a pair of independent standard normal variates,[17]

$$S_1 = (-2\ln U_1)^{\frac{1}{2}}\cos 2\pi U_2, \quad S_2 = (-2\ln U_1)^{\frac{1}{2}}\sin 2\pi U_2.$$

Table 9.3 lists equations to generate random variates for common probability density functions, given random values of standard uniform (R_U), generated on the interval (0,1), and standard normal (R_N) distributions. Most engineering parameters fall into either the normal, lognormal, or exponential distributions, and are listed in the table.

In the next section, a procedure is used to generate a time–history from a power spectrum of a random process.

---

[17] A proof is given on p. 86 of Rubinstein.

9.11.3 Generating a Time–History for a Random Process Defined by a Power Spectral Density

When computing the vibratory response of a machine or component in a random environment, it is sometimes necessary to work in the time–domain rather than in the frequency–domain. This is especially true when the differential equation governing structural behavior is *nonlinear*. Such systems are solved by more specialized procedures, generally numerically in time. Therefore, we are met with the challenge of utilizing the information stored in a power spectral density for a time–domain analysis. The procedure developed here is by Borgman[18] for ocean engineering application; we will do the same.

A primary characteristic of the ocean, used in offshore structural engineering, is the wave elevation or wave height $\eta(x,t)$, which is a function of position x and time t. Where this parameter is modeled realistically, it is done so assuming that $\eta(x,t)$ is a random process in time and that the possible wave heights are defined by a spectral density $S_\eta(\omega)$. A commonly used spectrum is the *Pierson–Moskowitz*,

$$S_\eta(\omega) = \frac{A_0}{\omega^5} e^{-B/\omega^4},$$

where the constants A_0 and B depend on location. See Equation 5.32 for additional details.

Therefore, instead of using a simple harmonic representation of the wave height, use

$$\eta(x,t) = \int_0^\infty \sin(kx - \omega t + \varepsilon)\sqrt{S_\eta(\omega)d\omega},$$

where the wave contains a range of frequency components ω, k is the wave number, and ε is a random phase angle selected from a list of random numbers uniformly distributed over the interval between 0 and 2π. The wave height $\eta(x,t)$ has units of length and the quantity $\sqrt{S_\eta(\omega)d\omega}$ has units of $\sqrt{(\text{length}^2 \text{sec})\cdot(\text{rad/sec})}$ = length. Physically, the integral represents a sum of harmonics weighted by a measure, $S_\eta(\omega)$, of how much of the total wave energy is within a particular energy band $d\omega$.

For computational purposes, we would like to discretize the integral over the range of ω where there is significant energy, say $\omega_0 < \omega < \omega_N$. This discretization results in *equal areas* rather than equal spectrum bands. Such a procedure avoids the presence of periodicities in the resulting time history. This partition is

$$\omega_0 < \omega_1 < \omega_2 < \cdots < \omega_N,$$

---

[18] L.E. Borgman, "Ocean Wave Simulation for Engineering Design," *J. Waterways and Harbors Division, ASCE*, **95** (WW4), 557–583, Nov. 1969.

and for $\omega < \omega_0$ and $\omega > \omega_N$, for practical purposes $S_\eta(\omega) = 0$. Over each band, averaged quantities are used. Define

$$\Delta\omega_i = \omega_i - \omega_{i-1}$$
$$\bar{\omega}_i = \frac{\omega_i + \omega_{i-1}}{2}, \quad i = 1, 2, \ldots, N,$$

where the overbar denotes average. For averaged quantities,

$$\bar{\eta}(x,t) = \sum_{i=1}^{N} \sin(\bar{k}_i x - \bar{\omega}_i t + \varepsilon_i)\sqrt{S_\eta(\bar{\omega}_i)\Delta\omega_i},$$

where $\bar{\omega}_i^2 = \bar{k}_i g$ for deep water. Let the cumulative area under the density function be given by

$$\sum_{i=1}^{j \leq N} S_\eta(\bar{\omega}_i)\Delta\omega_i \equiv S(\omega_j). \qquad (9.49)$$

Then

$$S_\eta(\bar{\omega}_i)\Delta\omega_i \approx S(\omega_i) - S(\omega_{i-1}),$$

which is equal to a constant, say a^2, as stipulated by the discretization procedure. The expression for a^2 is derived next.

Using the above, the average wave height is given by

$$\bar{\eta}(x,t) = \sum_{i=1}^{N} \sin(\bar{k}_i x - \bar{\omega}_i t + \varepsilon_i)\sqrt{S(\omega_i) - S(\omega_{i-1})}$$
$$= a \sum_{i=1}^{N} \sin(\bar{k}_i x - \bar{\omega}_i t + \varepsilon_i).$$

The total area of all the discretized regions is

$$Na^2 = S(\omega_N) \approx S(\infty) = \int_0^\infty S_\eta(\omega)d\omega. \qquad (9.50)$$

In words, this equation signifies that the total area is equal to the number of discretized bands N times the area of each band a^2. This total area equals the cumulative area $S(\omega_N)$ according to Equation 9.49 for $j = N$. Since most of the energy is at or below ω_N, this area is approximately equal to $S(\infty)$, which for the ocean wave equals the total area under the Pierson–Moskowitz spectrum $S_\eta(\omega)$.

Substitute the expression for the Pierson–Moskowitz into the last expression in Equation 9.50 and integrate to find

$$\int_0^\infty \frac{A_0}{\omega^5} e^{-B/\omega^4} d\omega = \frac{A_0}{4B} e^{-B/\omega^4} \bigg|_0^\infty = \frac{A_0}{4B}.$$

Then,
$$Na^2 = \frac{A_0}{4B},$$
and
$$a^2 = \frac{A_0}{4BN}.$$

Finally, we determine the positions of the partition frequencies as follows. For ω_N,
$$S(\omega_N) = \frac{A_0}{4B} e^{-B/\omega_N^4}.$$

Because of equipartition,
$$S(\omega_i) = \frac{i}{N} S(\omega_N)$$
$$\frac{A_0}{4B} e^{-B/\omega_i^4} = \frac{i}{N} \frac{A_0}{4B} e^{-B/\omega_N^4}.$$

Solving this last equation for ω_i, we find
$$\omega_i = \left(\frac{B}{\ln(N/i) + (B/\omega_N^4)} \right)^{1/4}.$$

This leads us to the needed time history,
$$\bar{\eta}(x,t) = \sqrt{\frac{A_0}{4BN}} \sum_{i=1}^{N} \sin(\bar{k}_i x - \bar{\omega}_i t + \varepsilon_i),$$

where $\bar{\omega}_i = (\omega_i + \omega_{i-1})/2$. Figure 9.21 shows the time history of a sample wave derived using the above procedure. This wave profile can now be used to generate wave force time histories via the Morison equation.

9.12 Concepts Summary

Special classes of problems introduced have been structures with periodicity and the inverse vibration problem, in both cases with and without uncertainty. Briefly mentioned are the important but more advanced problems of dynamic stability and the vibration of structures with internal fluid.

The advanced topic of random vibration of multi degree of freedom structures is introduced and studied via a modal analysis approach for a two degree of freedom structure.

Finally, as an introduction to approximate and computational techniques, Rayleigh's quotient is derived and applied for estimating the natural frequencies of a structure. The Monte Carlo method is introduced as

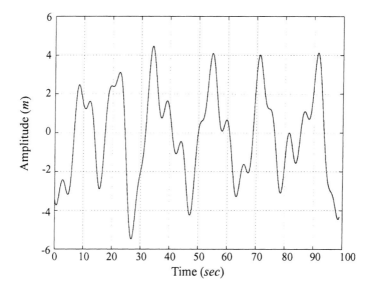

Figure 9.21: Sample wave profile derived by Borgman's method applied to the Pierson-Moskowitz wave height spectrum for wind velocity $V = 25$ m/sec.

9.13. PROBLEMS

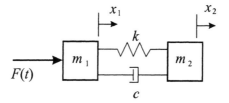

Figure 9.22: Robotic arm.

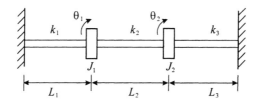

Figure 9.23: Two disk shaft systems.

a numerical probabilistic tool of importance. A useful tool by Borgman to generate time histories from power spectra is derived and detailed for the ocean wave height, needed in ocean engineering.

9.13 Problems

Problems for Section 9.3 — Unrestrained Systems

1. Figure 9.22 may be used as a simple model of a robotic arm undergoing forced motion. The forcing is at the base of the arm.

   Derive the matrix equation of motion, and solve in general for the response if (i) $F(t) = u(t)$ N, and (ii) $F(t) = \cos 2t$ N. Solve this problem in two ways: (a) using a constraint matrix, and (b) directly without a constraint matrix.

2. The system shown in Figure 9.23 undergoes rotational motion. Derive the equations of motion and solve for the response in terms of the initial conditions. Let the initial conditions be given by $\theta_1(0)$, $\dot{\theta}_1(0)$, $\theta_2(0)$, and $\dot{\theta}_2(0)$. Solve this problem in two ways: (i) using a constraint matrix, and (ii) directly without a constraint matrix.

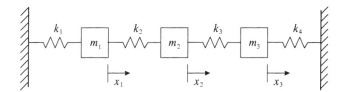

Figure 9.24: Three degree of freedom periodic system.

3. Two identical disks are connected by an elastic shaft of stiffness K. The moment of inertia of each disk with respect to the axis of the shaft is J. The system is at rest when a constant torque M is instantaneously applied to one of the disks. Determine the motion of the system assuming that the mass of the disks is much greater than that of the shaft. Solve this problem in two ways: *(i)* using a constraint matrix, and *(ii)* directly without a constraint matrix.

4. Solve *Example* 9.1, where $k_1 = 3$ Nm/rad, $k_2 = 10$ Nm/rad, and $I_1 = 5$ kg-m^2/rad, $I_2 = 2I_1$, $I_3 = I_2/2$.

5. For *Example* 9.2, show all the steps in the derivation of the three modes of vibration.

Problems for Section 9.5 – Periodic Structures

6. *(i)* For the three degree of freedom system shown in Figure 9.24, derive the equations of motion and solve for the frequencies and modes of vibration for the arbitrary stiffnesses k_1, k_2, k_3, k_4 and masses m_1, m_2, m_3. Then simplify these results for the case where all stiffnesses equal $k = 1$ N/m and all masses equal $m = 1$ kg. In this instance, plot the modes and responses for $x_1(0) = 0.5$ cm and all other initial conditions equal zero.

   (ii) For the previous system suppose that $k_1 = k_3 = k_4 = k = 1$ N/m, $m_1 = m_2 = m_3 = m = 1$ kg, and $k_2 = (1+\epsilon)k = (1+0.05)k$, where ϵ signifies a stiffness imperfection. Solve for the frequencies and modes of vibration, and plot the modes and responses for $x_1(0) = 0.5$ cm and all other initial conditions equal zero.

   Compare the results with and without imperfections. How do these results compare with those discussed in the section. Discuss.

7. For the previous problem, in part *(ii)*, solve assuming $k_1 = k_2 = k_3 = k_4 = k$ and $m_1 = m_3 = m$ and $m_2 = m(1+\epsilon)$. Use the same parameter values. Discuss.

9.13. PROBLEMS

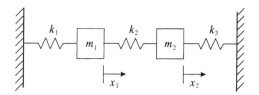

Figure 9.25: Two degree of freedom system with spring constraint.

Problems for Section 9.6 – Inverse Vibration: Estimation of Mass and Stiffness

8. Derive Equations 9.23–9.26.

9. Derive Equation 9.28.

10. Suppose that for the two degree of freedom system of Figure 9.10 we obtain the data: $f_1 = 4$ Hz, $f_2 = 10$ Hz, $f_3 = 8$ Hz, and $m_1 + m_2 = 20$ kg. Solve for the system parameters k_1, k_2, m_1, m_2.

11. In the inverse problem with uncertainty, suppose that instead of a fixed constraint, we take an elastic constraint in order to generate an additional equation, as shown in Figure 9.25. The benefit of using a spring as a constraint, rather than just fixing mass m_2 as in Figure 9.11, is that k_3 can be varied to represent a free end using $k_3 = 0$, or a fixed end using $k_3 = \infty$.

    Derive the ratios needed to estimate the mean values and variances of the structural parameters. We will need the ratios R_i as well as the expressions for the stiffnesses and masses in terms of these ratios. Assume that each mass is exactly 20 kg.

12. In *Example* 9.4, formulate expressions for the mean value and variance of R_3.

Problems for Section 9.9 – Stochastic Response of a Linear Multi–Degree of Freedom System

13. Reduce the methodology of this section to the specific case of a two degree of freedom system.

14. Follow the methodology of this section to study a forced vibrating system with the following property matrices,

$$[m] = \begin{bmatrix} 1 & 0 \\ 0 & 1 \end{bmatrix}, \quad [k] = \begin{bmatrix} 2 & -1 \\ -1 & 1 \end{bmatrix},$$

where each component force vector $\{F_1(t)\ F_2(t)\}^T$ is white noise. Therefore, transform the coupled equations of motion into modal coordinates, and proceed with an analysis of this two degree of freedom system. Discuss any assumptions needed as the steps of the analysis are made.

Problems for Section 9.10 – Rayleigh's Quotient

15. For a two disk torsional system (see Figure 9.23), follow the procedure of *Example* 9.6 to evaluate the two frequencies of oscillation approximately. Explain the procedure used to guess the respective mode shapes. Check that the approximate fundamental frequency is above the exact frequency, and that the second frequency is below the exact.

16. How would Rayleigh's quotient need to be altered for an unrestrained system? Apply the method to a two degree of freedom undamped system in unrestrained rectilinear motion.

9.14 Mini–Projects

The following papers from the recent published literature use multi–degree of freedom models to study problems of interest to the vibration community. They may be used as bases for mini–projects for interested students that would appreciate the opportunity to explore more advanced applications of the material they learned in this chapter.

1. *Transient Nozzle Excitation of Mistuned Bladed Discs*, R. Rzadkowski, *J. Sound and Vibration* (1996) **190**(4), 629–643.

    A numerical technique to compute the transient response of a mistuned bladed disc (non–identical blades) is presented on the basis of an extended beam theory including bending–torsional vibration and moderately thick plate theory. By using the extended Hamilton's principle and the Ritz method, the equations of motion are obtained. The best and worst distribution of detuned blades from the stress point of view is shown.

2. *Asymmetric Effects on Mode Localization of Symmetric Structures*, H.–J. Liu, L.–C. Zhao, J.–Y. Chen, *J. Sound and Vibration* (1996) **194**(4), 645–651.

    In structural engineering analysis, it is common to assume that the structure can be idealized as regular, even though in reality there are always imperfections. However, mistuning may induce localization in systems such as chains of interconnecting pendula, mistuned blade/disc systems, irregular space structures and multi–span beams. The present work is devoted to

9.14. MINI–PROJECTS

mode localization of a symmetric structure with components such that the symmetry condition is violated.

3. *Nonlinear Response of Two Disordered Pendula*, A.A. Tjavaras, M.S. Triantafyllou, *J. Sound and Vibration* (1996) **190**(1), 65–76.

   The effect of nonlinearities on the response of a localized system is investigated by studying the forced response of two coupled disordered pendula.

4. *Forced Response of Uniform n–Mass Oscillators and an Interesting Series*, M. Gurgoze, A. Ozer, *J. Sound and Vibration* (1994) **173**(2), 283–288.

   The vibration of a linear discrete mechanical system of n degrees of freedom are governed in physical space by a matrix differential equation of nth order. This means, in general, the solution of an eigenvalue problem of dimension n. For $n \geq 3$, the eigenvalue problems can generally be solved only numerically. Only in special cases is it possible to determine the eigencharacteristics analytically. A uniform oscillator with n equal masses m and n equal springs is an example of such a system. These can be thought of as a simplified discretized model for the longitudinal vibration of a clamped–free beam.

5. *Control of Maglev Suspension Systems*, Y. Cai, S.S. Chen, *J. of Vibration and Control*, **2**: 349–368, 1996.

   This study investigates alternative designs for control of maglev vehicle suspension systems. Active and semiactive control law designs are introduced into primary and secondary suspensions of maglev vehicles.

6. *Modelling and Simulation of the Golf Stroke*, A.R. Whittaker, *Computer Modeling and Simulation in Engineering*, **1**: 31–45, 1996.

   The aim of this article is to create a realistic model of a golf club that can be used to simulate the downswing and the impact between club and ball, thus predicting the resulting motion of the ball.

7. *Critical Cross Power Spectral Density Functions and the Highest Response of Multi–Supported Structures Subjected to Multi–Component Earthquake Excitations*, A. Sarkar, C.S. Manohar, *Earthquake Engineering and Structural Dynamics*, Vol. 25, 303–315 (1996).

   The highest response of multi–supported structures subjected to partially specified multi–component earthquake support motions is considered. Conditions for maximum response are analyzed and applied to a three–span suspension cable bridge and to a piping structure of a nuclear power plant.

8. *Adaptive Passive Vibration Control*, M.A. Franchek, M.W. Ryan, R.J. Bernhard, *J. Sound and Vibration* (1995) **189**(5), 565–585.

There are four classifications of vibration control. The three most common classifications are *passive*, *active*, and *hybrid* control. Passive control involves the use of reactive or resistive devices that either load the transmission path of the disturbing vibration or absorb vibrational energy. Active control also loads the transmission path but achieves this loading through the use of force actuators requiring external energy. Hybrid control integrates a passive approach with an active control structure. The hybrid approach is intended to reduce the amount of external power necessary to achieve control. Here, the fourth classification of adaptive–passive control is discussed, which combines a tunable passive device with a tuning strategy such that optimal performance is guaranteed.

9. *Simple Active Tuned Mass Damper Control Methodology for Tall Buildings Subject to Wind Loads*, S. Ankireddi, H.T.Y. Yang, *J. Structural Engineering*, Vol. 122, No. 1, Jan. 1996, 83–91.

   A procedure for the design of an active tuned mass damper for vibration control in tall buildings subject to wind loads is presented. The building motions are modeled by the first mode of the response, and it is assumed that the excitation is white noise.

10. *Role of Control–Structure Interaction in Protective System Design*, S.J. Dyke, B.F. Spencer, Jr., P. Quast, M.K. Sain, *J. Engineering Mechanics*, Vol. 121, No. 2, Feb. 1995, 322–338.

    The importance of including control–structure interaction when modeling a control system is discussed, and a general framework within which one can study its effect on protective systems is presented.

11. *Partial State Feedback Control with Application to Aircraft Wing Vibration Damping*, Y.D. Song, *J. Sound and Vibration*, (1995) **188**(3), 455–460.

    The aircraft wing vibration control problem is investigated, with emphasis on suppression of bending and torsion deformation. Such deformations may become significant when a flight vehicle travels at a high speed and/or undergoes an external disturbance.

12. *Parametric Study of Active Mass Dampers for Wind–Excited Tall Buildings*, Y.L. Xu, *Engineering Structures*, Vol. 18, No. 1, 64–76, 1996.

    A method for selecting design parameters of active mass dampers and estimating motion reduction of wind–excited tall buildings is proposed, based on aeroelastic model tests of uncontrolled tall buildings.

13. *Semi–Active Suspensions with Adaptive Capability*, I. Youn, A. Hac, *J. Sound and Vibration*, (1995) **180**(3), 475–495.

This paper deals with the synthesis of control laws for a two degree of freedom model of a vehicle with a semi-active suspension. The suspension system consists of a continuously variable damper and a spring that can vary the suspension stiffness among three different levels.

14. *Nonlinear Friction-Induced Vibration in Water-Lubricated Bearings*, T.A. Simpson, R.A. Ibrahim, *J. Vibration and Control*, **2**: 87–113, 1996.

    The problem of friction-induced vibration and squeal in water-lubricated shipboard bearings has received extensive studies. Here, the nonlinear aspects are studied using a nonlinear two degree of freedom model.

15. *Zones of Chaotic Behavior in the Parametrically Excited Pendulum*, S.R. Bishop, M.J. Clifford, *J. Sound and Vibration* (1996) **189**(1), 142–147.

    A driven pendulum is considered which can freely move in a plane and where the pivot point is subjected to a vertical periodic driving force. The so-called parametric pendulum can display a variety of dynamical behavior including stable equilibria, periodic oscillations, continuous rotations, subharmonic solutions and chaotic motions. These are studied here.

16. *Controlled Semiactive Hydraulic Vibration Absorber for Bridges*, W.N. Patten, R.L. Sack, Q. He, *J. of Structural Engineering*, Vol. 122, No. 2, February 1996, 187–192.

    Heavy truck traffic on highway bridges is known to produce impact loads that cause large vibrations. This phenomenon can reduce the expected service life of highway bridges. A means of mitigating those deleterious vibrations using an automatic control system that requires no pumps or line power is proposed.

17. *Effects of Severed Tethers in Space*, F. Angrilli, R. Da Forno, F. Reccanello, A. Zago, and G. Fanti, *J. Spacecraft and Rockets*, Vol. 34, No. 2, March–April 1997, 239–245.

    Experimental results and mathematical analyses are compared for determining the velocity of a severed tether and, consequently, its effects on the orbiter.

18. *The Motion of a Damped n Degree of Freedom System With a Single Natural "Frequency,"* E.V. Wilms, *J. Sound and Vibration* (1997) **204**(4), 679–680.

19. *Modal Sensitivity Analysis of Coupled Acoustic-Structural Systems*, J. Luo, H.C. Gea, *J. Vibration and Acoustics*, Vol. 119, October 1997, 545–550.

    Modal analysis of coupled acoustic-structural systems leads to an unsymmetric eigenproblem of special form which introduces different left and right eigenvectors. In this paper, the relation between left and right eigenvectors is established by symmetrization of the unsymmetric eigensystem.

20. *On the Zeros of Structural Frequency Response Functions and Their Sensitivities*, J.E. Mottershead, *Mechanical Systems and Signal Processing* (1998) **12**(5), 591–597.

    This paper is concerned with the information introduced by measured zeros from frequency response functions and its application to model assessment and updating.

21. *Mode Localization in Simply Supported Two–Span beams of Arbitrary Span Lengths*, D–O. Kim, I–W. Lee, *J. Sound and Vibration* (1998) **213**(5), 952–961.

    The objective of this study is to show the possibility of drastic occurrences of mode localization in non–periodic structures. Free vibration analysis of simply supported two–span beams of arbitrary span lengths is theoretically investigated.

22. *Twenty Years of Structural Dynamic Modification – A Review*, P. Avitabile, *Sound and Vibration*, January 2003.

    Structural dynamic modification techniques have been available as a design tool for several decades. This article reviews the methodology of the technique including both proportional and complex mode formulations. Issues pertaining to limitations on their use and truncation effects of unmeasured modes are discussed.

23. *Vortex–Induced Vibration of a Cylinder with Two Degrees of Freedom*, N. Jauvtis, C.H.K. Williamson, *J. Fluids and Structures*, 17 (2003) 1035–1042.

    In this work is studied the response of an elastically mounter cylinder, which is free to move in two degrees of freedom in a fluid flow, and which has low mass and damping.

Chapter 10

Continuous Models for Vibration

"Where reality is approached"

Continuous models of vibrating systems are viewed as more realistic because the structural properties are distributed rather than being concentrated at discrete points. The price to pay for increased realism is increased complexity. The governing equations of motion go from being ordinary differential equations for the discrete models, to partial differential equations for the continuous ones, since displacement is a function of both time and position. For elementary applications, deriving and solving the partial differential equations does not pose a great challenge. But, when nonuniform structures with varying cross–sections and material properties need to be modeled, approximate techniques of the type we began to study in the last chapter are needed. In this chapter, the focus is the study of strings, beams, membranes and plates. These are components in most structures and machines, and understanding how they behave is of great use to modeling more complex systems. Direct and modal solutions are primarily explored.

10.1 Continuous Limit of a Discrete Formulation

One can derive the continuous governing equations of motion by starting with a discrete version of the system, and then taking appropriate limits. In this way, the continuous model is understood to be a further modeling refinement. As an example of this limiting procedure, examine the derivation

CHAPTER 10. CONTINUOUS MODELS FOR VIBRATION

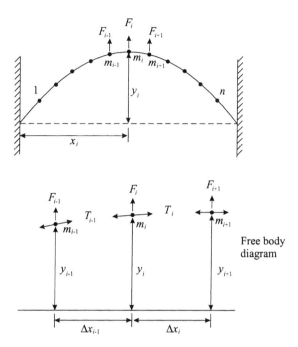

Figure 10.1: Massless string with point masses.

of the equation of motion of a string. Section 10.2 examines the vibrating string in more detail.

Start with a massless string to which point masses are fixed at equidistant locations, as shown in Figure 10.1. The point mass m_i has force F_i and tensions T_{i-1} and T_i acting on it. These are functions of time. Applying Newton's second law of motion in the vertical y direction results in the difference equation of motion,

$$T_i \frac{y_{i+1} - y_i}{\Delta x_i} - T_{i-1} \frac{y_i - y_{i-1}}{\Delta x_{i-1}} + F_i = m_i \frac{d^2 y_i}{dt^2},$$

where the assumption of small slopes leads to the approximation $\sin\theta \approx \tan\theta$. Define $\Delta y_i = y_{i+1} - y_i$ and $\Delta y_{i-1} = y_i - y_{i-1}$, with the result that Equation ?? becomes

$$T_i \frac{\Delta y_i}{\Delta x_i} - T_{i-1} \frac{\Delta y_{i-1}}{\Delta x_{i-1}} + F_i = m_i \frac{d^2 y_i}{dt^2}, \quad i = 1, 2, \ldots, n,$$

for all the point masses. The first two terms on the left hand side represent a change in the vertical force on point mass m_i. This realization may be

10.2. VIBRATION OF STRINGS

reflected in the equation as

$$\Delta\left(T_i \frac{\Delta y_i}{\Delta x_i}\right) + F_i = m_i \frac{d^2 y_i}{dt^2}, \qquad i = 1, 2, \ldots, n.$$

Divide both sides by Δx_i,

$$\frac{\Delta}{\Delta x_i}\left(T_i \frac{\Delta y_i}{\Delta x_i}\right) + \frac{F_i}{\Delta x_i} = \frac{m_i}{\Delta x_i} \frac{d^2 y_i}{dt^2}, \qquad i = 1, 2, \ldots, n.$$

Let the number n of point masses m_i increase, $n \to \infty$, while the space between them decreases to zero, and let their masses decrease such that the average mass per unit length remains constant. Position x_i becomes x, and in the limit as $\Delta x_i \to 0$, the difference equation becomes

$$\frac{\partial}{\partial x}\left[T(x)\frac{\partial y(x,t)}{\partial x}\right] + f(x,t) = m(x)\frac{\partial^2 y(x,t)}{\partial t^2}, \qquad (10.1)$$

which governs over the length of the string, $0 \leq x \leq L$, where $f(x,t)$ is the limiting value of the force per unit length and $m(x)$ is the limiting value of the mass per unit length. This discussion has served to demonstrate that discrete and continuous systems are really different idealizations of the same structure. Therefore, techniques developed for discrete systems can be applied to continuous ones, adjusting for the obvious differences. Taking the limit $n \to \infty$ implicitly expresses the fact that the continuous system is *infinite dimensional*. It has an infinite number of degrees of freedom.

Taking the sum of the forces in the horizontal direction and setting this to zero, since there are no external forces in this direction, results in

$$T_{i+1} \cos \theta_{i+1} - T_i \cos \theta_i = 0.$$

Using the small angle assumption, that $\cos \theta_{i+1} \approx \cos \theta_i \approx 1$, forms the basis for the approximation $T_{i+1} \approx T_i$. The tension is constant due to these assumptions, which are a result of small displacements.

Equation 10.1 will be derived again using the free body diagram of a *continuous* element in the next section.

10.2 Vibration of Strings

A string is a valuable model for understanding the dynamic behavior of continuous systems. Strings have been used for simplified models of telephone wires, conveyor belts, even models of human DNA. When a string is stretched from its equilibrium position, the tension that is created internally

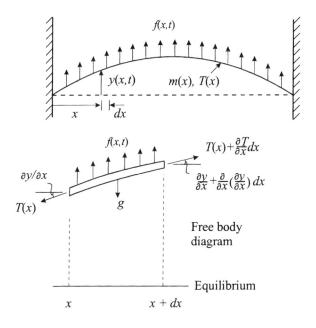

Figure 10.2: String and free body.

acts as the restoring force to bring the string back to its original undeformed position. The transversely vibrating *beam* derives its restoring force from bending stiffness as shown in Section 10.5.1. The string cannot transmit a bending moment but can resist axial tension.

It is straightforward to derive the governing equation of motion using the free body diagram for a section of the displaced string, as per Figure 10.2. The variables introduced are $T(x)$, the tension in the string, $f(x,t)$, the applied transverse force per unit length, and $m(x)$, the mass per unit length. All are functions of position x. Using Newton's second law of motion, and assuming small displacements such that $\sin\theta \approx \theta$, the sum of the forces in the transverse direction y are set equal to the mass times acceleration of the string element in the y direction,

$$\left(T(x) + \frac{\partial T(x)}{\partial x}dx\right)\left(\frac{\partial y}{\partial x} + \frac{\partial^2 y}{\partial x^2}dx\right) + f(x,t)dx - T(x)\frac{\partial y}{\partial x} = m(x)dx\frac{\partial^2 y}{\partial t^2}.$$

Expand the products on the left hand side and ignore second order terms $(dx)^2$ on the assumption that for linear vibration these terms are not significant.[1] Divide by dx, with the result being the governing equation of linear

---

[1] Any and all assumptions must be subsequently verified numerically, experimentally,

10.2. VIBRATION OF STRINGS

motion,

$$\frac{\partial}{\partial x}\left[T(x)\frac{\partial y}{\partial x}\right] + f(x,t) = m(x)\frac{\partial^2 y}{\partial t^2}. \tag{10.2}$$

The free vibration problem that is used to solve for system frequencies and modes is obtained by setting $f(x,t) = 0$,

$$\frac{\partial}{\partial x}\left[T(x)\frac{\partial y}{\partial x}\right] = m(x)\frac{\partial^2 y}{\partial t^2}. \tag{10.3}$$

Note that for the case of constant tension, $T(x) = T$, and the result is the *wave equation*

$$\frac{\partial^2 y}{\partial x^2} = \frac{1}{c^2}\frac{\partial^2 y}{\partial t^2},$$

where $c = \sqrt{T/m}$ is the velocity of wave propagation with units of length/time. The solution to this problem is found using a wave propagation solution. Then, the wave equation is derived using Hamilton's principle.

There are two general approaches to solving continuous system equations, the modal or *standing wave* approach and the *wave propagation* approach. The standard solution technique for linear systems of finite length is based on a modal approach. It is instructive, however, to first consider the wave propagation solution, since the two methods are complementary, provide us with two perspectives, and of course, lead to the same solution. The modal solution is developed afterwards in Section 10.2.4.

10.2.1 Wave Propagation Solution

Wave propagation solutions are particularly useful where there is an *infinite* or *semi–infinite* dimension in a structure. An example is the modeling of ground vibration for earthquake engineering applications. However, finite length structures can also be studied in this way.

Assume that the string undergoing small transverse vibration has constant tension T, constant mass per unit length m, and is governed by the classical wave equation, rewritten here using a more compact notation,[2]

$$y_{,xx} = \frac{1}{c^2}y_{,tt}, \tag{10.4}$$

---

or by comparison to more exact analytical solutions.

[2] The shorthand notation is, for example

$$y_{,tt} \equiv \frac{\partial^2 y}{\partial t^2}, \quad y_{,xt} \equiv \frac{\partial^2 y}{\partial x \partial t}.$$

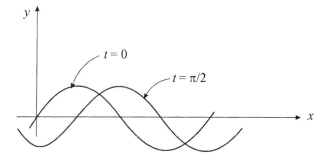

Figure 10.3: Example of a right-going wave.

where $c = \sqrt{T/m}$. The general solution begins with the assumption that response $y(x,t)$ equals the sum of two different waves traveling in opposite directions,

$$y(x,t) = F_1(x - ct) + F_2(x + ct),$$

where F_1 and F_2 are two arbitrary functions that can be differentiated twice with respect to x and t. Take the following derivatives,

$$\begin{aligned} y_{,tt} &= c^2 F_{1,tt} + c^2 F_{2,tt} \\ y_{,xx} &= F_{1,xx} + F_{2,xx} \end{aligned}$$

and substitute them into governing Equation 10.4,

$$F_{1,xx} + F_{2,xx} = \frac{1}{c^2}(c^2 F_{1,tt} + c^2 F_{2,tt}), \qquad (10.5)$$

to find that it is satisfied for *any* two different functions of $x \mp ct$. To demonstrate this, assume the harmonic solution

$$y(x,t) = A \sin \frac{2\pi}{\lambda}(x - ct) + B \cos \frac{2\pi}{\lambda}(x + ct). \qquad (10.6)$$

Differentiating this harmonic solution with respect to x and t as necessary, we see that the wave equation and thus Equation 10.5 are satisfied.

To better understand each term in solution Equation 10.6, consider the position of the harmonic function $y(x,t) = \sin(x - t)$ at two instances of time, $t = 0$ sec and $t = \pi/2$ sec, as shown in Figure 10.3. Reading the value of y in each case for several values of x will show that the wave has traveled *to the right* a distance of $\pi/2$ rad in this period of time. Similarly, one can show that $\sin(x + t)$ will travel to the left.

10.2. VIBRATION OF STRINGS

Consider the right–going wave, $y(x,t) = A\sin\frac{2\pi}{\lambda}(x-ct)$. The *period* P can be defined for any location x as the amount of time for the value of y to repeat. It can be related to the frequency ω rad/sec by $P = 2\pi/\omega$. For the right–going wave,

$$\omega = \frac{2\pi c}{\lambda}$$

and therefore $P = \lambda/c$, or the *wavelength* $\lambda = Pc$, which has length units. The number of waves per unit length, known as the *wavenumber* κ, equals the inverse of the wavelength, $\kappa = 1/\lambda$ with units of length^{-1}. With these new parameters, the right–going wave can be written as

$$y(x,t) = A\sin(2\pi\kappa x - \omega t). \tag{10.7}$$

Suppose that two waves of equal amplitude A and frequency ω travel in opposite directions. Then the wave propagation solution is the sum of the two waves,

$$\begin{aligned} y(x,t) &= A\sin(2\pi\kappa x - \omega t) + A\sin(2\pi\kappa x + \omega t) \\ &= 2A\sin 2\pi\kappa x \cos\omega t. \end{aligned} \tag{10.8}$$

This equation is no longer a traveling wave, but rather a *standing wave*, that is, one with a profile that is *oscillating* rather than propagating.

For this standing wave, there are *nodes* where $y = 0$, at $\sin 2\pi\kappa x = 0$, and *anti–nodes*, or maximum amplitudes where $\sin 2\pi\kappa x = \pm 1$. At this stage in the analysis, the above holds for *any* ω. But once the boundary conditions are specified, for example the string is fixed at both ends, only certain frequencies will satisfy the above conditions for nodes and maxima. We will learn much more about satisfying the boundary conditions for standing waves later in this chapter. But next we examine how traveling waves in bounded media can satisfy these conditions as well.

For the standing wave Equation 10.8, assume that the displacements are fixed for all time at the locations $x = 0$ and $x = L$,

$$\begin{aligned} y(0,t) &= 0 \\ y(L,t) &= 0 = 2A\sin 2\pi\kappa L \cos\omega t. \end{aligned}$$

The first equation is satisfied identically and, therefore, provides us with no new information. The second equation can only be valid for all t if $\sin 2\pi\kappa L = 0$. Then, $2\pi\kappa L = r\pi$ for integer r, providing the necessary condition

$$\kappa = \frac{r}{2L}. \tag{10.9}$$

But

$$\kappa = \frac{1}{\lambda} = \frac{\omega}{2\pi c}. \tag{10.10}$$

CHAPTER 10. CONTINUOUS MODELS FOR VIBRATION

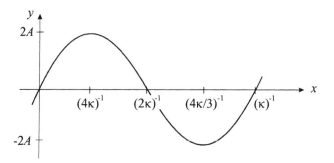

Figure 10.4: Superposition of traveling waves.

See Figure 10.4 for a sketch of this standing wave. Equate Equations 10.9 to 10.10 and since r is an integer, define $\omega = \omega_r$,

$$\omega_r = \frac{r\pi c}{L} = \frac{r\pi}{L}\sqrt{\frac{T}{m}}, \quad r = 1, 2, \ldots, \tag{10.11}$$

where ω_r are the *allowable vibration frequencies* for the string, for the given boundary conditions. We will revisit these results using the standing wave approach in the next section. The purpose of showing both approaches is to demonstrate that problems can be cast in different forms and different approaches can sometimes provide useful alternate perspectives.

If the two amplitudes leading to Equation 10.8 are unequal, there exists, in addition to the standing wave, a wave traveling in the direction of the stronger partial wave, with an amplitude equal to the difference of the amplitudes of the components, as the following example shows.

Example 10.1 Addition of Two Waves with Different Amplitudes
Two waves of different amplitudes traveling in opposite directions are defined by

$$\begin{aligned} y_1(x,t) &= A_1 \exp[i(2\pi\kappa x - \omega t)] \\ y_2(x,t) &= A_2 \exp[i(2\pi\kappa x + \omega t)]. \end{aligned}$$

If $A_2 = A_1 + \varepsilon$ represents the relation between the two amplitudes, then the sum of the two waves is

$$\begin{aligned} y_1(x,t) + y_2(x,t) &= A_1 e^{i(2\pi\kappa x - \omega t)} + A_1 e^{i(2\pi\kappa x + \omega t)} + \varepsilon e^{i(2\pi\kappa x + \omega t)} \\ &= 2A_1 \cos 2\pi\kappa x \cos \omega t + \varepsilon \cos(2\pi\kappa x + \omega t), \end{aligned}$$

10.2. VIBRATION OF STRINGS

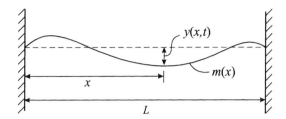

Figure 10.5: Schematic of string for Hamilton's derivation.

where $\varepsilon = A_2 - A_1$ and the real part of the complex exponential is retained. The extra ε term denotes an extra propagating wave in the direction y_2 with amplitude ε. This extra term may be written as

$$\varepsilon \cos(2\pi\kappa x + \omega t) = \varepsilon \frac{y_2}{A_2}$$
$$= y_2 \left(1 - \frac{A_1}{A_2}\right),$$

where the real part of y_2 is implied. The extra term can be seen to be some fraction of propagating wave y_2. ∎

Therefore, the vibration of continuous structures can be viewed as a study of standing waves or modes. The modal approach to the vibrating string begins in Section 10.2.4.

10.2.2 The Wave Equation via Hamilton's Principle

Continuing our practice of examining different formulations of governing equations, the wave equation is derived for a string utilizing Hamilton's principle. Figure 10.5 depicts the problem and defines the necessary variables. To avoid confusion, let $T(x)$ represent the string internal tension and $\tau(t)$ represent the kinetic energy required for Hamilton's principle.

As before, the string mass per unit length is denoted by $m(x)$ and the transverse displacement by $y(x,t)$. The kinetic energy for an element of string of length dx is

$$\frac{1}{2}m(x)dx \left[\frac{\partial y(x,t)}{\partial t}\right]^2,$$

and for the string of length L, the kinetic energy is

$$\tau(t) = \frac{1}{2}\int_0^L m(x) \left[\frac{\partial y(x,t)}{\partial t}\right]^2 dx.$$

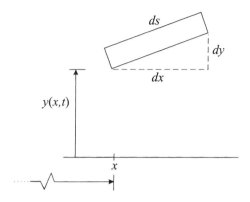

Figure 10.6: Schematic for string elongation.

The potential energy $V(t)$ is proportional to the increase in string length ds when compared to the string at rest. The change in potential energy equals the work done by the internal tension $T(x)$ in elongating the string an amount Δ, that is, $V(t) = \int_0^L T(x)d\Delta$, where Δ will be shown to be a function of x. See Figure 10.6.

For the element of length dx, the increase in length[3] is $d\Delta = ds - dx$,

$$\begin{aligned} ds - dx &= \left\{1 + \left[\frac{\partial y(x,t)}{\partial x}\right]^2\right\}^{1/2} dx - dx \\ &\approx \frac{1}{2}\left[\frac{\partial y(x,t)}{\partial x}\right]^2 dx, \end{aligned}$$

where the last approximation is valid for *small slopes*, using the *binomial series* for the square root. Therefore, the potential energy for the entire string is

$$V(t) = \frac{1}{2}\int_0^L T(x)\left[\frac{\partial y(x,t)}{\partial x}\right]^2 dx.$$

Substituting these expressions into Hamilton's principle must render the

---

[3] We simply use the right triangle with hypotenuse ds, along with the two sides that represent the difference in coordinates between the two ends of the segment, as shown in Figure 10.6 where $ds^2 = dx^2 + dy^2$, or $ds = \sqrt{dx^2 + dy^2} = dx\sqrt{1 + (dy/dx)^2}$.

10.2. VIBRATION OF STRINGS

integral below stationary, that is, $\delta I = 0$, where

$$\begin{aligned} I &= \int_{t_1}^{t_2} [T(t) - V(t)] dt \\ &= \frac{1}{2} \int_{t_1}^{t_2} \int_0^L \left[m(x) \left(\frac{\partial y}{\partial t} \right)^2 - T(x) \left(\frac{\partial y}{\partial x} \right)^2 \right] dx\, dt, \quad (10.12) \end{aligned}$$

and the variation is on displacement y subject to the conditions that at $t = t_1$ and $t = t_2$,

$$\delta y(x, t) = 0. \quad (10.13)$$

To proceed further, we do the following:

- interchange the order of integration with respect to t and x;
- interchange δ and ∂;
- integrate by parts;
- when integrating the time integrals by parts, recall that variations vanish at the end–times;

where operators δ, $\partial/\partial t$, and $\partial/\partial x$ are commutative, and $\int dt$ and $\int dx$ interchange due to the uniform convergence of the limiting processes that define the integrals.

Consider separately the variation of each term on the right hand side above, noting that the first term is a time derivative and the second is a space derivative. Integrate the variation of the first term by parts with respect to t,

$$\begin{aligned} \delta \int_{t_1}^{t_2} \frac{1}{2} m(x) \left(\frac{\partial y}{\partial t} \right)^2 dt &= \int_{t_1}^{t_2} m(x) \frac{\partial y}{\partial t} \delta \left(\frac{\partial y}{\partial t} \right) dt \\ &= \int_{t_1}^{t_2} m(x) \frac{\partial y}{\partial t} \frac{\partial (\delta y)}{\partial t} dt \\ &= \left(m(x) \frac{\partial y}{\partial t} \right) \delta y \Big|_{t_1}^{t_2} - \int_{t_1}^{t_2} \frac{\partial}{\partial t} \left(m(x) \frac{\partial y}{\partial t} \right) \delta y\, dt \\ &= - \int_{t_1}^{t_2} m(x) \frac{\partial^2 y}{\partial t^2} \delta y\, dt, \quad (10.14) \end{aligned}$$

where use has been made of the conditions in Equation 10.13.

Looking at the second term in the integrand of Equation 10.12, but now taking the variation of the integration with respect to x, yields

$$\begin{aligned}\delta \int_0^L \frac{1}{2}T(x)\left(\frac{\partial y}{\partial x}\right)^2 dx &= \int_0^L T(x)\frac{\partial y}{\partial x}\delta\left(\frac{\partial y}{\partial x}\right) dx \\ &= \int_0^L T(x)\frac{\partial y}{\partial x}\frac{\partial(\delta y)}{\partial x} dx \\ &= \left(T(x)\frac{\partial y}{\partial x}\right)\delta y \bigg|_0^L - \int_0^L \frac{\partial}{\partial x}\left(T(x)\frac{\partial y}{\partial x}\right)\delta y\, dx.\end{aligned}$$
(10.15)

Substitute Equations 10.14 and 10.15 into Hamilton's principle,

$$\int_{t_1}^{t_2}\left[\int_0^L \left(\frac{\partial}{\partial x}\left[T(x)\frac{\partial y}{\partial x}\right] - m(x)\frac{\partial^2 y}{\partial t^2}\right)\delta y\, dx - \left(T(x)\frac{\partial y}{\partial x}\right)\delta y\bigg|_0^L\right] dt = 0.$$
(10.16)

Since δy is arbitrary except for the requirement that it vanishes at the ends, then over the domain $0 \leq x \leq L$,

$$\frac{\partial}{\partial x}\left[T(x)\frac{\partial y}{\partial x}\right] - m(x)\frac{\partial^2 y}{\partial t^2} = 0. \tag{10.17}$$

This is the wave equation previously derived using the free body diagram of forces along with Newton's second law of motion. The second term in Equation 10.16 must vanish at the ends due to the condition that factor δy equals zero at the boundaries,

$$\left(T(x)\frac{\partial y}{\partial x}\right)\delta y\bigg|_0^L = 0.$$

If an end is fixed, then $\delta y = 0$ at that end. If one end of the string is free, then $T(x)\partial y/\partial x = 0$ at that end. We next study the solution of the wave equation.

10.2.3 The Boundary Value Problem for the String

To find the natural modes of vibration of a string, make the most elementary assumption regarding the solution of a partial differential equation; its solution is *separable*,

$$y(x,t) = Y(x)F(t). \tag{10.18}$$

This solution implies that behavior in time and space are independent. This is the same approach used in Chapter 8 when the eigenvalue problem was first developed.

10.2. VIBRATION OF STRINGS

Since Equation 10.18 must satisfy the governing equation, take the appropriate derivatives,[4] $y' = Y'F$, $\ddot{y} = Y\ddot{F}$, and substitute them into governing Equation 10.3 or 10.17 to find

$$\frac{d}{dx}[T(x)Y'(x)F(t)] = m(x)Y(x)\ddot{F}$$

$$\underbrace{\frac{1}{m(x)Y(x)}\frac{d}{dx}[T(x)Y'(x)]}_{I} = \underbrace{\frac{\ddot{F}}{F}}_{II} = -\omega^2.$$

In the last equation, time dependent variables and space dependent variables have been placed on opposite sides of the equal sign. Since part I is only a function of x, and part II is only a function of t, it must be that each is equal to the same constant, say $-\omega^2$. When the equation in time, $\ddot{F} + \omega^2 F = 0$, is solved, the result is simple harmonic motion with frequency ω.

The equation in space is

$$-\frac{d}{dx}[T(x)Y'(x)] = \omega^2 m(x)Y(x), \qquad (10.19)$$

which must be satisfied in the domain $0 \leq x \leq L$. In addition, the *boundary conditions* must be specified[5] to completely define the problem. Equation 10.19 is mathematically known as a *Sturm–Liouville* eigenvalue problem.[6]

10.2.4 Modal Solution for Fixed–Fixed Boundary Conditions

In this section, we derive the natural frequencies and mode shapes for the uniform string that is fixed at both ends. The natural frequencies and mode shapes are a result of an eigenvalue/eigenvector analysis.

---

[4] Prime denotes differentiation with respect to x and overdot differentiation with respect to t.

[5] Note that the key to the solutions is satisfying the boundary conditions, and the behavior of the system is quite different for the various boundary conditions. Boundary conditions may be grouped into two basic types: *geometric boundary conditions* and *natural boundary conditions*. Geometric (or imposed) boundary conditions are deflections or rotations. Natural (or dynamic) boundary conditions are a result of moment or shear force balance.

[6] In this book, we study vibratory systems that are *self-adjoint*, a property that leads to the symmetry of the property matrices. Therefore, think of the self-adjoint property as the counterpart to the symmetry of the mass and stiffness matrices of the previous chapter. A good text on matrices or linear algebra can provide you with details.

Eigenvalues – Natural Frequencies

Consider the uniform vibrating string using a modal or standing wave approach, assuming constant tension, where the string is fixed at both ends. A string fixed at end $x = 0$ does not displace and $y(0,t) = Y(0)F(t) = 0$, or $Y(0) = 0$. A uniform string is one where the geometry and material properties do not vary along its length. The constant tension approximation is valid for a string oscillating with small amplitudes.

Equation 10.19 simplifies considerably for constant T. The eigenvalue problem becomes

$$Y'' + \beta^2 Y = 0, \quad \beta^2 = \frac{\omega^2 m}{T}, \tag{10.20}$$

with the boundary conditions $Y(0) = 0$ and $Y(L) = 0$, and constant m. Recall that the wave speed c is given by $c = \sqrt{T/m}$ and, therefore, $\beta = \omega/c$. The second order constant coefficient differential Equation 10.20 has a periodic solution in x,

$$Y(x) = C_1 \sin \beta x + C_2 \cos \beta x, \tag{10.21}$$

and satisfaction of the boundary conditions requires that the following two equations are solved,

$$Y(0) = 0 = C_2$$
$$Y(L) = 0 = C_1 \sin \beta L.$$

Since $C_1 = 0$ results in the trivial solution $Y(x) = 0$, then $\sin \beta L = 0$ and, therefore,

$$\beta_r L = r\pi, \quad r = 1, 2, \ldots.$$

Substituting for β yields the allowable frequencies of oscillation,

$$\omega_r = r\pi \sqrt{\frac{T}{mL^2}}, \quad r = 1, 2, \ldots,$$

which is the same expression as was derived for the traveling wave solution, Equation 10.11. The ratio $\sqrt{T/mL^2}$ has units of 1/sec. It is interesting to note that the string has an infinite number of discrete natural frequencies. This corresponds to the understanding of the continuous system as having an infinite number of degrees of freedom.

The frequency distribution depends on the boundary conditions and the geometric and material properties. This equation also signifies that the higher the string tension, the higher will be the set of frequencies. Further, if the string is of a higher mass per unit length or if it is longer, the set of frequencies will be lower, in proportion to $1/\sqrt{m}$ and $1/L$. We have all

10.2. VIBRATION OF STRINGS

experienced such variation in behavior, whether playing a guitar or just with a rubber band; a tighter guitar string plays a higher frequency musical note; a heavier string plays a lower note. Also, since all the frequencies are integer multiples of the fundamental frequency ($r = 1$), the vibration of the string will yield a musical tone as in the case of a violin or piano string.

Eigenfunctions – Mode Shapes

Once the natural frequencies of the system are known, the mode shapes are given by Equation 10.21,

$$Y_r(x) = C_r \sin \omega_r \sqrt{\frac{m}{T}} x = C_r \sin \frac{r \pi x}{L}.$$

The term *eigenfunction* is used for continuous systems since the mode is a continuous function. $Y_r(x)$ is accurate only to a constant value since, if $Y_r(x)$ is a solution of Equation 10.20, then so is $C_r Y_r(x)$, where C_r is any constant. To specify C_r, a normalization is carried out according to the rule

$$\int_0^L m Y_r^2(x) dx = 1, \quad r = 1, 2, \ldots.$$

Therefore,

$$C_r^2 \int_0^L m \sin^2 \frac{r \pi x}{L} dx = 1$$

$$C_r^2 = \frac{2}{mL},$$

and the modes have the specific form

$$Y_r(x) = \sqrt{\frac{2}{mL}} \sin \frac{r \pi x}{L}, \quad r = 1, 2, \ldots,$$

as shown in Figure 10.7 for an Aluminum string of 1 in^2 cross section and $m = 0.1$ lb-sec^2/in^2 and $L = 120$ in. The above normalization is customary and is useful so that different analysts can obtain the same numerical results. Note that each mode in this case has the same peak amplitude, $\sqrt{2/mL}$, since C_r is independent of r.

Complete Solution

Finally, the time history for each mode r can be written using Equation 10.18,

$$\begin{aligned} y_r(x,t) &= Y_r(x) F_r(t) \\ &= \sqrt{\frac{2}{mL}} \sin \frac{r \pi x}{L} (A_r \sin \omega_r t + B_r \cos \omega_r t), \end{aligned}$$

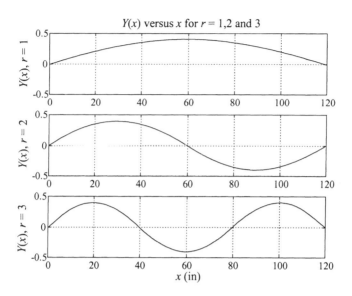

Figure 10.7: Modes of string for $m = 0.1$ lbsec^2/in^2 and $L = 120$ in.

with $\omega_r = r\pi\sqrt{T/mL^2} = r\pi c/L$. The complete solution is found by applying the *expansion theorem*, as for discrete systems, and summing all modal components,

$$y(x,t) = \sum_{r=1}^{\infty} \sqrt{\frac{2}{mL}} \sin \frac{r\pi x}{L} \left(A_r \sin \frac{r\pi c}{L} t + B_r \cos \frac{r\pi c}{L} t \right). \quad (10.22)$$

It is good practice to occasionally check that the units of derived equations make sense. Look closer at Equation 10.22. The response $y(x,t)$ is a displacement with length units. Therefore, the right hand side must also have such units. In the factor $\sqrt{2/mL}$, m has units of mass/length and L units of length. Then, $\sqrt{2/mL}$ has units of $1/\sqrt{\text{mass}}$. This means that the coefficients A_r and B_r must have units of $\sqrt{\text{mass}} \times$ length as verified in Equation 10.24. Therefore, Equation 10.22 is dimensionally correct. Doing this provides a good additional check on a mathematical derivation. Use this procedure to verify the equality of units in the following several equations that satisfy the initial conditions.

The arbitrary constants A_r and B_r are functions of the initial conditions, which have not yet been specified. Assume the initial displacement of the string be given by $y(x,0) = f(x)$ and the initial velocity by $\dot{y}(x,0) = 0$. These initial values are used with Equation 10.22, as follows. Satisfy the

10.2. VIBRATION OF STRINGS

initial velocity first, to find

$$\dot{y}(x,0) = 0 = \sum_{r=1}^{\infty} \sqrt{\frac{2}{mL}} \omega_r \sin \frac{r\pi x}{L} (A_r \cos \omega_r t - B_r \sin \omega_r t)|_{t=0}$$

$$= \sum_{r=1}^{\infty} \sqrt{\frac{2}{mL}} A_r \omega_r \sin \frac{r\pi x}{L}.$$

The only way that this equality can hold for all x is if $A_r = 0$ for all r. Satisfying the initial displacement condition yields the relation

$$y(x,0) = f(x) = \sum_{r=1}^{\infty} \sqrt{\frac{2}{mL}} B_r \sin \frac{r\pi x}{L}. \tag{10.23}$$

From the theory of Fourier series, this is a Fourier sine expansion,[7] with the coefficients B_r given by

$$\sqrt{\frac{2}{mL}} B_r = \frac{2}{L} \int_0^L f(x) \sin \frac{r\pi x}{L} dx,$$

or

$$B_r = \sqrt{\frac{2m}{L}} \int_0^L f(x) \sin \frac{r\pi x}{L} dx. \tag{10.24}$$

Therefore, the complete solution for the given initial conditions is

$$y(x,t) = \sum_{r=1}^{\infty} \sqrt{\frac{2}{mL}} \sin \frac{r\pi x}{L} \left(\sqrt{\frac{2m}{L}} \int_0^L f(x) \sin \frac{r\pi x}{L} dx \right) \cos \omega_r t$$

$$= \frac{2}{L} \sum_{r=1}^{\infty} \sin \frac{r\pi x}{L} \left(\int_0^L f(x) \sin \frac{r\pi x}{L} dx \right) \cos \omega_r t.$$

This equation can be solved once the initial displacement function $f(x)$ is known. Note that a units check of Equation 10.24 verifies that coefficients A_r and B_r have units of $\sqrt{\text{mass}} \times \text{length}$. Figure 10.8 shows the complete solution is approached by adding partial sums. Here, $m = 0.1$ lb sec^2/in^2, $L = 120$ in,

$$f(x) = \frac{x}{L}\left(\frac{x}{L} - \frac{1}{3}\right)\left(\frac{x}{L} - \frac{3}{4}\right)\left(\frac{x}{L} - 1\right) \text{ in,}$$

and $\omega_r = r\pi\sqrt{T/mL^2}$, where $T = 100$ lb.

---

[7] B_r can be obtained by multiplying both sides of Equation 10.23 by $\sin(r\pi x/L)$ and integrating over the domain x from 0 to L. Such a procedure makes use of the orthogonality of the sine function.

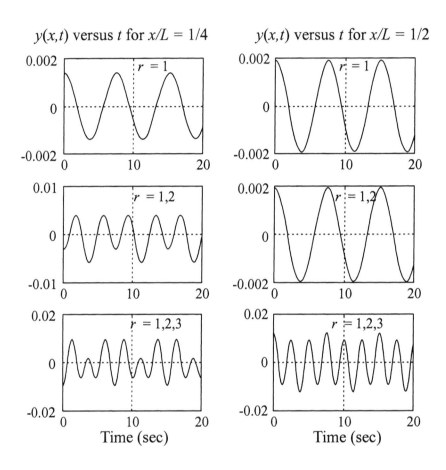

Figure 10.8: Partial sum solutions for the string: $r = 1$, $r = 1, 2$, $r = 1, 2, 3$.

10.3. LONGITUDINAL (AXIAL) VIBRATION OF BEAMS

We now have two options regarding the next topic of study, the beam. From a structural behavior perspective, given that we have just completed studying the transversely vibrating string, it makes sense to proceed to the transversely vibrating beam. From a mathematical perspective, and from a pedagogical perspective, it makes sense to study the beam in axial vibration and in torsional vibration since the governing equations are identical to that of the string. Also, the governing equation and boundary conditions are less difficult to formulate and solve than those for the bending beam. Therefore, we proceed with the axially vibrating beam.

10.3 Longitudinal (Axial) Vibration of Beams

Beams[8] are fundamental components in structures and machines. They also prove to be useful models for more complex structural behavior. We will derive, study, and solve the governing equations for beams undergoing longitudinal, torsional, and transverse motion. We assume small amplitude motion for linear behavior. The equation of motion is formulated using both Newton's second law of motion, and by the application of Hamilton's principle.

10.3.1 Newton's Approach to the Governing Equation

Consider the schematic of the beam in Figure 10.9(a). Displacements, strains, and stresses are assumed uniform at a given cross section. Newton's second law of motion is a statement of the force balance for any free body element of the beam. From the figure, force P acts to the left and P plus an undetermined increment dP acts to the right. Had this been a static problem, $dP = 0$. For the dynamic problem, the sum of the forces equals the product of mass and acceleration. Let the element have a mass per unit length of $m(x)$. Alternatively $m(x) = \rho(x)A(x)$, where $\rho(x)$ is the density and $A(x)$ is the area of the cross section at x). Then, by Newton's second law of motion for an element of length dx,

$$[P + dP](x, t) - P(x, t) = m(x)dx\frac{\partial^2 u(x,t)}{\partial t^2}.$$

---

[8] We have found the terminology *beam* interchanged with the alternatives *bar* and *rod*. Numerous texts have been examined, and there appears to be no accepted standard, even within the same text! It appears that *beam* is used for transverse motion, *rod* for axial motion, and *shaft* for rotational motion. We use beam in our discussions except for the case of a beam under torsional vibration. Such instances are likely to be in application to rotating machinery, and therefore we will adopt the conventional *shaft* for such cases.

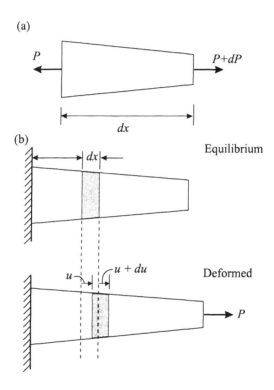

Figure 10.9: Axial deformation due to load P.

10.3. LONGITUDINAL (AXIAL) VIBRATION OF BEAMS

From the strength of materials, $P = AE\varepsilon = AE\partial u/\partial x$, and

$$dP(x,t) = \frac{\partial P(x,t)}{\partial x}dx$$
$$= \frac{\partial}{\partial x}\left(A(x)E\frac{\partial u(x,t)}{\partial x}\right)dx,$$

and

$$\frac{\partial}{\partial x}\left(A(x)E\frac{\partial u(x,t)}{\partial x}\right) = m(x)\frac{\partial^2 u(x,t)}{\partial t^2}.$$

If the beam is uniform, then $m(x) = m = \rho A$ and the area A can be canceled from both sides of the equation, leading to the wave equation,

$$c^2\frac{\partial^2 u(x,t)}{\partial x^2} = \frac{\partial^2 u(x,t)}{\partial t^2},$$

where $c^2 = E/\rho$ has units of speed squared. Before solving this equation, let us rederive it using Hamilton's principle.

10.3.2 Hamilton's Approach to the Governing Equation

We next use Hamilton's principle to derive the governing equation of motion for the longitudinal vibration of a beam where x is the direction of motion. Considering the deformed beam of Figure 10.9(b), in response to load P, the change in length dx is

$$\Delta(dx) = du = \frac{\partial u}{\partial x}dx = \varepsilon\, dx,$$

where $u(x,t)$ is the axial displacement, and ε is the strain. We require expressions for the potential and kinetic energies.

The strain energy associated with this deformation is

$$dV = \frac{1}{2}P\,du = \frac{1}{2}P\varepsilon\,dx.$$

The axial stress at a particular location x is given by $\sigma = P/A$, and by $\sigma = E\varepsilon$. The potential energy of deformation for the element is then

$$dV = \frac{1}{2}\frac{P^2}{EA(x)}dx = \frac{1}{2}EA(x)\varepsilon^2 dx.$$

For the whole beam, the potential energy is

$$V = \frac{1}{2}\int_0^L \frac{P^2}{EA(x)}dx = \frac{1}{2}\int_0^L EA(x)\varepsilon^2 dx = \frac{1}{2}\int_0^L EA(x)\left(\frac{\partial u}{\partial x}\right)^2 dx,$$

where $EA(x)$ is the axial stiffness.

The kinetic energy is given by

$$T = \frac{1}{2}\int_0^L m(x)\left(\frac{\partial u}{\partial t}\right)^2 dx,$$

where $m(x)$ is the mass per unit length.

Now apply and Hamilton's principle. Substitute V and T,

$$\delta \int_{t_1}^{t_2}(T-V)dt$$

$$= \delta \int_{t_1}^{t_2}\left[\frac{1}{2}\int_0^L m(x)\left(\frac{\partial u}{\partial t}\right)^2 dx - \frac{1}{2}\int_0^L EA(x)\left(\frac{\partial u}{\partial x}\right)^2 dx\right]dt$$

$$= \int_{t_1}^{t_2}\left[\int_0^L m(x)\frac{\partial u}{\partial t}\delta\left(\frac{\partial u}{\partial t}\right)dx - \int_0^L EA(x)\frac{\partial u}{\partial x}\delta\left(\frac{\partial u}{\partial x}\right)dx\right]dt = 0.$$

Perform the usual interchanges and integration by parts. Recall that δu is taken to vanish at $t = t_1$ and $t = t_2$, and find the resulting expression to be

$$\int_{t_1}^{t_2}\left\{\int_0^L\left[\frac{\partial}{\partial x}\left(EA(x)\frac{\partial u}{\partial x}\right) - m(x)\frac{\partial^2 u}{\partial t^2}\right]\delta u\, dx \right.$$
$$\left. - \left|\left(EA(x)\frac{\partial u}{\partial x}\right)\delta u\right|_0^L\right\}dt = 0.$$

This equation must be satisfied for those δu that vanish at the ends of the beam, at $x = 0$ and $x = L$. Since δu is arbitrary in $0 < x < L$, the factor in the square braces must equal zero,

$$\frac{\partial}{\partial x}\left(EA(x)\frac{\partial u}{\partial x}\right) - m(x)\frac{\partial^2 u}{\partial t^2} = 0. \tag{10.25}$$

This is the equation of motion. In addition, the possible boundary conditions are given by

$$\left(EA(x)\frac{\partial u}{\partial x}\right)\delta u\bigg|_0^L = 0. \tag{10.26}$$

For example, a clamped end at $x = 0$ results in a zero displacement δu over all time t, that is, $u(0,t) = 0$. A free end at $x = L$ implies that a moment cannot be resisted and $EA(L)\partial u(L)/\partial x = 0$. Note that Equation 10.25 has the same form as the governing equation for the vibrating string.[9]

---

[9]In Equation 10.2, replace y by u, and T by EA.

10.3.3 Simplified Eigenvalue Problem

The assumed solution for the fixed–free longitudinal vibration problem is $u(x,t) = U(x)F(t)$. $U(x)$ is the mode of vibration. Differentiate this assumed solution and substitute into Equation 10.25 to find

$$\frac{d}{dx}\left[EA(x)\frac{dU(x)}{dx}\right] = m(x)U(x)\frac{\ddot{F}}{F} = -\omega^2 m(x)U(x), \qquad (10.27)$$

where the assumption of harmonic behavior results in the replacement of $\ddot{F}/F$ by $-\omega^2$. The fixed–free boundary conditions to be satisfied are

$$U(0) = 0$$
$$EA(L)\frac{dU(L)}{dx} = 0.$$

One further assumption is made, that the beam is uniform, $A(x) = A$, $m(x) = m$, and define $\beta^2 = \omega^2/c^2$, where $c = \sqrt{EA/m}$ is the wave speed. Equation 10.27 is thus transformed into the following simple ordinary differential equation,

$$\frac{d^2U}{dx^2} + \beta^2 U = 0,$$

with the general modal solution

$$U(x) = C_1 \sin\beta x + C_2 \cos\beta x. \qquad (10.28)$$

C_1 and C_2 are determined from the given boundary conditions,

$$U(0) = 0 = C_2$$
$$U'(L) = 0 = C_1\beta\cos\beta L.$$

Two possible solutions exist for the second boundary equation: $C_1 = 0$ or $\beta\cos\beta L = 0$. The first results in the trivial solution $U(x) = 0$, which is of no use. The second equation defines β, that is, $C_1 \neq 0$ and $\beta\cos\beta L = 0$. Therefore, $\beta_r L = (2r-1)\pi/2$, and

$$\beta_r = (2r-1)\frac{\pi}{2L}, \quad r = 1, 2, \ldots.$$

From the definition of β, solve for ω,

$$\omega_r^2 = \beta_r^2 \frac{EA}{m}$$

$$\omega_r = (2r-1)\frac{\pi}{2}\sqrt{\frac{EA}{mL^2}}.$$

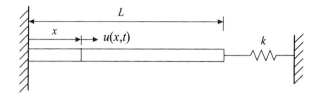

Figure 10.10: Uniform beam with elastic restraint in longitudinal motion.

Equation 10.28 for the rth mode becomes

$$U_r(x) = C_r \sin \frac{(2r-1)\pi}{2L} x, \quad r = 1, 2, \ldots.$$

Since the modes can only be specified to a constant, it is customary to normalize them according to

$$\int_0^L m U_r^2 dx = 1,$$

and find $C_r = \sqrt{2/mL}$.

The complete solution is then

$$\begin{aligned} u(x,t) &= \sum_{r=1}^{\infty} C_r \sin \beta_r x (A_r \sin \omega_r t + B_r \cos \omega_r t) \\ &= \sum_{r=1}^{\infty} \sqrt{\frac{2}{mL}} \sin \frac{(2r-1)\pi}{2L} x \left(A_r \sin \frac{(2r-1)\pi}{2L} \sqrt{\frac{EA}{m}} t \right. \\ &\quad \left. + B_r \cos \frac{(2r-1)\pi}{2L} \sqrt{\frac{EA}{m}} t \right). \end{aligned} \qquad (10.29)$$

Coefficients A_r and B_r are found by satisfying the initial conditions. Suppose that the initial displacement is $u(x,0)$ and the initial velocity $\dot{u}(x,0)$. Then A_r and B_r can be found by using Equation 10.29 and its time derivative.

Example 10.2 Axial Vibration of Uniform Beam with Elastic Boundary

As a more interesting example of an axially vibrating beam, consider the schematic in Figure 10.10, where the right end of the beam is restrained elastically.

10.3. LONGITUDINAL (AXIAL) VIBRATION OF BEAMS

We know the differential equation of motion for the uniform beam in longitudinal vibration to be

$$EA\frac{\partial^2 u}{\partial x^2} = m\frac{\partial^2 u}{\partial t^2},$$

along with the respective eigenvalue problem

$$EA\frac{d^2 U}{dx^2} = -\omega^2 mU. \tag{10.30}$$

The boundary conditions are

$$U(0) = 0$$
$$EA(L)U'(L) = -kU(L),$$

where the last condition is a force balance equation at the interface between the right end of the beam and the left end of the spring.

Eigenvalue problem Equation 10.30 can be written as

$$\frac{d^2 U}{dx^2} + \beta^2 U = 0,$$

with $\beta^2 = \omega^2 m/EA$. We have already solved this problem, except here there is an elastic boundary condition at the end $x = L$. Therefore, the modal equation is $U(x) = C_1 \sin \beta x + C_2 \cos \beta x$, with $U(0) = 0 = C_2$ and

$$U'(L) = -\frac{k}{EA}U(L), \quad \text{or}$$
$$C_1\beta \cos \beta L = -\frac{k}{EA}C_1 \sin \beta L.$$

This expression can be solved for β using

$$\tan \beta L = -\frac{\beta EA}{k}.$$

This is an *implicit* relation for β. For an arbitrary set of parameter values, say, $k = EA$ and $L = 1$, we have $\tan \beta = -\beta$, which is plotted in Figure 10.11.

The first few positive roots of this equation are 0, 2.029, 4.913, 7.980, ..., and the roots are anti–symmetric about the vertical axis.[10] Once β is evaluated, the frequencies can be found using

$$\omega_r^2 = \beta_r^2 \frac{EA}{m}.$$

---

[10] Another easy way to find the roots is graphically. Superimpose the curves $\tan \beta$ and $-\beta$, and the intersections are the roots.

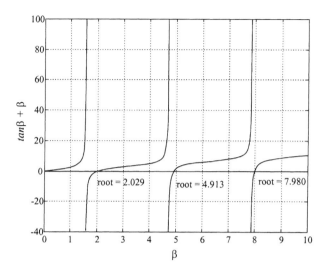

Figure 10.11: Plot of $\tan\beta + \beta = 0$.

Now the modes can be written explicitly, leading to the complete solution in time and space. ∎

10.3.4 Orthogonality of the Normal Modes

Suppose a uniform beam is loaded axially by some force $F(x,t)$, as shown in Figure 10.12. We would like to demonstrate how this problem can be solved using modal techniques very similar to those used for multi degree of freedom vibration problems.

The equation of motion is given by

$$EA\frac{\partial^2 u}{\partial x^2} - m\frac{\partial^2 u}{\partial t^2} + F(x,t) = 0. \tag{10.31}$$

Assume the product solution for $u(x,t)$ with the resulting eigenvalue problem,

$$\frac{d^2 U}{dx^2} + \beta^2 U = 0, \tag{10.32}$$

where $\beta^2 = \omega^2 m/EA$, with solution

$$U(x) = C_1 \sin\beta x + C_2 \cos\beta x. \tag{10.33}$$

10.3. LONGITUDINAL (AXIAL) VIBRATION OF BEAMS

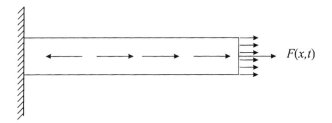

Figure 10.12: Beam under general forcing $F(x,t)$.

The frequencies ω_j and $U_j(x)$ can be obtained as usual by satisfying the boundary conditions.

By way of this example, we demonstrate the orthogonality of the normal modes. We motivate the more formal discussion of Section 10.5.4 on the orthogonality of the modes by tackling this solution with an expansion of the solution and the forcing in terms of the modes, as follows,

$$u(x,t) = \sum_{j=1}^{\infty} u_j(t) U_j(x) \qquad (10.34)$$

$$F(x,t) = \sum_{j=1}^{\infty} F_j(t) U_j(x). \qquad (10.35)$$

Equations 10.32 and 10.33 show that the modes have a harmonic character. Such harmonic functions have orthogonality properties that can be used to evaluate the expansion functions $u_j(t)$ and $F_j(t)$. We have seen this approach in Chapter 8. Multiply both sides of Equation 10.34 by $mU_k(x)$ and integrate over the x domain from 0 to L, as follows,

$$\int_0^L mu(x,t) U_k(x) dx = \sum_{j=1}^{\infty} mu_j(t) \int_0^L U_j(x) U_k(x) dx.$$

Since $U_j(x)$ is harmonic, integrals such as those on the right hand side are zero except where $k = j$. It is customary to use the normalization $\int_0^L mU_j(x)^2 dx = 1$. Doing this, we find

$$u_j(t) = \int_0^L mu(x,t) U_j(x) dx$$

$$F_j(t) = \int_0^L mF(x,t) U_j(x) dx. \qquad (10.36)$$

Substituting expansion Equations 10.34 and 10.35 into the Equation of motion 10.31, leads to the summation

$$\sum_{j=1}^{\infty}\left[EAu_j(t)\frac{d^2U_j}{dx^2} - m\frac{d^2u_j}{dt^2}U_j + F_j(t)U_j\right] = 0. \tag{10.37}$$

From the eigenvalue problem we have

$$-EA\frac{d^2U_j}{dx^2} = \omega_j^2 mU_j,$$

which can be substituted into Equation 10.37, resulting in the equation

$$\sum_{j=1}^{\infty}\left[-\omega_j^2 mu_j(t) - m\frac{d^2u_j}{dt^2} + F_j(t)\right]U_j = 0.$$

Since $U_j(x) \neq 0$, the terms in the square brackets must add to zero,

$$\ddot{u}_j + \omega_j^2 u_j = \frac{F_j(t)}{m}, \quad j = 1, 2, \ldots.$$

The solution for the jth term is

$$u_j(t) = A_j \cos\omega_j t + B_j \sin\omega_j t + \frac{1}{m\omega_j}\int_0^t F_j(\tau)\sin\omega_j(t-\tau)d\tau,$$

where any force $F(x,t)$ can be substituted into Equation 10.36 for the evaluation of $F_j(t)$. The complete solution is the infinite sum,

$$u(x,t) = \sum_{j=1}^{\infty} u_j(t)U_j(x).$$

The number of terms needed for an accurate representation depends on the particular loading characteristics. The initial conditions are satisfied, as usual, by specifying A_j and B_j.

The next section continues with the torsional vibration of a shaft. The mathematics is the same, but the physical interpretation is new.

10.4 Torsional Vibration of Shafts

The torsional vibration of a shaft has numerous applications. Shafts are the backbone of power transmission systems, and other rotor bearing systems. For example, a pump is often connected to a motor with a cooling fan

10.4. TORSIONAL VIBRATION OF SHAFTS

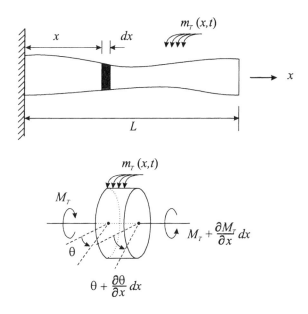

Figure 10.13: Shaft with free body diagram.

through a coupling, and a turbopump in a space shuttle main engine is driven by a gas turbine. Such rotor–bearing systems are often modeled as flexible rotors subjected to torques. Forces may act upon the rotating shafts in the longitudinal direction. The analysis and design of such systems is a specialized discipline, but we can begin to understand the torsional properties of the shaft given our current state of knowledge.

Assume that there is no *warping* of the cross–sectional planes of the shaft. Warping occurs for non–uniform or non–circular shafts, where the plane cross–section distorts out of plane. Consider the sketch of a shaft, an element of its length, and the free body diagram for that element in Figure 10.13.

Define the following properties at section x: shear modulus G, polar moment of inertia for circular cross–section $J(x)$, angle of twist $\theta(x,t)$, internal twisting moment $M_T(x,t)$, mass polar moment of inertia per unit length $I(x)$, and external twisting moment per unit length $m_T(x,t)$. From strength of materials,

$$M_T = GJ(x)\frac{\partial \theta}{\partial x}. \qquad (10.38)$$

Apply Newton's second law to the free body diagram and find the torsional

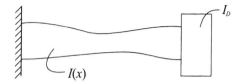

Figure 10.14: Shaft with rigid disk attached at end.

equation of motion

$$[M_T + \frac{\partial M_T}{\partial x}dx] + m_T(x,t)dx - M_T = I(x)dx\frac{\partial^2 \theta}{\partial t^2},$$

or

$$\frac{\partial M_T}{\partial x} + m_T(x,t) = I(x)\frac{\partial^2 \theta}{\partial t^2}. \quad (10.39)$$

Substitute Equation 10.38 into Equation 10.39 to obtain the torsional vibration equation of motion,

$$\frac{\partial}{\partial x}\left[GJ(x)\frac{\partial \theta}{\partial x}\right] + m_T(x,t) = I(x)\frac{\partial^2 \theta}{\partial t^2}.$$

For free torsional vibration, set $m_T(x,t) = 0$. As examples of possible boundary conditions, a clamped end at $x = 0$ permits no rotation, $\theta(0,t) = 0$; a free end at $x = L$ implies no moment resistance,

$$M_T(L,t) = GJ(L)\theta'(L) = 0.$$

Note that the above equations have the same mathematical form as those for a *string* and for the longitudinal vibration of a *beam*.[11] The torsion problem is solved in exactly the same way as the string and beam problems. Next study torsional motion with a more interesting and useful boundary condition.

10.4.1 Torsion of Shaft with Rigid Disk at End

Consider the torsional oscillations of a shaft that has a rigid disk attached at the free end, as sketched in Figure 10.14. The disk has a mass moment of inertia I_D. This model can be used to represent the torsional oscillations of power transmission shafts in machinery.

---

[11] If $I(x)$ is replaced by $mJ(x)$, the governing equation is the wave equation with torsional wave speed $c = \sqrt{G/m}$.

10.4. TORSIONAL VIBRATION OF SHAFTS

All the previous analysis is still valid except for the boundary condition at the rigid disk, which acts as an applied moment,

$$GJ(L)\frac{\partial \theta(L,t)}{\partial x} = -I_D(L)\frac{\partial^2 \theta(L,t)}{\partial t^2}. \tag{10.40}$$

Solution of the eigenvalue problem is via separation of variables,

$$\theta(x,t) = \Theta(x)F(t),$$

where $F(t)$ is harmonic. The eigenvalue problem becomes

$$\frac{d}{dx}[GJ(x)\Theta'(x)] = -\omega^2 I(x)\Theta(x), \tag{10.41}$$

with boundary conditions $\Theta(0) = 0$ and, from Equation 10.40,

$$GJ(L)\Theta'(L) = \omega^2 I_D(L)\Theta(L).$$

Note that the second boundary condition is a function of frequency ω. This is due to the existence of the inertia at the right end and is known as a *frequency dependent boundary condition*.

To solve eigenvalue Equation 10.41, assume the shaft to be uniform, $J(x) = J$, $I(x) = I$, and define $\beta^2 = \omega^2 I/GJ$. Making these substitutions results in the simplified eigenvalue problem

$$\frac{d^2\Theta}{dx^2} + \beta^2\Theta = 0,$$

with solution

$$\Theta(x) = C_1 \sin \beta x + C_2 \cos \beta x.$$

Applying the two boundary conditions results in the two equations to find that $\Theta(0) = 0 = C_2$ and therefore

$$GJC_1\beta \cos \beta L = \omega^2 I_D C_1 \sin \beta L. \tag{10.42}$$

Substitute $\omega^2 = \beta^2 GJ/I$ into Equation 10.42, and simplify to find

$$\tan \beta L = \frac{I}{I_D \beta},$$

or

$$\tan \beta L = \frac{IL}{I_D}\frac{1}{\beta L}. \tag{10.43}$$

This is a transcendental equation in βL that must be solved recursively for βL. There is an infinity of solutions for β_r and for frequency,

$$\omega_r^2 = \frac{\beta_r^2 GJ}{I}, \quad r = 1, 2, \ldots.$$

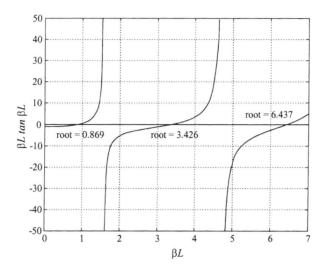

Figure 10.15: Graph of $\beta L \tan \beta L = 1$.

ω_r is not an integer multiple of the fundamental frequency.

Figure 10.15 shows a graph of Equation 10.43 for the ratio $IL/I_D = 1$, that is, $\beta L \tan \beta L = 1$, which has the following first three roots: $\beta_r L = 0.860$, 3.426, and 6.437. Once $\beta_r L$ is evaluated, ω_r can be found, and the solution completed as in previous problems.

Now we are ready to proceed with the study of the transversely vibrating beam. This is a new class of problems that is governed by a fourth-order partial differential equation. However, the solution technique is similar to that used for the earlier problems.

10.5 Transverse Vibration of Beams

The modeling of a transversely vibrating beam includes the effects of shear distortion and bending moment. Timoshenko was the first to include *shear distortion effects* in the modeling, and therefore his name is attached to such models. The formulation and basic solutions along with simple boundary conditions are studied in this section. In Section 10.6, special loading and boundary conditions are explored and possible applications mentioned.[12]

---

[12] An extensive discussion of various beam theories can be found in *Dynamics of Transversely Vibrating Beams Using Four Engineering Theories*, S.M. Han, H. Benaroya, T. Wei., J. Sound and Vibration (1999) **225**(5), 935–988.

10.5. TRANSVERSE VIBRATION OF BEAMS

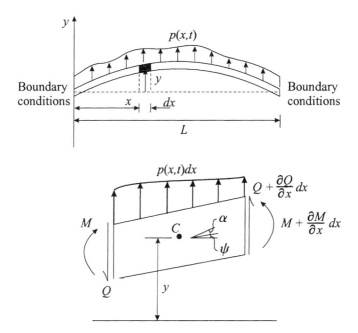

Figure 10.16: Transverse deflection of a beam and its free body diagram.

When rotary inertia and shear effects are ignored, the beam is called the *Bernoulli–Euler* beam. It is, of course, a special case of the more general Timoshenko beam.

10.5.1 Derivation of the Equations of Motion for the Beam with Shear Distortion: The *Timoshenko* Beam

The equation governing the transverse vibration of a beam of length L is derived, with the following properties at section x; $m(x)$ is the mass per unit length, $A(x)$ is the cross–sectional area, and $I(x)$ is the moment of inertia. Assume small deflections $y(x,t)$ and rotations $\partial y/\partial x$, and include the bending $M(x,t)$ and shear $Q(x,t)$ effects.

Consider a free body of a section of length dx as shown in Figure 10.16. Its slope is due to a bending component ψ and to a *shear distortion* component α,

$$\frac{\partial y}{\partial x} = \psi(x,t) + \alpha(x,t).$$

From elementary beam theory, the bending moment[13] is related to the slope by $M = EI\, \partial\psi/\partial x$, and the shear is related to the slope by $Q = s\alpha\, A(x)G$, where s is a number[14] that depends on the shape of the cross section, and G is the constant shear modulus of the material. Q may then be rewritten as

$$Q = s\left(\frac{\partial y}{\partial x} - \psi\right) A(x)G.$$

$s\,A(x)$ is called the *reduced section* and is computed from classical beam theory. For example, $s = 5/6$ for a plane rectangular cross section and $s = 1/1.175$ for a plane circular cross section.

Hamilton's principle, $\int_{t_1}^{t_2}(\delta T + \delta W)dt = 0$, is used to derive the boundary value problem and proceed to derive the kinetic and potential energies, and their variations. The kinetic energy due to translation and rotation for the whole beam is given by

$$T(t) = \frac{1}{2}\int_0^L m(x)\left[\frac{\partial y}{\partial t}\right]^2 dx + \frac{1}{2}\int_0^L J(x)\left[\frac{\partial \psi}{\partial t}\right]^2 dx,$$

where

$$J(x) = \rho(x)I(x) = \frac{m(x)}{A(x)}I(x) = k^2(x)m(x). \qquad (10.44)$$

$J(x)$ is the mass moment of inertia per unit length, and $k(x)$ is the radius of gyration, both about the neutral bending axis. Note also that in the second integral above for the rotary inertia, *only the effect of bending is included since the shear only results in a distortion, not a rotation.*

Using the chain rule, the variation in kinetic energy δT is

$$\delta T(t) = \int_0^L m(x)\frac{\partial y}{\partial t}\delta\left(\frac{\partial y}{\partial t}\right)dx + \int_0^L k^2(x)m(x)\frac{\partial \psi}{\partial t}\delta\left(\frac{\partial \psi}{\partial t}\right)dx. \qquad (10.45)$$

Before proceeding with the virtual work done on the structure, δW, partition the work into a *conservative* part that is equal to the change in potential

---

[13] The exact relation is
$$\frac{M}{EI} = \frac{y''}{[1+(y')^2]^{3/2}}.$$
For small displacements and slopes, y' is very small when compared to 1 and can be ignored.

[14] See **Vibration Problems in Engineering**, 5th Edition, by W. Weaver, S. Timoshenko, and D. Young, McGraw-Hill, 1990, pp.434–436, for additional values of s. This simplified relation between Q and α is introduced to make the equations analytically tractable, while accounting for the fact that shear is parabolically distributed on a cross section. Therefore, the equations of this section are valid for thin beams, where the length to depth is at least 10:1.

10.5. TRANSVERSE VIBRATION OF BEAMS

and strain energy, and a *non–conservative* part that includes the work done by external forces $p(x,t)$,

$$\begin{aligned}\delta W(t) &= \delta W_c(t) + \delta W_{nc}(t) \\ &= -\delta V(t) + \int_0^L p(x,t)\delta y(x,t)dx.\end{aligned} \qquad (10.46)$$

The change in potential energy equals the work done by the conservative actions due to the moment and the shear force,

$$\begin{aligned}V(t) &= \frac{1}{2}\int_0^L M(x,t)\frac{\partial \psi}{\partial x}dx + \frac{1}{2}\int_0^L Q(x,t)\alpha\, dx \\ &= \frac{1}{2}\int_0^L EI(x)\left(\frac{\partial \psi}{\partial x}\right)^2 dx + \frac{1}{2}\int_0^L s\alpha^2 A(x)G\, dx,\end{aligned}$$

where the second term is due to the shear deformation effect. The variation $\delta V(t)$ is given by

$$\delta V(t) = \int_0^L EI(x)\frac{\partial \psi}{\partial x}\delta\left(\frac{\partial \psi}{\partial x}\right)dx + \int_0^L sA(x)G\alpha\,\delta\alpha\, dx,$$

where the number of variables is reduced next by substituting $\delta\alpha = \delta(\frac{\partial y}{\partial x} - \psi)$. Introduce Equations 10.45 and 10.46 into Hamilton's variational principle,

$$\begin{aligned}\int_{t_1}^{t_2}\Bigg\{&\left[\int_0^L m(x)\frac{\partial y}{\partial t}\delta\left(\frac{\partial y}{\partial t}\right)dx + \int_0^L k^2(x)m(x)\frac{\partial \psi}{\partial t}\delta\left(\frac{\partial \psi}{\partial t}\right)dx\right] \\ &-\left[\int_0^L EI(x)\frac{\partial \psi}{\partial x}\delta\left(\frac{\partial \psi}{\partial x}\right)dx + \int_0^L sA(x)G\left(\frac{\partial y}{\partial x}-\psi\right)\delta\left(\frac{\partial y}{\partial x}-\psi\right)dx\right] \\ &+ \int_0^L p(x,t)\delta y(x,t)dx\Bigg\}dt = 0,\end{aligned}$$

where the variation δ operates on the function that immediately follows.

Perform the usual interchanges and integration by parts and combine terms to find

$$\begin{aligned}\int_{t_1}^{t_2}\Bigg[&\int_0^L\left\{\frac{\partial}{\partial x}\left[sGA(x)\left(\frac{\partial y}{\partial x}-\psi\right)\right]-m(x)\frac{\partial^2 y}{\partial t^2}+p(x,t)\right\}\delta y\, dx \\ &+\int_0^L\left\{\frac{\partial}{\partial x}\left(EI(x)\frac{\partial \psi}{\partial x}\right)+sGA(x)\left(\frac{\partial y}{\partial x}-\psi\right)-k^2(x)m(x)\frac{\partial^2 \psi}{\partial t^2}\right\}\delta\psi\, dx \\ &-\left(EI(x)\frac{\partial \psi}{\partial x}\right)\delta\psi\bigg|_0^L - \left[sGA(x)\left(\frac{\partial y}{\partial x}-\psi\right)\right]\delta y\bigg|_0^L\Bigg]dt = 0. \qquad (10.47)\end{aligned}$$

To proceed note that $\delta\psi$ and δy are arbitrary for $0 < x < L$. It then follows that the governing equations of motion for the vibration of a Timoshenko beam are

$$\frac{\partial}{\partial x}\left[sGA(x)\left(\frac{\partial y}{\partial x}-\psi\right)\right] - m(x)\frac{\partial^2 y}{\partial t^2} + p(x,t) = 0 \quad (10.48)$$

$$\frac{\partial}{\partial x}\left(EI(x)\frac{\partial\psi}{\partial x}\right) + sGA(x)\left(\frac{\partial y}{\partial x}-\psi\right) - k^2(x)m(x)\frac{\partial^2\psi}{\partial t^2} = 0, \quad (10.49)$$

with possible boundary conditions defined by

$$\left[EI(x)\frac{\partial\psi}{\partial x}\right]\delta\psi\bigg|_0^L = 0 \quad (10.50)$$

$$\left[sGA(x)\left(\frac{\partial y}{\partial x}-\psi\right)\right]\delta y\bigg|_0^L = 0. \quad (10.51)$$

These boundary conditions are interpreted to mean that *either* the term in the square brackets *or* the variation parameter equals zero at $x = 0$ and $x = L$. The term in the square brackets in Equation 10.50 is a moment and the term $\delta\psi$ is a rotation. Therefore, the relevant boundary condition is that *either* the moment *or* the rotation equals zero at an end. Similarly, in Equation 10.51, the term in the square brackets is the shear force and δy is a deflection. Then, *either* the shear force *or* the deflection equals zero at a boundary. The boundary conditions are applied in the following sections. The free vibration problem can be solved[15] beginning with the assumed harmonic responses, $y(x,t) = Y(x)\exp(i\omega t)$ and $\psi(x,t) = \Psi(x)\exp(i\omega t)$.

Before proceeding with particular examples and boundary conditions, it is of interest to consider the uniform beam, where $A(x) = A$, $m(x) = m$, $k(x) = k$, and $I(x) = I$. Equations 10.48 and 10.49 can now be combined into a single equation of motion. For the uniform beam, the governing equations reduce to

$$sGA\frac{\partial}{\partial x}\left(\frac{\partial y}{\partial x}-\psi\right) - m\frac{\partial^2 y}{\partial t^2} + p(x,t) = 0 \quad (10.52)$$

$$EI\frac{\partial^2\psi}{\partial x^2} + sGA\left(\frac{\partial y}{\partial x}-\psi\right) - k^2 m\frac{\partial^2\psi}{\partial t^2} = 0. \quad (10.53)$$

We choose to eliminate ψ in combining these two equations, leading to an equation governing $y(x,t)$. First differentiate Equation 10.53 with respect

---

[15] See solution beginning on page 197 of **Mechanical Vibrations: Theory and Applications to Structural Dynamics**, M. Géradin, D. Rixen, John Wiley and Sons, 1994.

10.5. TRANSVERSE VIBRATION OF BEAMS

to x,

$$EI\frac{\partial^3 \psi}{\partial x^3} + sGA\left(\frac{\partial^2 y}{\partial x^2} - \frac{\partial \psi}{\partial x}\right) - k^2 m\frac{\partial^3 \psi}{\partial x \partial t^2} = 0. \tag{10.54}$$

From Equation 10.52, solve for $\partial \psi/\partial x$, as well as the other derivatives required in Equation 10.54,

$$\frac{\partial \psi}{\partial x} = \frac{\partial^2 y}{\partial x^2} - \frac{m}{sGA}\frac{\partial^2 y}{\partial t^2} + \frac{p(x,t)}{sGA} \tag{10.55}$$

$$\frac{\partial^3 \psi}{\partial x^3} = \frac{\partial^4 y}{\partial x^4} - \frac{m}{sGA}\frac{\partial^4 y}{\partial x^2 \partial t^2} + \frac{1}{sGA}\frac{\partial^2 p}{\partial x^2} \tag{10.56}$$

$$\frac{\partial^3 \psi}{\partial x \partial t^2} = \frac{\partial^4 y}{\partial x^2 \partial t^2} - \frac{m}{sGA}\frac{\partial^4 y}{\partial t^4} + \frac{1}{sGA}\frac{\partial^2 p}{\partial t^2}. \tag{10.57}$$

Substitute Equations 10.55, 10.56, and 10.57 into 10.54, and after a bit of algebra, find the governing equation to be

$$EI\frac{\partial^4 y}{\partial x^4} + m\frac{\partial^2 y}{\partial t^2} - \left(k^2 m + \frac{EIm}{sGA}\right)\frac{\partial^4 y}{\partial x^2 \partial t^2} + \frac{(km)^2}{sGA}\frac{\partial^4 y}{\partial t^4}$$
$$= p(x,t) + \frac{k^2 m}{sGA}\frac{\partial^2 p}{\partial t^2} - \frac{EI}{sGA}\frac{\partial^2 p}{\partial x^2}, \tag{10.58}$$

where the external loading $p(x,t)$ has been retained, even though for the eigenvalue problem it is set to zero. Each term in Equation 10.58 represents a different physical aspect of beam behavior. If rotary effects can be neglected, set $k = 0$ in Equation 10.58. However, if the shear effect can be neglected, set $s = 0$ in Equations 10.52 and 10.53, which results in $Q = 0$. The much relied upon and basic Bernoulli–Euler beam equation that is used for many simplified studies assumes $k = 0$ and $s = 0$,

$$EI\frac{\partial^4 y}{\partial x^4} + m\frac{\partial^2 y}{\partial t^2} = p(x,t). \tag{10.59}$$

10.5.2 Boundary Conditions

We summarize the possible boundary conditions that are derivable from Equations 10.50 and 10.51. These are used in subsequent problems.

Fixed End

A fixed end is defined as one where no displacement or rotation can take place for all time. For location $x = 0$,

$$y(0,t) = 0$$
$$\psi(0,t) = 0.$$

Hinged End

A hinged end does not allow displacement, and has no moment resistance and therefore permits rotation. Thus, for location $x = 0$,

$$y(0,t) = 0$$

$$M(0,t) = \left| EI(x)\frac{\partial \psi}{\partial x} \right|_{x=0} = 0.$$

Free End

A free end has no moment or shear resistance. Thus, for location $x = 0$,

$$M(0,t) = \left| EI(x)\frac{\partial \psi}{\partial x} \right|_{x=0} = 0$$

$$Q(0,t) = \left| sGA(x)\left(\frac{\partial y}{\partial x} - \psi\right) \right|_{x=0} = 0.$$

These are the basic boundary conditions. In practical applications, the designer might choose to mix these properties. We consider hybrid boundary conditions in Section 10.6. Of course, the above boundary conditions can be written in the same form for a boundary at $x = L$.

10.5.3 Simplified Eigenvalue Problem

In order to be able to analytically tackle the eigenvalue problem, some reasonable simplifying assumptions are needed. When the cross-sectional dimensions are much smaller than the length (the rule of thumb is 1:10), the shear distortion effect and the rotary inertia effect are reasonably neglected.[16] Also, for the eigenvalue problem, external forces $p(x,t)$ are set equal to zero.

Therefore, Equations 10.48 and 10.49 become

$$\frac{\partial}{\partial x}\left[sGA(x)\left(\frac{\partial y}{\partial x} - \psi\right)\right] - m(x)\frac{\partial^2 y}{\partial t^2} = 0 \quad (10.60)$$

$$\frac{\partial}{\partial x}\left(EI(x)\frac{\partial \psi}{\partial x}\right) + sGA(x)\left(\frac{\partial y}{\partial x} - \psi\right) = 0. \quad (10.61)$$

Solve Equation 10.61 for $sGA(x)(\partial y/\partial x - \psi)$, and substitute this into Equation 10.60. Due to the above assumption of no shear distortion, $\partial y/\partial x = \psi(x,t) + \alpha(x,t)$ can be written as $\partial \psi/\partial x = \partial^2 y/\partial x^2$ since $\alpha = 0$.

---

[16] $k(x)$, which is related to $J(x)$, is approximately equal to zero in Equation 10.44 and in subsequent equations.

10.5. TRANSVERSE VIBRATION OF BEAMS

The resulting governing equation for $y(x,t)$ is the Bernoulli–Euler beam with variable properties,

$$\frac{\partial^2}{\partial x^2}\left[EI(x)\frac{\partial^2 y}{\partial x^2}\right] = -m(x)\frac{\partial^2 y}{\partial t^2}. \tag{10.62}$$

Assume the product solution,

$$y(x,t) = Y(x)F(t), \tag{10.63}$$

differentiate and substitute this solution into the governing equation to obtain

$$\frac{d^2}{dx^2}[EI(x)Y''(x)F(t)] = -m(x)Y(x)\ddot{F}$$

$$\frac{1}{m(x)Y(x)}\frac{d^2}{dx^2}[EI(x)Y''(x)] = -\frac{\ddot{F}}{F}. \tag{10.64}$$

Using the same argument as for the vibrating string, equate each side of Equation 10.64 to the constant ω^2 so that the solution in time is harmonic,

$$\ddot{F} + \omega^2 F = 0, \tag{10.65}$$

and $Y(x)$ is governed by

$$\frac{d^2}{dx^2}[EI(x)Y''(x)] = \omega^2 m(x)Y(x). \tag{10.66}$$

This is the eigenvalue problem. Equation 10.65 requires two initial conditions and Equation 10.66 requires four boundary conditions for a complete solution.

The eigenvalue problem must be solved for a particular set of boundary conditions, resulting in expressions for the eigenfunctions $Y(x)$ and frequencies ω which the structure can accommodate in free vibration. The solution to Equation 10.65 is harmonic, $F(t) = A\cos(\omega t - \phi)$. A and ϕ are found by satisfying the initial conditions. Possible idealized boundary conditions include the following sets.

Fixed End

At a fixed or clamped end, deflection and rotation cannot occur, and the boundary conditions are

$$\begin{aligned} y(0,t) &= Y(0)F(t) = 0 \\ y(L,t) &= Y(L)F(t) = 0 \\ y'(0,t) &= Y'(0)F(t) = 0 \\ y'(L,t) &= Y'(L)F(t) = 0. \end{aligned}$$

Therefore, $Y(0) = Y(L) = 0$ and $Y'(0) = Y'(L) = 0$, for all t.

Hinged End

A hinged end cannot deflect and cannot resist a rotation. The zero deflection equations have been already provided. For zero moment resistance, the moment equation, $M = EI(x)\partial^2 y/\partial x^2 = EI(x)Y''(x)F(t)$, yields the following conditions,

$$EI(0)Y''(0)F(t) = 0$$
$$EI(L)Y''(L)F(t) = 0.$$

EI can be divided out of the above expressions, and $Y''(0) = Y''(L) = 0$ for all t.

Free End

A free end has no resistance to moment and shear force. The moment conditions have been provided above. For zero shear resistance, the shear equation,

$$Q = -\frac{\partial}{\partial x}\left[EI(x)\frac{\partial^2 y}{\partial x^2}\right] = -\frac{d}{dx}[EI(x)Y''(x)]F(t),$$

yields the following conditions,

$$\frac{d}{dx}[EI(0)Y''(0)]F(t) = 0$$
$$\frac{d}{dx}[EI(L)Y''(L)]F(t) = 0.$$

EI can be divided out of the above expressions, and for all time

$$Y'''(0) = 0$$
$$Y'''(L) = 0,$$

for all t.

Actual boundary conditions are in between those of the above cases, since *no boundary is exactly hinged or completely fixed*, but is somewhere in between the extremes.

Example 10.3 Solution for a Beam That is Hinged (Simply Supported) at Both Ends

For the hinged beam, the boundary conditions are $Y(0) = Y(L) = 0$ and $Y''(0) = Y''(L) = 0$. In addition, assume that the beam is uniform:

$$I(x) = I, \quad m(x) = m,$$

10.5. TRANSVERSE VIBRATION OF BEAMS

and therefore Equation 10.66 becomes

$$EI \frac{d^4}{dx^4}[Y(x)] = \omega^2 m Y(x). \tag{10.67}$$

Define $\beta^4 = \omega^2 m/EI$, and the eigenvalue equation becomes

$$Y''''(x) - \beta^4 Y(x) = 0, \tag{10.68}$$

with eigenfunction solution

$$Y(x) = C_1 \sin \beta x + C_2 \cos \beta x + C_3 \sinh \beta x + C_4 \cosh \beta x. \tag{10.69}$$

Applying the $x = 0$ boundary conditions to Equation 10.69, we find

$$Y(0) = 0 = C_2 + C_4$$
$$Y''(0) = 0 = -C_2 + C_4,$$

leading to $C_2 = C_4 = 0$. Physically, this result makes sense since the nature of the cosine and the hyperbolic cosine is such that they cannot equal zero at a boundary, as required by the hinged condition. Using Equation 10.69 again, for the $x = L$ boundary conditions, leads to

$$Y(L) = 0 = C_1 \sin \beta L + C_3 \sinh \beta L \tag{10.70}$$
$$Y''(L) = 0 = -C_1 \beta^2 \sin \beta L + C_3 \beta^2 \sinh \beta L, \tag{10.71}$$

where the factor β^2 cancels out. Adding (and subtracting) Equations 10.70 and 10.71 yields the relations

$$2C_3 \sinh \beta L = 0$$
$$2C_1 \sin \beta L = 0.$$

There are several possible cases to consider here. The first is that $C_1 = C_3 = 0$, but this is the trivial solution. The second case is that $C_1 = 0$ and $C_3 \neq 0$, with $\sinh \beta L = 0$, which is only true for $\beta = 0$, which is again trivial and of no interest. We are then left with the case $C_3 = 0$ and $C_1 \neq 0$, implying that $\sin \beta L = 0$. For this last equality to hold,

$$\beta_r L = r\pi, \quad r = 1, 2, \ldots.$$

Therefore, from the definition of β,

$$\omega_r^2 = \beta_r^4 \frac{EI}{m}$$
$$= \left(\frac{r\pi}{L}\right)^4 \frac{EI}{m}$$
$$\omega_r = \left(\frac{r\pi}{L}\right)^2 \sqrt{\frac{EI}{m}}, \quad r = 1, 2, \ldots. \tag{10.72}$$

Equation 10.72 specifies the frequencies that can be accommodated in free vibration for the given boundary conditions.[17] The eigenfunction becomes

$$Y_r(x) = C_1 \sin\left(\frac{r\pi x}{L}\right). \tag{10.73}$$

To fix the value of C_1, Equation 10.73 is generally normalized according to the rule

$$\int_0^L mY_r^2(x)\,dx = 1$$

$$\int_0^L C_1^2 m \sin^2\left(\frac{r\pi x}{L}\right) dx = 1$$

$$\Rightarrow C_1 = \sqrt{\frac{2}{mL}}.$$

C_1 is independent of index r. The mode shapes are then

$$Y_r(x) = \sqrt{\frac{2}{mL}} \sin\left(\frac{r\pi x}{L}\right), \quad r = 1, 2, \ldots, \tag{10.74}$$

which are plotted in Figure 10.17 for $m = 0.1$ slug/in and $L = 120$ in.

The response for each mode, with index r, is obtained by forming the product $F_r(t)Y_r(x)$, as per Equation 10.63,

$$y_r(x,t) = A_r \sqrt{\frac{2}{mL}} \cos(\omega_r t - \phi_r) \sin\left(\frac{r\pi x}{L}\right), \quad r = 1, 2, \ldots.$$

Summing over r, the complete response is

$$y(x,t) = \sum_{r=1}^{\infty} \left[A_r \sqrt{\frac{2}{mL}} \cos(\omega_r t - \phi_r) \sin\left(\frac{r\pi x}{L}\right) \right], \tag{10.75}$$

where A_r and ϕ_r are determined by satisfying the initial conditions. As a practical matter, one retains as many terms as necessary in Equation 10.75 to attain sufficient accuracy. How sufficient depends on the particular problem and the parameter values of the system.

We complete this problem by considering the specific initial displacement $y(x, 0) = Cx(1 - x/L)$ and zero initial velocity $\dot{y}(x, 0) = 0$. These initial conditions can be satisfied by solving for A_r and ϕ_r, for which we need the orthogonality properties of the modes, a topic to be discussed in Section 10.5.4. Therefore, this example will be concluded in *Example* 10.6,

---

[17]Note that $\omega_r = r^2 \omega_1$.

10.5. TRANSVERSE VIBRATION OF BEAMS

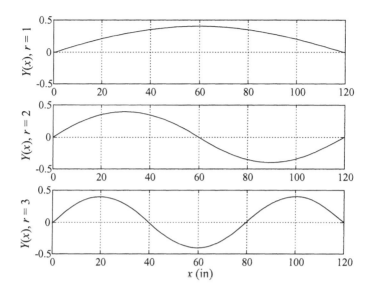

Figure 10.17: Mode shapes for hinged–hinged beam.

where vibration–induced stresses are also discussed. In the following example alternate boundary conditions are considered. ∎

Example 10.4 Solution for the Fixed–Free Case
Consideration of alternate boundary conditions requires picking up the analysis of the last example at Equation 10.69, $Y(x) = C_1 \sin \beta x + C_2 \cos \beta x + C_3 \sinh \beta x + C_4 \cosh \beta x$. For a beam fixed at $x = 0$ and the other end free at $x = L$, the boundary conditions for the eigenfunctions are

$$Y(0) = 0, \qquad Y'(0) = 0,$$
$$Y''(L) = 0, \qquad Y'''(L) = 0.$$

At the fixed end, the displacement and slope are zero, and at free end, the moment and shear force are zero. Applying the boundary conditions for $x = 0$, we find

$$C_2 + C_4 = 0$$
$$C_1 + C_3 = 0,$$

so that the eigenfunction is reduced to

$$Y(x) = C_1(\sin \beta x - \sinh \beta x) + C_2(\cos \beta x - \cosh \beta x). \tag{10.76}$$

Applying the two conditions at $x = L$, results in the relations

$$C_1(\sin\beta L + \sinh\beta L) + C_2(\cos\beta L + \cosh\beta L) = 0 \quad (10.77)$$
$$C_1(\cos\beta L + \cosh\beta L) - C_2(\sin\beta L - \sinh\beta L) = 0. \quad (10.78)$$

Solve 10.78 for C_2 in terms of C_1, and substitute into 10.76 to find

$$Y(x) = \frac{C_1}{\sin\beta L - \sinh\beta L}[(\sin\beta L - \sinh\beta L)(\sin\beta x - \sinh\beta x) + (\cos\beta L + \cosh\beta L)(\cos\beta x - \cosh\beta x)], \quad (10.79)$$

where C_1 is eventually incorporated into the complete solution, as shown here. To establish the frequencies that the beam can sustain in free vibration, rewrite Equations 10.77 and 10.78 in matrix form,

$$\begin{bmatrix} (\sin\beta L + \sinh\beta L) & (\cos\beta L + \cosh\beta L) \\ (\cos\beta L + \cosh\beta L) & -(\sin\beta L - \sinh\beta L) \end{bmatrix} \begin{Bmatrix} C_1 \\ C_2 \end{Bmatrix} = \begin{Bmatrix} 0 \\ 0 \end{Bmatrix}.$$

The determinant of the matrix must be equal to zero for a non–trivial solution to exist, that is,

$$(\sin\beta L + \sinh\beta L)(\sin\beta L - \sinh\beta L) + (\cos\beta L + \cosh\beta L)^2 = 0.$$

This is the characteristic equation. Simplification results in the implicit equation

$$\cos\beta L \cosh\beta L = -1,$$

which must be solved for βL numerically. Recall that $\beta^4 = \omega^2 m/EI$, or $\omega = \beta^2\sqrt{EI/m}$. The first three roots are $\beta L = 1.875, 4.694, 7.855$, and therefore

$$\omega_1 = (1.875)^2\sqrt{\frac{EI}{mL^4}} \text{ rad/sec}$$

$$\omega_2 = (4.694)^2\sqrt{\frac{EI}{mL^4}} \text{ rad/sec}$$

$$\omega_3 = (7.855)^2\sqrt{\frac{EI}{mL^4}} \text{ rad/sec}.$$

Equation 10.79 for the modes is plotted in Figure 10.18 for $L = 120$ in, assuming $C_1 = 1$. For example, using Equation 10.79, for $\beta L = 1.875$, the first mode is given by

$$Y(x) = \frac{C_1}{-2.230}[-2.230(\sin\beta x - \sinh\beta x) + 3.037(\cos\beta x - \cosh\beta x)].$$

10.5. TRANSVERSE VIBRATION OF BEAMS

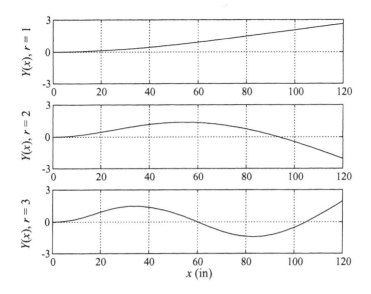

Figure 10.18: First three eigenfunctions for clamped–free beam; $C_1 = 1$.

The solution for each mode r is given by

$$\begin{aligned} y_r(x,t) &= B_r Y_r(x) F_r(t) \\ &= \overline{C}_r Y_r(x) \cos(\omega_r t - \phi_r), \end{aligned} \quad (10.80)$$

where $\overline{C}_r$ is the arbitrary constant obtained by multiplying the two constants from each solution $Y_r(x)$ and $F_r(t)$. In Equation 10.79, there is one eigenfunction $Y(x)$ for each root βL. The complete solution is then the sum

$$y(x,t) = \sum_{r=1}^{\infty} y_r(x,t),$$

where $\overline{C}_r$ and ϕ_r are determined by satisfying the initial conditions. ∎

In the next section, the orthogonality properties of the modes are further explored.

10.5.4 Orthogonality of the Normal Modes

The orthogonality of the eigenfunctions is demonstrated for the bending beam. The following procedure is applicable to all other vibration problems considered in this chapter.

The eigenvalue problem is defined by Equation 10.66, and each eigenfunction will satisfy this relation. Denote two distinct solutions by $Y_r(x)$ and $Y_s(x)$, so that

$$\frac{d^2}{dx^2}[EI(x)Y_r''(x)] = \omega_r^2 m(x) Y_r(x) \tag{10.81}$$

$$\frac{d^2}{dx^2}[EI(x)Y_s''(x)] = \omega_s^2 m(x) Y_s(x). \tag{10.82}$$

Multiply Equation 10.81 by $Y_s(x)$, and integrate the left hand side by parts twice over the domain $0 \le x \le L$ to find

$$\int_0^L Y_s(x) \frac{d^2}{dx^2}[EI(x)Y_r''(x)]\, dx = \left| Y_s(x)\frac{d}{dx}[EI(x)Y_r''(x)] \right|_0^L$$

$$- [Y_s'(x) EI(x) Y_r''(x)]_0^L + \int_0^L EI(x) Y_r''(x) Y_s''(x)\, dx$$

$$= \omega_r^2 \int_0^L m(x) Y_r(x) Y_s(x)\, dx. \tag{10.83}$$

Similarly, multiply Equation 10.82 by $Y_r(x)$, again integrate the left hand side by parts twice, and obtain an equation of the same form as Equation 10.83. Subtract this second equation from Equation 10.83 to find

$$(\omega_r^2 - \omega_s^2)\int_0^L m(x) Y_r(x) Y_s(x)\, dx = \left| Y_s(x) \frac{d}{dx}[EI(x)Y_r''(x)] \right|_0^L$$

$$- [Y_s'(x) EI(x) Y_r''(x)]_0^L - \left| Y_r(x) \frac{d}{dx}[EI(x) Y_s''(x)] \right|_0^L$$

$$+ [Y_r'(x) EI(x) Y_s''(x)]_0^L .$$

For any combination of fixed, hinged, or free ends, all the terms on the right hand side vanish, resulting in

$$(\omega_r^2 - \omega_s^2) \int_0^L m(x) Y_r(x) Y_s(x)\, dx = 0, \tag{10.84}$$

and since this derivation assumes $\omega_r \ne \omega_s$, the factor $(\omega_r^2 - \omega_s^2)$ can be divided out to find the first orthogonal relation,

$$\int_0^L m(x) Y_r(x) Y_s(x)\, dx = 0, \quad r \ne s. \tag{10.85}$$

In Equation 10.84, if $r = s$, then we cannot divide by $(\omega_r^2 - \omega_s^2)$. This suggests a normalization condition that includes Equation 10.85 as a special

10.5. TRANSVERSE VIBRATION OF BEAMS

case,
$$\int_0^L m(x)Y_r(x)Y_s(x)dx = \delta_{rs}, \quad r,s = 1,2,\ldots, \tag{10.86}$$

where the *Kronecker delta* is defined as

$$\delta_{rs} = \begin{cases} 1 & \text{if } r = s \\ 0 & \text{if } r \neq s. \end{cases}$$

Therefore, for $r = s$, Equation 10.86 becomes

$$\int_0^L m(x)Y_r^2(x)dx = 1,$$

naturally yielding the identity $(\omega_r^2 - \omega_r^2) \cdot 1 = 0$ for $r = s$ in Equation 10.84. Substituting Equation 10.86 into 10.83 results in the equation

$$\int_0^L Y_s(x)\frac{d^2}{dx^2}[EI(x)Y_r''(x)]\,dx = \omega_r^2 \delta_{rs}, \quad r,s = 1,2,\ldots. \tag{10.87}$$

After integration by parts, using the same procedure and assumptions as before, we find

$$\int_0^L EI(x)Y_r''(x)Y_s''(x)dx = \omega_r^2 \delta_{rs}, \quad r,s = 1,2,\ldots. \tag{10.88}$$

These orthogonality properties lead to a modal solution technique for the continuous–parameter vibration problem. This possibility may have been anticipated since the continuous system was shown to be the limit of the discrete system, as the number of degrees of freedom approach infinity.

The next example demonstrates how this process works and, at the same time, includes a viscous damping term and an external forcing term on the right hand side of the equation of motion.

Example 10.5 Transverse Vibration Response of a Forced and Damped Beam

Begin with Equation 10.62, and add viscous damping and external forcing. For a simply–supported beam with uniform properties, the equation of motion is

$$EI\frac{\partial^4 y}{\partial x^4} + c\frac{\partial y}{\partial t} + m\frac{\partial^2 y}{\partial t^2} = p(x,t), \tag{10.89}$$

where the natural frequencies are given by Equation 10.72 and the mode shapes by Equation 10.74. To solve this problem, use the normal mode method, which utilizes the modal orthogonality properties. First, expand the applied force $p(x,t)$ in terms of the modes, then do the same with the structural displacement,[18]

$$p(x,t) = \sum_{j=1}^{\infty} p_j(t) Y_j(x).$$

Multiply both sides of this equation by $mY_k(x)$ and then integrate over the beam span,

$$\int_0^L mp(x,t)Y_k(x)dx = \sum_{j=1}^{\infty} p_j(t) \int_0^L mY_j(x)Y_k(x)dx.$$

On the right hand side of this equation the orthogonality properties of the modes imply that the integral equals zero for $k \neq j$, and otherwise is normalized to 1 for $k = j$. Therefore,

$$\int_0^L mp(x,t)Y_j(x)dx = p_j(t). \tag{10.90}$$

Apply the same procedure to the displacement response

$$y(x,t) = \sum_{j=1}^{\infty} y_j(t) Y_j(x), \tag{10.91}$$

and multiply and integrate as above to find

$$\int_0^L my(x,t)Y_j(x)dx = y_j(t).$$

$y(x,t)$ can be differentiated with respect to x and t so that all the terms in the equation of motion Equation 10.89 can be put into the modal expansion form,

$$\sum_{j=1}^{\infty} \left[EIy_j(t)\frac{d^4Y_j}{dx^4} + c\frac{dy_j}{dt}Y_j + m\frac{d^2y_j}{dt^2}Y_j \right] = \sum_{j=1}^{\infty} p_j(t)Y_j(x). \tag{10.92}$$

The modes are defined by and satisfy Equation 10.67: $EId^4Y_j/dx^4 = \omega_j^2 mY_j$. Substituting this relation into Equation 10.92 leads to

$$\sum_{j=1}^{\infty} [\omega_j^2 my_j + c\dot{y}_j + m\ddot{y}_j - p_j(t)]Y_j(x) = 0.$$

---

[18] This procedure is the *method of eigenfunction expansion*. We have previously applied it to multi degree of freedom structures in Chapter 8. It is also known as a *modal expansion*. It is assumed here that the modes $Y_j(x)$ satisfying the boundary conditions of the problem have been obtained. For the case where non–harmonic, time–dependent boundary conditions exist, the method of Section 11.5.1 needs to be applied.

10.5. TRANSVERSE VIBRATION OF BEAMS

Since the eigenfunctions $Y_j(x)$ are not zero except for unrestrained motion, the expression in the square brackets must vanish for every j,

$$\ddot{y}_j + \frac{c}{m}\dot{y}_j + \omega_j^2 y_j = \frac{1}{m}p_j(t), \quad j = 1, 2, \ldots \quad (10.93)$$

As for single degree of freedom systems, let $c/m = 2\zeta_j\omega_j$, thus defining

$$\zeta_j = \frac{c}{2m\omega_j}.$$

Equation 10.93 is the damped, forced, harmonic oscillator, with the well-known convolution solution,

$$y_j(t) = \frac{1}{m}\int_0^t p_j(\tau)g_j(t-\tau)d\tau, \quad (10.94)$$

where $g_j(t)$ is the impulse response function for a damped oscillator. Once solved, $y_j(t)$ is substituted into the expansion equation

$$\begin{aligned}y(x,t) &= \sum_{j=1}^\infty y_j(t)Y_j(x) \\ &= \frac{1}{m}\sum_{j=1}^\infty Y_j(x)\int_0^t p_j(\tau)g_j(t-\tau)d\tau,\end{aligned}$$

and, substituting Equation 10.90 for $p_j(t)$, results in

$$y(x,t) = \sum_{j=1}^\infty Y_j(x)\int_0^t \left[\int_0^L p(\xi,\tau)Y_j(\xi)d\xi\right]g_j(t-\tau)d\tau. \quad (10.95)$$

This is the complete steady–state time domain solution. If the transient response to non–zero initial conditions is required, then it must be added to Equation 10.94. ∎

Bernoulli

Daniel Bernoulli was the son of Johann Bernoulli. He was born in Groningen on 8 Feb 1700 while his father held the chair of mathematics there. He died on 17 March 1782 in Basel, Switzerland. His older brother was Nicolaus(II) Bernoulli and his uncle was Jacob Bernoulli so he was born into a family of leading mathematicians but also into a family where there was unfortunate rivalry, jealousy and bitterness.

When Daniel was five years old the family returned to their native city of Basel where Daniel's father filled the chair of mathematics left vacant on the death of his uncle Jacob

Bernoulli. When Daniel was five years old his younger brother Johann(II) Bernoulli was born. All three sons would go on to study mathematics.

Daniel wanted to embark on an academic career like his father so he applied for two chairs at Basel. His application for the chair of anatomy and botany was decided by drawing of lots and he was unlucky in this game of chance. The next chair to fall vacant at Basel that Daniel applied for was the chair of logic, but again the game of chance of the final selection by drawing of lots went against him. Having failed to obtain an academic post, Daniel went to Venice to study practical medicine.

While in Venice, Daniel had also designed an hour glass to be used at sea so that the trickle of sand was constant even when the ship was rolling in heavy seas. He submitted his work on this to the Paris Academy and in 1725, the year he returned from Italy to Basel, he learnt that he had won the prize of the Paris Academy. Daniel had also attained fame through his work Mathematical exercises and on the strength of this he was invited to take up the chair of mathematics at St Petersburg. His brother Nicolaus(II) Bernoulli was also offered a chair of mathematics at St Petersburg so in late 1725 the two brothers travelled to St Petersburg.

Within eight months of their taking up the appointments in St Petersburg Daniel's brother died of fever. Daniel was left, greatly saddened at the loss of his brother and also very unhappy with the harsh climate. He thought of returning to Basel and wrote to his father telling him how unhappy he was in St Petersburg. Johann Bernoulli was able to arrange for one of his best pupils, Leonard Euler, to go to St Petersburg to work with Daniel. Euler arrived in 1727 and this period in St Petersburg, which Daniel left in 1733, was to be his most productive time.

One of the topics which Daniel studied in St Petersburg was that of vibrating systems. As Straub writes: "From 1728, Bernoulli and Euler dominated the mechanics of flexible and elastic bodies, in that year deriving the equilibrium curves for these bodies. ... Bernoulli determined the shape that a perfectly flexible thread assumes when acted upon by forces of which one component is vertical to the curve and the other is parallel to a given direction. Thus, in one stroke he derived the entire series of such curves as the velaria, linteraria, catenaria... "

While in St Petersburg he made one of his most famous discoveries when he defined the simple nodes and the frequencies of oscillation of a system. He showed that the movements of strings of musical instruments are composed of and infinite number of harmonic vibrations all superimposed on the string.

A second important work which Daniel produced while in St Petersburg was one on probability and political economy. Daniel makes the assumption that the moral value of the increase in a person's wealth is inversely proportional to the amount of that wealth. He then assigns probabilities to the various means that a person has to make money and deduces an expectation of increase in moral expectation. Daniel applied some of his deductions to insurance.

Undoubtedly the most important work which Daniel Bernoulli did while in St Petersburg was his work on hydrodynamics. Even the term itself is based on the title of

10.5. TRANSVERSE VIBRATION OF BEAMS

the work which he produced called *Hydrodynamica* and, before he left St Petersburg, Daniel left a draft copy of the book with a printer. However the work was not published until 1738 and although he revised it considerably between 1734 and 1738, it is more the presentation that he changed rather then the substance. This work contains for the first time the correct analysis of water flowing from a hole in a container. This was based on the principle of conservation of energy which he had studied with his father in 1720. Daniel also discussed pumps and other machines to raise water. One remarkable discovery appears in Chapter 10 of *Hydrodynamica* where Daniel discussed the basis for the kinetic theory of gases. He was able to give the basic laws for the theory of gases and gave, although not in full detail, the equation of state discovered by Van der Waals a century later.

In 1750, he was appointed to the chair of physics and taught physics at Basel for 26 years until 1776. He gave some remarkable physics lectures with experiments performed during the lectures. Based on experimental evidence he was able to conjecture certain laws which were not verified until many years later. Among these was Coulomb's law in electrostatics.

Daniel Bernoulli did produce other excellent scientific work during these many years back in Basel. In total he won the Grand Prize of the Paris Academy 10 times, for topics in astronomy and nautical topics. He won in 1740 (jointly with Euler) for work on Newton's theory of the tides; in 1743 and 1746 for essays on magnetism; in 1747 for a method to determine time at sea; in 1751 for an essay on ocean currents; in 1753 for the effects of forces on ships; and in 1757 for proposals to reduce the pitching and tossing of a ship in high seas.

Another important aspect of Daniel Bernoulli's work that proved important in the development of mathematical physics was his acceptance of many of Newton's theories and his use of these together with the more powerful calculus of Leibniz. Daniel worked on mechanics and again used the principle of conservation of energy which gave an integral of Newton's basic equations. He also studied the movement of bodies in a resisting medium using Newton's methods.

He also continued to produce good work on the theory of oscillations and in a paper he gave a beautiful account of the oscillation of air in organ pipes. His strengths and weaknesses are summed up by Straub "Bernoulli's active and imaginative mind dealt with the most varied scientific areas. Such wide interests, however, often prevented him from carrying some of his projects to completion. It is especially unfortunate that he could not follow the rapid growth of mathematics that began with the introduction of partial differential equations into mathematical physics. Nevertheless he assured himself a permanent place in the history of science through his work and discoveries in hydrodynamics, his anticipation of the kinetic theory of gases, a novel method for calculating the value of an increase in assets, and the demonstration that the most common movement of a string in a musical instrument is composed of the superposition of an infinite number of harmonic vibrations..."

Daniel Bernoulli was much honored in his own lifetime. He was elected to most of he leading scientific societies of his day including those in Bologna, St Petersburg, Berlin, Paris, London, Bern, Turin, Zurich and Mannheim.

Example 10.6 Continuation of *Example* 10.3: Evaluation of Arbitrary Constants and Vibration–Induced Stresses

We continue *Example* 10.3 by evaluating the arbitrary constants A_r and ϕ_r of Equation 10.75. Also of interest is the connection between vibration and a stress computation that can be used to estimate the fatigue life of the structure. Use the following form for the solution,

$$y(x,t) = \sum_{r=1}^{\infty} \sqrt{\frac{2}{mL}} [A_r \cos\omega_r t + B_r \sin\omega_r t] \sin\left(\frac{r\pi x}{L}\right),$$

and evaluate the constants[19] A_r and B_r. The initial conditions are $y(x,0) = Cx(1-x/L)$ and $\dot{y}(x,0) = 0$.

Apply the initial displacement,

$$y(x,0) = Cx\left(1 - \frac{x}{L}\right) = \sum_{r=1}^{\infty} \sqrt{\frac{2}{mL}} A_r \sin\frac{r\pi x}{L}, \qquad (10.96)$$

and initial velocity,

$$\dot{y}(x,0) = 0 = \sum_{r=1}^{\infty} \sqrt{\frac{2}{mL}} B_r \omega_r \sin\frac{r\pi x}{L}.$$

From the second condition, $B_r = 0$ for all r.

Multiply Equation 10.96 by $\sin p\pi x/L$, where p is another index, and integrate over the length of the beam,

$$\int_0^L Cx\left(1 - \frac{x}{L}\right) \sin\frac{p\pi x}{L} dx = \sum_{r=1}^{\infty} \sqrt{\frac{2}{mL}} \int_0^L A_r \sin\frac{r\pi x}{L} \sin\frac{p\pi x}{L} dx,$$

to take advantage of orthogonality. The integral on the right hand side equals zero except where $p = r$. Thus, performing the integrations leads to

$$\frac{2CL^2}{(r\pi)^3}(1 - \cos r\pi) = \sqrt{\frac{2}{mL}} A_r \frac{L}{2}.$$

The term $(1 - \cos r\pi)$ equals zero for r *even* and equals two for r *odd*. Therefore,

$$A_r = \frac{8C}{(r\pi)^3}\sqrt{\frac{mL^3}{2}}, \quad r = 1, 3, \ldots,$$

---

[19] This is not the same A_r as in *Example* 10.3.

10.5. TRANSVERSE VIBRATION OF BEAMS

and the response is given by

$$y(x,t) = \sum_{r=1,3,\ldots}^{\infty} \frac{8LC}{(r\pi)^3} \cos\omega_r t \sin\frac{r\pi x}{L}. \tag{10.97}$$

The displacement of the beam as a function of space and time can be related to the stresses that develop in a beam undergoing vibratory motion. Stress oscillations are directly related to the fatigue life and reliability of a structure and its components.

Elementary mechanics of materials relates the bending moment M to the second derivative of the slope by the equation

$$M = EI\frac{\partial^2 y}{\partial x^2},$$

and the stress σ_M on an element of beam a distance c from the neutral axis of the beam is

$$\sigma_M = \frac{Mc}{I} = Ee\frac{\partial^2 y}{\partial x^2}.$$

Then, using Equation 10.97,

$$\sigma_M(x,t) = Ee \sum_{r=1,3,\ldots}^{\infty} \frac{-8C}{r\pi L} \sin\frac{r\pi x}{L} \cos\omega_r t.$$

The stress can be evaluated for particular x and t. For a simply supported beam, the maximum stress occurs at midspan, $x = L/2$. Then, $\sin r\pi x/L = \sin r\pi/2$ which is non–zero for r odd,

$$\sigma_M\left(\frac{L}{2},t\right) = -Ee \sum_{r=1,3,\ldots}^{\infty} \frac{8C}{r\pi L} \sin\frac{r\pi}{2} \cos\omega_r t.$$

$\omega_r t$ can be written in terms of the fundamental frequency ω_1,

$$\begin{aligned}\omega_r &= r^2\left(\frac{\pi}{L}\right)^2 \sqrt{\frac{EI}{m}} \\ &= r^2\omega_1,\end{aligned}$$

and then $\omega_r t = r^2\omega_1 t$. The fundamental frequency is related to the fundamental period by $\omega_1 = 2\pi/\tau_1$. Therefore, at $t = \tau_1/2$, we have $\omega_r t = r^2\pi$, and $\cos\omega_r t = \cos r^2\pi = -1$ for r odd. This specific stress is then given[20] by

$$\sigma_M\left(\frac{L}{2},\frac{\tau_1}{2}\right) = Ee\sum_{r=1,3,\ldots}^{\infty}\frac{8C}{r\pi L}\sin\frac{r\pi}{2}$$

$$\sigma_M\left(\frac{L}{2},\frac{\tau_1}{2}\right) = \frac{2EeC}{L}, \qquad (10.98)$$

with the same units as the modulus of elasticity E, such as psi.

To conclude with numerical results, assume the beam is a pipe of length $L = 200$ in, and modulus of elasticity $E = 30 \times 10^6$ psi. Assume initial value constant $C = 1$, and $e = 2$ in. The stress calculation using Equation 10.98 yields

$$\sigma_M\left(\frac{L}{2},\frac{\tau_1}{2}\right) = \frac{2\times(30\times10^6)\times 2\times 1}{200}\text{ psi} = 0.6\times 10^6\text{ psi}.$$

Note that the stress is linearly related to the value of C. ∎

The study of transversely vibrating beams continues in the next section for beams with special boundary conditions and forces.

Timoshenko

Stephen P. Timoshenko was born in 1878 in Russia. His father Timofeyevich was a surveyor and his mother was a voracious reader. During his early childhood, Timoshenko enjoyed playing in piles of sand near building construction sites - he built fortresses, castles and especially rail roads. He joined The Petersburg Polytechnic in 1903 and in 1904 he travelled to Europe to become better acquainted with German technical school and their teaching methods. Timoshenko took interest in reading Rayleigh's book, *The Theory of Sound*, and he was particularly captivated by the approximate methods of calculating vibration frequencies of complex structures. In 1907–1908 he gave a course on strength of materials and later on published the notes in lithographic form. The books of Lord Raleigh exerted a large influence on the development of Timoshenko's scientific work.

He investigated a number of new problems involving the stability of a compressed bar. In connection with the Quebec Bridge disaster in Canada he started working on the theory of stability of composite beams and found simpler methods of solving problems.

---

[20]We make use of the identity

$$\sum_{r=1,3,\ldots,}^{\infty}\frac{1}{r}\sin\frac{r\pi}{2} = (1-\frac{1}{3}+\frac{1}{5}-\frac{1}{7}+\cdots) = \frac{\pi}{4}.$$

10.6. BEAM VIBRATION: SPECIAL PROBLEMS

In 1912 when he went to England and found that the lab facility at Cambridge University was poorer than those in German laboratories.

In 1913 Timoshenko became Professor of Ministry of Ways of Communications and Electrical Engineering Institute. Then he joined as Professor at Zagreb Polytechnic and he continued up to 1922 and for some time Timoshenko worked at Westinghouse. His lectures on Applied Mechanics at the University of Michigan attracted a large number of students from other departments and also young teachers. In Michigan a summer session in applied mechanics was instituted and Prandtl, Southwell, Westergaard and von Kármán participated. During this period he published a number of books such as *Strength of Materials*, *Theory of Elasticity* and *Elastic Stability*. In 1936 he moved to Stanford University and during this period he published *Engineering Mechanics*, *Theory of Plates and Shells*, *Theory of Structures* and *Advanced Dynamics*. He wrote his last book on the *History of Strength of Materials* where he traced the history from Leonardo da Vinci and Galileo to the present. The Stanford University laboratory of Engineering Mechanics was named after him. He was elected to the National Academy of Sciences and the Royal Society. He received many honorary doctoral degrees from various universities such as Lehigh, Zurich Tech Institute and Glasgow University. In 1935, American Society of Mechanical Engineers conferred upon him the Worcester Reed Warner Medal for achievement in the field of mechanics.

From the perspective of more than half a century, Timoshenko's great influence upon applied science and technology in America resulted less from his original, creative discoveries than from his ideals of engineering education, his superb skill as a teacher, and his highly developed pragmatic skill in using fragments of exact solutions for a variety of approximate solutions to difficult problems in applied mechanics. He died in 1972.

10.6 Beam Vibration: Special Problems

The following sections discuss special beam vibration problems. Special forces and boundary conditions are examined in order to demonstrate the power of these models to explain large classes of problems.

10.6.1 Transverse Vibration of Beam with Axial Force

Consider the lateral vibration of a beam, loaded both laterally and axially, as in Figure 10.19. Our interest lies with establishing the added effect of the axial force on the response. Whenever a beam or column is compressed, there is concern about its buckling. As the axial load is increased, a critical value is reached where a new (buckled) deformation configuration is possible. Since this configuration is generally undesirable, structural failure is assumed to occur at this critical load. Design codes generally assume a failure at some load less than the buckling load.

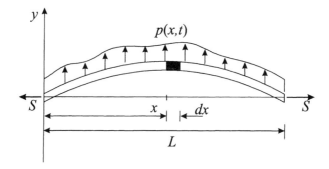

Figure 10.19: Beam with transverse and axial forces.

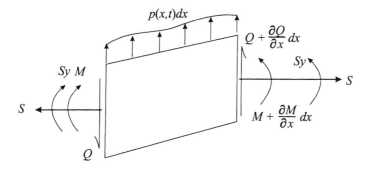

Figure 10.20: Free body for the transverse vibration of beam with axial force.

To formulate this problem, use the free body diagram of an arbitrary section of the beam and draw all external forces, noting that there is an additional moment term Sy due to the constant axial force S, where y is the deflection at the section under consideration. This is shown in Figure 10.20.

While the equation of motion in the vertical y direction remains the same (since S acts approximately perpendicular to y), the moment equation about the center of the cross–section now must include the moment due to S, that is Sy,

$$M + Sy = EI(x)\frac{\partial^2 y}{\partial x^2},$$

and governing Equation 10.62 becomes

$$\frac{\partial^2}{\partial x^2}\left[EI(x)\frac{\partial^2 y}{\partial x^2} - Sy\right] + p(x,t) = -m(x)\frac{\partial^2 y}{\partial t^2}.$$

10.6. BEAM VIBRATION: SPECIAL PROBLEMS

Making the uniformity assumption as before, setting $p(x,t) = 0$ for the eigenvalue problem, eigenvalue Equation 10.67 becomes

$$EI\frac{d^4}{dx^4}Y(x) - S\frac{d^2}{dx^2}Y(x) = \omega^2 m Y(x), \quad (10.99)$$

where the effect of axial force S is significant. For this problem assume

$$Y_r(x) = C_r \sin\frac{r\pi x}{L}, \quad r = 1, 2, \ldots,$$

take appropriate derivatives and substitute these into Equation 10.99, to find

$$EIC_r\left(\frac{r\pi}{L}\right)^4 \sin\frac{r\pi x}{L} + SC_r\left(\frac{r\pi}{L}\right)^2 \sin\frac{r\pi x}{L} = \omega_r^2 m C_r \sin\frac{r\pi x}{L},$$

where we note that the coefficients C_r can be cancelled. Thus,

$$\omega_r = \left(\frac{r\pi}{L}\right)^2 \sqrt{\frac{EI}{m}}\sqrt{1 + \frac{S}{EI}\left(\frac{L}{r\pi}\right)^2}, \quad r = 1, 2, \ldots.$$

Note that for a tensile axial force $+S$, the effect is an increase in the frequencies of free vibration.[21] Had a compressive force $-S$ been applied, the frequencies would be given by

$$\omega_r = \left(\frac{r\pi}{L}\right)^2 \sqrt{\frac{EI}{m}}\sqrt{1 - \frac{S}{EI}\left(\frac{L}{r\pi}\right)^2}, \quad (10.100)$$

resulting in lower natural frequencies. The question to ask here is, for what compressive load will the frequencies shift down so that the fundamental frequency becomes zero?

For $r = 1$, the term

$$\frac{S}{EI}\left(\frac{L}{r\pi}\right)^2$$

is the ratio of S to the Euler buckling load. If $SL^2/EI\pi^2 \to 1$, the lowest mode of vibration approaches a zero frequency and transverse buckling occurs for $S = EI\pi^2/L^2$. A plot of ω_1 as a function of S using Equation 10.100 is shown in Figure 10.21 for a solid steel column of circular cross section of radius 1 in, length $L = 10$ in, and mass per unit length of $m = 0.283\,\pi$ lb-sec^2/in^2.

---

[21] We had known this intuitively. This is comparable to the tension in the string, where higher tension results in a higher set of frequencies.

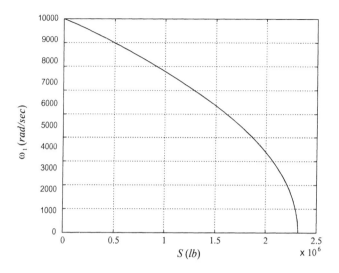

Figure 10.21: Natural frequency ω_1 as a function of axial force S.

10.6.2 Transverse Vibration of Beam with Elastic Restraints

To demonstrate again the importance of the boundary conditions, let us consider the effect of the degree of fixity of the supports. In previous problems, the boundaries were clearly defined as fixed, hinged, or free. Suppose, however, that the boundary properties are in reality somewhere between fixed and hinged. What would the effect be and how could this be modeled in the equation for the boundary condition? Physically, such a boundary would resist deflection and rotation, with properties that are a mix of the fixed and the hinged ends. One possibility for modeling such intermediate conditions is to restrain the ends of the beam using a combination of linear and torsional springs, as in Figure 10.22.

The eigenvalue problem and the general solution are identical to previous results, except that there are the following boundary conditions for shear force V and moment M,

$$\begin{aligned} V(0) &= EIY'''(0) = -k_1 Y(0) \\ V(L) &= EIY'''(L) = k_3 Y(L) \\ M(0) &= EIY''(0) = k_2 Y'(0) \\ M(L) &= EIY''(L) = -k_4 Y'(L). \end{aligned}$$

The sign convention used above takes positive shear to be up on the left

10.6. BEAM VIBRATION: SPECIAL PROBLEMS

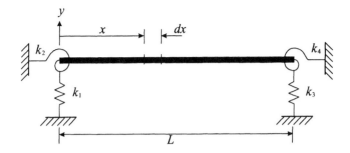

Figure 10.22: Beam with elastic restraints.

end and down on the right end of the beam, and a positive pair of bending moments results in a concave up beam.

Begin with general eigenfunction Equation 10.69,

$$Y(x) = C_1 \sin \beta x + C_2 \cos \beta x + C_3 \sinh \beta x + C_4 \cosh \beta x,$$

along with the appropriate derivatives,

$$\begin{aligned} Y'(x) &= \beta(C_1 \cos \beta x - C_2 \sin \beta x + C_3 \cosh \beta x + C_4 \sinh \beta x) \\ Y''(x) &= \beta^2(-C_1 \sin \beta x - C_2 \cos \beta x + C_3 \sinh \beta x + C_4 \cosh \beta x) \\ Y'''(x) &= \beta^3(-C_1 \cos \beta x + C_2 \sin \beta x + C_3 \cosh \beta x + C_4 \sinh \beta x). \end{aligned}$$

Substituting these into the equations for the boundary conditions leads to the following four simultaneous algebraic equations,

$$EI\beta^3 C_1 - k_1 C_2 - EI\beta^3 C_3 - k_1 C_4 = 0$$
$$(-EI\beta^3 \cos \beta L - k_3 \sin \beta L)C_1 + (EI\beta^3 \sin \beta L - k_3 \cos \beta L)C_2$$
$$+(EI\beta^3 \cosh \beta L - k_3 \sinh \beta L)C_3 + (EI\beta^3 \sinh \beta L - k_3 \cosh \beta L)C_4 = 0$$
$$-k_2 C_1 - EI\beta C_2 - k_2 C_3 + EI\beta C_4 = 0$$
$$(-EI\beta \sin \beta L + k_4 \cos \beta L)C_1 + (-EI\beta \cos \beta L - k_4 \sin \beta L)C_2$$
$$+(EI\beta \sinh \beta L + k_4 \cosh \beta L)C_3 + (EI\beta \cosh \beta L + k_4 \sinh \beta L)C_4 = 0.$$

As before, a non–trivial solution exists only if the 4x4 determinant of the coefficients of $C_1, \ldots, C_4$ is set equal to zero,

$$\begin{vmatrix} EI\beta^3 & -k_1 & -EI\beta^3 & -k_1 \\ k_{21} & k_{22} & k_{23} & k_{24} \\ -k_2 & -EI\beta & -k_2 & EI\beta \\ k_{41} & k_{42} & k_{43} & k_{44} \end{vmatrix} = 0. \qquad (10.101)$$

where

$$k_{21} = -EI\beta^3 \cos\beta L - k_3 \sin\beta L$$
$$k_{22} = EI\beta^3 \sin\beta L - k_3 \cos\beta L$$
$$k_{23} = EI\beta^3 \cosh\beta L - k_3 \sinh\beta L$$
$$k_{24} = EI\beta^3 \sinh\beta L - k_3 \cosh\beta L$$

and

$$k_{41} = -EI\beta \sin\beta L + k_4 \cos\beta L$$
$$k_{42} = -EI\beta \cos\beta L - k_4 \sin\beta L$$
$$k_{43} = EI\beta \sinh\beta L + k_4 \cosh\beta L$$
$$k_{44} = EI\beta \cosh\beta L + k_4 \sinh\beta L.$$

This results in the characteristic equation for the frequencies, which are then used to evaluate the modes of vibration.

Note that the fixed and free boundary conditions can be recovered here by setting the appropriate spring constants to zero or infinity. For example, letting $k_1 = k_2 \longrightarrow \infty$ and $k_3 = k_4 \longrightarrow 0$ results in the fixed–free beam.

10.6.3 Transverse Vibration of Beam on Elastic Foundation

All previous boundary conditions were at discrete points, the ends of the beam. It is possible that a continuous boundary effect occurs along the length of the beam. In such cases, the boundary condition becomes part of the governing equation. An important example of such a problem is the beam on an elastic foundation,[22] a schematic of which appears in Figure 10.23. Applications include machine vibration, the vibration of a structure on a foundation, and structural response to ground shock such as earthquakes and explosives.

Here the elastic restraint against transverse motion is distributed continuously along the length of the beam and damping effects are ignored. For element dx, the differential equation of motion, following Equation 10.62, is

$$\frac{\partial^2}{\partial x^2}\left[EI(x)\frac{\partial^2 y}{\partial x^2}\right]dx = -k_f(x)y\,dx - \varrho A(x)dx\frac{\partial^2 y}{\partial t^2},$$

where the restraining force is given by the term $k_f(x)y\,dx$. The parameter $k_f(x)$ is the constant stiffness per unit length of foundation, and $\varrho A(x) =$

---
[22] The example of the rotating shaft that is immersed in a lubricated sleeve is one where there is continuous damping along the length instead of continuous stiffness.

10.6. BEAM VIBRATION: SPECIAL PROBLEMS

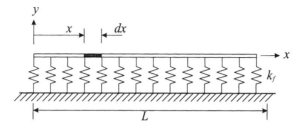

Figure 10.23: Beam on elastic foundation.

$m(x)$, the mass of the beam per unit length. For a *prismatic beam* (uniform cross–section),

$$EI\frac{\partial^4 y}{\partial x^4} + k_f y = -\varrho A \frac{\partial^2 y}{\partial t^2}.$$

It is clear from this equation how the foundation stiffness alters the mathematical character of the governing equation. Assuming a product solution and performing the usual derivatives and substitutions, the equation governing the modal variable $Y(x)$ is found to be

$$EI\frac{d^4 Y}{dx^4} - (\varrho A \omega^2 - k_f)Y = 0,$$

or

$$\frac{d^4 Y}{dx^4} + \left(\frac{-\varrho A}{EI}\omega^2 + \frac{k_f}{EI}\right)Y = 0. \tag{10.102}$$

Let $m = \varrho A$, redefine β so as to include the foundation stiffness, $\beta^4 = (\omega^2 m - k_f)/EI$, and rewrite Equation 10.102 in standard form,

$$Y''''(x) - \beta^4 Y(x) = 0. \tag{10.103}$$

This is the same as Equation 10.68, except for the definition of β, and, therefore, the equation for ω becomes, for the hinged–hinged case,

$$\omega_r = \left(\frac{r\pi}{L}\right)^2 \sqrt{\frac{EI}{m} + \frac{k_f}{m}\left(\frac{L}{r\pi}\right)^4}. \tag{10.104}$$

A conclusion we can draw is that the *elastic end restraints* of the previous section result in changes of both frequencies and mode shapes, whereas an *elastic foundation* changes only the frequencies. This is not a general conclusion and depends on the boundary conditions.

The importance of the term $k_f (L/r\pi)^4$ is problem–specific and by comparison to the value of EI. For most values of k_f, this term can be ignored

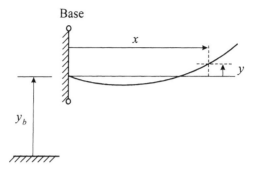

Figure 10.24: Beam on a moving support.

when compared with EI. The importance of the elastic foundation is primarily on the lowest frequencies.

10.6.4 Response of a Beam with a Moving Support

An interesting and important problem is that of a vibrating beam on a moving support, as shown schematically in Figure 10.24.

The governing equation for such a beam is Equation 10.62, except that $y(x,t)$ is replaced by $y(x,t) + y_b(t)$,

$$\frac{\partial^2}{\partial x^2}\left[EI(x)\frac{\partial^2(y+y_b)}{\partial x^2}\right] = -m(x)\frac{\partial^2(y+y_b)}{\partial t^2},$$

where $y_b(t)$ represents the motion of the base translating as a rigid body in addition to the displacement $y(x,t)$ due to bending motion. Since y_b is only a function of time, the governing equation can be simplified to resemble the base excited structure of Chapter 3,

$$\frac{\partial^2}{\partial x^2}\left[EI(x)\frac{\partial^2 y}{\partial x^2}\right] + m(x)\frac{\partial^2 y}{\partial t^2} = -m(x)\frac{\partial^2 y_b}{\partial t^2}.$$

The steady–state solution to this equation for a uniform beam is given by Equation 10.95, repeated here,

$$y(x,t) = \sum_{j=1}^{\infty} Y_j(x) \int_0^t \left[\int_0^L p(\xi,\tau)Y_j(\xi)d\xi\right] g_j(t-\tau)d\tau,$$

with $p(x,t) = -m(\partial^2 y_b/\partial t^2)$, where $g_j(t)$ is the impulse response function for the undamped oscillator. If the base excitation is harmonic, $y_b(t) = B\sin\Omega t$, then $p(x,t) = p(t) = mB\Omega^2 \sin\Omega t$. Once solved, the total motion equals $y(x,t) + y_b(t)$.

10.6. BEAM VIBRATION: SPECIAL PROBLEMS

Harmonic Loads as Boundary Conditions

When a continuous system is subjected to forces, the equation of motion becomes non–homogeneous. Sometimes, the applied forces can be treated as time–dependent boundary conditions. When these forces are harmonic, the method of separation of variables proceeds as usual. Otherwise, other techniques, such as those of Section 11.5.1, must be used.

Consider the procedure for the case of a harmonic force. A uniform cantilever beam of length L fixed at the left end is subjected on the right, free end, to a harmonic load acting down in the positive y direction equal to $P\cos\Omega t$. The transverse beam vibration equation is still valid, but with the following boundary conditions,

$$y(0,t) = 0 \qquad \frac{\partial^2 y(L,t)}{\partial x^2} = 0$$

$$\frac{\partial y(0,t)}{\partial x} = 0 \qquad EI\frac{\partial^3 y(L,t)}{\partial x^3} = -P\cos\Omega t.$$

The last boundary condition is a statement that the shear force on the right end of the beam is equal and opposite to that of the applied force.

Assuming a separable solution leads to the solution of the eigenvalue problem, Equation 10.69, reproduced here,

$$Y(x) = C_1 \sin\beta x + C_2 \cos\beta x + C_3 \sinh\beta x + C_4 \cosh\beta x.$$

The constants are determined by satisfying the four boundary conditions listed above. $Y(0) = 0$ leads to $C_2 + C_4 = 0$. $Y'(0) = 0$ leads to $C_1 + C_3 = 0$. The last two boundary conditions are used in the form $Y''(0) = 0$ and $EIY'''(L)F(t) = -P\cos\Omega t$.

Note how the harmonic loading appears in the boundary condition and that both the x and t variables appear on the left hand side. This last boundary condition can be rewritten as

$$EIY'''(L)(B_1 \cos\omega t + B_2 \sin\omega t) = -P\cos\Omega t,$$

where $F(t) = B_1 \cos\omega t + B_2 \sin\omega t$. This can be satisfied only if $B_2 = 0$ and $\omega = \Omega$. Then,

$$EIB_1 Y'''(L) = -P, \qquad (10.105)$$

giving us our fourth equation involving the coefficients. There are now the four equations needed for a solution of the constants C_i.

10.6.5 Response of a Beam to a Traveling Force

A brief introduction is provided here to the formulation of traveling force problems. A number of important applications can be cast in this form.

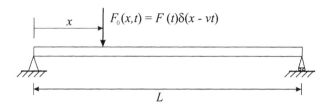

Figure 10.25: Traveling point force on a beam.

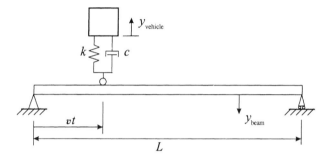

Figure 10.26: Traveling force due to single degree of freedom vehicle.

Vehicle–structure *interaction problems*[23] such as automobile–bridge, jet–aircraft carrier, structure–ocean wave, and train–track are familiar ones. High speed ground transportation systems based on magnetically levitated (*maglev*) vehicles result in electromagnetic coupling between vehicle and track. Figures 10.25 and 10.26 hint at how preliminary models may be formulated.

In the figures, the variable v denotes the velocity of the vehicle. $F_0(t)$ can be a harmonic function such as $\sin \Omega t$, and studies of the relation between beam vibration characteristics and parameters v and Ω are of importance. In Figure 10.26, an n degree of freedom model can be used for the vehicle, if necessary. The beam model may be as simple as the Bernoulli–Euler, or may include the additional effects of the Timoshenko beam. For a rail system, the beam may be on a foundation, and damping can be added. Many options exist, depending on application, and a realistic model can be built by adding some of the individual effects of the more basic models we have already studied.

---

[23] Interaction problems are generally those where two dissimilar systems are coupled in some way, and, therefore, their vibration characteristics must be solved simultaneously.

10.7 Concepts Summary

This has been another chapter of many new ideas and problems. Continuous systems are introduced as a limiting case for an n degree of freedom discrete system, in the limit as $n \to \infty$. The simplest continuous system is the string.

The string is modeled in a number of different ways, demonstrating that vibrating continuous systems can be viewed as either modal or wave systems. It is shown that either approach yields identical results of different mathematical form. The wave approach was left at this point to demonstrate how the equation of motion of a string, beam, rod, or shaft can be derived using Newton's second law of motion or Hamilton's principle. In all these models of one dimensional continuous systems, a modal approach is used to solve for primarily free vibration response. Proportional and non-proportional damping is introduced.

Some beam problems with special loading or boundary conditions are also derived and discussed.

10.8 Problems

Problems for Section 10.2 – Vibration of Strings

1. Solve for the string response, Equation 10.22, for the following initial conditions:

   (i) $y(x,0) = 0$, $\dot{y}(x,0) = g(x)$

   (ii) $y(x,0) = \sin x$, $\dot{y}(x,0) = 0$

   (iii) $y(x,0) = 0$, $\dot{y}(x,0) = \sin x$

   (iv) $y(x,0) = \sin x$, $\dot{y}(x,0) = \sin x$.

   In all cases, sketch the initial conditions and plot the response $y(x,t)$.

2. A string is stretched between $x = 0$ and $x = L$ and has a variable density $\rho = \rho_0 + \varepsilon x$, where ρ_0 and ε are constants. The initial displacement is $f(x)$, and the string is released from rest.

   (i) If the tension T is constant, then show that the governing equation of motion is

   $$T\frac{\partial^2 y}{\partial x^2} = \rho \frac{\partial^2 y}{\partial t^2},$$

   for $0 < x < L$, $t > 0$, with boundary conditions $y(0,t) = 0$, $y(L,t) = 0$, and initial conditions $y(x,0) = f(x)$, $dy(x,0)/dt = 0$.

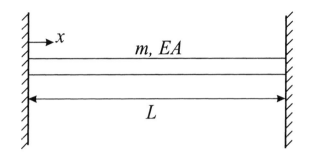

Figure 10.27: Fixed beam in longitudinal vibration.

(ii) Show that the frequencies of normal mode vibration are given by $f_n = \omega_n/2\pi$, where ω_n are the positive roots of

$$J_{1/3}(\alpha\omega)J_{-1/3}(\beta\omega) = J_{1/3}(\beta\omega)J_{-1/3}(\alpha\omega),$$

where

$$\alpha = \frac{2\rho_0}{3\varepsilon}\sqrt{\frac{\rho_0}{T}}$$

$$\beta = \frac{2(\rho_0 + \varepsilon L)}{3\varepsilon}\sqrt{\frac{\rho_0 + \varepsilon L}{T}}.$$

Problems for Section 10.3 – Longitudinal Vibration of Beams

3. For the uniform cantilever beam in longitudinal vibration, Figure 10.27, derive the expressions for the natural frequencies and the mode shapes. Sketch the first three modes.

4. For Equation 10.29, solve for the constants A_r and B_r by satisfying the initial conditions.

5. Consider the free axial motion of a beam with elastic boundary of *Example* 10.2, solve for the first four natural frequencies, modes, and the complete solution, for the following parameter cases:

   (i) $k = EA$ N/m, $L = 10$ m
   (ii) $k = 10EA$ N/m, $L = 1$ m
   (iii) $k = EA/10$ N/m, $L = 1$ m
   (iv) $k = EA/10$ N/m, $L = 10$ m.

10.8. PROBLEMS

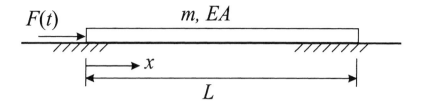

Figure 10.28: Beam on smooth surface.

6. A beam lies on a horizontal and smooth surface and at $t = 0$ is forced at one end by $F(t)$, as shown in the sketch of Figure 10.28. Derive the general response for any forcing function using the known eigenvalues and eigenfunctions.

   Then evaluate the specific steady state response for the following forcing functions:

   (i) $F(t) = F_0 u(t)$ N, where $u(t)$ is the unit step function
   (ii) $F(t) = \sin \Omega t$ N for the case where $\Omega = \omega_n/2$ rad/sec
   (iii) $F(t) = 1 - t/t_0$ N, $0 \leq t \leq t_0$ sec
   (iv) $F(t) = \delta(t)$ N, where $\delta(t)$ is the Dirac delta function.

7. A cantilever beam has its free end stretched uniformly so that the original length L becomes L_0, and then it is released at $t = 0$. See the sketch of Figure 10.29. Begin with the general solution for the axial response of a beam,

$$u(x,t) = \sum_{r=1}^{\infty} \sin \frac{(2r-1)\pi x}{2L} \left(A_r \sin \frac{(2r-1)\pi x}{2L} \sqrt{\frac{EA}{m}} t \right.$$
$$\left. + B_r \cos \frac{(2r-1)\pi x}{2L} \sqrt{\frac{EA}{m}} t \right),$$

   satisfy the boundary conditions to fix the arbitrary constants, and derive the particular response.

8. Solve for the first three modes and frequencies of the non–uniform longitudinally vibrating beam of Figure 10.30 beginning with the usual equation of motion. The following properties hold:

$$EA(x) = 2EA(1 - x/L),$$
$$\text{and } m(x) = 2m(1 - x/L).$$

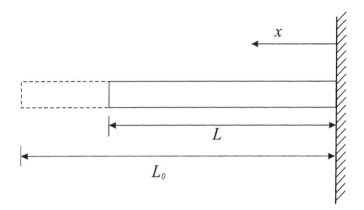

Figure 10.29: Stretched cantilever beam.

Solve this problem for arbitrary k, and then reduce the solution for the case where $k = 0$. How far can this problem be solved analytically?

Problems for Section 10.4 – Torsional Vibration of Shafts

9. The uniform shaft of Figure 10.31 is subjected to a torque at its free end, $T_0 \sin \omega t$. Find the steady state vibration response by solving the free vibration problem and including the forcing as a boundary condition. Can this problem be solved as a forced vibration problem where the free end is left unforced and the forcing is accounted for as an external loading? Do this.

10. For the torsion problem of a shaft with rigid disk at end of Section 10.4.1, solve for the frequencies, modes and total response for the cases:

    (i) $IL = I_D$;

    (ii) $IL = 10 I_D$;

    (iii) $IL = I_D/10$.

Problems for Section 10.5 – Transverse Vibration of Beams

11. Derive Equation 10.47.

12. For the simply–supported beam of Figure 10.32, derive the natural frequencies and the natural modes of vibration. Sketch the first three mode shapes. What changes if the beam is on a slope?

10.8. PROBLEMS

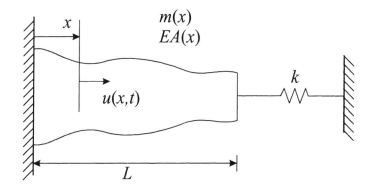

Figure 10.30: Elastic constraint of a longitudinally vibrating beam.

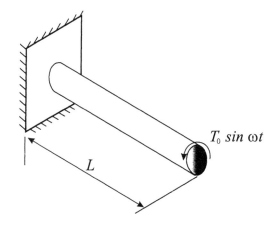

Figure 10.31: Torsional vibration of a shaft.

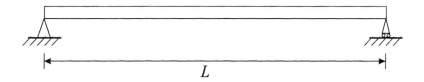

Figure 10.32: Simply-supported beam.

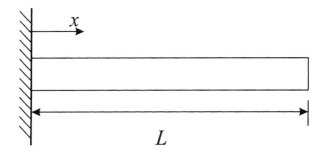

Figure 10.33: Transverse vibration of a cantilever beam.

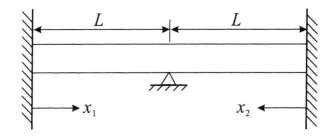

Figure 10.34: Two-span beam.

13. Derive the natural frequencies and mode shapes of the transversely vibrating beam of Figure 10.33. Sketch the first three mode shapes. What changes if the beam is vertical?

14. For the transverse vibration of a simply–supported uniform beam, solve for the transient response if the initial conditions are given by

$$y(x,0) = B\left(\frac{x}{L} - 3\frac{x^2}{L^2} + 2\frac{x^3}{L^3}\right)$$
$$\dot{y}(x,0) = 0.$$

15. A two–span beam may be used as a simple model of a bridge. For this model, sketched in Figure 10.34, derive the frequency equations.

16. The beam of Figure 10.35 is released and rotates at Ω rad/sec as a rigid body. Assume that upon impact the beam latches onto the support and there is no rebound and no loss in energy. At impact the beam will transfer all its rotational kinetic and potential energies into

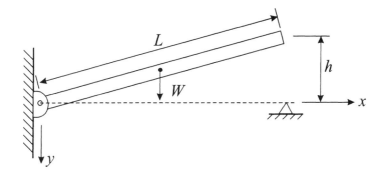

Figure 10.35: A freely falling rotating beam.

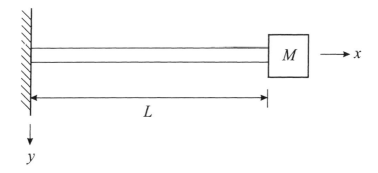

Figure 10.36: Transversely vibrating beam with end mass.

transverse vibrational motion. Derive the response after impact. How would the analysis proceed if there is a rebound?

17. List the boundary conditions for the transversely vibrating beams of Figures 10.36 and 10.37.

18. For the transversely vibrating beam of Figure 10.38, find the modes and the first three natural frequencies of vibration.

19. Equations 10.48 and 10.49 are solved using the assumed solutions $y(x,t) = Y(x)\exp(i\omega t)$ and $\psi(x,t) = \Psi(x)\exp(i\omega t)$. Set up the equations that are thus obtained.

20. In Equations 10.52 and 10.53, eliminate $y(x,t)$ and solve for $\psi(x,t)$. What physical motion does this equation govern?

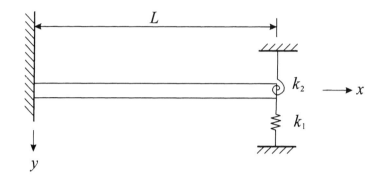

Figure 10.37: Transversely vibrating beam with elastic support.

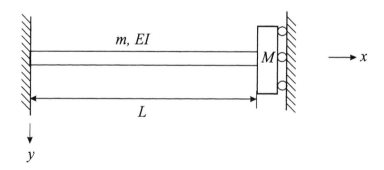

Figure 10.38: Transversely vibrating beam with end roller.

10.8. PROBLEMS

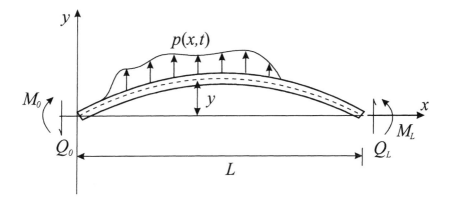

Figure 10.39: Transversely vibrating beam.

21. Derive Equation 10.58 making the necessary simplifications.

22. Derive Equation 10.59. State all assumptions.

23. Complete the solution of Equation 10.80 for the initial conditions $y(x,0) = d(x)$ and $\dot{y}(x,0) = v(x)$.

24. Derive Equation 10.88 starting with Equation 10.87.

25. Solve Equation 10.95 for zero initial conditions and for each of the following forcing functions:

    (i) $p(x,t) = u(t)$ N, where $u(t)$ is the unit step function,

    (ii) $p(x,t) = \sin \Omega t$ N,

    (iii) $p(x,t) = e^{-\varphi t}$ N, for $t \geq 0$ sec,

    (iv) $p(x,t) = \sin \Omega t$ N for $0 \leq \Omega \leq \pi/2$ rad/sec.

26. Derive the equation of motion and boundary conditions for the transversely vibrating beam using Hamilton's principle. Assume small deflections. Let I_p equal the mass moment of inertia per unit length of the beam about the neutral axis. Figure 10.39 provides a sketch of the beam along with positive sign conventions.

27. Solve for the response of the cantilever beam sketched in Figure 10.40.

28. In *Example* 10.6, solve for the case where $t = \tau_1/4$ sec.

Problems for Section 10.6 – Beam Vibration: Special Problems

560   CHAPTER 10. CONTINUOUS MODELS FOR VIBRATION

Figure 10.40: Beam with harmonic end load.

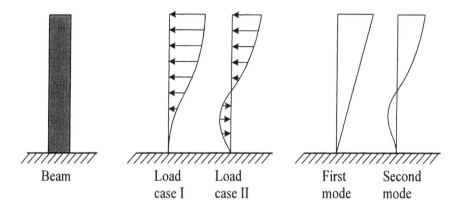

Figure 10.41: Modal participation factor.

29. Derive the equation of motion for the transversely vibrating beam with axial force applying Hamilton's principle.

30. Solve Equation 10.101 by setting the determinant of the coefficients equal to zero, resulting in a polynomial that is solved for the frequencies. Use the following parameter values: $E = 30 \times 10^6$ lb/in^2, $I = 10 \times 10^4$ in^4, $k_1 = 100$ lb/in, $k_2 = 1000$ lb/in, $k_3 = 500$ lb/in, $k_4 = 2000$ lb/in.

31. Complete the solution for the beam with harmonic end load beginning with Equation 10.105.

32. Discuss the modal participation factor for a vibrating beam that is subjected to loads as shown in Figure 10.41. Refer to Section 8.6.2 for a discussion of the modal participation factor.

10.9 Mini–Projects

The following *mini–projects* are provided as possible avenues for extending the basic studies of the chapter to more advanced and more applied studies. At this point, the student has the ability and background to begin to tackle such problems, with some guidance from the instructor. Each mini–project is intended for extended study, either by the individual student, or in small groups of two or three students.

1. *On the Bernoulli–Euler Beam Theory with Random Excitation*, J.L. Bogdanoff, J.E. Goldberg, *J. of the Aero/Space Sciences*, Vol. 27, No. 5, May 1960, 371–376.

   Mean square displacement and stress are calculated in a simply supported Bernoulli–Euler beam with distributed external viscous damping for several types of random excitation.

2. *Mean–Square Response of Beams to Nonstationary Random Excitation*, G. Ahmadi, M.A. Satter, *AIAA Journal*, Vol. 13, No. 8, Aug. 1975, 1097–1100.

   The response of an elastic beam under nonstationary random loading has been studied, resulting in the mean–square response of the beam.

3. *On Some General Properties of Combined Dynamical Systems*, E.H. Dowell, *J. Applied Mechanics*, 1978.

   The dynamics of combined systems composed of several component systems are studied.

4. *Vibrational Aspects of Rotating Turbomachinery Blades*, A. Leissa, *Applied Mechanics Reviews*, Vol. 34, No. 5, May 1981, 629–635.

   The blades of rotating turbomachinery are typically the most critical parts of the design. For reasons of efficiency involving gas dynamics and weight, they must be thin, yet they must operate in severe thermal environments and at high rotational speeds causing large centrifugal body forces. This is a review paper.

5. *The Inverse Problem for the Vibrating Beam*, G.M.L. Gladwell, *Proceedings of the Royal Society, London*, A **393**, 277–295 (1984).

   The problem of reconstructing the mass and stiffness distributions of a vibrating beam from frequency data is studied. The beam is modeled by a system of rigid rods joined together by rotational springs and with masses at the joints.

6. *Nonlinear Vibration of a Traveling Tensioned Beam*, J.A. Wickert, *Int. J. Nonlinear Mechanics*, Vol. 27, No. 3, 503–517, 1992.

   Free nonlinear vibration of an axially moving, elastic, tensioned beam is analyzed over the sub- and supercritical transport speed ranges.

7. *Dynamics and Control of a Translating Flexible Beam with a Prismatic Joint*, S.S.K. Tadikonda, H. Baruh, *J. Dynamic Systems, Measurement, and Control*, Sept. 1992, Vol. 114, 422–427.

   A dynamic model of a translating flexible beam, with tip mass at one end and emerging or retracting into a rigid base at the other is developed. Applications include the dynamics and control of flexible links in translational motion with a prismatic joint at one end.

8. *Modeling and Identification of Boundary Conditions in Flexible Structures*, H. Baruh, J.B. Boka, *Modal Analysis: the International J. of Analytical and Experimental Modal Analysis*, v.8, n.2, April 1993, 107–117.

   An approach is presented to model and to identify boundary conditions in structures. The unknown boundary conditions are modeled as axial and torsional springs at the boundaries.

9. *Forced Vibration of a Mass–Loaded Beam with Heavy Tip Body*, T.-P. Chang, *J. Sound and Vibration*, (1993) **164**(3), 471–484.

   The deterministic and random vibration response analysis of a model which simulates a robotic arm is presented. The model is a uniform, mass–loaded, hysteretically damped beam, the left end of which is attached by both translational and rotational springs, and the right end of which is free and carrying a heavy tip mass. Reliability of such structures is an application.

10. *Nonlinear Vibrations of a Floating Beam Under Harmonic Excitation*, F. Zhu, *J. Sound and Vibration* (1994) **172**(5), 689–696.

    The floating beam vibrates on a fluid that is assumed to be inviscid, irrotational, and incompressible.

11. *Analysis of Free Vibrations of Tall Buildings*, L. Qiusheng, C. Hong, L. Guiqing, *J. Engineering Mechanics*, Vol. 120, No. 9, Sept. 1994, 1861–1876.

    Frame, shear–wall, and frame–shear–wall tall buildings are analyzed using shear beam models with varying cross sections. Cantilever beams are utilized with various base boundary conditions to simulate different foundations.

12. *Free Vibration of Stepped Beams with Elastic Ends*, M.A. De Rosa, *J. Sound and Vibration*, (1994) **173**(4), 563–567.

    The most general case of a stepped beam with a single step has been solved, and the free vibration frequencies of a slender Euler–Bernoulli stepped beam with two elastic ends are calculated.

13. *Vibration of Rotating Turbomachine Blades with Flexible Roots*, M. Swaminadham, *J. Sound and Vibration*, (1994) **174**(2), 284–288.

    Turbine blades are modeled as uniform cantilever beams fixed to the disk. Individual as well as combined effects of root compliance and rotational speeds are analyzed using the first order central difference method.

14. *Exact Solution for the Free Vibration of a Hanging Cord with a Tip Mass*, R.I. Sujith and D.H. Hodges, *J. Sound and Vibration*, (1995) **179**(2), 359–361.

    An exact solution is presented for the free vibration characteristics of an inextensible cord with a tip mass, hanging vertically in a uniform gravitational field.

15. *Railway Noise – Can Random Sleeper Spacings Help?* M.A. Heckl, *Acoustica*, Vol. 81 (1995), 559–564.

    A mathematical model is used to predict the vibration of a rail excited by a rolling wheel. The track is modeled as an infinitely long beam (rail) supported over a finite section by discrete support systems.

16. *Parametric Instability of a Cantilevered Column Under Periodic Loads in the Direction of the Tangency Coefficient*, C–C. Chen, M–K. Yeh, *J. Sound and Vibration*, (1995) **183**(2), 253–267.

    The assumed modes method is used to convert the original continuous column to a lumped one for the instability analysis. Lagrange's equation is used to derive the equation of motion, which becomes a *Mathieu* equation for periodic loading. Transition curves are derived for regions of instability.

17. *Linear Vibration of a Coupled String–Rigid Bar System*, B. Yang, *J. Sound and Vibration* (1995) **183**(3), 383–399.

    Because of the dynamic interaction between the strings and the rigid bar, the coupled system is not self–adjoint and classical modal analysis cannot be applied. A eigenfunction expansion formulation is developed to analyze the equations. Applications of multiple flexible and rigid bodies include vehicles, airplanes, spacecraft, robots, computer storage systems, rotating machinery, highway structures, buildings,, bridges, and pipelines conveying fluids.

18. *Vibration Analysis of Timoshenko Beams with Non–Homogeneity and Varying Cross–Section*, X. Tong, B. Tabarrok, K.Y. Yeh, *J. Sound and Vibration*, (1995) **186**(5), 821–835.

    An analytic solution is presented for the free and forced vibration of stepped Timoshenko beams, and it is used for the approximate analysis of generally non–uniform Timoshenko beams. In the case of free vibration, the frequency equation is expressed in terms of some initial parameters at one end of the beam; while in the case of forced vibration, the solution may be obtained by solving a set of algebraic equations with only two unknowns.

19. *Free Vibration Characteristics of Variable Mass Rockets Having Large Axial Thrust/Acceleration*, A. Joshi, *J. Sound and Vibration*, (1995) **187**(4), 727–736.

    Modern–day rockets and satellite launch vehicles have a large thrust to weight ratio, as there is a continuing need to launch heavier satellites deeper into space. Simultaneously, there are efforts to reduce the cost of launching satellites in terms of cost per unit weight of payload, which results in more slender and flexible rocket structures. Furthermore, the mass of the launch vehicles depletes continuously due to propellant consumption as well as separation of various rocket stages. All of these factors result in structural vibration frequencies which are not only closer to, or even within, the control system bandwidth but also vary as a function of the trajectory. In addition, the presence of large thrust/acceleration leads to the development of significant axial forces, which also have an impact on the transverse vibration frequencies of beams. Such a system is studied.

20. *Effects of Vehicle Suspension Design on Dynamics of Highway Bridges*, M.F. Green, D. Cebon, D.J.Cole, *J. Structural Engineering*, Vol. 121, No. 2, Feb. 1995, 272–282.

    The effects of heavy vehicles with leaf–spring and air–spring suspensions on the dynamic response of short–span highway bridges are considered. The dynamic bridge responses are calculated by modeling the bridge and the vehicle separately and then combining the models with an iterative procedure.

21. *Vibration Modes and Frequencies of Timoshenko Beams with Attached Rigid Bodies*, M.W.D. White, G.R. Heppler, *J. Applied Mechanics*, March 1995, Vol. 62, 193–199.

    The equations of motion and boundary conditions for a free–free Timoshenko beam with rigid bodies attached at the endpoints are derived. The frequency equations for beams with attached rigid bodies are related to those for beams without rigid bodies.

10.9. MINI-PROJECTS

22. *Longitudinal Vibrations of a Prismatic Bar Suddenly Subjected to a Tensile Load at One End When the Other is Elastically Restrained*, C.A. Rossit, D.V. Bambill, P.A.A. Laura, *J. Sound and Vibration*, (1996) **188**(1), 145–148.

    The longitudinal motion of an elastically restrained bar responding to a suddenly applied load is used as a first approximation model of a drill string when a load is applied suddenly at one end while the other is embedded in a Winkler–type foundation.

23. *Stability Analysis of a Cracked Shaft Subjected to the End Load*, L–W. Chen, H–K. Chen, *J. Sound and Vibration*, (1995) **188**(4), 497–513.

    The stability of a rotating cracked shaft subjected to an end axial compressive load is studied, and the influences of the existing open crack on the natural whirling speeds of the shaft are discussed.

24. *On the Eigenfrequencies of a Cantilever Beam with Attached Tip Mass and a Spring–Mass System*, M. Gurgoze, *J. Sound and Vibration*, (1996) **190**(2), 149–162.

    The Lagrange multipliers method is used to derive the frequency equation of a clamped–free Euler–Bernoulli beam with tip mass where a spring–mass system is attached to it. Applications include robot arms.

25. *Vibration of a Free–Free Beam Under Tensile Axial Loads*, X.Q. Liu, R.C. Ertekin, H.R. Riggs, *J. Sound and Vibration*, (1996) **190**(2), 273–282.

    The free vibration of a free–free beam under tensile axial loads can be used as an appropriate model for pipeline towing problems in ocean engineering. The non–dimensional governing equation and boundary conditions for the in–plane vibration of a uniform, free–free beam subject to constant tension are derived.

26. *Free Vibration of a Complex Euler–Bernoulli Beam*, N. Popplewell, D. Chang, *J. Sound and Vibration*, (1996) **190**(5), 852–856.

    A unified treatment is proposed for finding the free vibration of a non–uniform beam having material or cross–sectional discontinuities, intermediate spring supports, or non–classical end supports.

27. *The Effect of a Moving Mass and Other Parameters on the Dynamic Response of a Simply Supported Beam*, G. Michaltsos, D. Sophianopoulos, A.N. Kounadis, *J. Sound and Vibration*, (1996) **191**(3), 357–362.

    The linear dynamic response of a simply supported uniform beam under a moving load of constant magnitude and velocity, including the effect of its mass, is studied.

28. *A "Direct" Solution for the Free Vibration of a Hanging Uniform Cord with a Particle Tip Mass*, S. Naguleswaran, *J. Sound and Vibration* (1996) **191**(3), 453–458.

    The mode shape differential equation is solved without any coordinate transformation and is therefore a direct method. The first three normalized mode shapes for typical values of the cord mass to tip mass ratio are presented graphically and the corresponding slopes at the point of suspension are tabulated.

29. *Free Vibration of Simply Supported Beam Partially Loaded with Distributed Mass*, K.T. Chan, T.P. Leung, W.O. Wong, *J. Sound and Vibration*, (1996) **191**(4), 590–597.

    The vibration of a simply supported beam, partially loaded with distributed mass, is studied. Frequencies and mode shapes are calculated.

30. *Dynamic Behavior of a Drill–String: Experimental Investigation of Lateral Instabilities*, A. Berlioz, J. Der Hagopian, R. Dufour, E. Draoui, *J. Vibration and Acoustics*, July 1996, Vol. 118, 292–298.

    This paper focuses on laboratory tests concerned with the lateral behavior of a rod that is representative of part of a drill–string used in rotary oil drilling.

31. *Transverse Vibration of a Moving String: I: A Physical Overview; II: A Comparison Between the Closed–Form Solution and the Normal–Mode Solution*, L. Lengoc, H. McCallion, *J. Systems Engineering*, (1996) 6:61–71, 72–78.

    Two physical explanations for the transverse oscillation of a moving string are presented: *(i)* transverse waves propagating longitudinally, and *(ii)* forces acting on a taut moving string due to oscillation. The first explanation leads to the closed–form solutions for the moving string directly from considerations of propagating transverse waves. The second explanation, based upon coupling of natural modes in time due to Coriolis actions, leads to a double–sine series solution of the equation of motion. The Galerkin method is shown to give matrices which are identical to those resulting from the coupled normal mode solution.

32. *Resonance Conditions and Deformable Body Coordinate Systems*, A.A. Shabana, *J. Sound and Vibration*, (1996) **192**(1), 389–298.

    It is demonstrated that resonance conditions are not absolute in the sense that different resonance frequencies can be obtained for the same system if the deformation is defined in different coordinate systems that can have arbitrary rigid body displacements. As a consequence, discussion of the

10.9. MINI-PROJECTS

resonance phenomenon must be accompanied with a specification of the coordinate system of the elastic body.

33. *Vibration Analysis and Diagnosis of a Cracked Shaft*, T.C. Tsai, Y.Z. Wang, J. Sound and Vibration, (1996) **192**(3), 607–620.

    A diagnostic method of determining the position and size of a transverse open crack on a stationary shaft without disengaging it from the machine system is investigated. The crack is modeled as a joint of a local spring. The position of the crack can be predicted by comparing the fundamental mode shapes of the shaft with and without a crack.

34. *Free Vibrations of a Linearly Tapered Cantilever Beam with Constraining Springs and Tip Mass*, N.M. Auciello, J. Sound and Vibration, (1996) **192**(4), 905–911.

    A method is used to calculate the frequencies of a cantilever beam with linearly varying breadth and height. Included are the eccentricity and rotary inertia of the tip mass. The transverse displacement is in two dimensions.

35. *Transverse Vibrations of a Linearly Tapered Cantilever Beam with Tip Mass of Rotary Inertia and Eccentricity*, N.M. Auciello, J. Sound and Vibration, (1996) **194**(1), 25–34.

    An exact analysis of free vibration of cantilever tapered beams with a mass at the tip and flexible constraint is presented. The rotary inertia of the concentrated mass is considered along with its eccentricity.

36. *Instability of a Clamped–Elastically Restrained Timoshenko Column Carrying a Tip Load, Subjected to a Follower Force*, K. Sato, J. Sound and Vibration, (1996) **194**(4), 623–630.

    It is of great importance to study the vibration and stability of various types of elastic columns, used widely as machine and airplane structures. The mathematical theory is developed using Hamilton's principle and some new results are given on the influence of the attached load on the instability of the system.

37. *Coupled Timoshenko Beam Vibration Equations for Free Symmetric Bodies*, A.N. Kathnelson, J. Sound and Vibration, (1996) **195**(2), 348–352.

    Despite the simplicity of the beam model, the Timoshenko beam vibration equations accounting for the shear deformation and rotary inertia give the correct eigenfrequencies even of short beams provided an appropriate value for the shear coefficient is used. Here, coupled equations are used to model structures such as plates.

38. *A Variational Approach to the Vibrations of Tapered Beams with Elastically Restrained Ends*, R.O. Grossi, B. del V. Arenas, *J. Sound and Vibration*, (1996) **195**(3), 507–511.

   This work extends earlier work on the same topic where uniform beams were studied. Here, tapered beams are considered.

39. *Lateral Free Vibration of a Single Pile With or Without an Axial Load*, U.B. Halabe, S.K. Jain, *J. Sound and Vibration*, (1996) **195**(3), 531–544.

   The results of free vibration analysis in lateral direction for a single pile with or without axial load are presented here. The soil is modeled by a linear homogeneous Winkler model. The boundary conditions treated are free– and fixed–head and free–tip.

40. *Approximate Formula for the Frequencies of a Rotating Timoshenko Beam*, V.T. Nagaraj, *J. of Aircraft*, Vol. 33, No. 3, March 1996, 637–639.

   It is well known that shear deformation and rotary inertia can have an important influence on the frequencies of non rotating beams. This influence can be especially significant for beams made of advanced composite materials because of the high ratio of Young's modulus to the shear modulus. For non rotating beams, these effects can be modeled by Timoshenko's equations for which exact solutions are available. Here, the influence of shear deformation and rotary inertia are approximated for rotating beams.

41. *Elastic Response of Columns After Sudden Loss of Bracing*, R.H. Plaut, R.-H. Yoo, *J. Engineering Mechanics*, Vol. 122, No. 4, April 1996, 383–384.

   A pinned column with an initial deflection and an internal brace are modeled. The brace is modeled as a translational spring. Axial compressive loads are applied at the brace and at the top of the column. Then the brace and the corresponding load are suddenly removed, and the ensuing motion of the column is analyzed.

42. *Transverse Vibration of a Timoshenko Beam Acted on by an Accelerating Mass*, H.P. Lee, *Applied Acoustics*, Vol. 47, No. 4, pp. 319–330, 1996.

   The transverse vibration of a Timoshenko beam acted on by an accelerating mass is analyzed and compared with the corresponding behavior of a Timoshenko beam subjected to an equivalent moving force neglecting the inertial effects of the mass.

43. *Axial Force Stabilization of Transverse Vibration in Pinned and Clamped Beams*, C.D. Rahn, C.D. Mote, Jr., *J. Dynamic Systems, Measurement, and Control*, June 1996, Vol. 118, 379–380.

   An axial force stabilizes the transverse vibration of a beam with pinned and/or clamped boundary conditions.

10.9. MINI–PROJECTS

44. *The Timoshenko Beam on an Elastic Foundation and Subject to a Moving Step Load, Part 1: Steady–State Response, Part 2: Transient Response*, S.F. Felszeghy, *J. Vibration and Acoustics*, July 1996, Vol. 118, 277–291.

    The response of a simply supported semi–infinite Timoshenko beam on an elastic foundation to a moving step load is determined. The response is found from summing the solutions to two mutually complementary sets of governing equations. The first solution is a particular solution to the forced equations of motion. The second solution is to a set of homogeneous equations of motion and nonhomogeneous boundary conditions so formulated as to satisfy the initial and boundary conditions of the actual problem when the two solutions are summed.

45. *In–Flight Flexure and Spin Lock–In for Antitank Kinetic Energy Projectiles*, A.G. Mikhail, *J. Spacecraft and Rockets*, Vol.33, No.5, Sept–Oct 1996, 657–664.

    A time–dependent analysis, coupling the rod vibration equation and pure roll motion equation, is formulated to numerically simulate the in–flight bending behavior of antitank kinetic energy projectile rods. The projectile is modeled as undergoing continuous, simulated planar pitching motion with the aerodynamic, spin, and structural forces included.

46. *Dynamic Response of an Axially Loaded Bending–Torsion Coupled Beam*, S.H.R. Eslimy-Isfahany, J.R. Banerjee, *J. of Aircraft*, Vol. 33, No. 3, May–June 1996, 601–607.

    A method is presented to predict the response of an axially loaded beam coupled in bending and torsion to deterministic and random loads. The beam is uniform and is assumed to carry a constant axial load. It is subject to time–dependent bending and/or torsional load, which can be either deterministic or random. Both concentrated and distributed loads are considered. The deterministic load is assumed to vary harmonically, whereas the random load is assumed to be Gaussian, having both stationary and ergodic properties. The theory developed is applied to a cantilever wind turbine blade for which there is substantial coupling between the bending and torsional modes of deformation.

47. *Prediction of the Dynamic Behavior of Non–Symmetric Coaxial Co– or Counter–Rotating Rotors*, G. Ferraris, V. Maisonneuve, M. Lalanne, *J. Sound and Vibration*, (1996) **195**(4), 649–666.

    This paper is concerned with the dynamic behavior of non–symmetric rotors and analyzes the dynamic behavior of a ducted propfan, which compares well with experimental results.

48. *Active Vibration Control of the Axially Moving String Using Space Feedforward and Feedback Controllers*, S. Ying, C.A. Tan, *J. Vibration and Acoustics*, Vol. 118, July 1996, 306–312.

   Closed–form results for the transverse response of both the uncontrolled and controlled string are given in the s domain. The proposed control law indicates that vibration in the region downstream of the control force can be canceled.

49. *Optimum Placement of Piezoelectric Sensor/Actuator for Vibration Control of Laminated Beams*, Y.K. Kang, H.C. Park, W. Hwang, K.S. Han, *AIAA Journal*, Vol. 34, No. 9, Sept. 1996, 1921–1926.

   Optimum placement is investigated numerically using the finite element method and verified experimentally for vibration control of laminated beams. Hamilton's principle is used to derive the equation of motion for the plate.

50. *Complex Modal Analysis of a Flexural Vibrating Beam with Viscous End Conditions*, G. Oliveto, A. Santini, E. Tripodi, *J. Sound and Vibration*, (1997) **200**(3), 327–345.

   The complex mode superposition method for the dynamical analysis of a simply supported beam with two rotational viscous dampers attached at its end is presented.

51. *A Comparison of Modal Expansion and Travelling Wave Methods for Predicting Energy Flow in Beam Structures*, J. Pan, J. Pan, *J. Sound and Vibration*, (1998) **214**(1), 1–15.

   Both modal expansion and travelling wave methods are commonly used for predicting the response and vibrational energy flow in structures. They describe the same structural wave motion problem from different viewpoints. In this paper, energy flows carried by the torsional and flexural waves in beam structures are predicted by these two methods and the results compared.

52. *Vibrations of a Cantilever Tapered Beam with Varying Section Properties and Carrying a Mass at the Free End*, N.M. Auciello, G. Nolè, *J. Sound and Vibration*, (1998) **214**(1), 105–119.

   The free vibration frequencies of a beam composed of two tapered beam sections with different physical characteristics with a mass at its end can be determined by either using the exact procedure, for which purpose the solution to the problem can be expressed using Bessel functions, or the approximate Rayleigh–Ritz treatment, with the assumption of orthogonal polynomials as test functions. The results and the numerical comparison between the two methods are provided in diagrams and tables. The effect of different materials on vibration modes is also provided.

53. *Effect of Crack Depth on the Natural Frequency of a Prestressed Fixed–Fixed Beam*, S. Masoud, M.A. Jarrah, M. Al-Maamory, *J. Sound and Vibration* (1998) **214**(2), 201–212.

    The effect of crack depth on the transverse vibrational characteristics of a prestressed fixed–fixed beam is investigated. The coupling effect between the crack depth and the axial load on the natural frequency of the system is also investigated.

54. *Vibration Analysis of Beams with a Two Degree of Freedom Spring–Mass System*, T-P. Chang, C-Y. Chang, *Int. J. Solids Structures*, Vol. 35, Nos. 5–6, pp. 383–401, 1998.

    In this paper, the natural frequencies and mode shapes of a Bernoulli–Euler beam with a two degree of freedom spring–mass system are determined by using Laplace transform with respect to the spatial variable. The deterministic and random vibration responses of the beam are obtained by using modal analysis.

55. *Natural Frequency and Mode Shape Sensitivities of Damped Systems: Part I, Distinct Natural Frequencies*, I-W. Lee, D-O. Kim, G-H. Jung, *J. Sound and Vibration*, (1999) **223**(3), 399–412.

    A procedure for determining the sensitivities of the eigenvalues and eigenvectors of damped vibratory systems with distinct eigenvalues is presented.

56. *Transverse Vibrations of a Flexible Beam Sliding Through a Prismatic Joint*, M. Gürgöze, S. Yüksel, *J. Sound and Vibration* (1999) 223(3), 467–482.

    In this paper, the vibrations of an axially moving flexible beam sliding through an arbitrarily driven prismatic joint, restricted to move on a horizontal plane, are investigated.

57. *The Influence of Damping on Waves and Vibrations*, L. Gaul, *Mechanical Systems and Signal Processing* (1999) **13**(1), 1–30.

    Wave propagation and vibration are associated with the removal of energy by dissipation or radiation. In mechanical systems damping forces causing dissipation are often small compared to restoring and inertia forces. However, their influence can be great and is discussed in the present survey paper together with the transmission of energy away from the system by radiation.

58. *Adaptation of the Concept of Modal Analysis to Time–Varying Structures*, K. Liu, M.R. Kujath, *Mechanical Systems and Signal Processing* (1999) **13**(3), 413–422.

   Most of the concepts of modal analysis are well established in the literature, and widely understood for linear time–invariant systems. This paper discusses an adaptation of modal analysis concepts to time–varying systems.

59. *On the Existence of a Non–Vibratory Mode Shape for Free–Free Straight Beams Under Compressive Loads*, G.C. Nihous, *J. Sound and Vibration* (1999) **220**(5), 948–953.

60. *Flow–Induced Vibration of an Euler–Bernoulli Beam*, X.Q. Wang, R.M.C. So, Y. Liu, *J. Sound and Vibration* (2001) **243**(2), 241–268.

   Flow–induced vibration of a fixed–fixed elastic cylinder with a large aspect ratio (≈ 58) is considered.

61. *Natural Frequencies and Mode Shapes of a Free–Free Beam with Large End Masses*, C.L. Kirk, S.M. Wiedemann, *J. Sound and Vibration* (2002) **254**(5), 939–949.

   An analytical solution is presented for the natural frequencies, mode shapes and orthogonality conditions, of a free-free beam with large offset masses connected to the beam by torsion springs. The study lays the foundation for investigations into the dynamics and vibration control of multi–link articulated systems such as the Space Shuttle Remote Manipulator.

62. *Modal Analysis of Rotating Composite Cantilever Plates*, H.H. Yoo, S.K. Kim, D.J. Inman, *J. Sound and Vibration* (2001) **258**(2), 238–246.

   A set of ordinary differential equations of motion for the plate is derived using the assumed modes method. The equations of motion include the coupling terms between the in–plane and the lateral motions as well as the motion–induced stiffness variation terms.

63. *Vortex–Induced Vibration of Two Side–by–Side Euler–Bernoulli Beams*, R.M.C. So, X.Q. Wang, *J. Sound and Vibration* (2003) **259**(3), 677–700.

   Vortex–induced vibration of two side–by–side elastic beams in a cross flow is numerically studied. The two beams are identical and fixed at both ends. In the numerical approach, the Euler–Bernoulli beam theory is used to model the beam vibration, and the laminar Navier–Stokes equations are solved to give the flow field. The flow equations are resolved using a finite element method and the flow–induced forces are calculated at every time step in order to correctly reflect the fluid–beam interaction.

10.9. MINI–PROJECTS

64. *Forced Vibrations of a Clamped–Free Beam with a Mass at the Free End with an External Periodic Disturbance Acting on the Mass with Applications in Ships' Structures*, D.V. Bambill, S.J. Escanes, C.A. Rossit, *Ocean Engineering* 30 (2003) 1065–1077.

    The Euler–Bernoulli model has been used to obtain an exact analytical solution to the title problem, of importance to the structural engineer in several fields of technology: ocean and naval engineering, aerospace applications, among others.

Chapter 11

Continuous Models for Vibration: Advanced Models

"Reality becomes more complex"

11.1 Vibration of Membranes

In this section and the next, modeling of membranes and plates is introduced, bringing the student to the point where advanced study is possible. Membranes and plates are the respective two–dimensional extensions of strings and beams. Similarities to earlier models appear as we derive and solve the equations of motion. Damping of membranes and plates are not considered in this text.

Membranes have numerous applications, for example, inflatable structures, parachutes, and biomechanical components such as the valve of an artificial heart.

11.1.1 Rectangular Membranes

Consider a *thin* and *uniform* rectangular membrane under constant tension along the boundary, which coincides with the x and y axes. The transverse displacement w is of interest and is assumed *small*. The governing equation of motion in this direction is derived along with given arbitrary initial conditions on displacements and velocities. The procedure is the same as for the string, except for the second dimension. There is no moment resistance.

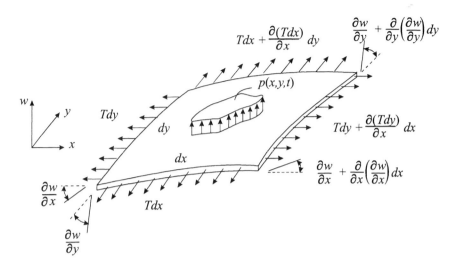

Figure 11.1: Free body diagram of a rectangular membrane element.

The free body diagram of a membrane element is shown in Figure 11.1. Summing the forces in the transverse direction yields

$$\left[Tdy + \frac{\partial T}{\partial x}dx\right]\left[\frac{\partial w}{\partial x} + \frac{\partial^2 w}{\partial x^2}dx\right] + p(x,y,t)dx\,dy - T\frac{\partial w}{\partial x}dy$$
$$+ \left[Tdx + \frac{\partial T}{\partial y}dy\right]\left[\frac{\partial w}{\partial y} + \frac{\partial^2 w}{\partial y^2}dy\right] - T\frac{\partial w}{\partial y}dx = m\,dx\,dy\,\frac{\partial^2 w}{\partial t^2},$$

where the left–hand side represents the forces on the free body and the right–hand side is the resulting inertia. m is the constant mass per unit area, and the assumption of small displacements permit the approximate replacement of the sine of an angle by the angle. Expand the products, simplify, and recall that T is constant, to find

$$Tdy\left[\frac{\partial w}{\partial x} + \frac{\partial^2 w}{\partial x^2}dx\right] - Tdy\frac{\partial w}{\partial x}$$
$$+ Tdx\left[\frac{\partial w}{\partial y} + \frac{\partial^2 w}{\partial y^2}dy\right] - Tdx\frac{\partial w}{\partial y} + p(x,y,t)dx\,dy = m\,dx\,dy\,\frac{\partial^2 w}{\partial t^2}.$$

Simplifying further leads to the final governing equation of motion,

$$\frac{T}{m}\left[\frac{\partial^2 w}{\partial x^2} + \frac{\partial^2 w}{\partial y^2}\right] + \frac{1}{m}p(x,y,t) = \frac{\partial^2 w}{\partial t^2}.$$

11.1. VIBRATION OF MEMBRANES

This is the *wave equation* in two dimensions. For the free vibration analysis, define $T/m = c^2$ and set $p(x,y,t) = 0$,

$$c^2 \left[\frac{\partial^2 w}{\partial x^2} + \frac{\partial^2 w}{\partial y^2} \right] = \frac{\partial^2 w}{\partial t^2}.$$

As we have done a number of times, assume the product solution to this partial differential equation,

$$w(x,y,t) = W(x,y)F(t),$$

and substitute as needed to find

$$c^2[W_{,xx} + W_{,yy}]F = W\ddot{F}.$$

The respective eigenvalue problem is

$$\nabla^2 W(x,y) + \beta^2 W(x,y) = 0, \quad \text{plus boundary conditions,} \qquad (11.1)$$

where

$$\nabla^2 \equiv \frac{\partial^2}{\partial x^2} + \frac{\partial^2}{\partial y^2}$$

$$\beta^2 = \left(\frac{\omega}{c}\right)^2 = \omega^2 \left(\frac{m}{T}\right).$$

∇ is the *Laplacian* operator called *nabla*.[1] To solve the eigenvalue problem, another separation of variables is required; substitute $W(x,y) = X(x)Y(y)$ and its derivatives into Equation 11.1 to find

$$\frac{X''}{X} + \frac{Y''}{Y} + \beta^2 = 0.$$

Using familiar arguments, set

$$\frac{X''}{X} = -\frac{Y''}{Y} - \beta^2 = -\alpha^2.$$

The negative sign is chosen for the constant α^2 on physical grounds in order that the solution is harmonic in x and y. Therefore, two equations must be solved,

$$X'' + \alpha^2 X = 0$$
$$Y'' + \gamma^2 Y = 0,$$

---
[1] Nabla comes from the Hebrew word *naval*, meaning harp; nun, bet, lamed.

with $\gamma^2 = \beta^2 - \alpha^2$. The solutions to these equations are

$$\begin{aligned} X(x) &= A_1 \sin \alpha x + A_2 \cos \alpha x \\ Y(y) &= A_3 \sin \gamma y + A_4 \cos \gamma y, \end{aligned}$$

and, therefore,

$$\begin{aligned} W(x,y) &= C_1 \sin \alpha x \sin \gamma y + C_2 \sin \alpha x \cos \gamma y \\ &+ C_3 \cos \alpha x \sin \gamma y + C_4 \cos \alpha x \cos \gamma y. \end{aligned} \quad (11.2)$$

The coefficients C_i are determined by satisfying the boundary conditions. To proceed further analytically, assume that the membrane is simply supported along all four sides. We anticipate that, of the four components in Equation 11.2, only the first term will be non–zero for the following simply–supported boundary conditions: $W(0,y) = 0$, $W(a,y) = 0$, and $W(x,0) = 0$, $W(x,b) = 0$. The conditions at each boundary must be examined;

(i) Along $x = 0$, $W(0,y) = C_3 \sin \gamma y + C_4 \cos \gamma y = 0$, which can be true for arbitrary y only if $C_3 = C_4 = 0$.

(ii) Along $x = a$, $W(a,y) = C_1 \sin \alpha a \sin \gamma y + C_2 \sin \alpha a \cos \gamma y = 0$, which yields two possible solutions. The first is that $C_1 = C_2 = 0$, but this is the well known trivial solution that is of no practical value. The other possible solution is $\sin \alpha a = 0$.

(iii) Along $y = 0$, $W(x,0) = C_2 \sin \alpha x + C_4 \cos \alpha x = 0$, which can be true for arbitrary x only if $C_2 = C_4 = 0$.

(iv) Along $y = b$, $W(x,b) = C_1 \sin \alpha x \sin \gamma b + C_3 \cos \alpha x \sin \gamma b = 0$, which also has two possible solutions. The trivial solution requires $C_1 = C_3 = 0$. The other possibility is that $\sin \gamma b = 0$.

Therefore, the only solution that can satisfy all the above is that $C_1 \neq 0$ along with the following *characteristic equations*,

$$\begin{aligned} \sin \alpha a &= 0 \quad \longrightarrow \quad \alpha_j a = j\pi \\ \sin \gamma b &= 0 \quad \longrightarrow \quad \gamma_k b = k\pi, \end{aligned}$$

where $j, k = 1, 2, \ldots$. Thus,

$$\beta_{jk} = (\alpha_j^2 + \gamma_k^2)^{1/2} = \pi \left[\left(\frac{j}{a}\right)^2 + \left(\frac{k}{b}\right)^2 \right]^{1/2},$$

and

$$\omega_{jk} = c\beta_{jk} = \sqrt{\frac{T}{m}} \beta_{jk}, \quad j, k = 1, 2, \ldots,$$

11.1. VIBRATION OF MEMBRANES

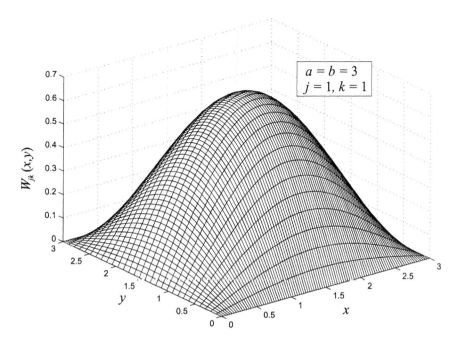

Figure 11.2: (a) Example mode for square membrane.

with the eigenfunctions or modes given by

$$W_{jk}(x,y) = C_{jk} \sin \frac{j\pi x}{a} \sin \frac{k\pi y}{b}, \quad j,k = 1,2,\ldots.$$

Two modes are drawn in Figures 11.2 for a square membrane with $a = 3$ and $b = 3$. Figure 11.2(a) is the mode for $j = k = 1$ and Figure 11.2(b) is the mode for $j = k = 2$. Note that it is possible for *two* modes to exist for the *same* frequency. This is the first time that we have such a situation. For example, if $a = b$ then $\beta_{jk} = \beta_{kj}$ and $W_{jk} = W_{kj}$.

To specify the values of coefficients C_{jk}, normalize according to

$$\int_0^a \int_0^b m C_{jk}^2 \sin^2 \frac{j\pi x}{a} \sin^2 \frac{k\pi y}{b} dy\, dx = 1,$$

with the result that, for all j and k,

$$C_{jk} = \frac{2}{\sqrt{mab}}.$$

with $\gamma^2 = \beta^2 - \alpha^2$. The solutions to these equations are

$$X(x) = A_1 \sin \alpha x + A_2 \cos \alpha x$$
$$Y(y) = A_3 \sin \gamma y + A_4 \cos \gamma y,$$

and, therefore,

$$W(x,y) = C_1 \sin \alpha x \sin \gamma y + C_2 \sin \alpha x \cos \gamma y$$
$$+ C_3 \cos \alpha x \sin \gamma y + C_4 \cos \alpha x \cos \gamma y. \quad (11.2)$$

The coefficients C_i are determined by satisfying the boundary conditions. To proceed further analytically, assume that the membrane is simply supported along all four sides. We anticipate that, of the four components in Equation 11.2, only the first term will be non–zero for the following simply–supported boundary conditions: $W(0,y) = 0$, $W(a,y) = 0$, and $W(x,0) = 0$, $W(x,b) = 0$. The conditions at each boundary must be examined;

(i) Along $x = 0$, $W(0,y) = C_3 \sin \gamma y + C_4 \cos \gamma y = 0$, which can be true for arbitrary y only if $C_3 = C_4 = 0$.

(ii) Along $x = a$, $W(a,y) = C_1 \sin \alpha a \sin \gamma y + C_2 \sin \alpha a \cos \gamma y = 0$, which yields two possible solutions. The first is that $C_1 = C_2 = 0$, but this is the well known trivial solution that is of no practical value. The other possible solution is $\sin \alpha a = 0$.

(iii) Along $y = 0$, $W(x,0) = C_2 \sin \alpha x + C_4 \cos \alpha x = 0$, which can be true for arbitrary x only if $C_2 = C_4 = 0$.

(iv) Along $y = b$, $W(x,b) = C_1 \sin \alpha x \sin \gamma b + C_3 \cos \alpha x \sin \gamma b = 0$, which also has two possible solutions. The trivial solution requires $C_1 = C_3 = 0$. The other possibility is that $\sin \gamma b = 0$.

Therefore, the only solution that can satisfy all the above is that $C_1 \neq 0$ along with the following *characteristic equations*,

$$\sin \alpha a = 0 \quad \longrightarrow \quad \alpha_j a = j\pi$$
$$\sin \gamma b = 0 \quad \longrightarrow \quad \gamma_k b = k\pi,$$

where $j, k = 1, 2, \ldots$. Thus,

$$\beta_{jk} = (\alpha_j^2 + \gamma_k^2)^{1/2} = \pi \left[\left(\frac{j}{a}\right)^2 + \left(\frac{k}{b}\right)^2 \right]^{1/2},$$

and

$$\omega_{jk} = c\beta_{jk} = \sqrt{\frac{T}{m}} \beta_{jk}, \quad j,k = 1, 2, \ldots,$$

CHAPTER 11. ADVANCED CONTINUOUS MODELS

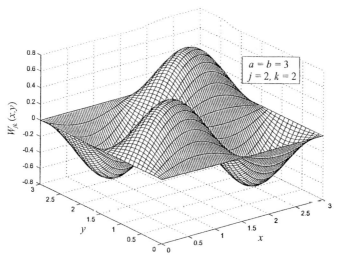

Figure 11.2: (b) Example mode for square membrane.

At this point, the complete solution in space and time can be recovered for each mode jk. First specify the time function,

$$F_{jk}(t) = B_{jk1} \cos \omega_{jk} t + B_{jk2} \sin \omega_{jk} t, \tag{11.3}$$

and then $w_{jk}(x, y, t) = W_{jk}(x, y) F_{jk}(t)$, or

$$\begin{aligned} w_{jk}(x, y, t) &= \frac{2}{\sqrt{mab}} \sin \frac{j\pi x}{a} \sin \frac{k\pi y}{b} (B_{jk1} \cos \omega_{jk} t + B_{jk2} \sin \omega_{jk} t), \\ j, k &= 1, 2, \ldots. \end{aligned} \tag{11.4}$$

The complete solution is the double sum

$$w(x, y, t) = \sum_{j=1}^{\infty} \sum_{k=1}^{\infty} w_{jk}(x, y, t).$$

We are now ready to satisfy the initial conditions by appropriately choosing the coefficients B_{jk1} and B_{jk2}. For generality, let the initial conditions be

$$\begin{aligned} w(x, y, 0) &= d(x, y) \\ \frac{\partial w(x, y, 0)}{\partial t} &= v(x, y), \end{aligned}$$

November 2004

Dear Customer:

Thank you for your purchase of *Mechanical Vibration: Analysis, Uncertainties, and Control, 2nd edition* (Cat. #DK801x) by Haym Benaroya.

Unfortunately, the author inadvertently left out the page that should have followed page 580. The missing page is on the reverse of this sheet. We sincerely regret any inconvenience this may have caused.

CRC Press

#DK801X / 0-8247-5380-1

11.1. VIBRATION OF MEMBRANES

where $d(x,y)$ is the initial displacement field, and $v(x,y)$ is the initial velocity field.

Applying the initial displacement field leads to the double Fourier series

$$d(x,y) = \sum_{j=1}^{\infty}\sum_{k=1}^{\infty} \frac{2}{\sqrt{mab}} B_{jk1} \sin\frac{j\pi x}{a} \sin\frac{k\pi y}{b}, \qquad (11.5)$$

where $d(x,y)$ must be continuous and have the derivatives $d_{,x}$, $d_{,y}$, and $d_{,xy}$. Examine each part of the double Fourier series in step. Define

$$K_j(y) = \sum_{k=1}^{\infty} \frac{2}{\sqrt{mab}} B_{jk1} \sin\frac{k\pi y}{b} \qquad (11.6)$$

so that

$$d(x,y) - \sum_{j=1}^{\infty} K_j(y) \sin\frac{j\pi x}{a}. \qquad (11.7)$$

For fixed y, Equation 11.7 is a Fourier sine series with

$$K_j(y) = \frac{2}{a}\int_0^a d(x,y) \sin\frac{j\pi x}{a} dx.$$

Equation 11.6 is also a Fourier sine series with

$$\frac{2}{\sqrt{mab}} B_{jk1} = \frac{2}{b}\int_0^b K_j(y) \sin\frac{k\pi y}{b} dy.$$

Therefore, combine terms and solve for B_{jk1},

$$B_{jk1} = 2\sqrt{\frac{m}{ab}}\int_0^b\int_0^a d(x,y) \sin\frac{j\pi x}{a} \sin\frac{k\pi y}{b} dx\,dy, \quad j,k=1,2,\ldots. \qquad (11.8)$$

Equation 11.8 is sometimes called the *generalized Euler formula*.

Following the same procedure applying the initial velocity field, leads to

$$B_{jk2} = \frac{2}{\omega_{jk}}\sqrt{\frac{m}{ab}}\int_0^b\int_0^a v(x,y) \sin\frac{j\pi x}{a} \sin\frac{k\pi y}{b} dx\,dy, \quad j,k=1,2,\ldots, \qquad (11.9)$$

where this is a frequency–dependent coefficient. Once initial displacement $d(x,y)$ and initial velocity $v(x,y)$ are specified, Equations 11.8 and 11.9 can be evaluated and substituted in Equation 11.4.

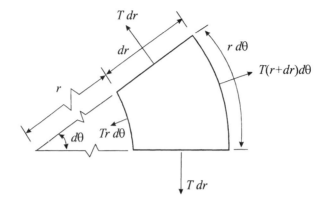

Figure 11.3: Free body diagram of a circular membrane.

11.1.2 Circular Membranes

It is of interest and important to consider next the circular membrane. Applications include specialized inflatable structures and circular elements of biomechanical devices such as heart valves. The free body diagram of Figure 11.3 depicts forces on a membrane sector, where *polar coordinates* are now best suited to represent the geometry.

In polar coordinates, the eigenvalue problem is

$$\nabla^2 W(r,\theta) + \beta^2 W(r,\theta) = 0, \quad \text{plus boundary conditions}, \qquad (11.10)$$

where

$$\nabla^2 \equiv \frac{\partial^2}{\partial r^2} + \frac{1}{r}\frac{\partial}{\partial r} + \frac{1}{r^2}\frac{\partial^2}{\partial \theta^2},$$

and $\beta = \omega/c$. Assume a solution of the form

$$W(r,\theta) = R(r)\Theta(\theta),$$

and substitute this along with its appropriate derivatives into Equation 11.10 to find

$$R''\Theta + \frac{1}{r}R'\Theta + \frac{1}{r^2}R\Theta'' + \beta^2 R\Theta = 0.$$

Collecting like variables,

$$\frac{r^2}{R}\left(R'' + \frac{1}{r}R' + \beta^2 R\right) = -\frac{\Theta''}{\Theta} = j^2. \qquad (11.11)$$

A positive sign is chosen for the constant j^2 so that Θ has a harmonic character. That is, $\Theta'' + j^2\Theta = 0$ is harmonic in θ, as required by the

11.1. VIBRATION OF MEMBRANES

circular geometry. Also note that j must be an integer to satisfy the requirement that $\theta = \theta + 2\pi n$, where n is an integer. In this way, continuity and uniqueness are maintained in the circumferential direction around the circular membrane. Thus,

$$\Theta_j(\theta) = A_{1j} \sin j\theta + A_{2j} \cos j\theta, \quad j = 0, 1, 2, \ldots,$$

where the coefficients A_{1j} and A_{2j} will be determined.

The equation in $R(r)$,

$$R'' + \frac{1}{r}R' + \left(\beta^2 - \frac{j^2}{r^2}\right)R = 0,$$

must be now solved. This equation has an expansion solution[2] known as the *Bessel function*,

$$R_j(r) = A_{3j} J_j(\beta r) + A_{4j} Y_j(\beta r),$$

where J_j is the Bessel function of order j of the *first kind*, and Y_j is the Bessel function of order j of the *second kind*. The general solution of the eigenvalue problem is

$$\begin{aligned} W_j(r,\theta) &= R_j(r)\Theta_j(\theta) \\ &= C_{1j} J_j(\beta r) \sin j\theta + C_{2j} J_j(\beta r) \cos j\theta \\ &+ C_{3j} Y_j(\beta r) \sin j\theta + C_{4j} Y_j(\beta r) \cos j\theta, \quad j = 0,1,2,\ldots. \end{aligned}$$

The coefficients C_{lj} are determined by satisfying the boundary conditions. For example, consider the membrane fixed on the rim at $r = a$, that is, $W_j(a, \theta) = 0$. Since there is a finite deflection at the center of the membrane, and since $Y_j(\beta r)$ has the property $Y_j(0) = \infty$, then for physical reasons it must be that $C_{3j} = C_{4j} = 0$. The modes of vibration are then given by

$$W_j(r,\theta) = C_{1j} J_j(\beta r) \sin j\theta + C_{2j} J_j(\beta r) \cos j\theta, \quad j = 0,1,2,\ldots. \quad (11.12)$$

To satisfy the boundary condition at the rim, set $r = a$,

$$W_j(a,\theta) = C_{1j} J_j(\beta a) \sin j\theta + C_{2j} J_j(\beta a) \cos j\theta = 0,$$

for all θ, and $j = 0, 1, 2, \ldots$. Since the solution $C_{1j} = C_{2j} = 0$ is trivial, it must be that

$$J_j(\beta a) = 0, \quad j = 0, 1, 2, \ldots. \quad (11.13)$$

---

[2]Assume an expansion solution of the general form: $y = a_0 + a_1 x + a_2 x^2 + \ldots$. Substitute this into the governing equation, and find the power series solution defined as the *Bessel function*.

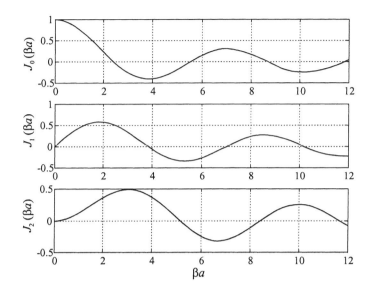

Figure 11.4: Roots of the Bessel function $J_j(\beta a) = 0$.

This equation specifies the allowable frequencies of free vibration. Equations 11.13 are the characteristic equations for this problem. There are an *infinity of characteristic equations*, one for each j, and, for each of these characteristic equations, the Bessel function has an *infinity of roots*, or frequencies.

For example, the first three roots of $J_0(\beta a)$ are 2.405, 5.520, and 8.654, and the first three roots of $J_1(\beta a)$ are 3.832, 7.016, and 10.173. See Figure 11.4.

In order to signify this multiplicity of roots, use the notation

$$\beta_{jk} = \frac{\omega_{jk}}{c}.$$

For each frequency there are two distinct modes, and there are fewer frequencies than modes. Such a case is known as a *degenerate* case, and Equation 11.12 for the modes becomes

$$W_{1jk}(r, \theta) = C_{1jk} J_j(\beta_{jk} r) \sin j\theta, \qquad (11.14)$$
$$W_{2jk}(r, \theta) = C_{2jk} J_j(\beta_{jk} r) \cos j\theta, \qquad (11.15)$$
$$j, k = 0, 1, 2, \ldots.$$

Note that for $j = 0$, $\sin j\theta = 0$ and $\cos j\theta = 1$. This reduced case is considered separately when evaluating the coefficients C_{ljk}. Following

11.1. VIBRATION OF MEMBRANES

Meirovitch,[3] the natural modes can be normalized according to the rule

$$\int_D mW_{0k}^2 \, dD = 1,$$

where D is a two dimensional domain of integration. For the $j = 0$ case,

$$\int_0^{2\pi} \int_0^a mC_{0k}^2 J_0^2\left(\frac{\omega_{0k}}{c}r\right) r \, dr \, d\theta = 1$$

$$\implies C_{0k}^2 = \frac{1}{\pi m a^2 J_1^2[\omega_{0k}a/c]}. \tag{11.16}$$

For all other j, the natural modes are degenerate

$$\int_0^{2\pi} \int_0^a mC_{1jk}^2 J_j^2\left(\frac{\omega_{jk}}{c}r\right) \sin^2(j\theta) r \, dr \, d\theta = 1$$

$$\implies C_{1jk}^2 = \frac{2}{\pi m a^2 J_{j+1}^2[\omega_{jk}a/c]}, \tag{11.17}$$

$$\int_0^{2\pi} \int_0^a mC_{2jk}^2 J_j^2\left(\frac{\omega_{jk}}{c}r\right) \cos^2(j\theta) r \, dr \, d\theta = 1$$

$$\implies C_{2jk}^2 = \frac{2}{\pi m a^2 J_{j+1}^2[\omega_{jk}a/c]}. \tag{11.18}$$

The normal modes are then

$$W_{0k}(r,\theta) = \frac{1}{\sqrt{\pi m a} J_1[\omega_{0k}a/c]} J_0\left(\frac{\omega_{0k}}{c}r\right), \quad k = 1, 2, \ldots$$

$$W_{1jk}(r,\theta) = \frac{\sqrt{2}}{\sqrt{\pi m a} J_{j+1}[\omega_{jk}a/c]} J_j\left(\frac{\omega_{jk}}{c}r\right) \sin j\theta, \quad j, k = 1, 2, \ldots$$

$$W_{2jk}(r,\theta) = \frac{\sqrt{2}}{\sqrt{\pi m a} J_{j+1}[\omega_{jk}a/c]} J_j\left(\frac{\omega_{jk}}{c}r\right) \cos j\theta, \quad j, k = 1, 2, \ldots.$$

---

[3] **Analytical Methods in Vibrations**, Macmillan, 1967, 176–177.

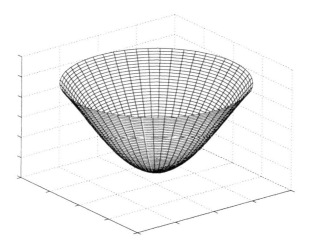

Figure 11.5: (a) Modes of a circular membrane: $j = 0, k = 1$.

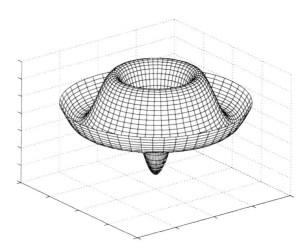

Figure 11.5: (b) Modes of a circular membrane: $j = 0, k = 3$.

11.1. VIBRATION OF MEMBRANES

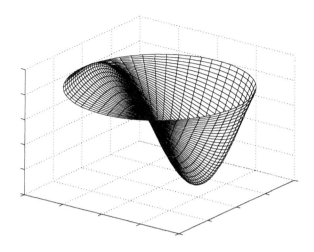

Figure 11.5: (c) Modes of a circular membrane: $j = 1, k = 1$.

Nodal lines of zero displacement are given by circles of constant radius r, and by diametrical lines of constant angle θ. Representative modes (exaggerated scale) are shown in Figures 11.5.

Equations 11.14 and 11.15 can now be written as the sum

$$W_{jk}(r,\theta) = \begin{cases} W_{0k} \text{ for } j = 0 \\ W_{1jk} + W_{2jk} \text{ for } j, k = 1, 2, 3, \ldots, \end{cases}$$

and the complete response is given by

$$w(r,\theta,t) = \sum_{j=1}^{\infty} \sum_{k=1}^{\infty} w_{jk}(r,\theta,t),$$
$$\text{where } w_{jk}(r,\theta,t) = W_{jk}(r,\theta) F_{jk}(t).$$

$F_{jk}(t)$ is a harmonic function such as Equation 11.3, with arbitrary coefficients that can be established by satisfying the initial conditions.

Bessel

Friedrich Wilhelm Bessel lived during the years 1784–1846 and was born Minden, Germany. He died in Konigsberg, Germany. Bessel, as a boy, dreamed of travel, so he studied languages, geography and the principles of navigation. This led him to study astronomy and mathematics. In 1804, he wrote a paper on Halley's comet, calculating the orbit from data from observations made in 1607. He sent it to Olbers and it was published.

588   CHAPTER 11. ADVANCED CONTINUOUS MODELS

Olbers offered Bessel a post as assistant at the Lilienthal Observatory and, after some thought, Bessel left the affluence of his commercial job for the poverty of the Observatory post.

At the age of 26, Bessel was appointed director of Frederick William III of Prussia's new Konigsberg Observatory and he was granted the title of doctor by the University of Göttingen. Bessel remained at Konigsberg for the rest of his life, and it was there that he undertook his monumental task of determining the positions and proper motions of stars which led to the discovery in 1838 of the parallax of 61 Cygni. Bessel also worked out a method of mathematical analysis involving what is now known as the Bessel function. He introduced this in 1817 in his study of the problem of determining the motion of three bodies moving under mutual gravitation. In 1824, he developed Bessel functions more fully in a study of planetary perturbations.

11.2 Vibration of Plates

We endeavor next to develop and analyze the equations governing the dynamic behavior of plates.[4] Plates are useful models not only for the obvious structural components, but also for internal combustion engine components such as pistons and cylinder heads, diaphragms used in artificial hearts, pumping devices, and lenses used in a variety of optical devices. We begin by looking at the most basic plate model, one of rectangular cross–section and uniform thickness. We will also learn how to formulate and solve the most basic plate vibration problems of the kind alluded to in the above applications. For an introduction to the damping of plates, the reader is referred to the cited texts.

The plate is more complex than the membrane because plates can resist bending moments. Therefore, such moments are part of the free body diagram and affect the final form of the governing equations of motion. The same analogy held for the restoring forces of the string as compared to the moment resistance of the beam.

The following assumptions are made *a priori* so that a simplified but useful model can be approached analytically. Assume

---

[4] An advanced book on the vibration of plates is **The Bending and Stretching of Plates**, Second Edition, by E.H. Mansfield, Cambridge University Press, 1989. It includes studies of plates of various shapes, variable rigidity problems, approximate methods, and large–deflection theory. Another book, **Vibrations of Elastic Plates**, by Y.–Y. Yu, Springer–Verlag, 1996, provides an introduction to linear and nonlinear dynamical modeling of sandwich plates, laminated composites, and piezoelectric layered plates. Two excellent starting points on plates and shells are the monographs **Theory of Plates and Shells**, by S. Timoshenko and S. Woinowsky–Krieger, McGraw–Hill, 1959, and **Vibrations of Plates and Shells**, by W. Soedel, Marcel Dekker, 1981.

11.2. VIBRATION OF PLATES

1. plate thickness h is small when compared with other dimensions (a rule of thumb is a ratio of less than 1:10),

2. there is no strain along the middle surface during bending,

3. plane sections do not warp and therefore remain plane,

4. normal stresses in the direction transverse to the plate can be ignored due to the small relative thickness,

5. the load is normal to the surface,

6. deflections are small, and

7. the influence of shear and rotary inertia are neglected. More general problems require approximate and computational models.

11.2.1 Derivation of the Equation of Motion for Rectangular Plates

Consider the free body diagrams, Figure 11.6 for load and shear forces and Figure 11.7 for moments. Take the positive coordinate direction as down in the figures, and use the right hand rule for positive moments. Our development will make use of results from solid mechanics[5] as applied to the free body diagrams and the resulting equations of motion.

For the plate in question, the net shear forces are

$$Q_x = \int_{-h/2}^{h/2} \tau_{xz} dz, \quad Q_y = \int_{-h/2}^{h/2} \tau_{yz} dz.$$

Q_x is the net shear force on the face with a perpendicular in the x direction. Similarly for Q_y. Next, use Newton's second law of motion, setting the sum of forces in the z direction (perpendicular to the xy plane of the plate) equal to the inertia force in the z direction,

$$\frac{\partial Q_x}{\partial x} + \frac{\partial Q_y}{\partial y} + p(x, y, t) = \rho h \frac{\partial^2 w}{\partial t^2}, \qquad (11.19)$$

where $w(x, y, t)$ is the deflection of the middle plane[6] in the z direction, $p(x, y, t)$ is the load intensity per unit area, ρ is the volume density, and ρh is the mass per unit area. In Equation 11.19, $dx\,dy$ has been divided out

---
[5] See, for example, the following excellent texts: **Foundations of Solid Mechanics** by Y.C. Fung, Prentice–Hall, 1965, and **Applied Elasticity** by C.-T. Wang, McGraw–Hill, 1953.

[6] Note that w is not a function of z, but only x, y and time.

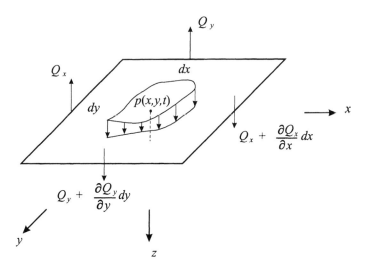

Figure 11.6: Free body diagram for rectangular plate: shear.

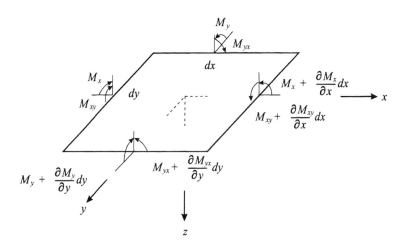

Figure 11.7: Free body diagram for rectangular plate: moments.

11.2. VIBRATION OF PLATES

of all expressions. Newton's second law of motion is applied next to the sum of the moments about the x and y axes. The intermediate equation shown next is of the sum of the moments about the x axis, neglecting rotary inertia[7],

$$M_y dx - \left(M_y + \frac{\partial M_y}{\partial y}dy\right)dx - M_{xy}dy$$
$$+ \left(M_{xy} + \frac{\partial M_{xy}}{\partial x}dx\right)dy + Q_y\frac{dy}{2}dx + \left(Q_y dx + \frac{\partial Q_y}{\partial y}dx\, dy\right)\frac{dy}{2} = 0$$

or $\quad -\dfrac{\partial M_y}{\partial y}dy\, dx + \dfrac{\partial M_{xy}}{\partial x}dx\, dy + Q_y dx\, dy = 0.$

Higher order products such as $(dx)^2$ are neglected based on our earlier assumptions. The same procedure is used for the moments about the y axis. Having done this, the moment equations about the x and y axes are, respectively,

$$\frac{\partial M_{xy}}{\partial x} - \frac{\partial M_y}{\partial y} + Q_y = 0 \qquad (11.20)$$

$$\frac{\partial M_x}{\partial x} + \frac{\partial M_{yx}}{\partial y} - Q_x = 0. \qquad (11.21)$$

The notation M_y, for example, indicates the bending moment due to normal stress σ_y, and M_{xy} indicates a twisting moment due to shear stress τ_{xy}. Perform the appropriate differentiations of Equations 11.20 and 11.21 in order to obtain expressions for $\partial Q_x/\partial x$ and $\partial Q_y/\partial y$, which are then substituted into Equation 11.19, to find

$$\frac{\partial^2 M_x}{\partial x^2} + \frac{\partial^2 M_y}{\partial y^2} - 2\frac{\partial^2 M_{xy}}{\partial x \partial y} = -p(x,y,t) + \rho h \frac{\partial^2 w}{\partial t^2}, \qquad (11.22)$$

where $M_{xy} = -M_{yx}$. The moments need to be related to the plate deflection w. To do this, let u and v be components of displacement at any point in the plate, parallel to the xy axis,

$$u = -z\frac{\partial w}{\partial x}, \quad v = -z\frac{\partial w}{\partial y}.$$

See Figure 11.8 for a schematic of the displacement u.

---

[7]Details of this derivation can be found in **Mechanics of Elastic Structures** by J.T. Oden and E.A. Ripperger, McGraw–Hill 1981.

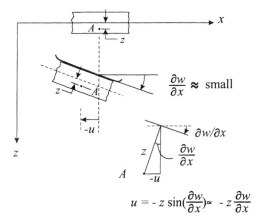

Figure 11.8: Schematic of strain–displacement for a plate.

The strains are related to the displacements in the following way,

$$\varepsilon_x = \frac{\partial u}{\partial x} = -z\frac{\partial^2 w}{\partial x^2}$$
$$\varepsilon_y = \frac{\partial v}{\partial y} = -z\frac{\partial^2 w}{\partial y^2}$$
$$\gamma_{xy} = \frac{\partial u}{\partial y} + \frac{\partial v}{\partial x} = -2z\frac{\partial^2 w}{\partial x \partial y}.$$

Hooke's law for plane stress relates these strains to the stress resultants,

$$\sigma_x = \frac{E}{1-\nu^2}(\varepsilon_x + \nu\varepsilon_y) = -\frac{Ez}{1-\nu^2}\left(\frac{\partial^2 w}{\partial x^2} + \nu\frac{\partial^2 w}{\partial y^2}\right)$$
$$\sigma_y = \frac{E}{1-\nu^2}(\varepsilon_y + \nu\varepsilon_x) = -\frac{Ez}{1-\nu^2}\left(\frac{\partial^2 w}{\partial y^2} + \nu\frac{\partial^2 w}{\partial x^2}\right)$$
$$\tau_{xy} = G\gamma_{xy} = -2Gz\frac{\partial^2 w}{\partial x \partial y} = -\frac{Ez}{1+\nu}\frac{\partial^2 w}{\partial x \partial y},$$

where ν is Poisson's ratio, E is Young's modulus of elasticity, and G is the shearing modulus.

Each stress resultant is multiplied by its respective moment arm, yielding

11.2. VIBRATION OF PLATES

the following moments,

$$M_x = \int_{-h/2}^{h/2} z\sigma_x dz = -D\left(\frac{\partial^2 w}{\partial x^2} + \nu \frac{\partial^2 w}{\partial y^2}\right) \quad (11.23)$$

$$M_y = \int_{-h/2}^{h/2} z\sigma_y dz = -D\left(\frac{\partial^2 w}{\partial y^2} + \nu \frac{\partial^2 w}{\partial x^2}\right) \quad (11.24)$$

$$M_{xy} = -\int_{-h/2}^{h/2} z\tau_{xy} dz = -M_{yx} = D(1-\nu)\frac{\partial^2 w}{\partial x \partial y}, \quad (11.25)$$

where $D = Eh^3/12(1-\nu^2)$ is called the *flexural rigidity* of the plate. Equations 11.23–11.25 relate moments to deflection w. Substitute the appropriate derivatives of these moments into Equation 11.22 to find[8]

$$D\left(\frac{\partial^4 w}{\partial x^4} + 2\frac{\partial^4 w}{\partial x^2 \partial y^2} + \frac{\partial^4 w}{\partial y^4}\right) + \rho h \frac{\partial^2 w}{\partial t^2} = p(x, y, t) \quad (11.26)$$

or

$$D\nabla^4 w + \rho h \ddot{w} = p(x, y, t), \quad (11.27)$$

where ∇^4 is called the *biharmonic* operator. Equation 11.26 or 11.27 is the equation of motion for the transverse vibration of a plate given the stated assumptions.

11.2.2 The Eigenvalue Problem

The eigenvalue problem is obtained by first setting $p(x, y, t) = 0$, and then assuming the solution to be separable in space and time,

$$w(x, y, t) = W(x, y)F(t).$$

Substituting the assumed solution and its derivatives into Equation 11.27,

$$DF(t)\nabla^4 W(x, y) + \rho h W(x, y)\ddot{F}(t) = 0,$$

---

[8] Use the following:

$$\frac{\partial^2 M_x}{\partial x^2} = -D\left(\frac{\partial^4 w}{\partial x^4} + \nu \frac{\partial^4 w}{\partial x^2 \partial y^2}\right)$$

$$\frac{\partial^2 M_y}{\partial y^2} = -D\left(\frac{\partial^4 w}{\partial y^4} + \nu \frac{\partial^4 w}{\partial x^2 \partial y^2}\right)$$

$$-2\frac{\partial^2 M_{xy}}{\partial x \partial y} = -2D(1-\nu)\frac{\partial^4 w}{\partial x^2 \partial y^2}.$$

or
$$\frac{D}{\rho h}\frac{\nabla^4 W}{W} = -\frac{\ddot{F}}{F} = \omega^2, \quad \text{a constant.}$$

The eigenvalue problem is an equation in W,
$$\nabla^4 W - \beta^4 W = 0, \tag{11.28}$$

where $\beta^4 = \omega^2 \rho h / D$. Equation 11.28 can be solved by recognizing its equivalent factored form,
$$(\nabla^2 + \beta^2)(\nabla^2 - \beta^2)W = 0.$$

There are two possible solutions to this equation, $W = W_1$ and $W = W_2$, each of which will satisfy the equality to zero, that is,
$$\begin{aligned}(\nabla^2 + \beta^2)W_1 &= 0 \\ (\nabla^2 - \beta^2)W_2 &= 0.\end{aligned} \tag{11.29}$$

The complete solution by linear superposition will be $W = W_1 + W_2$. Solution W_1 is that for a membrane, as is solution W_2 if Equation 11.29 is written as
$$(\nabla^2 + (i\beta)^2)W_2 = 0.$$

Equation 11.2 is the solution for W_1, reproduced here,
$$\begin{aligned}W_1(x,y) &= C_1 \sin\alpha x \sin\gamma y + C_2 \sin\alpha x \cos\gamma y \\ &+ C_3 \cos\alpha x \sin\gamma y + C_4 \cos\alpha x \cos\gamma y.\end{aligned} \tag{11.30}$$

Similarly,[9]
$$\begin{aligned}W_2(x,y) &= C_5 \sinh\alpha_1 x \sinh\gamma_1 y + C_6 \sinh\alpha_1 x \cosh\gamma_1 y \\ &+ C_7 \cosh\alpha_1 x \sinh\gamma_1 y + C_8 \cosh\alpha_1 x \cosh\gamma_1 y.\end{aligned} \tag{11.31}$$

From the membrane solutions, recall that $\beta^2 = \alpha^2 + \gamma^2$ and $\beta_1^2 = \alpha_1^2 + \gamma_1^2$.

In order to proceed analytically,[10] assume the case of the simply supported plate. In this instance, all terms in Equations 11.30 and 11.31 must be zero except for the term $C_1 \sin\alpha x \sin\gamma y$. None of the other terms can satisfy the conditions of no moment resistance and no deflection at all boundaries. Therefore,
$$W(x,y) = C_1 \sin\alpha x \sin\gamma y. \tag{11.32}$$

---

[9] Use the identities $\sin(ix) = i\sinh x$ and $\cos(ix) = \cosh x$ in Equation 11.30 and incorporate any imaginary numbers into the arbitrary constants.

[10] More complex problems can, of course, be solved using approximate and numerical procedures.

11.2. VIBRATION OF PLATES

The set of displacement boundary conditions requires that
$$W(0,y) = W(a,y) = W(x,0) = W(x,b) = 0.$$
From Equation 11.32 this leads to
$$\sin \alpha a = 0 \implies \alpha_j a = j\pi$$
$$\sin \gamma b = 0 \implies \gamma_k b = k\pi, \quad j,k = 1,2,\ldots.$$
Therefore,
$$\omega_{jk} = \beta_{jk}^2 \sqrt{\frac{D}{\rho h}} = (\alpha_j^2 + \gamma_k^2)\sqrt{\frac{D}{\rho h}}$$
$$= \pi^2 \left[\left(\frac{j}{a}\right)^2 + \left(\frac{k}{b}\right)^2\right]\sqrt{\frac{D}{\rho h}},$$
and the modes are given by $W_{jk}(x,y) = C_{jk} \sin \alpha_j x \sin \gamma_k y$, or
$$W_{jk}(x,y) = C_{jk} \sin \frac{j\pi x}{a} \sin \frac{k\pi y}{b}, \tag{11.33}$$
where, by using the usual normalization procedure, $C_{jk} = 2/\sqrt{\rho h a b}$. The complete solution is then the double sum
$$w(x,y,t) = \sum_{j=1}^{\infty}\sum_{k=1}^{\infty} W_{jk}(x,y)F_{jk}(t), \tag{11.34}$$
where the remainder of the solution follows lines identical to those of the membrane.

Example 11.1 Forced, Simply–Supported Rectangular Plate
For a uniform, simply–supported plate that is loaded harmonically at one point (x_1,y_1), the loading function is given by
$$P(x,y,t) = \sin\Omega t\ \delta(x-x_1, y-y_1),$$
where the δ is the Kronecker delta function, defined as
$$\delta(m,n) = \begin{cases} 1 & \text{if } m=0 \text{ and } n=0 \\ 0 & \text{if } m \neq 0 \text{ and/or } n \neq 0. \end{cases}$$

From the previous discussion, the normal modes $W_{jk}(x,y)$ and frequencies ω_{jk} are known. Assuming zero initial conditions, the response is given by Equation 11.34, where
$$F_{jk}(t) = \int_0^t \left[\int_0^a \int_0^b P(\zeta,\eta,\tau) W_{jk}(\zeta,\eta)d\zeta d\eta\right] g_{jk}(t-\tau)d\tau.$$

See *Example* 10.5 for the analogous beam problem. When substituting the two dimensional Kronecker delta function for the force $P(x,y,t)$, the double integral is reduced to a single value at $x = x_1$ and $y = y_1$, with all other values equal to zero. Then,

$$F_{jk}(t) = \int_0^t \frac{2}{\sqrt{\rho hab}} \sin \Omega \tau \sin \frac{j\pi x_1}{a} \sin \frac{k\pi y_1}{b} g_{jk}(t-\tau)d\tau.$$

The function $g_{jk}(t)$ is the impulse response function, here for an undamped oscillator,

$$g_{jk}(t) = \frac{1}{\omega_{jk}} \sin \omega_{jk} t, \quad t \geq 0.$$

The complete solution is then

$$w(x,y,t) = \frac{4}{\rho hab} \sum_{j=1}^{\infty} \sum_{k=1}^{\infty} \frac{1}{\omega_{jk}} \sin \frac{j\pi x_1}{a} \sin \frac{k\pi y_1}{b} \sin \frac{j\pi x}{a} \sin \frac{k\pi y}{b}$$
$$\cdot \int_0^t \sin \Omega \tau \sin \omega_{jk}(t-\tau) d\tau.$$

Note that for certain values of x_1 and y_1, there are modes that cannot be excited by the given force. These are the modes where jx_1/a and ky_1/b are integers. If the loading function is more complicated, a more intricate algebraic procedure is needed to evaluate $F_{jk}(t)$. ∎

The problem of circular plates can be approached using the above techniques, but the algebra becomes more complicated due to the polar system of coordinates that must be used.

11.3 Random Vibration of Continuous Structures

While the general random vibration of a continuous structure is quite complex, we can begin by studying the random vibration of elastic and continuous structures using the modal approach. The algebra can become quite extensive, but it is instructive to carry out the procedures, since computational approaches will be based on similar ideas. The lateral vibration of a beam is used in this introduction, and we refer to *Example* 10.5 beginning with Equation 10.89 for several key equations. In order to avoid confusion between displacement and mode function, we use Z for displacement, and $\hat{Y}_j(\cdot)$ for mode function. Even though $\hat{Y}_j(\cdot)$ is capitalized, it is not a random function. We will draw on the probability concepts of Chapter 5.

11.3. RANDOM VIBRATION OF CONTINUOUS STRUCTURES

Assume that the loading $P(x,t)$ is random and represents an ensemble of functions of space and time. Only the first two statistics of the force are needed to proceed with the analysis of the response. Its mean value is denoted
$$E\{P(x,t)\} = \mu_P(x,t),$$
with autocorrelation
$$E\{P(x_1,t_1)P(x_2,t_2)\} = R_{PP}(x_1,x_2;t_1,t_2).$$
These statistics are available through data gathering experiments performed on the loading to be experienced by the structure. The goal in this analysis is to derive the statistics for the response, μ_Z and R_{ZZ}, which are functions of the statistics of the loading and the structural parameters.

For steady–state vibration problems, it is reasonable to assume that the statistics of the loading are stationary. Begin with Equation 10.95,
$$Z(x,t) = \sum_{j=1}^{\infty} \hat{Y}_j(x) \int_0^t \left[\int_0^L P(\xi, t-\tau)\hat{Y}_j(\xi)d\xi \right] g_j(\tau)d\tau,$$
where the shift τ has been placed within $P(x,t)$. Take the expectation of both sides,
$$E\{Z(x,t)\} = \sum_{j=1}^{\infty} \hat{Y}_j(x) \int_0^t \left[\int_0^L E\{P(\xi, t-\tau)\}\hat{Y}_j(\xi)d\xi \right] g_j(\tau)d\tau.$$
$g(t)$ is the impulse response function, and for stationary loading, the expected value is a constant, $E\{P(\xi, t-\tau)\} = \mu_P(\xi)$. In order to simplify the following analysis, assume that the loading has zero mean. This does not overly simplify the problem since a non–zero mean can be introduced later by simply shifting the response by the (constant) mean value. For zero mean loading, $\mu_P(x) = 0$ and the response is also zero mean, $E\{Z(x,t)\} = \mu_Z = 0$.

We need to relate the autocorrelation of the response to the autocorrelation of the loading. This requires several steps, as follows. For the loading, using Equation 10.90,
$$\begin{aligned}
R_{P_jP_k}(t_1,t_2) &= E\{P_j(t_1)P_k(t_2)\} \\
&= m^2 \int_0^L \int_0^L E\{P(x_1,t_1)P(x_2,t_2)\}\hat{Y}_j(x_1)\hat{Y}_k(x_2)dx_1dx_2 \\
&= m^2 \int_0^L \int_0^L R_{PP}(x_1,x_2;t_1,t_2)\hat{Y}_j(x_1)\hat{Y}_k(x_2)dx_1dx_2,
\end{aligned}$$
(11.35)

where, for a stationary loading, $R_{PP}(x_1, x_2; t_1, t_2) = R_{PP}(x_1, x_2; \tau)$, $\tau = t_2 - t_1$, and therefore $R_{P_j P_k}(t_1, t_2) = R_{P_j P_k}(\tau)$. Using Equation 10.91, the response autocorrelation is given by

$$\begin{aligned} R_{ZZ}(x_1, x_2; t_1, t_2) &= E\{Z(x_1, t_1) Z(x_2, t_2)\} \\ &= \sum_{j=1}^{\infty} \sum_{k=1}^{\infty} E\{Z_j(t_1) Z_k(t_2)\} \hat{Y}_j(x_1) \hat{Y}_k(x_2) \\ &= \sum_{j=1}^{\infty} \sum_{k=1}^{\infty} R_{Z_j Z_k}(t_1, t_2) \hat{Y}_j(x_1) \hat{Y}_k(x_2), \quad (11.36) \end{aligned}$$

where,[11] using Equation 10.94,

$$\begin{aligned} R_{Z_j Z_k}(t_1, t_2) &= \frac{1}{m^2} E \left\{ \int_{-\infty}^{\infty} P_j(t_1 - \theta) g_j(\theta) d\theta \int_{-\infty}^{\infty} P_k(t_2 - \kappa) g_k(\kappa) d\kappa \right\} \\ &= \frac{1}{m^2} \int_{-\infty}^{\infty} \int_{-\infty}^{\infty} E\{P_j(t_1 - \theta) P_k(t_2 - \kappa)\} g_j(\theta) g_k(\kappa) d\theta \, d\kappa \\ &= \frac{1}{m^2} \int_{-\infty}^{\infty} \int_{-\infty}^{\infty} R_{P_j P_k}(t_1 - \theta, t_2 - \kappa) g_j(\theta) g_k(\kappa) d\theta \, d\kappa. \end{aligned}$$

Since $P(x, t)$ is stationary, $R_{P_j P_k}(t_1 - \theta, t_2 - \kappa) = R_{P_j P_k}(t_2 - t_1 + \theta - \kappa)$ and $R_{Z_j Z_k}(t_1, t_2) = R_{Z_j Z_k}(t_2 - t_1)$. Therefore,

$$R_{Z_j Z_k}(t_2 - t_1) = \frac{1}{m^2} \int_{-\infty}^{\infty} \int_{-\infty}^{\infty} R_{P_j P_k}(t_2 - t_1 + \theta - \kappa) g_j(\theta) g_k(\kappa) d\theta \, d\kappa, \quad (11.37)$$

and

$$R_{ZZ}(x_1, x_2; t_2 - t_1) = \sum_{j=1}^{\infty} \sum_{k=1}^{\infty} R_{Z_j Z_k}(t_2 - t_1) \hat{Y}_j(x_1) \hat{Y}_k(x_2).$$

Generally, for linear systems, it is of value to have the frequency domain solution. Here, for the random vibration problem, the frequency domain is

---

[11] The convolution equation is

$$y_j(t) = \frac{1}{m} \int_0^t p_j(t - \tau) g_j(\tau) d\tau,$$

where $g_j(\tau)$, the impulse response function, is identically zero for $t < \tau$, that is, for the time instants preceding the excitation of the system. Hence, advancing the upper limit of the integral to infinity does not affect the value of the integral. Similarly, for $t < 0$, $p(t) = 0$ and therefore, advance the lower limit to minus infinity. The response can then be written as

$$y_j(t) = \frac{1}{m} \int_{-\infty}^{\infty} p_j(t - \tau) g_j(\tau) d\tau.$$

This property is known as *causality* and has the simple physical meaning that there is no response before the load acts on the structure.

11.3. RANDOM VIBRATION OF CONTINUOUS STRUCTURES

the spectral density of the response, $S_{ZZ}(\omega)$. Formally, this is given by the Fourier transform of $R_{ZZ}(\tau)$,

$$S_{ZZ}(\omega) = \sum_{j=1}^{\infty}\sum_{k=1}^{\infty}\left[\int_{-\infty}^{\infty} R_{Z_j Z_k}(\tau)e^{-i\omega\tau}d\tau\right]\hat{Y}_j(x_1)\hat{Y}_k(x_2), \qquad (11.38)$$

where $\tau = t_2 - t_1$, and the term in the square brackets equals $S_{Z_j Z_k}(\omega)$, which now needs to be evaluated in terms that have already been derived.

By definition, the Fourier transform relation exists,

$$S_{Z_j Z_k}(\omega) = \int_{-\infty}^{\infty} R_{Z_j Z_k}\tau e^{-i\omega\tau}d\tau.$$

$R_{Z_j Z_k}$ can be evaluated beginning with Equation 11.37,

$$R_{Z_j Z_k}(\tau) = \frac{1}{m^2}\int_{-\infty}^{\infty}\int_{-\infty}^{\infty} R_{P_j P_k}(\tau + \theta - \kappa)g_j(\theta)g_k(\kappa)d\theta\, d\kappa.$$

Take the Fourier transform of both sides, letting $\lambda = \tau + \theta - \kappa$,

$$S_{Z_j Z_k}(\omega) = \int_{-\infty}^{\infty} e^{-i\omega(\lambda-\theta+\kappa)}\frac{1}{m^2}\left[\int_{-\infty}^{\infty}\int_{-\infty}^{\infty} R_{P_j P_k}(\lambda)g_j(\theta)g_k(\kappa)d\theta\, d\kappa\right]d\lambda, \qquad (11.39)$$

where τ has been replaced by $\lambda - \theta + \kappa$ and $d\tau$ by $d\lambda$. Rewrite Equation 11.39 in a more useful form by separating the integrals according to dummy variables,[12]

$$S_{Z_j Z_k}(\omega) = \frac{1}{m^2}\int_{-\infty}^{\infty} g_j(\theta)e^{i\omega\theta}d\theta \int_{-\infty}^{\infty} g_k(\kappa)e^{-i\omega\kappa}d\kappa \int_{-\infty}^{\infty} R_{P_j P_k}(\lambda)e^{-i\omega\lambda}d\lambda. \qquad (11.40)$$

The Fourier transform of the impulse response function $g(t)$ is the frequency response function $H(i\omega)$. Therefore, Equation 11.40 becomes

$$S_{Z_j Z_k}(\omega) = \frac{1}{m^2}H_j^*(i\omega)H_k(i\omega)S_{P_j P_k}(\omega), \qquad (11.41)$$

where $H_j(i\omega) = [\omega_j^2 - \omega^2 + 2i\zeta_j\omega_j\omega]^{-1}$, $H_j^*(i\omega)$ is the complex conjugate of $H_j(i\omega)$, and $S_{P_j P_k}(\omega)$, the spectral density of the modal force components, is derived assuming that the Fourier transform for $P_j(t)$ exists, as it does for most physical processes,

$$P_j(t) = \frac{1}{2\pi}\int_{-\infty}^{\infty} P_j(\omega)e^{i\omega t}d\omega.$$

---

[12] Recall the similar procedures from Chapter 6, similar integrals existed, see Equation 5.44, when deriving the fundamental relation in random vibration, Equation 5.45. Also, from Chapter 9, the sequence of Equations 9.42–9.45 parallel our development here.

Take the Fourier transform of Equation 11.35,

$$S_{P_j P_k}(\omega) = m^2 \int_0^L \int_0^L S_{PP}(\omega) \hat{Y}_j(x_1) \hat{Y}_k(x_2) dx_1 dx_2, \quad (11.42)$$

where $S_{PP}(\omega)$ is the spectral density of the loading, a quantity that is estimated from data. With some straightforward substitutions, we find

$$S_{ZZ}(x_1, x_2; \omega) = \frac{1}{m^2} \sum_{j=1}^{\infty} \sum_{k=1}^{\infty} H_j^*(i\omega) H_k(i\omega) S_{P_j P_k}(\omega) \hat{Y}_j(x_1) \hat{Y}_k(x_2).$$

One value of having such an equation is that the mean square (MS) displacement can be evaluated,

$$Z_{\text{MS}}(x) = R_{ZZ}(x, x; 0) = \int_{-\infty}^{\infty} S_{ZZ}(x, x; \omega) d\omega.$$

Recall that if $\mu_Z(x) = 0$ then $Z_{\text{MS}}(x) = \sigma_Z^2(x)$, the variance.

The derivations are now complete, but what do they mean, and how do they help the designer? One of the functions of a probabilistic analysis is to help the designer bound uncertainties so that it is possible to understand how randomness in the forcing results in a scatter of possible structural responses. Furthermore, this scatter is not haphazard, but is defined by a standard deviation, and possibly a density function. It is the variance that is used to bound the mean value response. These are subjects for more advanced study, which are now accessible to us.

11.4 Approximate Methods

In Chapter 9 the approximate method known as Rayleigh's quotient was examined. This method is also applicable to the continuous systems of this chapter. Also introduced in this section is the more general *Rayleigh–Ritz* method, which is an extension of the Rayleigh quotient, and the *Galerkin* method, one of the most general approximate methods. Many computational approaches for the computer are based on these methods.

11.4.1 Rayleigh's Quotient

Recall that for the discrete systems Rayleigh's quotient was derived beginning with the eigenvalue problem. Both sides of the equation was pre–multiplied by the transpose of the eigenvector, and then solved for the natural frequency. In this equation for the frequency, a guess for the eigenvector was substituted and the equation solved for the estimated frequency.

11.4. APPROXIMATE METHODS

The same approach is used here for the continuous parameter problem. The procedure is demonstrated for the transverse vibration of a beam, beginning with the eigenvalue statement, Equation 10.83, where the orthogonality of the normal modes was already demonstrated.[13] For Rayleigh's quotient, we work with one mode and thus let $r = s$. Then,

$$\int_0^L Y_r(x) \frac{d^2}{dx^2}[EI(x)Y_r''(x)]dx = \omega_r^2 \int_0^L m(x)[Y_r(x)]^2 dx.$$

Solving this for ω_r^2 yields Rayleigh's quotient,

$$\omega_r^2 = \frac{\int_0^L Y_r(x) \frac{d^2}{dx^2}[EI(x)Y_r''(x)]dx}{\int_0^L m(x)[Y_r(x)]^2 dx}. \tag{11.43}$$

The numerator is proportional to the strain energy and the denominator to the kinetic energy of the mode if harmonic motion is assumed.[14] Usually, one can only guess at the function $Y_r(x)$. Let $\mathcal{Y}_1(x)$ represent the guess. The better the guess, the closer (from above) will the approximate frequency be to the actual value. For the first mode, we can obtain a reasonable approximation for ω_1. For the cantilever beam, guess for $Y_1(x)$ a function $\mathcal{Y}_1(x)$ that equals zero at $x = 0$, that is, $\mathcal{Y}_1(0) = 0$. At the free end, the deflection and slope must be not zero, and the guess $\mathcal{Y}_1(x)$ must be such that $\mathcal{Y}_1(L) \neq 0$ and $\mathcal{Y}_1'(L) \neq 0$. Several possible guesses for the eigenfunction are possible, but a simple one is

$$\mathcal{Y}_1(x) = a\left(1 - \cos\frac{\pi x}{2L}\right),$$

where a is a constant, and the 2 in the denominator guarantees that the phasing in the cosine results in the appropriate values for $\mathcal{Y}_1(x)$ at the boundaries.[15] Assume that the beam is uniform[16] and substitute the expression for $\mathcal{Y}_1(x)$ into Equation 11.43 to find the approximation for ω_1, denoted here by Ω_1, to be

$$\Omega_1^2 = \frac{\int_0^L EI\left(1 - \cos\frac{\pi x}{2L}\right)\left(\left[\frac{\pi}{2L}\right]^2 \cos\frac{\pi x}{2L}\right)dx}{\int_0^L m\left(1 - \cos\frac{\pi x}{2L}\right)^2 dx}$$

$$\Omega_1 = \frac{3.667}{L^2}\sqrt{\frac{EI}{m}}.$$

---
[13] The same procedure would be followed for axial vibration and torsional vibration.
[14] This is clearer when the Rayleigh–Ritz procedure is discussed in the next section.
[15] Try evaluating $\mathcal{Y}_1(0)$, $\mathcal{Y}_1(L)$, $\mathcal{Y}_1'(0)$, and $\mathcal{Y}_1'(L)$, and compare with the actual boundary conditions.
[16] Actually, the benefits of such approximate techniques are for problems where an exact solution is not possible, for example, components with non–uniform sections. Here, our purpose is to demonstrate technique.

The exact result is the same as above except with a coefficient of 3.516. As expected, the approximate value is above the actual value since the approximate mode is always stiffer than the actual mode. Note that if the beam is not uniform, then the expression for this variation with x, $EI(x)$, is retained within the second derivative, and included in the integration with respect to x.

To estimate ω_2, a guess for $Y_2(x)$ is needed, but this becomes a difficult process and a more effective approach is available using the Rayleigh–Ritz method.

11.4.2 Rayleigh–Ritz Method

The *Rayleigh–Ritz* method may be viewed as an extension of Rayleigh's quotient, providing a method to estimate higher frequencies. As with Rayleigh's quotient, an assumption is made regarding the deformation, while satisfying the geometric boundary conditions. The deformation is approximated by a finite sum,

$$Y_r(x) = \sum_{i=1}^{n} c_i \mathcal{Y}_i(x),$$

where $\mathcal{Y}_i(x)$ are a set of assumed *trial* functions,[17] such as eigenfunctions, and c_i are to be determined in a manner described below.

Begin again with Equation 10.83 for any two modes of a transversely vibrating beam with vanishing boundary conditions,

$$\int_0^L EI(x) Y_r''(x) Y_s''(x) dx = \omega_r^2 \int_0^L m(x) Y_r(x) Y_s(x) dx,$$

or in Rayleigh's quotient form,

$$\omega_r^2 = \frac{\int_0^L EI(x) Y_r''(x) Y_s''(x) dx}{\int_0^L m(x) Y_r(x) Y_s(x) dx} \equiv \frac{N}{D}, \qquad (11.44)$$

where N and D represent the equations in the numerator and denominator, respectively. Equation 11.44 is an exact expression for the frequency if the

---

[17] When selecting a trial function for use in an approximate method, it is recognized that only the eigenfunction will yield the exact results. Any other function leads to an approximate solution. As we know, the eigenfunctions satisfy the governing differential equation and all the boundary conditions. If the trial function satisfies all geometric (displacements and slopes) and natural (forces and moments) boundary conditions, it is called a *comparison function*. These have to be differentiable as many times as the order of the governing equation. If the trial function only satisfies geometric boundary conditions, then it is called an *admissible function*. It is easier to find functions that satisfy only the geometric conditions and not the natural conditions.

11.4. APPROXIMATE METHODS

eigenfunctions are substituted on the right hand side. But if a guess is used with the finite sum, then the above equation for the frequency becomes

$$\Omega_r^2 = \frac{\int_0^L EI(x)[\sum_{i=1}^n c_i \mathcal{Y}_i''][\sum_{j=1}^n c_j \mathcal{Y}_j''] dx}{\int_0^L m(x)[\sum_{i=1}^n c_i \mathcal{Y}_i][\sum_{j=1}^n c_j \mathcal{Y}_j] dx}$$

$$= \frac{\sum_{i=1}^n \sum_{j=1}^n c_i c_j \int_0^L EI(x) \mathcal{Y}_i'' \mathcal{Y}_j'' dx}{\sum_{i=1}^n \sum_{j=1}^n c_i c_j \int_0^L m(x) \mathcal{Y}_i \mathcal{Y}_j dx},$$

$r = 1, 2, \ldots, n$, where Ω_r represents the approximate value of the frequency. An n-term finite sum leads to estimates of the first n frequencies, as we see below. To simplify the notation, as well as to better understand the physical nature of the equations, define the following,

$$m_{ij} = \int_0^L m(x) \mathcal{Y}_i \mathcal{Y}_j dx \qquad (11.45)$$

$$k_{ij} = \int_0^L EI(x) \mathcal{Y}_i'' \mathcal{Y}_j'' dx. \qquad (11.46)$$

Then,

$$\Omega_r^2 = \frac{N}{D} = \frac{\sum_{i=1}^n \sum_{j=1}^n c_i c_j k_{ij}}{\sum_{i=1}^n \sum_{j=1}^n c_i c_j m_{ij}}. \qquad (11.47)$$

Our task is to determine the optimal values of c_i so that the best estimate is found for the frequency. Knowing that the exact frequency is always smaller than the approximate value, *minimize* Ω_r^2, the approximate value of ω_r^2 given by Equation 11.47, by differentiating it with respect to c_i and setting the resulting expression equal to zero, that is,

$$\frac{\partial}{\partial c_i}\left(\frac{N}{D}\right) = \frac{D \partial N/\partial c_i - N \partial D/\partial c_i}{D^2} = 0, \quad i = 1, 2, \ldots, n.$$

The only way this equation can equal zero is if the numerator equals zero, since D is never equal to zero. The numerator can be put into the more useful form

$$\frac{\partial N}{\partial c_i} - \frac{N}{D}\frac{\partial D}{\partial c_i} = 0, \quad i = 1, \ldots, n, \qquad (11.48)$$

where $\Omega_r^2 = N/D$, and n is the number of terms in the approximate solution. The infinite degree of freedom system has been replaced by an n degree of freedom system.

To proceed further, assume functions for $\mathcal{Y}_i(x)$. From the above equations, this function must be twice differentiable and still be a function of x

after the second derivative. Use the two–term approximation for the fixed–free beam

$$Y_r(x) = c_1 \mathcal{Y}_1(x) + c_2 \mathcal{Y}_2(x)$$
$$= c_1 \left(\frac{x}{L}\right)^2 + c_2 \left(\frac{x}{L}\right)^3.$$

N and D can be evaluated as follows,

$$N = \sum_{i=1}^{2}\sum_{j=1}^{2} c_i c_j k_{ij}$$
$$= \sum_{i=1}^{2} [c_i c_1 k_{i1} + c_i c_2 k_{i2}]$$
$$= c_1^2 k_{11} + c_1 c_2 k_{12} + c_2 c_1 k_{21} + c_2^2 k_{22}$$
$$= c_1^2 k_{11} + 2 c_1 c_2 k_{12} + c_2^2 k_{22}.$$

Similarly,

$$D = \sum_{i=1}^{2}\sum_{j=1}^{2} c_i c_j m_{ij}$$
$$= c_1^2 m_{11} + 2 c_1 c_2 m_{12} + c_2^2 m_{22},$$

where use has been made of the symmetries of m_{ij} and k_{ij}. Next evaluate the derivatives needed in Equation 11.48, for $n = 2$,

$$\frac{\partial N}{\partial c_1} = 2 c_1 k_{11} + 2 c_2 k_{12} \qquad (11.49)$$

$$\frac{\partial N}{\partial c_2} = 2 c_1 k_{12} + 2 c_2 k_{22} \qquad (11.50)$$

$$\frac{\partial D}{\partial c_1} = 2 c_1 m_{11} + 2 c_2 m_{12} \qquad (11.51)$$

$$\frac{\partial D}{\partial c_2} = 2 c_1 m_{12} + 2 c_2 m_{22}, \qquad (11.52)$$

11.4. APPROXIMATE METHODS

where

$$m_{11} = \int_0^L m\mathcal{Y}_1^2 dx = \frac{mL}{5}$$

$$m_{12} = \int_0^L m\mathcal{Y}_1\mathcal{Y}_2 dx = \frac{mL}{6} = m_{21}$$

$$m_{22} = \int_0^L m\mathcal{Y}_2^2 dx = \frac{mL}{7}$$

$$k_{11} = \int_0^L EI[\mathcal{Y}_1'']^2 dx = \frac{4EI}{L^3}$$

$$k_{12} = \int_0^L EI\mathcal{Y}_1''\mathcal{Y}_2'' dx = \frac{6EI}{L^3} = k_{21}$$

$$k_{22} = \int_0^L EI[\mathcal{Y}_2'']^2 dx = \frac{12EI}{L^3}.$$

Therefore, Equation 11.48 for $i = 1, 2$, in conjunction with Equations 11.49–11.52, can be written in matrix form as

$$\begin{bmatrix} k_{11} - \Omega_r^2 m_{11} & k_{12} - \Omega_r^2 m_{12} \\ k_{12} - \Omega_r^2 m_{12} & k_{22} - \Omega_r^2 m_{22} \end{bmatrix} \begin{Bmatrix} c_1 \\ c_2 \end{Bmatrix} = \begin{Bmatrix} 0 \\ 0 \end{Bmatrix}, \quad (11.53)$$

or in general matrix notation as

$$[[k] - \Omega_r^2[m]]\{c\} = \{0\}. \quad (11.54)$$

This is the same matrix eigenvalue problem we solved many times when considering multi degree of freedom systems. The values of c_1 and c_2 can be found by solving the 2×2 determinant $|[k] - \Omega_r^2[m]| = 0$. The evaluation of this determinant provides us with estimates of the two natural frequencies Ω_1^2 and Ω_2^2, since a two term approximate solution is taken, resulting in a two degree of freedom approximating system. Then, using the two natural frequencies, proceed to find the values c_{1r} and c_{2r}, $r = 1, 2$ in Equation 11.53. The complete solution is then given by

$$y(x,t) = \sum_{r=1}^{2} Y_r(x)F_r(t)$$
$$= [c_{11}\mathcal{Y}_1(x) + c_{21}\mathcal{Y}_2(x)]F_1(t) + [c_{12}\mathcal{Y}_1(x) + c_{22}\mathcal{Y}_2(x)]F_2(t),$$

where $F_1(t)$ is harmonic in ω_1 and $F_2(t)$ is harmonic in ω_2. This approach replaces an infinite dimensional continuous system by an n dimensional discrete system. The discrete system can then be solved using the techniques of Chapter 8.

Example 11.2 Physical Meaning of Equation 11.47
Explore Equation 11.47 further to have a better physical understanding of this procedure. Use the bending beam as the vehicle for our understanding.

The strain energy of a bending beam is given by

$$V(t) = \frac{1}{2}\int_0^L EI(x)\left(\frac{\partial^2 y}{\partial x^2}\right)^2 dx,$$

and the kinetic energy by

$$T(t) = \frac{1}{2}\int_0^L m(x)\left(\frac{\partial y}{\partial t}\right)^2 dx.$$

Using the product solution, $y(x,t) = Y(x)F(t)$, $V(t)$ and $T(t)$ become

$$V(t) = \frac{1}{2}F^2(t)\int_0^L EI(x)[Y''(x)]^2 dx$$

$$T(t) = \frac{1}{2}\dot{F}^2 \int_0^L m(x)Y^2(x) dx.$$

If $F(t)$ is harmonic, say $A\cos\omega t$, then the maximum $V(t)$ and the maximum $T(t)$ are given by

$$V_{max} = \frac{1}{2}A^2 \int_0^L EI(x)[Y''(x)]^2 dx$$

$$T_{max} = \frac{1}{2}A^2\omega^2 \int_0^L m(x)Y^2(x) dx.$$

The kinetic energy for $\omega = 1$ rad/sec is customarily defined as T^*_{max}. Then,

$$T_{max} = \omega^2 T^*_{max}.$$

For a system with no dissipation, such as those due to friction or damping, the maximum potential energy equals the maximum kinetic energy,

$$V_{max} = \omega^2 T^*_{max},$$

or

$$\omega^2 = \frac{N}{D} = \frac{V_{max}}{T^*_{max}},$$

11.4. APPROXIMATE METHODS

from Equation 11.47.

Therefore, Rayleigh's quotient, and its extension, the Rayleigh–Ritz procedure, are essentially statements on the ratio between potential energy and kinetic energy. Physically, it makes sense that this ratio is related to the frequency of oscillation, since the rate at which energy is being exchanged between kinetic and potential can be obtained by knowing the maxima of each, and the time it takes to go between maxima.

If $Y_i(x)$ is not exactly known, then approximate this function by $\mathcal{Y}_i(x)$. To obtain the minimum estimate of the frequency, differentiate the approximate expressions for V_{max} and T^*_{max},

$$\frac{\partial N}{\partial c_i} = \frac{\partial V_{max}}{\partial c_i}$$
$$\frac{\partial D}{\partial c_i} = \frac{\partial T^*_{max}}{\partial c_i}.$$

Consider the uniform cantilever beam via the approach motivated above. Assume $\mathcal{Y}(x) = c_1 x^2 + c_2 x^3$. The maximum potential energy is

$$V_{max} = \frac{1}{2} EI \int_0^L (2c_1 + 6c_2 x)^2 dx$$
$$= \frac{1}{2} EI (4c_1^2 L + 12 c_1 c_2 L^2 + 12 c_2^2 L^3).$$

Then,

$$\frac{\partial V_{max}}{\partial c_1} = EI(4c_1 L + 6c_2 L^2) = \sum_{j=1}^2 2k_{1j} c_j$$

$$\frac{\partial V_{max}}{\partial c_2} = EI(6c_1 L^2 + 12 c_2 L^3) = \sum_{j=1}^2 2k_{2j} c_j,$$

where the second set of equalities are from Equations 11.49 and 11.50. Thus, $k_{11} = 2EIL$, $k_{12} = 3EIL^2$, $k_{21} = 3EIL^2$, and $k_{22} = 6EIL^3$. Similarly for T^*_{max},

$$T^*_{max} = \frac{m}{2} \int_0^L (c_1 x^2 + c_2 x^3)^2 dx$$
$$= \frac{m}{2} \left(c_1^2 \frac{L^5}{5} + 2c_1 c_2 \frac{L^6}{6} + c_2^2 \frac{L^7}{7} \right).$$

Then

$$\frac{\partial T^*_{max}}{\partial c_1} = \frac{mL^5 c_1}{5} + \frac{mL^6 c_2}{6} = \sum_{j=1}^{2} 2m_{1j} c_j$$

$$\frac{\partial T^*_{max}}{\partial c_2} = \frac{mL^6 c_1}{6} + \frac{mL^7 c_2}{7} = \sum_{j=1}^{2} 2m_{2j} c_j,$$

where, from Equation 11.51 and 11.52, $m_{11} = mL^5/10$, $m_{12} = m_{21} = mL^6/12$, and $m_{22} = mL^7/14$.

Combine these terms into the eigenvalue formulation,

$$\begin{bmatrix} 2EIL - \Omega^2 mL^5/10 & 3EIL^2 - \Omega^2 mL^6/12 \\ 3EIL^2 - \Omega^2 mL^6/12 & 6EIL^3 - \Omega^2 mL^7/14 \end{bmatrix} \begin{Bmatrix} c_1 \\ c_2 \end{Bmatrix} = \begin{Bmatrix} 0 \\ 0 \end{Bmatrix}.$$

Taking the determinant of the matrix and setting it equal to zero yields

$$\Omega^4 - 1,224 \frac{EI}{mL^4} \Omega^2 + 15,120 \left(\frac{EI}{mL^4}\right)^2 = 0.$$

The two frequencies are estimated as

$$\Omega_1 = \frac{3.533}{L^2}\sqrt{\frac{EI}{m}} \text{ rad/sec}, \quad \Omega_2 = \frac{34.81}{L^2}\sqrt{\frac{EI}{m}} \text{ rad/sec},$$

with the exact numerical coefficients being 3.516 and 22.03, respectively. The estimate for ω_1 is very close, but the one for ω_2 is poor. It is generally found that accurate estimates are obtained for the first $n/2$ frequencies, where n equals the number of terms in the approximate solution. If a better approximation is necessary for ω_2, then a four term approximation is necessary. ∎

11.4.3 The Galerkin Method

While the Rayleigh–Ritz is a powerful technique, the *Galerkin* method is more general and powerful, being able to model non–self–adjoint problems such as nonlinear partial differential equations.[18] In this approach, an ap-

---

[18] In particular, the Rayleigh–Ritz procedure and the Galerkin method are essentially the same if the system is *self–adjoint*. But for non–self–adjoint systems, only the Galerkin method can be used. An excellent discussion of these and other related issues is provided by H.H.E. Leipholz, *On Some Developments in Direct Methods of the Calculus of Variations*, Applied Mechanics Reviews, Vol. 40, No. 10, Oct. 1987, 1379–1392. Some of the material is within your reach. Give it a try!

11.4. APPROXIMATE METHODS

proximate solution is assumed

$$u_a(x,t) = \sum_{i=1}^{n} \mathcal{Y}_i(x) q_i(t),$$

where the functions $\mathcal{Y}_i(x)$ are assumed to satisfy the boundary conditions. Since the approximate solution does not exactly satisfy the governing equation of motion, an error results. The procedure by which this error is minimized is known as the Galerkin method.

Using the axial vibration of a uniform beam as an example, substitute the approximate solution $u_a(x,t)$ into the governing equation,

$$EA\frac{\partial^2 u_a}{\partial x^2} - m\frac{\partial^2 u_a}{\partial t^2} = R_{error}, \qquad (11.55)$$

where R_{error}, known as the *residual error*, is due to the fact that u_a is not the exact solution. If $u_a(x,t)$ is the exact solution, then the residual error equals zero. In the Rayleigh–Ritz procedure, the expression for the natural frequency is minimized. Here, an expression that includes the residual error is minimized. However, since the error depends on the quality of the function $\mathcal{Y}_i(x)$ chosen above, the minimization equations are to be with respect to these functions.

Multiply Equation 11.55 by the variation of the assumed solution,

$$\delta u_a(x,t) = \sum_{i=1}^{n} \mathcal{Y}_i(x) \delta q_i(t),$$

and integrate the resulting equation over the domain,

$$\int_0^L \left[EA\frac{\partial^2 u_a}{\partial x^2} - m\frac{\partial^2 u_a}{\partial t^2}\right] \delta u_a dx = \int_0^L R_{error} \delta u_a dx.$$

Substitute the expression for δu_a to find

$$\sum_{i=1}^{n} \int_0^L \left[EA\frac{\partial^2 u_a}{\partial x^2} - m\frac{\partial^2 u_a}{\partial t^2}\right] \mathcal{Y}_i(x) \delta q_i(t) dx = \sum_{i=1}^{n} \int_0^L R_{error} \mathcal{Y}_i(x) \delta q_i(t) dx,$$

or

$$\sum_{i=1}^{n} \left\{ \int_0^L \left[EA\frac{\partial^2 u_a}{\partial x^2} - m\frac{\partial^2 u_a}{\partial t^2} - R_{error}\right] \mathcal{Y}_i(x) dx \right\} \delta q_i(t) = 0.$$

Since the variation $\delta q_i(t)$ is arbitrary, the term in the braces must equal zero, or

$$\int_0^L \left[EA\frac{\partial^2 u_a}{\partial x^2} - m\frac{\partial^2 u_a}{\partial t^2}\right] \mathcal{Y}_i(x) dx = \int_0^L R_{error} \mathcal{Y}_i(x) dx, \quad i = 1, 2, \ldots, n.$$
$$(11.56)$$

To minimize the error, the integral of the weighted residual error must be minimized. The criteria in selecting $\mathcal{Y}_i(x)$ is that the integral on the right hand side be as close to zero as possible. The $\mathcal{Y}_i(x)$ act as *weighting functions* selected such that the residual error is minimum over the domain, as specified in Equation 11.56. Such techniques are part of a group of approximate methods called *methods of weighted residuals*. Finite elements are such a method. Physically, Equation 11.56 equals the work done per unit length by the force R_{error} undergoing a deflection $\mathcal{Y}_i(x)$. For the exact solution, this work equals zero since, then, $R_{error} = 0$. Therefore, we actually solve

$$\int_0^L \left[EA\frac{\partial^2 u_a}{\partial x^2} - m\frac{\partial^2 u_a}{\partial t^2} \right] \mathcal{Y}_i(x) dx = 0, \quad i = 1, 2, \ldots, n,$$

where the approximate mode is used.[19] Note again that the term in the square brackets equals the residual error, R_{error}, as per Equation 11.55.

Example 11.3 Second Order Galerkin's Equations for the Axially Vibrating Beam
Assume a two–term approximation for the axial displacement,

$$u_a(x,t) = \mathcal{Y}_1(x)q_1(t) + \mathcal{Y}_2(x)q_2(t). \qquad (11.57)$$

The residual error is given by

$$R_{error} = EA\frac{\partial^2 u_a}{\partial x^2} - m\frac{\partial^2 u_a}{\partial t^2}. \qquad (11.58)$$

Multiply both sides of this equation by $\mathcal{Y}_i(x)$ and integrate over the beam length so that

$$\int_0^L R_{error} \mathcal{Y}_i(x) dx = 0, \quad i = 1, 2. \qquad (11.59)$$

Therefore, the two equations to solve are

$$\int_0^L R_{error} \mathcal{Y}_1(x) dx = 0$$
$$\int_0^L R_{error} \mathcal{Y}_2(x) dx = 0.$$

---

[19] Had we included an external force, it would have appeared on the right hand side of this equation.

11.4. APPROXIMATE METHODS

Specifically, by substituting Equation 11.57 into 11.58, and then into 11.59, these two equations become

$$EA \int_0^L [\mathcal{Y}_1'' q_1 + \mathcal{Y}_2'' q_2]\mathcal{Y}_1 dx - m \int_0^L [\mathcal{Y}_1 \ddot{q}_1 + \mathcal{Y}_2 \ddot{q}_2]\mathcal{Y}_1 dx = 0$$

$$EA \int_0^L [\mathcal{Y}_1'' q_1 + \mathcal{Y}_2'' q_2]\mathcal{Y}_2 dx - m \int_0^L [\mathcal{Y}_1 \ddot{q}_1 + \mathcal{Y}_2 \ddot{q}_2]\mathcal{Y}_2 dx = 0.$$

In matrix form, we have

$$\begin{bmatrix} m_{11} & m_{12} \\ m_{21} & m_{22} \end{bmatrix} \begin{Bmatrix} \ddot{q}_1 \\ \ddot{q}_2 \end{Bmatrix} + \begin{bmatrix} k_{11} & k_{12} \\ k_{21} & k_{22} \end{bmatrix} \begin{Bmatrix} q_1 \\ q_2 \end{Bmatrix} = \begin{Bmatrix} 0 \\ 0 \end{Bmatrix}, \quad (11.60)$$

where

$$m_{11} = -\int_0^L m\mathcal{Y}_1^2 dx$$

$$m_{12} = m_{21} = -\int_0^L m\mathcal{Y}_1\mathcal{Y}_2 dx$$

$$m_{22} = -\int_0^L m\mathcal{Y}_2^2 dx$$

$$k_{11} = \int_0^L EA\mathcal{Y}_1''\mathcal{Y}_1 dx = EA\mathcal{Y}_1'\mathcal{Y}_1 \Big|_0^L - \int_0^L EA\mathcal{Y}_1'\mathcal{Y}_1' dx$$

$$k_{12} = \int_0^L EA\mathcal{Y}_2''\mathcal{Y}_1 dx = EA\mathcal{Y}_2'\mathcal{Y}_1 \Big|_0^L - \int_0^L EA\mathcal{Y}_2'\mathcal{Y}_1' dx$$

$$k_{21} = \int_0^L EA\mathcal{Y}_1''\mathcal{Y}_2 dx = EA\mathcal{Y}_1'\mathcal{Y}_2 \Big|_0^L - \int_0^L EA\mathcal{Y}_1'\mathcal{Y}_2' dx$$

$$k_{22} = \int_0^L EA\mathcal{Y}_2''\mathcal{Y}_2 dx = EA\mathcal{Y}_2'\mathcal{Y}_2 \Big|_0^L - \int_0^L EA\mathcal{Y}_2'\mathcal{Y}_2' dx,$$

where the boundary values are assumed to equal zero, following Equation 10.26. At this point we can select $\mathcal{Y}_1$ and $\mathcal{Y}_2$ that satisfy the boundary conditions. Then solve the eigenvalue problem based on matrix Equation 11.60 to find $\omega_{1,2}$ and $q_{1,2}(t)$, as we have done many times before. k_{ij} have been left in their general form. ∎

The real power of such approximate methods is for problems that have no exact analytical solution, which include most problems of practical interest. For example, the analysis of the tapered cantilever beam of Figure 11.9 can be approached using the approximate methods of this section.

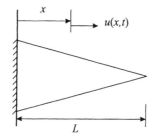

Figure 11.9: Tapered beam in axial vibration.

Depending on the degree of the taper, one has

$$EA(x) = EA\left(1 - \alpha(x)\frac{x}{L}\right)$$
$$m(x) = m\left(1 - \beta(x)\frac{x}{L}\right),$$

where $\alpha(x)$ and $\beta(x)$ represent the geometrical and mass distributions, respectively, and $\alpha(L) = 1$ and $\beta(L) = 1$. For a uniform fixed–free beam, the first mode is given by $\mathcal{Y}_1(x) = \sin(\pi x/2L)$, which satisfies the above boundary conditions, and can be used in an approximate model.

Galerkin

Boris Grigorievich Galerkin came from a poor family and had a harder time through his years of education than would otherwise have been the case. He attended secondary school in Minsk, then in 1893 he entered the Petersburg Technological Institute. Here he studied mathematics and engineering but he needed to make money to survive so at first he took on private tutoring, then from 1896 he worked as a designer. After graduating from the Technological Institute in 1899 he worked at the Kharkov Locomotive Plant. In 1903 Galerkin went to St Petersburg and there he became engineering manager at the Northern Mechanical and Boiler Plant.

Beginning in 1909 Galerkin began to study building sites and construction works throughout Europe. In the same year he began teaching at the Petersburg Technological Institute. His first publication on longitudinal curvature also appeared in 1909, following earlier work by Euler. This paper was highly relevant to his study of construction sites since the results were applied to the construction of bridges and frames for buildings.

His visits around European construction sites ended around 1914 but his academic work then turned to the area for which he is today best known, namely the method of approximate integration of differential equations known as the Galerkin method. He published his finite element method in 1915. In 1920 Galerkin was promoted to Head

11.5. WHERE VARIABLES DO NOT SEPARATE 613

of Structural Mechanics at the Petersburg Technological Institute. By this time he also held two chairs, one in elasticity at the Leningrad Institute of Communications Engineers and one in structural mechanics at Leningrad University.

In 1921 the St Petersburg Mathematical Society was reopened (it had closed in 1917 due to the Russian Revolution) as the Petrograd Physical and Mathematical Society. Galerkin played a major role in the Society along with Steklov, Bernstein, Friedmann and others. Other work for which Galerkin is also well known for his work on thin elastic plates. His major monograph on this topic *Thin Elastic Plates* was published in 1937. From 1940 until his death, Galerkin was head of the Institute of Mechanics of the Soviet Academy of Sciences.

A.T. Grigorian describes other work: "Galerkin's scientific research in the theory of casing (1934-45) revealed its broad application in industrial construction. His works in the field constitute a new direction in this important area. Galerkin was a consultant in the planning and building of many of the Soviet Union's largest hydrostations. In 1929, in connection with the building of the Dnepr dam and hydroelectric station, Galerkin investigated stresses in dams and retaining walls with trapezoidal profile. His results were used in planning the dam."

11.5 Where Variables Do Not Separate

Every solution of a partial differential equation in this chapter has used the method of separation of variables. There are numerous problems that cannot be solved with this approach. Separation of variables has worked as a solution for the cases where the time function turned out to be harmonic. Therefore, the product of the time function and the space function for the complete solution resulted in synchronous motion, later also called modal behavior. The lack of synchronous motion implies a wave propagation solution.

This section formulates and discusses two classes of problems that do not have separable solutions. The first is a vibration with *time–dependent boundary conditions*, where the time–dependence is *not* harmonic; separation of variables will fail in the boundary conditions. The second class of problems includes the vibration of a pipe with internal flow, and the motion of a moving taut string. Here, separation of variables fails in the equation of motion due to a mixed derivative term.

11.5.1 Response to Nonharmonic, Time–Dependent Boundary Conditions

In the problems considered earlier in this chapter, a variety of boundary conditions were considered, including harmonic loading that was incorporated into the boundary conditions. For non–harmonic boundary conditions, the separation of variables fails not in the equation of motion but, rather, in the boundary conditions. One could use the Laplace transform method,[20] but since we did not emphasize this approach for continuous systems, an approach is chosen that transforms the equation to another coordinate system where the separation of variables succeeds. We follow the development given in Clark (based on the method of Mindlin and Goodman),[21] who demonstrates the technique for a bending beam.

Before proceeding, consider the axially vibrating cantilever beam with a force $P(t)$ acting on its free end in the axial direction. The boundary condition on the free end must satisfy $EA\partial u(L,t)/\partial x = P(t)$. Assuming separation of variables $u(x,t) = X(x)F(t)$, this boundary condition becomes $EAX'(L)F(t) = P(t)$. If the loading is harmonic then the analysis can proceed as usual by separation of variables, but if $P(t)$ is some arbitrary time function, another technique is required.

The *Mindlin–Goodman* technique transforms the original homogeneous differential equation with non–homogeneous boundary conditions into a non–homogeneous differential equation with homogeneous boundary conditions that can then be solved by the usual modal analysis techniques. The approach works, as well, for a forced system, where the transformation results in a more complicated forcing.

Consider the transverse vibration of a uniform Bernoulli–Euler beam subjected to arbitrary forcing given by $q(x)p(t)$. The governing equation of motion is

$$EI\frac{\partial^4 y}{\partial x^4} + m\frac{\partial^2 y}{\partial t^2} = q(x)p(t),$$

where $m = \rho A$, with all notation defined earlier. This equation has four

---

[20] For an introduction to the application of Laplace transform techniques to partial differential equations, see Chapter 12 of **Elementary Applied Partial Differential Equation**, Second Edition, by R. Haberman, Prentice–Hall, 1987. The text by R.V. Churchill, **Operational Mathematics**, Second Edition, McGraw–Hill, 1958, is devoted to this subject. Finally, other transform techniques are fully discussed in **Fourier Transforms**, by I.N. Sneddon, McGraw–Hill, 1951. All three books discuss applications to problems of mechanical vibration, and are highly recommended.

[21] See Chapter 11 of **Dynamics of Continuous Elements**, by S.K. Clark, who references R.D. Mindlin, L.E. Goodman, "Beam Vibrations with Time–Dependent Boundary Conditions," *J. Applied Mechanics*, 17, 377–380, 1950, for originally developing the method.

11.5. WHERE VARIABLES DO NOT SEPARATE

boundary conditions, depicted generally for $X = 0$ or L as

$$\begin{aligned} y(X,t) &= f_0(t) \\ \frac{\partial y(X,t)}{\partial x} &= f_1(t) \\ \frac{\partial^2 y(X,t)}{\partial x^2} &= f_2(t) \\ \frac{\partial^3 y(X,t)}{\partial x^3} &= f_3(t), \end{aligned}$$

respectively, where X in the boundary conditions may be either $X = 0$ or L. There are sixteen possible combinations of boundary conditions. Some of the $f_i(t)$ may be zero, but at least one must be an arbitrary function of time. If the functions are harmonic, a standard separation of variables will succeed. In addition to the boundary conditions, there are two initial conditions,

$$\begin{aligned} y(x,0) &= y_0(x) \\ \dot{y}(x,0) &= v_0(x). \end{aligned}$$

A direct separation of variables will not work due to the arbitrary time function $f_i(t)$, and therefore, we proceed to examine the possibility that the response variable $y(x,t)$ can be transformed into a new domain where homogeneity of the boundary conditions can be enforced. This benefit will come at the cost of a more complicated forcing function than the original. Assume the following relation

$$y(x,t) = z(x,t) + \sum_{i=0}^{3} f_i(t) g_i(x), \qquad (11.61)$$

where $z(x,t)$ is the transformed displacement, $f_i(t)$ are the general boundary conditions, and $g_i(x)$ are to be chosen so that the boundary conditions of $z(x,t)$ will be homogeneous and, therefore, separable. An example is

$$g_i(x) = \alpha_i + \beta_i x + \gamma_i x^2 + \delta_i x^3 + \epsilon_i x^4.$$

We discover below how to connect the number of terms necessary in such a polynomial to the boundary conditions of the specific problem.

To proceed, substitute Equation 11.61 for $y(x,t)$ into the governing equation of motion, using the additional relations,

$$\begin{aligned} \ddot{y} &= \ddot{z} + \sum_{i=0}^{3} \ddot{f_i} g_i(x) \\ y^{iv} &= z^{iv} + \sum_{i=0}^{3} f_i(t) g_i^{iv}(x), \end{aligned}$$

to find

$$EIz^{iv} + m\ddot{z} = q(x)p(t) - \sum_{i=0}^{3}[EIf_i(t)g_i^{iv}(x) + m\ddot{f}_i g_i(x)], \qquad (11.62)$$

where superscripts denote partial differentiations of that order with respect to x, that is, $z^{iv} = \partial^4 z(x,t)/\partial x^4$.

The transformation equation must also satisfy all boundary and initial conditions, where, again, $X = 0$ or $X = L$,

$$y(X,t) = f_0(t) = z(X,t) + \sum_{i=0}^{3} f_i(t)g_i(X) \qquad (11.63)$$

$$\frac{\partial y(X,t)}{\partial x} = f_1(t) = \frac{\partial z(X,t)}{\partial x} + \sum_{i=0}^{3} f_i(t)\frac{dg_i(X)}{dx} \qquad (11.64)$$

$$\frac{\partial^2 y(X,t)}{\partial x^2} = f_2(t) = \frac{\partial^2 z(X,t)}{\partial x^2} + \sum_{i=0}^{3} f_i(t)\frac{d^2 g_i(X)}{dx^2} \qquad (11.65)$$

$$\frac{\partial^3 y(X,t)}{\partial x^3} = f_3(t) = \frac{\partial^3 z(X,t)}{\partial x^3} + \sum_{i=0}^{3} f_i(t)\frac{d^3 g_i(X)}{dx^3}, \qquad (11.66)$$

and

$$y(x,0) = y_0(x) = z(x,0) + \sum_{i=0}^{3} f_i(0)g_i(x)$$

$$\dot{y}(x,0) = v_0(x) = \dot{z}(x,0) + \sum_{i=0}^{3} \dot{f}_i(0)g_i(x).$$

The procedure will be to select $g_i(x)$ such that all boundary conditions on $z(x,t)$ vanish. That is, with appropriate choices of $g_i(x)$, we can obtain the homogeneous boundary conditions $z(X,t) = 0$, $z'(X,t) = 0$, $z''(X,t) = 0$, and $z'''(X,t) = 0$. Based on Equations 11.63–11.66, $g_i(x)$ must be a polynomial in x to an order equal to the highest derivative. There is no unique choice of functions and all lead to the same solution.

Having selected the $g_i(x)$ that satisfy the boundary conditions on z, then for $X = 0$ or L, all z and derivative terms drop out and Equations 11.63–

11.5. WHERE VARIABLES DO NOT SEPARATE

11.66 become

$$f_0(t) - \sum_{i=0}^{3} f_i(t) g_i(X) = 0$$

$$f_1(t) - \sum_{i=0}^{3} f_i(t) \frac{dg_i(X)}{dx} = 0$$

$$f_2(t) - \sum_{i=0}^{3} f_i(t) \frac{d^2 g_i(X)}{dx^2} = 0$$

$$f_3(t) - \sum_{i=0}^{3} f_i(t) \frac{d^3 g_i(X)}{dx^3} = 0.$$

These equations ensure that the boundary conditions for $z(x,t)$ are homogeneous.

So far, $g_i(x)$ are defined and now solve for $z(x,t)$, which can be accomplished via separation of variables, based on the above discussion,

$$z(x,t) = \sum_{n=1}^{\infty} X_n(x) T_n(t), \qquad (11.67)$$

where $X_n(x)$ will be seen to be the modes, over $[0, L]$, of the transformed problem. As in earlier problems, we can expand the force factor[22] $q(x)$ in terms of the modes,

$$q(x) = \sum_{n=1}^{\infty} Q_n X_n(x), \qquad (11.68)$$

where Q_n are numbers evaluated by orthogonality considerations using[23]

$$Q_n = \frac{\int_0^L q(x) X_n(x) dx}{\int_0^L X_n^2(x) dx}. \qquad (11.69)$$

---

[22] We could have defined the loading as $p(x,t)$ instead of $q(x)p(t)$ and proceeded as we did in *Example* 10.5.

[23] Note that the integral relation in the denominators can be normalized to equal one, as we have done in the previous problems. Here, they have been left not normalized to demonstrate that such normalization is arbitrary in such problems, whereas in modal analysis the normalization is accepted.

Similarly, expand $g_i(x)$ and $g_i^{iv}(x)$ in terms of the modes,

$$g_i(x) = \sum_{n=1}^{\infty} G_{in} X_n(x),$$

$$G_{in} = \frac{\int_0^L g_i(x) X_n(x) dx}{\int_0^L X_n^2(x) dx}; \qquad (11.70)$$

$$g_i^{iv}(x) = \sum_{n=1}^{\infty} \Gamma_{in} X_n(x),$$

$$\Gamma_{in} = \frac{\int_0^L g_i^{iv}(x) X_n(x) dx}{\int_0^L X_n^2(x) dx}, \qquad (11.71)$$

where G_{in} and Γ_{in} are constants. Substitute all these expressions into governing Equation 11.62 to find

$$EI \sum_{n=1}^{\infty} X_n^{iv} T_n + m \sum_{n=1}^{\infty} X_n \ddot{T}_n$$
$$= \sum_{n=1}^{\infty} X_n p(t) Q_n - \sum_{i=0}^{3} [EI f_i(t) \sum_{n=1}^{\infty} \Gamma_{in} X_n + m \ddot{f}_i \sum_{n=1}^{\infty} G_{in} X_n],$$

where the equality must hold for each n independently,

$$EI X_n^{iv} T_n + m X_n \ddot{T}_n = X_n p(t) Q_n - \sum_{i=0}^{3} [EI f_i(t) \Gamma_{in} X_n + m \ddot{f}_i G_{in} X_n].$$

Divide both sides by $X_n T_n$ to find

$$EI \frac{X_n^{iv}}{X_n} + m \frac{\ddot{T}_n}{T_n} = \frac{p(t)}{T_n} Q_n - \sum_{i=0}^{3} \frac{EI f_i(t) \Gamma_{in} + m \ddot{f}_i G_{in}}{T_n}.$$

The time–dependent functions can be separated from the spatially–dependent functions in the governing equation

$$\frac{EI}{m} \frac{X_n^{iv}}{X_n} = -\frac{\ddot{T}_n}{T_n} + \frac{p(t)}{m T_n} Q_n - \sum_{i=0}^{3} \frac{(EI/m) f_i(t) \Gamma_{in} + \ddot{f}_i G_{in}}{T_n} = \omega_n^2,$$

which leads to the equations

$$X_n^{iv} - \frac{m \omega_n^2}{EI} X_n = 0 \qquad (11.72)$$

$$\ddot{T}_n + \omega_n^2 T_n = \frac{p(t)}{m} Q_n - \sum_{i=0}^{3} \left[\frac{EI f_i(t) \Gamma_{in}}{m} + \ddot{f}_i G_{in} \right], \qquad (11.73)$$

$$n = 1, 2, \ldots.$$

11.5. WHERE VARIABLES DO NOT SEPARATE

Define the right hand side of Equation 11.73 as $P(t)$ to ease the subsequent algebra. The solution to eigenvalue Equation 11.72 is

$$X_n(x) = C_{1n}\cos\frac{m_n x}{L} + C_{2n}\sin\frac{m_n x}{L} + C_{3n}\cosh\frac{m_n x}{L} + C_{4n}\sinh\frac{m_n x}{L}, \tag{11.74}$$

where

$$\frac{m_n}{L} \equiv \left(\frac{\omega_n^2 m}{EI}\right)^{1/4}.$$

Harmonic Equation 11.73 is satisfied by

$$T_n(t) = A_n \cos\omega_n t + B_n \sin\omega_n t + \frac{1}{\omega_n}\int_0^t P(\tau)\sin\omega_n(t-\tau)d\tau.$$

The complete solution of the transformed problem is given by Equation 11.67. To satisfy the initial conditions $T_n(0)$ and $\dot{T}_n(0)$, solve the following,

$$z(x,0) = \sum_{n=1}^{\infty} A_n X_n(x)$$

$$\dot{z}(x,0) = \sum_{n=1}^{\infty} B_n \omega_n X_n(x).$$

Using the orthogonality properties of the modes, as done before in similar circumstances, the constants are found to be

$$A_n = \frac{\int_0^L z(x,0) X_n(x) dx}{\int_0^L X_n^2(x) dx}$$

$$= \frac{\int_0^L \left(y_0(x) - \sum_{i=0}^{3} f_i(0) g_i(x)\right) X_n(x) dx}{\int_0^L X_n^2(x) dx} \tag{11.75}$$

$$B_n = \frac{\int_0^L \dot{z}(x,0) X_n(x) dx}{\omega_n \int_0^L X_n^2(x) dx}$$

$$= \frac{\int_0^L \left(v_0(x) - \sum_{i=0}^{3} \dot{f}_i(0) g_i(x)\right) X_n(x) dx}{\omega_n \int_0^L X_n^2(x) dx}. \tag{11.76}$$

Now we have all the parameters that we need to set up the solution $y(x,t) = z(x,t) + \sum_{i=0}^{3} f_i(t) g_i(x)$.

Apply the above procedure to an example in order to clarify any possible vagueness.

620   CHAPTER 11. ADVANCED CONTINUOUS MODELS

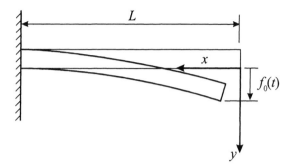

Figure 11.10: Cantilever beam subject to non–harmonic boundary condition.

Example 11.4 Cantilever Beam Subjected to an Arbitrary Transverse Time–Dependent Displacement at the Free End

Consider the cantilever beam shown in Figure 11.10 with the following boundary conditions,

$$y(0,t) = f_0(t) \quad \text{prescribed displacement at free end}$$
$$\frac{\partial^2 y(0,t)}{\partial x^2} = 0 \quad \text{no moment resistance at free end}$$
$$\frac{\partial y(L,t)}{\partial x} = 0 \quad \text{zero slope at fixed end}$$
$$y(L,t) = 0 \quad \text{no displacement at fixed end.}$$

For ease of algebra, assume zero initial conditions,

$$y(x,0) = 0, \quad \frac{\partial y(x,0)}{\partial t} = 0.$$

We can now proceed with the general transformation,

$$y(x,t) = z(x,t) + \sum_{i=0}^{3} f_i(t) g_i(x),$$

11.5. WHERE VARIABLES DO NOT SEPARATE

except that in this case, only $f_0(t) \neq 0$, and therefore,

$$z(0,t) = f_0(t) - \sum_{i=0}^{3} f_i(t)g_i(0) \tag{11.77}$$

$$z(L,t) = -\sum_{i=0}^{3} f_i(t)g_i(L) \tag{11.78}$$

$$\frac{\partial z(L,t)}{\partial x} = -\sum_{i=0}^{3} f_i(t)\frac{dg_i(L)}{dx} \tag{11.79}$$

$$\frac{\partial^2 z(0,t)}{\partial x^2} = -\sum_{i=0}^{3} f_i(t)\frac{d^2 g_i(0)}{dx^2}. \tag{11.80}$$

Given that the boundary conditions are up to the second order, assume

$$g_i(x) = \alpha_i + \beta_i x + \gamma_i x^2 + \delta_i x^3.$$

The above can be significantly simplified by recognizing that the only remaining $g_i(x)$ will also be for $i = 0$. In Equations 11.77–11.80, set all boundary conditions on $z(x,t)$ to zero. Therefore,

$$f_0(t) - \sum_{i=0}^{3} f_i(t)g_i(0) = 0 \tag{11.81}$$

$$-\sum_{i=0}^{3} f_i(t)g_i(L) = 0 \tag{11.82}$$

$$-\sum_{i=0}^{3} f_i(t)\frac{dg_i(L)}{dx} = 0 \tag{11.83}$$

$$-\sum_{i=0}^{3} f_i(t)\frac{d^2 g_i(0)}{dx^2} = 0. \tag{11.84}$$

This allows us to evaluate the constants needed to define $g_0(x)$, as follows. From Equation 11.81, we have

$$f_0(t) = f_0(t)g_0(0),$$

where $g_0(0) = \alpha_0$ and therefore, $\alpha_0 = 1$. From Equation 11.84,

$$0 = -f_0(t)(2\gamma_0),$$

or $\gamma_0 = 0$. From Equation 11.83,

$$0 = -f_0(t)(\beta_0 + 3\delta_0 L^2),$$

and from Equation 11.82,
$$0 = -f_0(t)(1 + \beta_0 L + \delta_0 L^3),$$
from which we find $\beta_0 = -3/2L$ and $\delta_0 = 1/2L^3$, giving us
$$g_0 = 1 - \frac{3}{2L}x + \frac{1}{2L^3}x^3.$$

Now use general solution Equation 11.74 for $X_n(x)$ to evaluate constants C_{in} by satisfying the boundary conditions. This is accomplished by recognizing that all boundary conditions for $z(x,t)$ are zero:
$$\begin{aligned} z(0,t) &= 0 \\ z'(L,t) &= 0 \\ z''(0,t) &= 0 \\ z(L,t) &= 0. \end{aligned}$$

Therefore, $X_n(0) = 0$, $X_n(L) = 0$, $dX_n(L)/dx = 0$, and $d^2 X_n(0)/dx^2 = 0$. Applying this to Equation 11.74 leads to the following equations,
$$\begin{aligned} C_{1n} &= 0, \quad C_{3n} = 0 \\ 0 &= C_{2n} \sin m_n + C_{4n} \sinh m_n & (11.85) \\ 0 &= C_{2n} \cos m_n + C_{4n} \cosh m_n. & (11.86) \end{aligned}$$

These last two equations define the characteristic equation from which we find $\tan m_n = -\tanh m_n$, leading to the values of m_n that can be evaluated recursively or via plotting the two sides of the equation against each other to find the intersections (roots). Once the m_n are evaluated, the modes are given by
$$X_n(x) = C_{2n} \sin \frac{m_n x}{L} + C_{4n} \sinh \frac{m_n x}{L}.$$

Use either Equation 11.85 or 11.86 to relate the two remaining coefficients. Use 11.85 to find
$$C_{4n} = -C_{2n} \frac{\sin m_n}{\sinh m_n},$$
and then
$$X_n(x) = \frac{C_{2n}}{\sinh m_n} \left(\sinh m_n \sin \frac{m_n x}{L} - \sin m_n \sinh \frac{m_n x}{L} \right). \quad (11.87)$$

We can define $C'_{2n} = C_{2n}/\sinh m_n$.

Next, evaluate all the coefficients and parameters that are part of the solution. Since the external load acts at the end of the beam, there is no

11.5. WHERE VARIABLES DO NOT SEPARATE

spatial variation, and therefore, $Q_n = 0$ and $q(x) = 0$, in Equations 11.69 and 11.68, respectively.

From Equation 11.70,

$$G_{0n} = \frac{\int_0^L g_0(x) X_n(x) dx}{\int_0^L X_n^2(x) dx},$$

where $g_0(x)$ and $X_n(x)$ have both been evaluated. Substituting, integrating, and simplifying leads to

$$G_{0n} = \frac{2}{m_n C'_{2n}(\sinh m_n - \sin m_n)}.$$

Since all the other $g_i(x) = 0$, we have $G_{2n} = G_{3n} = G_{4n} = 0$. Similarly, from Equation 11.71, $\Gamma_{in} = 0$.

To satisfy initial conditions, using Equations 11.75 and 11.76, coefficients A_n and B_n are needed, where $y_0(x) = 0$ and $v_0(x) = 0$,

$$A_n = \frac{\int_0^L \left(-\sum_{i=0}^3 f_i(0) g_i(x) X_n(x)\right) dx}{\int_0^L X_n^2(x) dx} = -f_0(0) G_{0n}$$

$$B_n = \frac{\int_0^L \left(-\sum_{i=0}^3 \dot{f}_i(0) g_i(x) X_n(x)\right) dx}{\omega_n \int_0^L X_n^2(x) dx} = -\frac{\dot{f}_0(0) G_{0n}}{\omega_n}.$$

The complete solution is

$$\begin{aligned} y(x,t) &= z(x,t) + \sum_{i=0}^3 f_i(t) g_i(x) \\ &= \sum_{n=1}^\infty X_n(x) \left[-f_0(0) G_{0n} \cos \omega_n t - \frac{\dot{f}_0(0) G_{0n}}{\omega_n} \sin \omega_n t \right. \\ &\quad \left. + \frac{1}{\omega_n} \int_0^t P(\tau) \sin \omega_n (t-\tau) d\tau \right] + f_0(t) \left(1 - \frac{3x}{2L} + \frac{x^3}{2L^3}\right), \end{aligned}$$

where

$$P(t) = \frac{p(t) Q_n}{m} - \sum_{i=0}^3 \left[\frac{EI}{m} f_i(t) \Gamma_{in} + \ddot{f}_i G_{in}\right] = -\ddot{f}_0(t) G_{0n}.$$

For the zero initial conditions, we have $f_0(0) = 0$ and $\dot{f}_0(0) = 0$, leading to

$$\begin{aligned} y(x,t) &= f_0(t) \left(1 - \frac{3x}{2L} + \frac{x^3}{2L^3}\right) \\ &\quad - \sum_{n=1}^\infty \frac{2 X_n(x)}{m_n C'_{2n}(\sinh m_n - \sin m_n)} \frac{1}{\omega_n} \int_0^t \ddot{f}_0(\tau) \sin \omega_n(t-\tau) d\tau, \end{aligned}$$

624   CHAPTER 11. ADVANCED CONTINUOUS MODELS

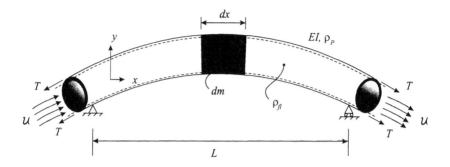

Figure 11.11: Pipe section with internal flow.

where the factor C'_{2n} in the denominator cancels with that factor within $X_n(x)$ in the numerator, as per Equation 11.87. The convolution integral can be integrated by parts twice so that

$$\frac{1}{\omega_n}\int_0^t \ddot{f}_0(\tau)\sin\omega_n(t-\tau)d\tau = f_0(t) - \omega_n \int_0^t f_0(\tau)\sin\omega_n(t-\tau)d\tau.$$

The problem is solved except for the substitution of a particular function for $f_0(t)$. ∎

Transformations such as those accomplished in the last problem are useful analytical tools. We often see problems transformed where a complicating feature is moved to another part of the equation where it is more easily handled. One example from earlier studies is the Laplace transform. The differential equation is transformed to an algebraic equation that is much easier to solve. But, the inverse transformation from the s domain into the t domain is generally difficult.

11.5.2 Flow in a Pipe with Constant Tension

Consider the schematic of the pipe with internal flow shown in Figure 11.11, where flow velocity $\mathcal{U}$ is assumed constant, the fluid has a constant mass per unit length of pipe ρ_{fl}, the tension in the pipe is a constant T, the pipe has constant bending rigidity EI, and constant mass per unit length ρ_P. Assume that linear vibration theory holds.

For an arbitrary pipe section, the vector diagram of velocities is shown

11.5. WHERE VARIABLES DO NOT SEPARATE

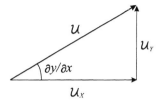

Figure 11.12: Velocity vector diagram.

in Figure 11.12. From this diagram, the flow velocity components are

$$\mathcal{U}_X = \mathcal{U}\cos\left(\frac{\partial y}{\partial x}\right) \simeq \mathcal{U}$$
$$\mathcal{U}_Y = \mathcal{U}\sin\left(\frac{\partial y}{\partial x}\right) \simeq \mathcal{U}\frac{\partial y}{\partial x}.$$

These velocities are for a static *slightly* bent beam. They must be added to the beam vibratory motion to find the complete flow velocity.

The coordinate x in these equations refers to the element of fluid dm_{fl} moving through the position which is a distance x from the fixed left support at time t. Thus, any derivatives that are taken with respect to time must be *total* or *material* derivatives.[24] For arbitrary function $f(x,t)$, the total derivative is given by

$$\frac{D}{Dt}f(x,t) = \frac{\partial f(x,t)}{\partial t} + \frac{\partial f(x,t)}{\partial x}\frac{\partial x}{\partial t}$$
$$= \frac{\partial f(x,t)}{\partial t} + \dot{x}\frac{\partial f(x,t)}{\partial x},$$

or using comma–subscript[25] notation for partial derivatives,

$$\frac{D}{Dt}f(x,t) = f_{,t} + \dot{x}f_{,x}$$

---

[24] The total derivative Df/Dt arises from the chain rule of higher–dimensional calculus. The total differential of a function Df of two or more variables, by the chain rule, accounts for all the contributions from each of the variables. Physically, such derivatives arise where a property changes with time and space, as in a flowing medium.

[25] With the comma–subscript notation, confusion may arise if I retain the style used throughout the text where equations are properly part of a sentence. That is, if an equation is at the end of a sentence, it is ended with a period. If an equation is a clause in a sentence, it is followed by a comma. Such rules are suspended where extra periods and commas would only add confusion.

Therefore, the velocity of the fluid in the transverse y direction is given by

$$\frac{D}{Dt}y(x,t) = v = \frac{\partial y(x,t)}{\partial t} + \frac{\partial y(x,t)}{\partial x}\frac{\partial x}{\partial t}$$
$$= y_{,t} + \mathcal{U}y_{,x} \qquad (11.88)$$

where y is the position of the fluid and $\mathcal{U}y_{,x} = \mathcal{U}_Y$. Physically, the velocity of the fluid in the transverse direction equals the sum of the flow velocity and the structural velocity in the transverse direction. It is implicitly assumed that the flow is uniform and there is no *cavitation*, regions within the pipe where the fluid has separated from the enclosing structure. It is also assumed that the pipe has no motion along the axial direction.

By Newton's second law of motion for element dm of pipe and internal fluid, the sum of the forces in the transverse direction is

$$F_y = dm_{fl}\frac{dv}{dt} + dm_P\frac{d}{dt}(y_{,t}), \qquad (11.89)$$

where $dm_{fl} = \rho_{fl}dx$, and $dm_P = \rho_P dx$ are the respective element masses. Since $v = v(x,t)$,

$$\frac{dv}{dt} = \frac{\partial v}{\partial t} + \frac{\partial v}{\partial x}\frac{\partial x}{\partial t}$$
$$= \frac{\partial v}{\partial t} + \frac{\partial v}{\partial x}\mathcal{U}$$
$$= \left(\frac{\partial}{\partial t} + \mathcal{U}\frac{\partial}{\partial x}\right)v.$$

Now substitute for v from Equation 11.88,

$$\frac{dv}{dt} = \left(\frac{\partial}{\partial t} + \mathcal{U}\frac{\partial}{\partial x}\right)(y_{,t} + \mathcal{U}y_{,x})$$
$$= y_{,tt} + 2\mathcal{U}y_{,tx} + \mathcal{U}^2 y_{,xx} \qquad (11.90)$$

Next, find the explicit expression for F_y in Equation 11.89. Draw the free body diagram of the pipe as shown in Figure 11.13. A similar diagram for the beam bending problem was viewed earlier in this chapter. The initial assumptions of constant tension and small angle bending leads to the following approximations,

$$\frac{dT}{dx} \approx 0$$
$$\sin\theta \approx \theta$$
$$\sin(\theta + \theta_{,x}dx) \approx \theta + \theta_{,x}dx.$$

11.5. WHERE VARIABLES DO NOT SEPARATE

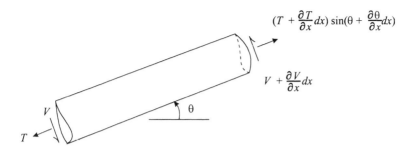

Figure 11.13: Free body diagram of pipe.

The net tension in the pipe is approximated by

$$F_T \approx T(\theta + \theta_{,x}dx) - T\theta = T\theta_{,x}dx,$$

and, using $\theta = \partial y/\partial x$,

$$F_T \approx Ty_{,xx}dx.$$

The net shear in the pipe is approximated by

$$F_S \approx -\left(V + \frac{\partial V}{\partial x}\right)dx + Vdx.$$

Make the substitutions $M = EI\partial^2 y/\partial x^2$ and $V = \partial M/\partial x$ to find

$$F_S \approx -EI\frac{\partial^4 y}{\partial x^4}dx.$$

Substituting the sum of the tension and shear forces on the left hand side of Equation 11.89 and using Equation 11.90, we have

$$-EIy_{,xxxx} + Ty_{,xx} = \rho_{fl}(y_{,tt} + 2\mathcal{U}y_{,tx} + \mathcal{U}^2 y_{,xx}) + \rho_P y_{,tt}$$

Combining terms, the governing equation of motion is

$$EIy_{,xxxx} - (T - \rho_{fl}\mathcal{U}^2)y_{,xx} + 2\rho_{fl}\mathcal{U}y_{,tx} + (\rho_{fl} + \rho_P)y_{,tt} = 0. \quad (11.91)$$

If $\mathcal{U} = 0$, then $T \to 0$ since T is the tension in the pipe due to fluid flow. T may be interpreted as the effective tension or compression of the pipe. As the fluid speed approaches zero, this tension does so as well, and we regain the Bernoulli–Euler beam equation. When $\rho_{fl}\mathcal{U}^2 = T$, the term $y_{,xx}$ drops out. $\rho_{fl}\mathcal{U}^2$ is a *centrifugal force* arising from the curvature of the pipe. This term and T together affect the system in a similar way, as a compressive load. The cross–derivative term $y_{,tx}$ is a *Coriolis force* and the last term $y_{,tt}$ is the inertia term. The term $EIy_{,xxxx}$ is due to pipe bending rigidity.

The centrifugal and Coriolis forces are generally called *fictitious forces*. The centrifugal is due to rotational motion and the Coriolis is present when a body is in motion relative to a rotating coordinate system.[26]

Attempting a separation of variables solution will not work due to the cross–derivative term, as we will see after the brief aside on the traveling string.

Traveling String: A Subsidiary Problem

If the bending rigidity is taken to be very small, as for a string or a fire hose, $EIy_{,xxxx} \approx 0$, and if we then let the pipe *disappear* and replace the whole cross–section including the fluid inside by a string or cable, then the equation of motion for a moving string with zero rigidity is obtained,

$$-(T - \rho_{string}\mathcal{U}^2)y_{,xx} + 2\rho_{string}\mathcal{U}y_{,tx} + \rho_{string}y_{,tt} = 0.$$

Arranging terms in a different way results in

$$y_{,tt} + 2\mathcal{U}y_{,tx} - \left(\frac{T}{\rho_{string}} - \mathcal{U}^2\right)y_{,xx} = 0,$$

where T/ρ_{string} equals the wave speed squared, c^2. The stability of motion of the moving string depends on the sign of the difference in the parenthesis.

Non–Separable Solution

Proceed as though we did not know that the term $y_{,tx}$ prevents a separation of variables. Assuming $y(x,t) = Y(x)F(t)$, Equation 11.91 becomes

$$EIY_{,xxxx}F - (T - \rho_{fl}\mathcal{U}^2)Y_{,xx}F + 2\rho_{fl}\mathcal{U}Y_{,x}F_{,t} + (\rho_{fl} + \rho_P)YF_{,tt} = 0.$$

Dividing by $Y(x)F(t)$ results in the equation

$$EI\frac{Y_{,xxxx}}{Y} - (T - \rho_{fl}\mathcal{U}^2)\frac{Y_{,xx}}{Y} + 2\rho_{fl}\mathcal{U}\frac{Y_{,x}F_{,t}}{YF} + (\rho_{fl} + \rho_P)\frac{F_{,tt}}{F} = 0,$$

which is not separable due to the third term in the expression. The lack of synchronous motion, or modes, means that any two points on the beam will vibrate *out of phase* with each other, and the beam behaves as a *wave*. This is similar to the non–proportional damping problem.

---

[26] Check any good reference on classical mechanics or applied dynamics for details. One is **Classical Mechanics – A Modern Perspective**, Second Edition, by V.D. Barger, M.G. Olsson, McGraw–Hill, 1995.

11.5. WHERE VARIABLES DO NOT SEPARATE

Therefore, a wave solution of the form Equation 10.7 is used,

$$y(x,t) = A\sin(2\pi\kappa x - \omega t).$$

Take the appropriate derivatives, substitute these into Equation 11.91, cancel common factors A and $\sin(2\pi\kappa x - \omega t)$, and obtain

$$EI(2\pi\kappa)^4 + (T - \rho_{fl}\mathcal{U}^2)(2\pi\kappa)^2 + 2\rho_{fl}\mathcal{U}(2\pi\kappa)\omega - (\rho_{fl} + \rho_P)\omega^2 = 0. \quad (11.92)$$

Recall that $2\pi\kappa = \omega/c$, so that there are several possible forms Equation 11.92 can take. With the rigidity term EI retained, this equation can be solved only numerically, with a solution that would see wave–like beam deflections propagating back and forth between the pipe supports rather than the synchronous modal behavior to which we are accustomed.

To proceed analytically, assume that the rigidity of the pipe is very small, as it would be in a water hose used in fire fighting. In this case, Equation 11.92 becomes

$$(T - \rho_{fl}\mathcal{U}^2)\kappa^2 + \frac{\rho_{fl}\mathcal{U}\omega}{\pi}\kappa - \frac{(\rho_{fl} + \rho_P)\omega^2}{(2\pi)^2} = 0.$$

This quadratic equation for wave number κ can be solved for its two roots,

$$\kappa_{1,2} = \frac{1}{2(T - \rho_{fl}\mathcal{U}^2)}\left[-\frac{\rho_{fl}\mathcal{U}\omega}{\pi} \pm \sqrt{\left(\frac{\rho_{fl}\mathcal{U}\omega}{\pi}\right)^2 + \frac{(T - \rho_{fl}\mathcal{U}^2)(\rho_{fl} + \rho_P)\omega^2}{\pi^2}}\right], \quad (11.93)$$

where $(\omega/\pi)^2$ can be factored out in the numerator. Since the number of waves per unit length, κ, is a real number, the argument of the square root above must be positive semi–definite, that is,

$$T(\rho_{fl} + \rho_P) \geq \rho_{fl}\rho_P\mathcal{U}^2.$$

No boundary conditions have yet been imposed. The pipe (or string) is fixed to prevent transverse displacement at two points a distance L apart: $y(0,t) = 0$ and $y(L,t) = 0$. Using the general solution for the two values κ_1 and κ_2,

$$y(x,t) = A_1\sin(2\pi\kappa_1 x - \omega t) + A_2\sin(2\pi\kappa_2 x - \omega t),$$

the first boundary condition leads to $A_1 = -A_2$. The second boundary condition leads to

$$\sin(2\pi\kappa_1 L - \omega t) = \sin(2\pi\kappa_2 L - \omega t)$$
$$= \sin(2\pi\kappa_2 L - \omega t + 2n\pi).$$

The second equality is just a statement of the 2π–periodicity of the sine function. Then, $2\pi\kappa_1 L = 2\pi\kappa_2 L + 2n\pi$, or

$$\kappa_1 = \kappa_2 + \frac{n}{L}, \quad n = 1, 2, \ldots.$$

Using Equation 11.93,

$$\kappa_1 - \kappa_2 = \frac{n}{L} = \frac{\omega\sqrt{T(\rho_{fl} + \rho_P) - \rho_{fl}\rho_P \mathcal{U}^2}}{\pi(T - \rho_{fl}\mathcal{U}^2)},$$

from which the allowable propagation frequencies are found to be

$$\omega_n = \frac{n\pi(T - \rho_{fl}\mathcal{U}^2)}{L\sqrt{T(\rho_{fl} + \rho_P) - \rho_{fl}\rho_P \mathcal{U}^2}}.$$

As $\mathcal{U} \to \sqrt{T/\rho_{fl}}$, frequency $\omega_n \to 0$. This limit may be interpreted as a stability limit.

The respective equation for a string, with $\rho_P = 0$ and $\rho_{fl} = \rho_{string}$, is

$$\begin{aligned}\omega_n &= \frac{n\pi}{L}\frac{(T - \rho_{string}\mathcal{U}^2)}{\sqrt{T\rho_{string}}} \\ &= \frac{n\pi}{L}\frac{(T/\rho_{string} - \mathcal{U}^2)}{\sqrt{T/\rho_{string}}} \\ &= \frac{n\pi}{L}\frac{(c^2 - \mathcal{U}^2)}{c},\end{aligned}$$

where $c^2 = T/\rho_{string}$, and as $\mathcal{U} \to c$, frequency $\omega_n \to 0$.

While these have been simplified results, they provide us with some sense of how such systems behave.

11.6 Concepts Summary

The two dimensional membrane and plate models are derived and solved modally. The membrane can be viewed as a two dimensional string, and the plate as a two dimensional beam. Some special problems are introduced. Rayleigh's approximate method is re–introduced, and the more powerful Rayleigh–Ritz and Galerkin methods introduced and applied to simple example problems. A modal analysis for a randomly loaded beam is outlined. The problems of non–harmonic boundary conditions and of pipe (beam) vibration with internal flow are also outlined as special cases of where separation of variables does not work as a solution method. The pipe problem results in a wave solution.

11.7 Problems

Problems for Section 11.1 – Vibration of Membranes

1. Look around your world and locate applications where a membrane model could be useful.

2. Derive Equation 11.5.

3. Derive Equation 11.8.

4. Derive Equation 11.9.

5. Derive Equation 11.11. Explain the need for $\theta = \theta_0 + 2\pi j$.

6. Look up Bessel's equation in your differential equation textbook and derive it and solve it in general.

7. Derive Equations 11.16, 11.17, and 11.18. This would be a good exercise to do with a symbolic manipulation code such as MAPLE.

8. For the simply supported rectangular membrane of Section 11.1.1, suggest possible initial displacements $d(x,y)$. Pick one and solve for the complete solution. Assume zero initial velocity.

9. Do the last problem over for an assumed initial velocity $v(x,y)$, with zero initial displacement.

10. Convert the wave equation from Cartesian coordinates,
$$\frac{\partial^2 u}{\partial t^2} = c^2 \left(\frac{\partial^2 u}{\partial x^2} + \frac{\partial^2 u}{\partial y^2} \right),$$
to polar coordinates,
$$\frac{\partial^2 u}{\partial t^2} = c^2 \left(\frac{\partial^2 u}{\partial r^2} + \frac{1}{r}\frac{\partial u}{\partial r} + \frac{1}{r^2}\frac{\partial^2 \theta}{\partial \theta^2} \right),$$
using the transformation $x = r\cos\theta$ and $y = r\sin\theta$.

Problems for Section 11.2 – Vibration of Plates

11. Look around your world and provide applications where a plate model could be useful.

12. Derive the equation of motion of a square plate Hamilton's principle.

13. Derive Equations 11.20 and 11.21.

14. Derive Equation 11.22.

15. Derive Equation 11.26.

16. Show that Equation 11.32 must be true for the given boundary conditions.

17. Derive coefficients $C_{jk} = 2/\sqrt{\rho ab}$ using the normalization procedure given for W_{jk} by Equation 11.33.

Problems for Section 11.3 – Random Vibration of Continuous Structures

18. Describe three applications where a continuous structure is loaded by a random force.

19. Suppose the simply supported beam is forced by a random load with mean value μ_F and autocorrelation $R_{FF}(\tau) = \exp(-\alpha\tau)$, where α is a constant. Solve for the mean, autocorrelation, and spectral density of the response.

20. For the previous problem, suppose calculations need to be simplified. Can the autocorrelation of the force be replaced by a triangular spike around $\tau = 0$? Try this replacement, justify the simplification, and solve. Compare with the above result.

21. What complications occur in the derivations of this section if the force cannot be assumed to be stationary?

Problems for Section 11.4 – Approximate Methods for Continuous Structures

22. Derive Rayleigh's quotient for an longitudinally vibrating beam, and estimate the fundamental frequency for the tapered beam of Figure 11.14. For this beam the following properties hold:

$$m(x) = m(1 - x/L)$$
$$\text{and } EA(x) = EA(1 - x/L).$$

Compare your result to the exact value of $\omega_1 = 2.40\sqrt{EA/mL^2}$.

23. For the previous problem, estimate the first two natural frequencies using the Rayleigh–Ritz procedure. Assume a trial function of the form $\mathcal{Y}(x) = c_1 x^2 + c_2 x^3$. Compare the fundamental frequencies estimated by the Rayleigh–Ritz and the Rayleigh quotient.

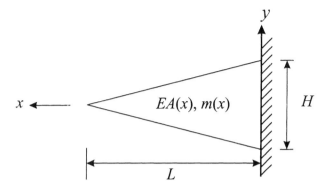

Figure 11.14: Tapered beam in longitudinal vibration.

24. For the tapered beam sketched in Figure 11.14, except now undergoing transverse vibration, estimate the first two natural frequencies using the Rayleigh–Ritz procedure. Assume a trial function of the form $\mathcal{Y}(x) = c_1 x^2 + c_2 x^3$. In this problem, use the following beam properties,

$$m(x) = \rho A(x) = \rho H(1 - x/L)$$
$$\text{and } EI(x) = (E/12)[H(1 - x/L)]^3.$$

Problems for Section 11.5 – Where Variables Do Not Separate

25. In *Example* 11.4, derive the general solution for $f_0(t) = x\exp(-t/L)$.

26. Solve the problem of *Example* 11.4 where the only nonzero boundary condition is $f_1(t) = \partial^2 y(0,t)/\partial x^2$.

27. Formulate the equation of motion for the fluid–conveying pipe using Lagrange's equation. Assume constant tension in the pipe, constant fluid velocity $\mathcal{U}$, linear material and vibration. Then simplify the equation for the moving string equation.

   Solution   Some intermediate results follow. The potential energy is due to bending strain energy, and to work done by the tension:

$$V_b = \int_0^L \frac{1}{2} EI y_{,xx}^2 \, dx$$
$$V_t = \int_0^L T \Delta s = \int_0^L T(ds - dx).$$

Refer to Figure 10.6 to relate ds to dx and dy, approximate the square root, to find
$$ds - dx \simeq \frac{1}{2} y_{,x}^2 dx.$$

The total potential is
$$V = V_b + V_t = \frac{1}{2} \int_0^L (EI y_{,xx}^2 + T y_{,x}^2) dx.$$

The kinetic energy is due to the vibrating pipe and to the motion of the fluid. The total kinetic energy is
$$T = T_{fl} + T_P = \frac{1}{2} \int_0^L \left[\rho_{fl} A_{fl} [\mathcal{U}^2 + (y_{,t} + \mathcal{U} y_{,x})^2] + \rho_P A_P y_{,t}^2 \right] dx.$$

These are substituted into Lagrange's equation to derive the equation of motion,
$$EI y_{,xxxx} - (T - \rho_{fl} A_{fl} \mathcal{U}^2) y_{,xx} + 2\rho_{fl} A_{fl} \mathcal{U} y_{,tx} + (\rho_{fl} A_{fl} + \rho_P A_P) y_{,tt} = 0.$$

Simplifying for the moving string, set $E = 0$, $\rho_P = \rho_{fl} = \rho$, $A_P = A_{fl} = A$, and the equation of motion becomes
$$y_{,tt} + 2\mathcal{U} y_{,tx} + (\mathcal{U}^2 - c^2) y_{,xx} = 0,$$

with $c^2 = T/\rho A$.

11.8 Mini–Projects

The following *mini–projects* are provided as possible avenues for extending the basic studies of the chapter to more advanced and more applied studies. At this point, the student has the ability and background to begin to tackle such problems, with some guidance from the instructor. Each mini–project is intended for extended study, either by the individual student, or in small groups of two or three students.

1. *Effect of Poisson's Ratio on the Fundamental Frequency of Transverse Vibration and Buckling Load of Circular Plates with Variable Profile*, P.A.A. Laura, V. Sonzogni, E. Romanelli, *Applied Acoustics*, Vol. 47, No. 3, 263–273, 1966.

   The problem described is solved by approximating the displacement function using simple polynomial coordinate functions which identically satisfy

11.8. MINI–PROJECTS

the boundary conditions. The numerical determinations of the eigenvalues under investigation are determined using the Rayleigh–Ritz method assuming that the plate thickness varies according to a particular functional relation.

2. *Response of Plates to Random Load*, M.M. Stanisic, *J. of the Acoustical Society of America*, Vol. 43, No. 6, 1968, 1351–1357.

   A technique has been developed to determine the covariance and spectral density functions for a plate of arbitrary boundary conditions subjected to loadings that are random in space and time.

3. *Wave Propagation in One–Dimensional Multi–Bay Periodically Supported Panels Under Supersonic Fluid Flow*, S. Mukherjee, S. Parthan, *J. Sound and Vibration*, (1995) **186**(1), 71–86.

   A simple method of vibration and stability of periodic multi–bay panels under supersonic flow is proposed. Floquet's principle, which has been used extensively for the analysis of vibration problems of periodic structures like equally spanned beams, stiffened plates and shells, has been extended here to predict the vibration characteristics of multi–bay periodic panels subjected to lengthwise supersonic airflow.

4. *Rayleigh–Ritz Method in Coupled Fluid–Structure Interacting Systems and its Applications*, F. Zhu, *J. Sound and Vibration*, (1995) **186**(4), 543–550.

   Based upon general thin shell theory and the basic equations of fluid mechanics, the Rayleigh–Ritz method for coupled fluid–structure free vibrations is developed for arbitrary tanks fully or partially filled with inviscid, irrotational and compressible or incompressible fluid, by means of the generalized orthogonality relations of wet modes and the associated Rayleigh quotients.

5. *Axisymmetric Dynamic Stability of an Annular Plate Subject to Pulsating Conservative Radial Loads*, H.P. Lee, T.Y. Ng, *Applied Acoustics*, **45** (1995), 167–179.

   The equations of motion of an annular plate subject to prescribed time–dependent radial compressive loads are formulated based on Hamilton's principle and the assumed mode method.

6. *Stochastic Vibration of an Elastic Beam Due to Random Moving Loads and Deterministic Axial Forces*, H.S. Zibdeh, *Engineering Structures*, Vol. 17, No. 7, 530–535, 1995.

   The random vibration of a simply supported elastic beam subjected to random loads moving with time–varying velocity is studied. The beam is also subjected to axial forces. Based on Euler–Bernoulli beam theory and stochastic methods, the governing partial differential equation is solved in closed form for the mean and variance of the response. Results are presented for different cases of speed, damping, and axial force parameters. The results show the effects of the variations of these parameters and interactions among them on random vibration characteristics of the beam.

7. *Vibration Control of Plates by Plate–Type Dynamic Vibration Absorbers*, T. Aida, K. Kawazoe, S. Toda, *J. Vibration and Acoustics*, Vol. 117, July 1995, 332–338.

A plate–type dynamic vibration absorber is presented for controlling the several predominant modes of vibration of a main plate under harmonic excitation, which consists of a dynamic absorbing plate under the same boundary conditions as the main plate with uniformly distributed connecting springs and dampers between the main and dynamic absorbing plates. Equations of motion of the system in the modal coordinates of the main plate become equal to those of a two degree of freedom system with two masses and three springs. An optimal plate type vibration absorber is derived.

8. *Some Remarks on the Identification of Damping Parameters at the Boundary of Vibrating Systems*, P. Hagedorn, U. Pabst, *Applied Mechanics Reviews*, Vol. 48, No. 11, Part 2, S107–S110, Nov. 1995.

In many cases, vibrating mechanical systems permit a reliable mathematical modeling with parameter values which are reasonably well known beforehand, except for the joints between different subsystems and at the boundaries. The boundary stiffness, which is often assumed as infinite, and the damping at the boundary, which is frequently ignored, are typically not well known. Here, the identification of the boundary stiffness and damping parameters from modal data is discussed. As an example, an elastic steel beam is treated, for which an experimental modal analysis has been carried out.

9. *Convergence of the Ritz Method*, A.W. Leissa, S.M. Shihada, *Applied Mechanics Reviews*, Vol. 48, No. 11, Part 2, S90–S95, Nov. 1995.

The Ritz method is widely used for the solution of problems in structural mechanics, especially eigenvalue problems where the free vibration frequencies or buckling loads are sought. Convergence is studied for the problem of free vibration of a cantilever beam, in particular with regard to the choice of displacement functions.

10. *The Modeling of Axially Translating Flexible Beams*, R.J. Theodore, J.H. Arakeri, A. Ghosal, *J. Sound and Vibration*, (1996) **191**(3), 363–376.

11.8. MINI-PROJECTS

The axially translating flexible beam with a prismatic joint can be modeled by using the Euler–Bernoulli beam equation together with the convective terms. In general, the method of separation of variables cannot be applied to solve this partial differential equation. A non–dimensional form of the Euler–Bernoulli beam equation is presented, obtained by using the concept of group velocity, and also the conditions under which separation of variables and assumed modes method can be used.

11. *Boundary Layer Induced Noise on Aircraft, Part I: The Flat Plate Model*, W.R. Graham, *J. Sound and Vibration*, (1996) **192**(1), 101–120.

    The importance of boundary layer contributions to aircraft cabin noise levels implies a need for a simple model capable of providing sufficient physical insight to address the problem at the design stage. This paper describes the initial form of such a model, which is based on the sound radiated by a single, flat, elastic plate under boundary layer excitation.

12. *An Approximate Analytical Solution of Beam Vibrations During Axial Motion*, B.O. Al–Bedoor, Y.A. Khulief, *J. Sound and Vibration*, (1996) **192**(1), 159–171.

    An approximate analytical solution for the transverse vibration of a beam during axial deployment is derived. The approach relies on removing the time dependency from the boundary conditions, and transferring it to the differential equation. The assumptions made are Euler beam theory and slow axial movement.

13. *Influence of the Material Constants on the Low Frequency Modes of a Free Guitar Plate*, A. Ezcurra, *J. Sound and Vibration*, (1996) **194**(4), 640–644.

    The sound production by non–electronic musical instruments is, from a physical point of view, a classical mechanics subject. Nevertheless, musical instruments are in general highly complex vibrating systems, although their components could be simple oscillators with many parameters having an influence on the final sound quality, in some cases in a way that is not well understood. In this work, the sensitivity of the eigenfrequencies of a guitar plate to different material parameter variations is established.

14. *Response of a Bending–Torsion Coupled Beam to Deterministic and Random Loads*, S.H.R. Eslimy-Isfahany, J.R. Banerjee, A.J. Sobey, *J. Sound and Vibration*, (1996) **195**(2), 267–283.

    The method of normal mode is used. The deterministic load is assumed to vary harmonically, whereas the random load is assumed to be Gaussian and ergodic. The analysis is applied to a cantilever aircraft wing for which there is substantial coupling between the bending and torsional modes of deformation.

15. *Dynamic Analysis of Nonuniform Beams with Time–Dependent Elastic Boundary Conditions*, S.Y. Lee, S.M. Lin, *J. Applied Mechanics*, Vol. 63, June 1996, 474–478.

    A specialized beam theory is used to study the dynamic response of a nonuniform beam with time–dependent elastic boundary conditions.

16. *Vibration of Rectangular Membranes with Linearly Varying Tension and Low Flexural Rigidity*, D.J. Gorman, R.K. Singhal, *Modal Analysis: the International J. of Analytical and Experimental Modal Analysis*, v.11, n.1, July 1996, 106–115.

    Analytical solutions are obtained for the free vibration of rectangular membranes with one–directional, linearly varying tension and light flexural rigidity. The linear variation in tension is a result of gravitational forces. A solution is obtained by the Rayleigh–Ritz energy method.

17. *Effects of Internal Flow on Vortex–Induced Vibration and Fatigue Life of Submarine Pipelines*, S. Zhonghan, Z. Qiang, *China Ocean Engineering*, Vol. 10, No. 3, 251–260, 1996.

    With the rapid development of offshore industries, submarine oil/gas pipelines have become widely used. The dynamic characteristics of pipelines show some new features due to the existence of both internal and external flows. The paper investigates the vortex-induced vibration of suspended pipeline exposed to steady submarine flow. The effects of internal flow are taken into account. Its influences on the amplitude of pipeline response, and then on fatigue life, are given in terms of the velocity of the internal flow.

18. *A Technique for the Systematic Choice of Admissible Functions in the Rayleigh–Ritz Method*, M. Amabili, R. Garziera, *J. Sound and Vibration* (1999) **224**(3), 519–539.

    A simple and systematic choice of admissible functions, which are the eigenfunctions of the closest, simple problem extracted from the one considered, is proposed.

19. *Vibration of Circular Plates Resting on a Sloshing Liquid: Solution of the Fully Coupled Problem*, M. Amabili, *J. Sound and Vibration* (2001) **245**(2), 261–283.

    The fully–coupled problem between sloshing modes of the free surface and bulging modes of the plate is solved by using the Rayleigh–Ritz method. The sloshing boundary condition is inserted into the eigenvalue problem.

20. *Linear Vibration Analysis of Cantilever Plates Partially Submerged in Fluid*, A. Ergin, B. Uğurlu, *J. Fluids and Structures*, 17 (2003) 927–939.

11.8. MINI–PROJECTS

In the analysis of the fluid–structure system, it is assumed that the fluid is ideal, and the fluid forces are associated with inertial effects of the surrounding fluid. Applications of such problems include vibration of ships and offshore platforms excited by wave impact, and dams and storage vessels under earthquake loading.

21. *Modelling of the Stiffness of Elastic Body*, S–J. Zhu, X–T. Weng, G. Chen, *J. Sound and Vibration* 262 (2003) 1–9.

    It is necessary for designing vibration isolation systems to know the components' static, dynamic and shock stiffnesses, which are currently obtained through experiments. This paper presents a new method for modeling the stiffness of elastic bodies with viscoelastic theory.

22. *On the Vibration Analysis of Rectangular Clamped Plates Using the Virtual Work Principle*, J.P. Arenas, *J. Sound and Vibration* 266 (2003) 912–918.

23. *Frequencies of beams carrying multiple masses: Rayleigh estimation versus eigenanalysis solutions*, K.H. Low, *J. Sound and Vibration* 268 (2003) 843–853.

Chapter 12

Nonlinear Oscillation

Nonlinearities in governing equations of motion can occur due to a variety of physical causes, material or geometric, or loading. By nonlinearity we mean functions for which the principle of linear superposition does not apply, for example, x^p, $\cos\phi$, $\exp(-xt)$. In all these instances, the functional relationship between displacement and damping and/or spring force is not linear. Because of the complexity of nonlinear behavior, there is no single overarching principle that governs the solution of nonlinear equations. There are, however, general approaches that can be utilized for the solution of certain classes of nonlinear differential equations. An excellent and readable monograph on nonlinear vibration is by Stoker.[1]

The study of nonlinear system behavior can be broadly categorized as either qualitative or quantitative. Qualitative approaches are concerned less with response time histories and more with the stability characteristics of the system in the neighborhood of an equilibrium. Quantitative approaches, on the other hand, are devoted to the derivation of usually approximate solutions to the governing nonlinear equation of motion. Perturbation methods[2] are a quantitative method used to approximate the response of systems with small nonlinearity. Of particular interest in nonlinear vibration[3,4] are systems that have periodic solutions. A nonlinear equation may have periodic solutions as well as nonperiodic solutions. Where the nonlinearity

---

[1] **Nonlinear Vibrations in Mechanical and Electrical Systems**, J.J. Stoker, original edition 1950, Wiley Classics Library Edition reprinted 1992.

[2] **Perturbation Techniques in Mathematics, Physics, and Engineering**, R. Bellman, Holt, Rinehart and Winston, New York, 1966.

[3] **Nonlinear Oscillations**, A.H. Nayfeh, D.T. Mook, Wiley–Interscience, New York, 1979.

[4] **Nonlinear Ordinary Differential Equations**, D.W. Jordan, P. Smith, Oxford University Press, Second Edition, 1988.

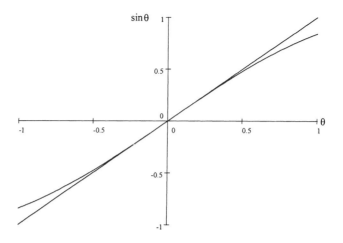

Figure 12.1: $\sin\theta$ versus its approximation θ rad.

is large, the analyst must use numerical methods to find the time history. The origin of quantitative methods lies in the early history of astronomical calculations, where it received the name *perturbation method*.

In this chapter, the focus is on single degree of freedom oscillators where the key concepts can be introduced. Higher order nonlinear systems are certainly important, but they are primarily tackled using numerical methods.

12.1 Examples of Nonlinear Vibration

Some examples of physical systems governed by nonlinear equations are given next. First consider systems where dissipation is ignored:

- *motion in a gravitational field:* A pendulum oscillates in a gravitational field. The governing equation of motion is

$$\ddot{\theta} + \frac{g}{l}\sin\theta = 0.$$

In linearized theory, for small oscillations, θ is small, and as an approximation we let $\sin\theta \simeq \theta$, resulting in the equation of the simple harmonic oscillator, $\ddot{\theta} + (g/l)\theta = 0$. Figure 12.1 shows the range over which the approximation looks good.

12.1. EXAMPLES OF NONLINEAR VIBRATION

- *restoring moments for floating bodies:* A floating body, such as a ship, oscillates in response to wave, current and wind loads that act on it. In general the motion is governed by a nonlinear equation.

- *elastic restoring forces:* The restoring force for a moored body is nonlinear as is the force-displacement relation for a shallow arch that can undergo snap-through buckling. While we generally may assume a linear elastic model for the relation between an applied force and the subsequent deformation, cases such as those mentioned above are elastic but nonlinear.

- *geometric nonlinearities:* As a body moves, nonlinear effects become significant. This can be true of a continuous process or a discontinuous process. The later may occur when additional forces come into play at discrete locations. For example, heat exchanger tubes with stops and ships with mooring lines and fenders have on-off constraints.

Consider the following nonlinear dissipative forces:

- *internal damping:* As a structure cycles through inelastic ranges of its constitutive relation, there is an energy loss through permanent deformation and the structure is said to be hysteretic.

- *interface damping or friction:* A body sliding on a surface experiences Coulomb friction.

- *flow-induced forces:* Drag between a fluid and structure results in a force $\sim v|v|$, where v is the relative flow velocity.

Example 12.1 Simple Pendulum

Consider the simple pendulum of Figure 12.2. Using the principles of earlier chapters, the equation of motion is given as

$$ml^2\ddot{\theta} + mgl\sin\theta = 0.$$

This problem was solved in Chapter 2 using the linear approximation $\sin\theta \simeq \theta$, with the result,

$$\begin{aligned}\theta &= A\sin(\omega_n t + \phi),\\ \omega_n &= \sqrt{g/l}.\end{aligned}$$

It is relatively straightforward to estimate the angles for which the approximation is no longer acceptable. For those cases, it is possible

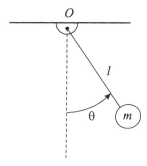

Figure 12.2: Simple pendulum.

to retain the first two terms of the series that the sine represents, $\sin\theta \simeq \theta - \theta^3/6$. The equation of motion is then

$$ml^2\ddot{\theta} + mgl\left(\theta - \frac{\theta^3}{6}\right) = 0,$$

or

$$\ddot{\theta} + \omega_n^2\left(\theta - \frac{\theta^3}{6}\right) = 0. \tag{12.1}$$

This is an oscillator with a nonlinear spring, $\ddot{\theta} + k(\theta) = 0$, where $k(\theta)$ is a general nonlinear function of θ. A linear spring behaves according to $k\theta$. The nonlinear spring can be categorized as a *soft spring*, where the slope decreases with increasing θ, or a *hard spring*, where the slope increases with θ, as seen in Figure 12.3.

Therefore,

$$\text{if } \frac{dk}{d\theta} = k \text{ then the spring is linear}$$
$$\text{if } \frac{dk}{d\theta} > k \text{ then the spring is hard}$$
$$\text{if } \frac{dk}{d\theta} < k \text{ then the spring is soft.}$$

By these categories we imply that the hard spring is strictly increasing and the soft spring is strictly decreasing, and k is some arbitrary stiffness (slope) that describes how the spring behaves for small θ. Of course, hybrid behavior is possible. This is generalized next. ∎

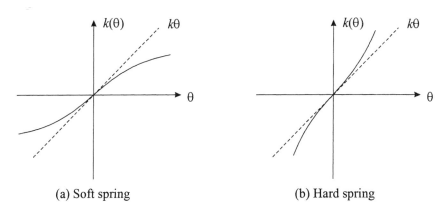

(a) Soft spring (b) Hard spring

Figure 12.3: (a) Soft spring results in larger displacement for same force increment. (b) Hard spring requires greater force for same displacement increment.

12.2 Fundamental Nonlinear Equations

A general nonlinear oscillator, where x is the displacement, can be represented by

$$m\ddot{x} + g(\dot{x}) + h(x) = F(t).$$

The inertia force is given by $m\ddot{x}$, the damping force by $-g(\dot{x})$, and the spring or restoring force by $-h(x)$. The external force or excitation is represented by the function $F(t)$. The reason for the negative signs is that by Newton's second law of motion, the sum of the forces on a free body is equated to the inertia force, that is, $m\ddot{x} = -g(\dot{x}) - h(x) + F(t)$. This general equation is specialized to two important classes of problems. One is the class where only the stiffness force is nonlinear, and the other where only the damping force is nonlinear.

The nonlinear stiffness class is governed by the equation

$$m\ddot{x} + c\dot{x} + h(x) = F(t).$$

The stiffness force $-h(x)$ is a nonlinear function of x, signifying how the restoring force acts in the different displacement regimes. Equations of this form are classified as the *Duffing* equation.

A class of nonlinear damping equations is governed by

$$m\ddot{x} + g(\dot{x}) + kx = F(t), \quad \text{where} \quad \begin{cases} \dot{x}g(\dot{x}) < 0 \text{ for small } \dot{x} \\ \dot{x}g(\dot{x}) > 0 \text{ for large } \dot{x}. \end{cases}$$

The damping force $-g(\dot{x})$ is a nonlinear function of velocity $\dot{x}$. Note the interesting behavior of this oscillator depending on whether the damping force acts in the direction of motion or acts opposite to the direction of velocity. These are called *self-excited* oscillators because the "damping" force is in the direction of velocity for small velocities so that the state of rest is unstable and motion develops from the rest position under the slightest disturbance even if the external force is zero. A special case of this equation is known as the *van der Pol* equation. We will study these equations in more detail later in this chapter.

We note two classes of equations:

Definition 1 *An **autonomous** equation is one where time t does not appear explicitly.*

Definition 2 *A **non-autonomous** equation is one where t appears explicitly, for example, in the forcing term.*

The difference between the non-autonomous and the autonomous systems is that the solutions of the former have the period of the external excitation, or more generally, are in a rational ratio to this period, while the period of the latter are determined by the parameters of the differential equation itself.

Also in an autonomous differential equation, t can be replaced by $t + t_0$, where t_0 is the phase, and still have the same solution. This means that the time axis can be arbitrarily translated and that the origin can be selected so that the initial velocity $dx/dt = 0$. Even so, sometimes for non-autonomous equations, additional conditions such as the zero initial velocity are imposed. The resulting analysis is less general, but greatly simplified and sometimes still useful. On occasion we make such additional assumptions in this chapter.

12.3 The Phase Plane

The phase plane[5] has already been introduced as an alternate view of the oscillatory system. An advantage of the phase plane is that the *trajectory*

---

[5] A number of definitions are in order here.

1. Given an n degree of freedom system, governed by second order differential equations, defined by n generalized coordinates, the n dimensional space is called the *configuration space*.

2. The n second order differential equations can be converted into $2n$ first order differential velocity (or momenta) equations. The $2n$ dimensional space of n generalized displacements and n generalized velocities is known as the *phase space*. The phase plane refers to the single degree of freedom case where a velocity is

12.3. THE PHASE PLANE

paths indicate the nature of the oscillation and motion. In a *qualitative* way, one can read the trajectories and observe the nature of system damping, whether the system is stable, oscillatory, periodic or otherwise. In a qualitative method, the differential equation does not have to be solved and yet it is possible to ascertain much about the properties of the system.

The *configuration space* or *plane* is the geometrical domain of the governing equations. A single degree of freedom system is governed by a second order differential equation, that is transformed into two first order differential equations in two generalized coordinates. For example, the second order linear differential equation $\ddot{x} + 2\zeta\omega_n \dot{x} + \omega_n^2 x = 0$ becomes

$$\dot{x} = y$$
$$\dot{y} = -\omega_n^2 x - 2\zeta\omega_n y.$$

Here the phase plane is plotted as velocity y vs. displacement x, or $\dot{x}$ vs. x. To uniquely determine the behavior of a particle at a point, the velocity as well as the location is needed. See Figure 12.4. A system with n degrees of freedom has $2n$ state equations, and this fact leads to the generalization of the two dimensional phase plane to the $2n$ dimensional phase or configuration space. Some call these $2n$ generalized coordinates, even though in Lagrangian dynamics the generalized coordinates are interpreted to mean the minimum set of coordinates that define the system configuration.

The phase plane cannot be used for a non–autonomous equation since the vector field at a point changes in time. The system may, however, be made autonomous by increasing its dimension by one. For a single non-autonomous equation then, the phase plane becomes a three–dimensional construction, thereby including the variation with time.

We have already plotted the phase plane for a linear oscillator. Consider the example of an undamped pendulum.

Example 12.2 Phase Plane for an Undamped Nonlinear Pendulum
Derive the trajectories of an undamped and nonlinear pendulum. Plot the trajectories for a variety of energy levels.
Solution The equation of motion is

$$\ddot{\theta} + \frac{g}{l}\sin\theta = 0$$
$$\text{or} \quad \ddot{\theta} + \omega_n^2 \sin\theta = 0.$$

---

plotted against a displacement in the plane.

3. The set of $2n$ first order equations are known as *state equations*, and their integrals (displacements) represent *state variables*. These variables define the $2n$ dimensional space called the *state space*.

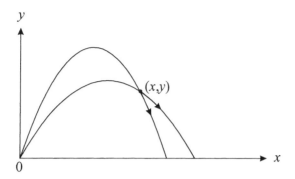

Figure 12.4: At coordinate (x, y) the velocity vector is needed in addition to the magnitude of the velocity y, in order to define the curve uniquely.

Define $x = \theta$ and $y = \dot{x} = \dot{\theta}$, then

$$\dot{x} = y$$
$$\dot{y} = -\omega_n^2 \sin x.$$

These two equations can be combined into an equation for the trajectory, where time is implicit,

$$\frac{dy}{dx} = -\frac{\omega_n^2 \sin x}{y},$$

or

$$y\,dy = -\omega_n^2 \sin x\, dx.$$

Integrate both sides assuming that at the end of a swing cycle $\dot{x} = 0$ and $x = x_0$, to find

$$y^2 = 2\omega_n^2 \left(\cos x - \cos x_0\right).$$

To simplify the appearance of the relation, let $z = y/\omega_n$,

$$z^2 = 2\left(\cos x - \cos x_0\right).$$

Figure 12.5 shows a few representative trajectories. Trajectories that intersect $x = \pm n\pi$ and those within represent simple harmonic motion. These are circles because the vertical axis has been normalized to the natural frequency. Otherwise they would be ellipses. Trajectories on the outside are not oscillatory. They may represent a pendulum that rotates like a propeller rather than oscillate about the static equilibrium. ∎

Example 12.3 Phase Plane of an Undamped Nonlinear Oscillator

12.3. THE PHASE PLANE

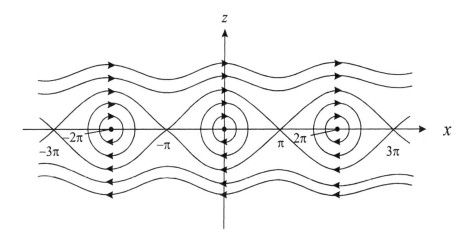

Figure 12.5: Trajectories for an undamped nonlinear pendulum where $z = y/\omega_n = \dot{\theta}/\omega_n$.

Consider the nonlinear equation

$$\ddot{x} + \omega_n^2 \left(x - 2\alpha x^3 \right) = 0.$$

This is the linear oscillator with the added nonlinear restoring term $-2\alpha\omega_n^2 x^3$. Define $x = \theta$ and $y = \dot{x} = \dot{\theta}$, then

$$\begin{aligned} \dot{x} &= y \\ \dot{y} &= -\omega_n^2 \left(x - 2\alpha x^3 \right), \end{aligned}$$

and

$$\frac{dy}{dx} = -\frac{\omega_n^2 \left(x - 2\alpha x^3 \right)}{y},$$

or

$$y\,dy = -\omega_n^2 \left(x - 2\alpha x^3 \right) dx.$$

Integrate both sides assuming that at the end of a swing cycle $\dot{x} = 0$ and $x = x_0$, to yield

$$z^2 + x^2 - \alpha x^4 = A^2, \tag{12.2}$$

where $z = y/\omega_n$ and $A^2 = x_0^2 \left(1 - \alpha x_0^2 \right)$ is a constant. The intersection points of the curves that separate oscillatory behavior from unstable behavior are given by solving Equation 12.2 for α with $z = 0$. Figure 12.6 depicts a number of trajectories for several values of α. For $\alpha < 1/(4A^2)$, the motion is periodic. Note that the vertical axis is normalized to ω_n. The

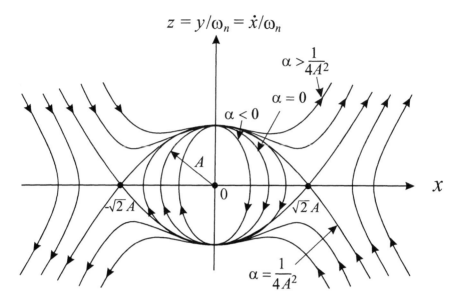

Figure 12.6: Trajectories for a nonlinear oscillator for different values of α.

circle for $\alpha = 0$ represents simple harmonic motion. For $\alpha = 1/(4A^2)$, the trajectories are given by the parabolas,

$$\frac{y}{\omega_n} = \pm\left(A - \frac{x^2}{2A}\right),$$

where the points $(x, y) = (\pm\sqrt{2}A, 0)$ are unstable equilibria. ∎

The above two examples can be generalized to the nonlinear governing equation

$$\ddot{x} + f(x, \dot{x}) = 0.$$

Define $\dot{x} = y$ and then $\dot{y} = -f(x, y)$. The slope of a trajectory is given by

$$\frac{dy}{dx} = \frac{(dy/dt)}{(dx/dt)} = -\frac{f(x, y)}{y} \equiv \varphi(x, y). \tag{12.3}$$

If $\varphi(x, y)$ is not indeterminate (due to a zero denominator), then there is a unique slope of the trajectory at every point (x, y) in the phase plane. If $y = 0$ and $f(x, y) \neq 0$, the point lies on the x axis and the slope of the trajectory is infinite. This means that all trajectories cross the x axis at a right angle. If $y = 0$ and $f(x, 0) = 0$, the point is called a *singular*

12.3. THE PHASE PLANE

point and the slope is indeterminate. Such a point represents an equilibrium position, that is, $x =$ constant and $y = \dot{x} = 0$, the velocity is zero. Therefore $\dot{y} = -f(x,y) = 0$ and there is no force on the system. The nature of the stability of the singular point requires additional work. In Figure 12.6 the points at $x = \pm\sqrt{2}A$ are unstable singular points.

12.3.1 Stability of Equilibria

As an introduction to the study of dynamic stability, a single degree of freedom nonlinear dynamic system is studied. It can be represented by the two first order differential equations

$$\frac{dx}{dt} = g_1(x,y) \qquad (12.4)$$

$$\frac{dy}{dt} = g_2(x,y), \qquad (12.5)$$

where g_1 and g_2 are nonlinear functions of x and y. The slope of the trajectories in the phase plane is given by

$$\frac{dy}{dx} = \frac{(dy/dt)}{(dx/dt)} = \frac{g_2(x,y)}{g_1(x,y)}. \qquad (12.6)$$

Equation 12.6 is a more general version of Equation 12.3. Define the singular or equilibrium point as (x_0, y_0) so that the slope has the indeterminate value $0/0$:

$$g_1(x_0, y_0) = g_2(x_0, y_0) = 0.$$

Example 12.4 The General Phase Plane
The simple phase plane discussed earlier, resulting in slope Equation 12.3, is useful for certain mechanical systems. More complex, or realistic mechanical, biological and geometrical problems are modeled using the more general first order system Equations 12.4 and 12.5. Because the slope Equation 12.6 is more general, the trajectories do not necessarily cross the x axis perpendicularly. An example of such a general first order system is

$$\dot{x} = 3x + 2y$$
$$\dot{y} = -2x - 2y.$$

Try plotting this set of curves, using Equation 12.6 to find the slope. These trajectories may cross the x axis at a slope other than $\pi/2$. ∎

It is useful to examine the behavior of the trajectories in the neighborhood of the singular points. Certain questions about the stability of the equilibrium can be answered by studying this neighborhood. This is most easily accomplished by expanding the nonlinear functions g_1 and g_2 in a Taylor series about a singular point. There may be many singular points in general. Each has to be examined individually. It can be assumed that $(x_0, y_0) = (0, 0)$ is such a singular point, since the slope of the trajectories does not vary with translation,

$$\begin{aligned} x_1 &= x - x_0 \\ y_1 &= y - y_0 \\ \frac{dy_1}{dx_1} &= \frac{dy}{dx}. \end{aligned}$$

Then, via the Taylor series of a function of two variables,

$$\begin{aligned} \dot{x} &= g_1(x, y) = \left.\frac{\partial g_1}{\partial x}\right|_{(0,0)} x + \left.\frac{\partial g_1}{\partial y}\right|_{(0,0)} y + \text{higher order terms} \\ \dot{y} &= g_2(x, y) = \left.\frac{\partial g_2}{\partial x}\right|_{(0,0)} x + \left.\frac{\partial g_2}{\partial y}\right|_{(0,0)} y + \text{higher order terms}. \end{aligned}$$

In the neighborhood of the singular point, the higher order terms can be ignored. Then, in matrix form,

$$\left\{\begin{array}{c} \dot{x} \\ \dot{y} \end{array}\right\} = \left[\begin{array}{cc} d_{11} & d_{12} \\ d_{21} & d_{22} \end{array}\right] \left\{\begin{array}{c} x \\ y \end{array}\right\}, \tag{12.7}$$

where d_{ij} represent the prior partial derivatives, respectively.

The solution to linearized matrix Equation 12.7 is geometrically similar to the solution of nonlinear Equations 12.4 and 12.5. Assume the solution

$$\left\{\begin{array}{c} x \\ y \end{array}\right\} = \left\{\begin{array}{c} X \\ Y \end{array}\right\} \exp(\lambda t), \tag{12.8}$$

where X, Y, and λ are constants. Substitute Equation 12.8 into Equation 12.7 and derive the eigenvalue problem,

$$\left[\begin{array}{cc} d_{11} - \lambda & d_{12} \\ d_{21} & d_{22} - \lambda \end{array}\right] \left\{\begin{array}{c} X \\ Y \end{array}\right\} = \left\{\begin{array}{c} 0 \\ 0 \end{array}\right\}.$$

The eigenvalues, $\lambda_{1,2}$, can be solved by evaluating the characteristic equation arising from the determinant of the characteristic matrix,

$$\left|\begin{array}{cc} d_{11} - \lambda & d_{12} \\ d_{21} & d_{22} - \lambda \end{array}\right| = 0,$$

12.3. THE PHASE PLANE

or

$$\lambda^2 - p\lambda + q = 0$$
$$p = (d_{11} + d_{22})$$
$$q = d_{11}d_{22} - d_{12}d_{21}.$$

Then, $\lambda_{1,2} = \left(p \pm \sqrt{p^2 - 4q}\right)/2$ with corresponding eigenvectors

$$\left\{\begin{array}{c} X \\ Y \end{array}\right\}_1, \left\{\begin{array}{c} X \\ Y \end{array}\right\}_2.$$

For the case $\lambda_1 \neq \lambda_2$ and $\lambda_1 \neq 0, \lambda_2 \neq 0$, the general solution is

$$\left\{\begin{array}{c} x \\ y \end{array}\right\} = C_1 \left\{\begin{array}{c} X \\ Y \end{array}\right\}_1 \exp(\lambda_1 t) + C_2 \left\{\begin{array}{c} X \\ Y \end{array}\right\}_2 \exp(\lambda_2 t),$$

where C_1 and C_2 are arbitrary constants. We note that these results are identical to those derived when studying the second order differential equation governing a damped harmonic oscillator. Classes of behavior can be defined in terms of the discriminant,

if $(p^2 - 4q) < 0$, the motion is oscillatory
if $(p^2 - 4q) > 0$, the motion is exponentially decaying.

Regarding stability, consider the expression

$$\exp(\lambda_1 t) = \exp\left(\frac{p}{2}\right) \exp\left(\frac{1}{2}\sqrt{p^2 - 4q}\right),$$

then

if $p > 0$, the system is unstable
if $p < 0$, the system is stable.

Therefore, the character of the trajectory depends on the value of p as well as the relative values of p^2 and $4q$. The following cases are described:

1. Case *(i)*: λ_1 and λ_2 are real and distinct; $p^2 > 4q$.

    (a) If λ_1 and λ_2 are of the same sign and $q > 0$, the equilibrium point is called a *node*. If $\lambda_2 < \lambda_1 < 0$ and $p < 0$, all the trajectories tend to the origin as $t \longrightarrow \infty$, and the origin is called a *stable node*. See Figure 12.7 (a). If $\lambda_2 > \lambda_1 > 0$ and $p > 0$, all the trajectories tend in the opposite direction as $t \longrightarrow \infty$, and the origin is called a *unstable node*. See Figure 12.7 (b).

(b) If λ_2 and λ_1 are of opposite signs and $q < 0$, with any sign for p, one solution tends to the origin and the other tends to infinity. The origin is called a *saddle point* and it corresponds to an *unstable equilibrium*. See Figure 12.7 (c).

2. Case *(ii)*: λ_1 and λ_2 are real and equal; $p^2 > 4q$.

   The trajectories are straight lines passing through the origin, which is the equilibrium point. The origin is a *stable node* if $\lambda_{1,2} < 0$ and an *unstable node* if $\lambda_{1,2} > 0$. See Figure 12.7 (d) for the case of a stable node. If the node is unstable, then the arrows point in the opposite direction.

3. Case *(iii)*: λ_1 and λ_2 are complex conjugates; $p^2 < 4q$.

   These trajectories are *logarithmic spirals*. The equilibrium point is called a *focus* or a *spiral point*.

   (a) If $p < 0$ and $q > 0$, the motion is asymptotically stable and thus the focus is stable. See Figure 12.7 (e).

   (b) If $p > 0$ and $q > 0$, the motion is unstable and thus the focus is unstable. See Figure 12.7 (f).

   (c) If $p = 0$, the trajectories reduce to circles if the vertical axis is normalized. Otherwise the trajectories are ellipses. The equilibrium point is called a *center* or a *vertex point*, and the motion is periodic and thus stable. See Figure 12.7 (g).

Statements about the stability of nonlinear systems based on their linearized counterparts may lead to significant behavior or critical behavior. If the linearized system is found to be asymptotically (as t increases) stable or unstable, it is said to possess *significant behavior*. In this case the stability characteristics of the linearized system are the same as of the nonlinear system. On the other hand, if the linearized system is found to be stable, the linearized system possesses *critical behavior*, and the stability conclusions about it do not necessarily extend to the full nonlinear system. In this instance an analysis of the nonlinear system is required.

Refer to the discussion in Section 6.3.1 on the more specific case of the stability of a damped oscillator.

12.4 Perturbation or Expansion Methods

A variety of approximation techniques are available to model and solve nonlinear governing equations. One quantitative approach is by a series

12.4. PERTURBATION METHODS

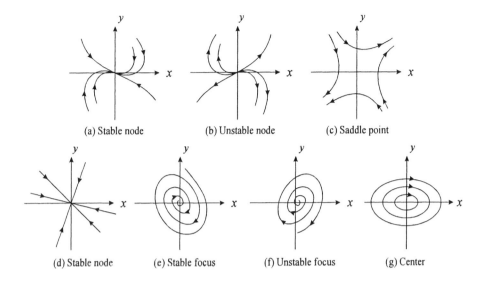

Figure 12.7: Stability curves.

expansion method known as a *perturbation* of the system. Inherent in this approach, as in all approximate analytical approaches, is that the nonlinearity is small. Perturbation methods[6] are useful as first steps in a nonlinear analysis to get a general behavior of the system. The method becomes cumbersome for more realistic models. Once the general behavior is understood other methods are useful.

The essence of the perturbation method as it is used to locate periodic solutions of nonlinear oscillators is the following. Given a harmonic oscillator, there exists an infinity of periodic solutions depending on the two constants of integration. The question arises, what will happen if the linear system $\ddot{x} + x = 0$ is perturbed by adding a small term, say, $\varepsilon f(t, x, \dot{x})$? A general statement is not possible but depends on the function $f(t, x, \dot{x})$ and the smallness of ε. For example, if $\varepsilon f(t, x, \dot{x}) = \varepsilon c \dot{x}$, then the trajectories become logarithmic spirals, stable if $c > 0$ and unstable if $c < 0$. The value of ε must be sufficiently small to ensure that the series solution converges. The perturbation method attempts to ascertain under what conditions the perturbed equation has periodic solutions. This is sometimes called the *problem of Poincaré*, who studied such problems rigorously.

Consider the unforced nonlinear pendulum as a motivating example.

---

[6]**Perturbations, Theory and Methods**, J.A. Murdock, Wiley–Interscience, 1991. This is an excellent book, the only one referenced here that provides a rigorous discussion of perturbation methods, their use and their mathematical basis.

Example 12.5 Nonlinear Pendulum

The approximate equation for an undamped nonlinear pendulum, Equation 12.1,

$$\ddot{\theta} + \omega_n^2 \left(\theta - \frac{\theta^3}{6}\right) = 0,$$

can be written in the more general form

$$\ddot{\theta} + \omega_n^2 \theta + \varepsilon \theta^3 = 0, \tag{12.9}$$

where $\omega_n^2 = \sqrt{g/l}$ and $\varepsilon = -\omega_n^2/6$. Equation 12.9 is known as an unforced *Duffing* equation. It is reasonable to assume that the nonlinearity is weak, that is, ε is small. In this way, the perturbation solution is given by

$$\theta(t) = \theta_0(t) + \varepsilon \theta_1(t) + \varepsilon^2 \theta_2(t) + \cdots,$$

where $\theta_i(t)$ are to be determined for all i. Take a two-term approximation for $\theta(t)$, and substitute this into the equation of motion,

$$\left(\ddot{\theta}_0 + \varepsilon \ddot{\theta}_1\right) + \omega_n^2 (\theta_0 + \varepsilon \theta_1) + \varepsilon (\theta_0 + \varepsilon \theta_1)^3 = 0,$$

and expand and group according to the order of the perturbation parameter ε,

$$(\ddot{\theta}_0 + \omega_n^2 \theta_0) + \varepsilon \left(\ddot{\theta}_1 + \omega_n^2 \theta_1 + \theta_0^3\right) + \varepsilon^2 \left(3\theta_0^2 \theta_1\right)$$
$$+ \varepsilon^3 \left(3\theta_0 \theta_1^2\right) + \varepsilon^4 \theta_1^3 = 0.$$

Since ε is assumed to be small, terms of order ε^2, ε^3, and ε^4 are higher than the order of the assumed solution and therefore truncated. The solution obtained in this way is said to be correct to order ε. Following the general procedure, each order of the solution is satisfied independently,[7]

$$\varepsilon^0 \; : \; \ddot{\theta}_0 + \omega_n^2 \theta_0 = 0 \tag{12.10}$$
$$\varepsilon^1 \; : \; \ddot{\theta}_1 + \omega_n^2 \theta_1 = -\theta_0^3. \tag{12.11}$$

θ_0 is known as the *generating solution* for the sequence of equations.

The Duffing equation requires two initial conditions. These can be specified in full generality, but assume that $\theta(0) = C$ and $\dot{\theta}(0) = 0$. If the initial velocity is other than zero, only the phase is changed but not the character of the solution for autonomous Equation 12.9. It is easiest to stipulate that

---

[7] When substituting the expansion into the differential equation, we obtain a power series in ε that must vanish *identically* in ε. This is why each order in ε results in a differential equation that must be satisfied.

12.4. PERTURBATION METHODS

the ε^0 order equation satisfy the initial conditions. Then the remaining equations of higher order will satisfy zero initial conditions.

As we know, these equations are solved in sequence. Equation 12.10 is solved for θ_0, which is then substituted into the right hand side of Equation 12.11 so that the solution θ_1 can be obtained. Therefore,

$$\theta_0(t) = A \sin(\omega_n t + \phi),$$

where satisfying the initial conditions results in $A = C$ and $\phi = \pi/2$. Equation 12.11 becomes[8]

$$\begin{aligned}\ddot{\theta}_1 + \omega_n^2 \theta_1 &= -C^3 \sin^3(\omega_n t + \pi/2) \\ &= -C^3 \left[\frac{3}{4}\sin(\omega_n t + \pi/2) - \frac{1}{4}\sin 3(\omega_n t + \pi/2)\right].\end{aligned}$$

The solution to this equation is

$$\theta_1(t) = \frac{3}{8\omega_n} t C^3 \cos(\omega_n t + \pi/2) - \frac{C^3}{32\omega_n^2} \sin 3(\omega_n t + \pi/2). \quad (12.12)$$

The two–term approximate solution is then $\theta(t) = \theta_0(t) + \varepsilon\theta_1(t)$. We immediately see a problem with the solution so obtained in the first term on the right hand side of Equation 12.12. This is a *secular term*, that is, an expression that grows without bounds, here due to the factor t multiplying the cosine. We know that the solution to Equation 12.9 should be periodic for small ε. The problem with the method as prescribed so far is that the truncation has removed terms that would balance, and in the limit, result in a periodic solution. This effect is demonstrated by expanding the following sine function,

$$\begin{aligned}\sin(\omega_n + \varepsilon)t &= \sin\omega_n t \cos\varepsilon t + \cos\omega_n t \sin\varepsilon t \\ &= \left(1 - \frac{1}{2!}\varepsilon^2 t^2 + \frac{1}{4!}\varepsilon^4 t^4 - \cdots\right) \sin\omega_n t \\ &+ \left(\varepsilon t - \frac{1}{3!}\varepsilon^3 t^3 + \frac{1}{5!}\varepsilon^5 t^5 - \cdots\right) \cos\omega_n t.\end{aligned}$$

Suppose we approximated the expression $\sin(\omega_n + \varepsilon)t$ by two terms of the expansion on the right hand side. The sin is harmonic and therefore the secular terms should not be there and must be removed in a logical way. This problem is corrected with Lindstedt's method derived in the next section. ∎

---

[8] Of course, $C\sin(\omega_n t + \pi/2) = C\cos\omega_n t$.

12.4.1 Lindstedt–Poincaré Method

An examination of the straightforward expansion of Example 12.5 shows that the solution is constrained to oscillate at the constant value ω_n. Linear harmonic systems are characterized by a constant period of oscillation, regardless of initial conditions. The linear response ($\varepsilon = 0$) is harmonic and of period $T_n = 2\pi/\omega_n$. Nonlinear quasi–harmonic systems, however, have periods, and thus frequencies, that are functions of the nonlinearity and the initial conditions. The breakdown of the above method, leading to secular terms, is due to ignoring this nonlinearity. In the presence of nonlinear terms the response is periodic and of period $T = 2\pi/\omega$, where ω is an unknown fundamental frequency depending on ε and initial conditions.

The *Lindstedt-Poincaré* method gets around this problem by expanding ω, the response frequency, as well as θ, in powers of a small parameter. Recall that the initial conditions of linear and nonlinear systems are coupled to the solution through the constants of integration. We therefore expect that the expansion in ω will be a function of an integration constant. Since ω does not appear in the equation of motion, only ω_n appears, the transformation $\tau = \omega t$ is first introduced. Then, $dt = d\tau/\omega$, and

$$\frac{d}{dt} = \omega \frac{d}{d\tau}$$
$$\frac{d}{dt}\left(\frac{d}{dt}\right) = \omega \frac{d}{d\tau}\left(\omega \frac{d}{d\tau}\right)$$
$$\text{or} \quad \frac{d^2}{dt^2} = \omega^2 \frac{d^2}{d\tau^2},$$

where τ is in radians. The essential feature of this method is emphasized by Minorsky.[9] There appears to be an arbitrariness in the approximations since two expansions have been introduced in the same differential equations. This arbitrariness enables us to dispose of the available constants so as to gradually eliminate the secular terms in the subsequent approximations.

Equation 12.9, $\ddot{\theta} + \omega_n^2 \theta + \varepsilon \theta^3 = 0$, becomes

$$\omega^2 \theta'' + \omega_n^2 \theta + \varepsilon \theta^3 = 0, \qquad (12.13)$$

where primes denote differentiation with respect to τ, $\omega_n^2 = \sqrt{g/l}$, and $\varepsilon = -\omega_n^2/6$. For $\varepsilon < 0$, the spring is soft and for $\varepsilon > 0$, the spring is hard. One can assume the initial conditions $\theta(\varepsilon, 0) = a_0$ and $\theta'(\varepsilon, 0) = 0$. For an autonomous system, we can allow the first order solution $\theta_0(\tau)$ to satisfy these conditions: $\theta_0(\varepsilon, 0) = a_0$ and $\theta_0'(\varepsilon, 0) = 0$ with the remaining orders

---
[9]**Nonlinear Oscillations**, N. Minorsky, Krieger Publishing Company, 1987. Reprint of original edition of 1962.

12.4. PERTURBATION METHODS

satisfying zero initial conditions $\theta_i(\varepsilon, 0) = a_0$ and $\theta'_i(\varepsilon, 0) = 0$, $i = 1, 2, \ldots$. For non–autonomous systems, the arbitrary constants of integration must be certain values so that periodic solutions exist. Not all initial conditions lead to periodic solutions.

We next expand θ and ω, and consider a linear approximation (only ε^1 terms are retained in addition to the linear terms),

$$\theta(\varepsilon, \tau) = \theta_0(\tau) + \varepsilon \theta_1(\tau) + \cdots$$
$$\omega = \omega_n + \varepsilon \omega_1 + \cdots.$$

Note that the oscillation frequency reduces to ω_n for $\varepsilon = 0$. Substitute these expansions into Equation 12.13, to find

$$(\omega_n + \varepsilon\omega_1)^2 (\theta_0 + \varepsilon\theta_1)'' + \omega_n^2 (\theta_0 + \varepsilon\theta_1) + \varepsilon (\theta_0 + \varepsilon\theta_1)^3 = 0.$$

This equation is expanded and grouped according to powers of ε,

$$\varepsilon^0 \;:\; \omega_n^2 \theta_0'' + \omega_n^2 \theta_0 = 0 \tag{12.14}$$
$$\varepsilon^1 \;:\; \omega_n^2 \theta_1'' + \omega_n^2 \theta_1 = -\theta_0^3 - 2\omega_n \omega_1 \theta_0'' \tag{12.15}$$
$$\vdots$$

We only retain the first two equations since a two term expansion is utilized. Note how such sets of equations are solved iteratively. Equation 12.14 is solved for $\theta_0(\tau)$ and then its powers and derivatives substituted into Equation 12.15 which is then solved for $\theta_1(\tau)$. Also note that each equation is linear and the nonlinear effects are on the right hand side, acting as inputs to the system. Linear theory is used to solve this sequence of equations. Simplify Equations 12.14 and 12.15,

$$\theta_0'' + \theta_0 = 0$$
$$\theta_1'' + \theta_1 = -\frac{1}{\omega_n^2}\theta_0^3 - 2\frac{\omega_1}{\omega_n}\theta_0''.$$

The solution to the first equation is $\theta_0(\tau) = a_0 \cos\tau$, after satisfying the initial conditions. The second equation becomes[10]

$$\theta_1'' + \theta_1 = \left(-\frac{3}{4}\frac{a_0^3}{\omega_n^2} + 2\frac{\omega_1 a_0}{\omega_n}\right)\cos\tau - \frac{a_0^3}{4\omega_n^2}\cos 3\tau.$$

---
[10] Use the trigonometric identity

$$\cos^3\tau = \frac{3}{4}\cos\tau + \frac{1}{4}\cos 3\tau.$$

To remove secular terms, the coefficient of the resonant loading $\cos\tau$ must be set to zero. This leads to an equation for ω_1,

$$\omega_1 = \frac{3}{8}\frac{a_0^2}{\omega_n}.$$

Therefore,

$$\theta_1(\tau) = a_1\cos\tau + b_1\sin\tau + \frac{a_0^3}{32\omega_n^2}\cos 3\tau.$$

a_1 and b_1 are found by applying the zero initial conditions for $\theta_1(\tau)$, leading to $a_1 = -a_0^3/32\omega_n^2$ and $b_1 = 0$. The approximate solution for $\theta(\varepsilon,\tau)$ is then

$$\theta(\varepsilon,\tau) = a_0\cos\tau - \varepsilon\frac{a_0^3}{32\omega_n^2}(\cos\tau - \cos 3\tau) + O(\varepsilon^2).$$

Transforming back to the t domain,

$$\theta(\varepsilon,t) = a_0\cos\omega t - \varepsilon\frac{a_0^3}{32\omega_n^2}(\cos\omega t - \cos 3\omega t) + O(\varepsilon^2)$$

$$\omega = \omega_n + \varepsilon\frac{3}{8}\frac{a_0^2}{\omega_n} + O(\varepsilon^2).$$

These equations provide a measure of the effect of the system nonlinearity on system response and frequency of response.

On When to Expand ω

There may be some ambiguity on when a second parameter expansion is needed. Sometimes we see the expansion in ω and other times not. Essentially, one expands parameters where their value is unknown. Of course, the oscillator response amplitude is unknown and expanded. For the autonomous oscillator the period and thus frequency are unknown, and ω is therefore expanded.

In the following sections, forced nonlinear oscillators are studied with the purpose of finding periodic oscillations at the forcing period or frequency. Thus there is no need to expand ω.

One of the interesting and challenging aspects of nonlinear equations is that they generally have numerous solutions. Depending on what the analyst needs to find out about the behavior of the nonlinear oscillator, different techniques may be appropriate. Here we are only interested in periodic solutions of slightly nonlinear oscillators.

12.4. PERTURBATION METHODS 661

Poincaré

Jules Henri Poincaré was born on 29 April 1854 in Nancy, Lorraine, France and died on 17 July 1912 in Paris, France. Poincaré entered the École Polytechnique in 1873, graduating in 1875. He was well ahead of all the other students in mathematics. His memory was remarkable and he retained much from all the texts he read but not in the manner of learning by rote, rather by linking the ideas he was assimilating particularly in a visual way. His ability to visualize what he heard proved particularly useful when he attended lectures since his eyesight was so poor that he could not see properly what his lecturers were writing on the blackboard.

Upon graduating from the École Polytechnique Poincaré continued his studies at the École des Mines. After completing his studies there, Poincaré spent a short while as a mining engineer at Vesoul while completing his doctoral work. As a student of Charles Hermite, Poincaré received his doctorate in mathematics from the University of Paris in 1879. His thesis was on differential equations and the examiners were somewhat critical of the work. In 1886 Poincaré was nominated for the chair of mathematical physics and probability at the Sorbonne. The intervention and support of Hermite was to ensure that Poincaré was appointed to this chair and to a chair at the École Polytechnique.

Before looking briefly at the many contributions that Poincaré made to mathematics and to other sciences, we should say a little about his way of thinking and working. He is considered as one of the great geniuses of all time. Poincaré kept very precise working hours. He undertook mathematical research for four hours a day, between 10 am and noon then again from 5 pm to 7 pm. He would read articles in journals later in the evening. An interesting aspect of Poincaré's work is that he tended to develop his results from first principles. For many mathematicians there is a building process with more and more being built on top of the previous work. This was not the way that Poincaré worked and not only his research, but also his lectures and books, were all developed carefully from basics.

Poincaré was a scientist preoccupied by many aspects of mathematics, physics and philosophy, and he is often described as the last universalist in mathematics. He made contributions to numerous branches of mathematics, celestial mechanics, fluid mechanics, the special theory of relativity and the philosophy of science. Much of his research involved interactions between different mathematical topics and his broad understanding of the whole spectrum of knowledge allowed him to attack problems from many different angles.

In applied mathematics he studied optics, electricity, telegraphy, capillarity, elasticity, thermodynamics, potential theory, quantum theory, theory of relativity and cosmology. In the field of celestial mechanics he studied the three-body-problem, as well as the theories of light and of electromagnetic waves. He is acknowledged as a codiscoverer, with Albert Einstein and Hendrik Lorentz, of the special theory of relativity.

After Poincaré achieved prominence as a mathematician, he turned his superb literary gifts to the challenge of describing for the general public the meaning and importance of science and mathematics. Poincaré's popular works include *Science and Hypothesis* (1901), *The Value of Science* (1905), and *Science and Method* (1908).

Poincaré achieved the highest honors for his contributions of true genius. He was elected to the Académie des Sciences in 1887 and in 1906 was elected President of the Academy. The breadth of his research led to him being the only member elected to every one of the five sections of the Academy, namely the geometry, mechanics, physics, geography and navigation sections. In 1908 he was elected to the Académie Francaise and was elected director in the year of his death. He was also made chevalier of the Légion d'Honneur and was honored by a large number of learned societies around the world. He won numerous prizes, medals and awards.

12.4.2 Forced Oscillations of Quasi–Harmonic Systems

Consider the forced Duffing equation,

$$\ddot{x} + \omega_n^2 x = \varepsilon \left[-\omega_n^2 \left(\alpha x + \beta x^3 \right) + F \cos \Omega t \right], \quad \varepsilon \ll 1, \qquad (12.16)$$

where α, β are given constant parameters, $\omega_n^2 = k/m$, and the harmonic forcing is $\varepsilon F \cos \Omega t$. In this case, the harmonic excitation is of small magnitude. It is expected therefore that the response will be almost or *quasi–harmonic*. Therefore, it is of interest to determine the circumstances under which response $x(t)$ is periodic with period $T = 2\pi/\Omega$. An expansion in the frequency of response is therefore not necessary.

This problem is approached using the perturbation method, with the following change of variables,

$$\Omega t = \tau + \phi$$
$$\frac{d}{dt} = \Omega \frac{d}{d\tau},$$

where τ is the new time variable and ϕ is a phase angle. This transformation results in a time scale with a period of oscillation of 2π. With these transformations, Equation 12.16 becomes

$$\Omega^2 x'' + \omega_n^2 x = \varepsilon \left[-\omega_n^2 \left(\alpha x + \beta x^3 \right) + F \cos \left(\tau + \phi \right) \right], \quad \varepsilon \ll 1, \qquad (12.17)$$

where primes denote differentiation with respect to τ. Recalling that secular terms must be removed, as they are non–physical, requires that the solution to Equation 12.17 be periodic, $x(\tau + 2\pi) = x(\tau)$. Assume also that $x(0) = C_0$, and for convenience that $x'(0) = 0$. The value of C_0 is not arbitrary, but is determined in terms of other system parameters below. The perturbation solution is based on the expansions of $x(\tau)$ and ϕ,

$$x(\tau) = x_0(\tau) + \varepsilon x_1(\tau) + \varepsilon^2 x_2(\tau) + \cdots$$
$$\phi = \phi_0 + \varepsilon \phi_1 + \varepsilon_2^2 \phi + \cdots,$$

12.4. PERTURBATION METHODS

where $x_i(\tau + 2\pi) = x_i(\tau)$, $x_i'(0) = 0$, for all i, and $x_0(0) = C_0$. These expansions are substituted into Equation 12.17, and equating the coefficients of like powers of ε, we find,

$$\Omega^2 x_0'' + \omega_n^2 x_0 = 0 \qquad (12.18)$$
$$\Omega^2 x_1'' + \omega_n^2 x_1 = -\omega_n^2 \left(\alpha x_0 + \beta x_0^3\right) + F\cos(\tau + \phi_0) \qquad (12.19)$$
$$\Omega^2 x_2'' + \omega_n^2 x_2 = -\omega_n^2 \left(\alpha x_1 + 3\beta x_0^2 x_1\right) + F\cos(\tau + \phi_1) \qquad (12.20)$$
$$\vdots$$

These equations are solved in sequence for $x_i(\tau)$, for all i, applying the periodicity and initial conditions. The solution to Equation 12.18 is

$$x_0(\tau) = C_0 \cos\frac{\omega_n}{\Omega}\tau,$$

where C_0 is a constant amplitude. 2π-periodicity in τ must be satisfied; this is possible only if $\Omega = \omega_n$. This substitution is made in the subsequent Equations 12.19 and 12.20. Also substitute for $x_0(\tau)$ in Equation 12.19 to find

$$x_1'' + x_1 = -\left(\alpha C_0 \cos\tau + \beta C_0^3 \cos^3\tau\right) + \frac{F}{\omega_n^2}\cos(\tau + \phi_0).$$

This expression can be simplified using the trigonometric relation $\cos^3\tau = (3\cos\tau + \cos 3\tau)/4$,

$$\begin{aligned}x_1'' + x_1 &= -\alpha C_0 \cos\tau - \beta C_0^3 \frac{1}{4}(3\cos\tau + \cos 3\tau)\\ &\quad + \frac{F}{\omega_n^2}(\cos\tau\cos\phi_0 - \sin\tau\sin\phi_0)\\ &= -\frac{F}{\omega_n^2}\sin\phi_0 \sin\tau - \left(\alpha C_0 + \frac{3}{4}\beta C_0^3 - \frac{F}{\omega_n^2}\cos\phi_0\right)\cos\tau\\ &\quad -\frac{1}{4}\beta C_0^3 \cos 3\tau.\end{aligned}$$

To avoid secular terms requires that the coefficients of $\sin\tau$ and $\cos\tau$ be set to zero, yielding conditions and relations for the respective parameters for periodicity. There are two sets of expressions that satisfy this condition: $\phi_0 = 0$, and $\phi_0 = \pi$,

$$\phi_0 = 0, \quad \alpha C_0 + \frac{3}{4}\beta C_0^3 - \frac{F}{\omega_n^2} = 0 \qquad (12.21)$$

$$\phi_0 = \pi, \quad \alpha C_0 + \frac{3}{4}\beta C_0^3 + \frac{F}{\omega_n^2} = 0, \qquad (12.22)$$

where C_0 is now determined, since α, β and F are all known. Equations 12.22 do not offer additional information beyond that of Equations 12.21. Equations 12.22 tell us that for phase $\phi_0 = \pi$, the response and the forcing are 180° out of phase. This is the same as being in phase with response of negative amplitude.

We proceed with Equations 12.21 to solve for $x_1(\tau)$,

$$x_1(\tau) = C_1 \cos \tau + \frac{1}{32} \beta C_0^3 \cos 3\tau. \tag{12.23}$$

The value of C_1 is determined based on the required periodicity of $x_2(\tau)$, just as C_0 was determined based on the periodicity of $x_1(\tau)$. Equation 12.23 is substituted into Equation 12.20, and utilizing a number of trigonometric identities, a set of expressions are derived that must be examined for the possibility of secular responses. Suppose only $x_1(\tau)$ is retained, then to order ε, or $O(\varepsilon)$,

$$\begin{aligned} x(\tau) &\simeq x_0(\tau) + \varepsilon x_1(\tau) \\ &= C_0 \cos \tau + \varepsilon \left(C_1 \cos \tau + \frac{1}{32} \beta C_0^3 \cos 3\tau \right), \end{aligned}$$

where $\phi \simeq \phi_0 + \varepsilon \phi_1 = 0$, since the ε^2 solution yields $\phi_1 = 0$. All the phase angle terms are zero since there is no damping in the system. This is recalled from our studies of linear undamped oscillators, and we could have avoided the ϕ expansion.

Equation 12.21 reminds us of the frequency response function $H(i\omega)$, Equation 3.23. Both functions are relations between response amplitude and forcing amplitude. Equation 12.21 is the analogous nonlinear relation. Introduce the relation

$$\omega_0^2 \equiv (1 + \varepsilon \alpha) \omega_n^2, \tag{12.24}$$

so that Equation 12.16 becomes

$$\ddot{x} + \omega_0^2 x + \varepsilon \omega_n^2 \beta x^3 = \varepsilon F \cos \Omega t, \quad \varepsilon \ll 1.$$

Solve Equation 12.24 for α and substitute into the second of Equations 12.21. Solve this equation for ω_n^2 to find

$$\omega_n^2 = \omega_0^2 \left(1 + \frac{3}{4} \varepsilon \beta C_0^2 \right) - \varepsilon \frac{F}{C_0}. \tag{12.25}$$

In deriving this equation, the following approximation is made

$$\left(1 - \frac{3}{4} \varepsilon \beta C_0^2 \right)^{-1} \simeq \left(1 + \frac{3}{4} \varepsilon \beta C_0^2 \right) \tag{12.26}$$

12.4. PERTURBATION METHODS

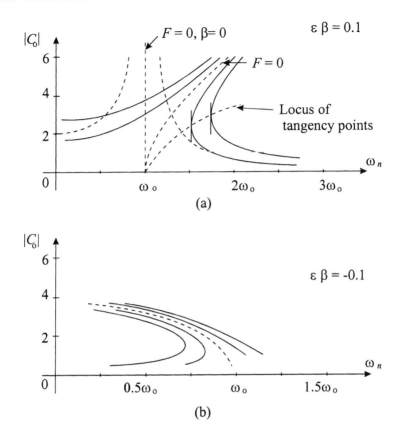

Figure 12.8: Amplitude curves for the Duffing equation for two values of $\varepsilon\beta$. In Figure (a), the dotted lines for $\beta = 0$ represent the linear problem.

based on the Taylor expansion for $1/(1+x) \simeq 1 - x$ for small x, and also a term of order ε^2 is dropped. Equation 12.25 can be plotted as $|C_0|$ versus ω_n, where ω_n is measured in units of ω_0. $\varepsilon\beta$ is given and εF is a parameter. Representative curves are shown in Figure 12.8 for $\varepsilon\beta = \pm 0.1$. Note that the parametric curves for a single case appear in pairs, on either side of the $F = 0$ curve.

Quite a wealth of information can be discerned from this set of curves. The sign of $\varepsilon\beta$ determines whether the curves slant to the right, $\varepsilon\beta > 0$, for a hardening spring, or to the left, $\varepsilon\beta < 0$, for a softening spring. Unlike for the linear case, there is no resonance condition where, for an undamped oscillator, the amplitudes grow ever larger around the natural frequency of the system.

Consider the line labeled as the locus of tangency points. A vertical line to the right of this line intersects the curves at two locations on one branch and at one location on the mirror branch. Any intersection represents a real root, and so this vertical line indicates that there are three possible amplitudes corresponding to a given excitation force amplitude. A vertical line to the left has only one real root, but also two complex roots. Following a representative curve as ω_0 increases results in ever larger amplitudes.

12.4.3 Jump Phenomenon

Of course, all systems have some damping, and it is interesting to consider a lightly damped Duffing oscillator to examine the significant changes that are predicted. Consider the governing equation

$$\ddot{x} + \omega_n^2 x = \varepsilon \left[-2\zeta \omega_n \dot{x} - \omega_n^2 \left(\alpha x + \beta x^3 \right) + F \cos \Omega t \right], \quad \varepsilon \ll 1,$$

where the small damping term $-2\varepsilon \zeta \omega_n \dot{x}$ has been added. (Note that α is dimensionless and β has units of x^{-2}.) Following the same procedure as before, the comparable equation to Equation 12.25 is found,

$$\left[\omega_0^2 \left(1 + \frac{3}{4} \varepsilon \beta C_0^2 \right) - \omega_n^2 \right]^2 + \left(2\varepsilon \zeta \omega_0^2 \right)^2 = \left(\frac{\varepsilon F}{C_0} \right)^2.$$

A plot of this equation shows similar, but significantly different curves. See Figure 12.9. The difference is that the two branches connect at some location. Compare with Figure 12.8.

This is significant because it means that an increase in the frequency does not always lead to an increase in amplitude. At some value, there will be a drop in amplitude to the other solution branch. This is clearly seen if one follows path 4 for increasing frequency, with a drop at point 1 to point 2. Similarly, for decreasing frequency, at location 3 there is a jump in amplitude to point 4. The path from point 1 to point 3 is unstable.

This behavior is called the *jump phenomenon*. It exists for nonlinear systems with damping. There are practical implications. For example, as a motor runs up to operating frequency, the designer will need to make sure that there are no jumps in amplitude.

12.4.4 Periodic Solutions of Non–Autonomous Systems

Consider the system governed by

$$\ddot{u} + \omega_n^2 u = F_0 \cos t + \varepsilon f \left(u, \dot{u}, t, \varepsilon \right),$$

12.4. PERTURBATION METHODS

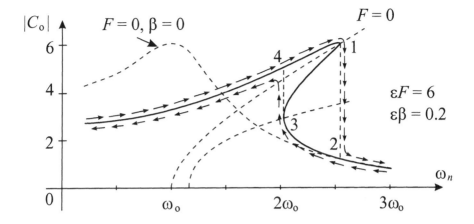

Figure 12.9: Jump phenomenon. The path between points 1 and 3 is unstable.

where the force is not small but the nonlinearities are of order ε. When $\varepsilon = 0$ and $F_0 = 0$, the nonlinear equation reduces to the simple harmonic oscillator with frequency ω_n and period $T = 2\pi/\omega_n$, while the forcing frequency is unity and of period 2π. For this case of $\varepsilon = 0$, the solution of the first order solution u_0 is

$$u_0 = C_0 \cos \omega_n (t - t_0) + \frac{F_0}{\omega_n^2 - 1} \cos t, \quad \omega_n^2 \neq 1.$$

C_0 and t_0 are arbitrary constants. There are two ways the solution can have the same 2π-period as the forcing term; ω_n must be an integer. Otherwise, if ω_n is an irrational number then the solution u_0 can be periodic only if $C_0 = 0$, in which case the period is automatically 2π, like the forcing.

Forced Pendulum

Nonlinear problems exhibit a coupling between the free and forced response and the initial conditions. The arbitrary constants of integration do not only specify the allowable initial conditions for periodic solutions, but are a result of enforcing periodicity.

Consider the *forced pendulum*, governed by the equation

$$\ddot{\theta} + \omega_n^2 \sin \theta = F \cos \omega t.$$

An approximate nonlinear equation that is representative of the fully nonlinear pendulum equation is

$$\ddot{\theta} + \omega_n^2 \left(\theta - \frac{\theta^3}{6}\right) = F \cos \omega t.$$

Consider a more general version of this equation. To do this first nondimensionalize the variables using the transformation $\tau = \omega t$. The nonlinear governing equation then becomes

$$\omega^2 \theta'' + \omega_n^2 \theta - \frac{1}{6}\omega_n^2 \theta^3 = \Gamma \omega^2 \cos \tau$$

$$\text{or} \quad \theta'' + \Omega^2 \theta - \frac{1}{6}\Omega^2 \theta^3 = \Gamma \cos \tau,$$

where $\Gamma \omega^2 = F$ and $\Omega = \omega_n/\omega$. To further generalize,[11] define $\varepsilon \equiv -\Omega^2/6$. Apply the perturbation expansion to

$$\theta'' + \Omega^2 \theta + \varepsilon \theta^3 = \Gamma \cos \tau, \tag{12.27}$$

reducing to the linear problem for $\varepsilon = 0$. We look for solutions of the same period as the driving force, which has 2π-periodicity.

The expansion is taken to be a function of ε and τ in the following way,

$$\theta(\varepsilon, \tau) \equiv \theta_0(\tau) + \varepsilon \theta_1(\tau) + \varepsilon^2 \theta_2(\tau) + \cdots, \tag{12.28}$$

where the periodicity condition $\theta_i(\tau + 2\pi) = \theta_i(\tau)$, $i = 0, 1, 2, \ldots$ removes any secular terms from the solution. If this approach is to work in a practical sense, it is necessary that $\varepsilon \ll 1$ such that $\theta_0(\tau) \gg \varepsilon \theta_1(\tau) \gg \varepsilon^2 \theta_2(\tau) \ldots$. Then we can retain the expansion to order ε or ε^2 and obtain a good approximation. What has been discussed so far makes no assumptions about the character of the forcing. An expansion in ω is not necessary here since oscillations with known period are sought.

The procedure is to substitute the expression for $\theta(\varepsilon, \tau)$ and its second derivative with respect to τ, $\theta''(\varepsilon, \tau)$, into Equation 12.27, with the result

$$\left(\theta_0'' + \varepsilon \theta_1'' + \varepsilon^2 \theta_2'' + \cdots\right) + \Omega^2 \left(\theta_0 + \varepsilon \theta_1 + \varepsilon^2 \theta_2 + \cdots\right)$$
$$+ \varepsilon \left(\theta_0 + \varepsilon \theta_1 + \varepsilon^2 \theta_2 + \cdots\right)^3 = \Gamma \cos \tau.$$

Note that when the cubic term is expanded, expressions of order ε^3 and ε^4 and higher are found.[12] Retain a solution to order ε^2 and drop all higher order terms, thus,

$$\varepsilon \left(\theta_0 + \varepsilon \theta_1 + \varepsilon^2 \theta_2\right)^3 \simeq \varepsilon \left(\theta_0^3 + 3\theta_0^2 \varepsilon \theta_1\right).$$

---

[11] We can define ε as a positive quantity and retain the negative sign in the equation of motion. Of course the results are the same, but sometimes retaining the physical meaning of ε as a frequency is desirable and then $\varepsilon > 0$.

[12] The expansion of the cubic term is

12.4. PERTURBATION METHODS

Since the θ_i are independent of ε, and the above must be satisfied for all ε, the expanded differential equation is actually a sequence of differential equations that can be identified by equating terms of the same order of ε,

$$\varepsilon^0 \;:\; \theta_0'' + \Omega^2 \theta_0 = \Gamma \cos \tau \qquad (12.29)$$
$$\varepsilon^1 \;:\; \theta_1'' + \Omega^2 \theta_1 = -\theta_0^3 \qquad (12.30)$$
$$\varepsilon^2 \;:\; \theta_2'' + \Omega^2 \theta_2 = -3\theta_0^2 \theta_1. \qquad (12.31)$$

The approximate solution is then given by Equation 12.28, with an error of $O\left(\varepsilon^3\right)$. We can immediately observe the major advantage to this expansion technique is the transformation of a nonlinear equation into a set of linear equations that are solved in sequence. The linear equations are all the same except for the input, and they can be solved using the convolution integral. Proceeding through this iterative solution, we will utilize our physical understanding of the system to help us in rejecting parts of the solution which do not fit our understanding. This will become clearer as we proceed to solve the example.

Solve Equations 12.29–12.31 in sequence. The solution to the ε^0 order governing equation is

$$\theta_{0h}(\tau) = a_0 \cos \Omega \tau + b_0 \sin \Omega \tau,$$

where the subscript h denotes homogeneous solution, and a_0 and b_0 are arbitrary constants the values of which are established below by enforcing the required periodicity. The particular solution is given by

$$\theta_{0p}(\tau) = \frac{\Gamma}{\Omega^2 - 1} \cos \tau.$$

The complete solution is the sum of the homogeneous and particular solutions. The particular solution is not valid at or near the resonant condition $\Omega = 1$. This case is solved separately.[13] We also are interested here, for the sake of brevity, only at solutions that have the period of the loading, that is $T = 2\pi/1$ sec. Other solutions exist, some of which are examined in the next section.

$$()^3 = \theta_0^3 + 3\theta_0^2 \varepsilon \theta_1 + 3\theta_0^2 \varepsilon^2 \theta_2 + 3\theta_0 \varepsilon^2 \theta_1^2$$
$$+ 6\theta_0 \varepsilon^3 \theta_1 \theta_2 + 3\theta_0 \varepsilon^4 \theta_2^2 + \varepsilon^3 \theta_1^3 + 3\varepsilon^4 \theta_1^2 \theta_2 + 3\varepsilon^5 \theta_1 \theta_2^2 + \varepsilon^6 \theta_2^3.$$

[13] Assume $\Omega \simeq 1$, and
$$\Omega^2 = 1 + \varepsilon \beta, \qquad (12.32)$$
with $\Gamma = \varepsilon \gamma$, and expand as before.

If the response must have the period 2π, then $a_0 = b_0 = 0$, and $\theta_0(\tau) = \theta_{0p}(\tau)$. This function is cubed and becomes the forcing for the equation governing $\theta_1(\tau)$,

$$\begin{aligned}\theta_0^3 &= \left(\frac{\Gamma}{\Omega^2 - 1}\cos\tau\right)^3 \\ &= \left(\frac{\Gamma}{\Omega^2 - 1}\right)^3 \left(\frac{3}{4}\cos\tau + \frac{1}{4}\cos 3\tau\right).\end{aligned}$$

The expansion of $\cos^3\tau$ is important because it helps identify the various harmonic components contained therein. Again ignoring the homogeneous solution, the particular solution is

$$\begin{aligned}\theta_{1p}(\tau) &= a_1\cos\tau + b_1\cos 3\tau; \quad \Omega \neq 1, 3, \\ a_1 &= \frac{3\Gamma^3}{4(\Omega^2 - 1)^4} \\ b_1 &= \frac{\Gamma^3}{4(\Omega^2 - 1)(\Omega^2 - 9)}.\end{aligned}$$

For the next term in the series, it is necessary to evaluate the input $3\theta_0^2\theta_1$. This step is omitted here. The truncated solution correct to order ε is

$$\theta(\varepsilon, \tau) \simeq \theta_0(\tau) + \varepsilon\theta_1(\tau) + O(\varepsilon^2),$$

where expressions for $\theta_0(\tau)$ and $\theta_1(\tau)$ are known, and $\varepsilon = -\Omega^2/6$. The variables can now be transformed back to the physical ones.

Arbitrary Forcing

Now consider the possibility that the forcing is general, $f(t)$. Begin with the general nonlinear governing equation

$$\ddot{\theta} + \alpha\dot{\theta} + \omega_n^2\theta + \varepsilon g\left(\theta, \dot{\theta}\right) = f(t),$$

with the expansion solution

$$\theta(\varepsilon, t) = \theta_0(t) + \varepsilon\theta_1(t) + \varepsilon^2\theta_2(t) + \ldots,$$

where $\theta_0(t)$ satisfies the linear differential equation. Before substituting the expansion into the governing equation, the nonlinear function $g\left(\theta, \dot{\theta}\right)$

12.4. PERTURBATION METHODS

is itself expanded, *about the linear solution* $\left(\theta_0, \dot{\theta}_0\right)$, so that we can keep track of the various orders in ε,

$$\begin{aligned}
g\left(\theta, \dot{\theta}\right) &\equiv g\left(\theta_0 + \varepsilon\theta_1 + \varepsilon^2\theta_2 + \ldots, \dot{\theta}_0 + \varepsilon\dot{\theta}_1 + \varepsilon^2\dot{\theta}_2 + \ldots\right) \\
&= g\left(\theta_0, \dot{\theta}_0\right) + \left(\varepsilon\theta_1 + \varepsilon^2\theta_2 + \ldots\right)\frac{\partial}{\partial \theta}g\left(\theta_0, \dot{\theta}_0\right) \\
&\quad + \left(\varepsilon\dot{\theta}_1 + \varepsilon^2\dot{\theta}_2 + \ldots\right)\frac{\partial}{\partial \dot{\theta}}g\left(\theta_0, \dot{\theta}_0\right) \\
&\quad + higher\ order\ terms.
\end{aligned}$$

Retaining terms to order ε^2, and equating terms of equal power in ε, the following sequence is generated,

$$\varepsilon^0 \ : \ \ddot{\theta}_0 + \alpha\dot{\theta}_0 + \omega_n^2\theta_0 = f(t)$$
$$\varepsilon^1 \ : \ \ddot{\theta}_1 + \alpha\dot{\theta}_1 + \omega_n^2\theta_1 = -g\left(\theta_0, \dot{\theta}_0\right)$$
$$\varepsilon^2 \ : \ \ddot{\theta}_2 + \alpha\dot{\theta}_2 + \omega_n^2\theta_2 = -\theta_1\frac{\partial}{\partial x}g\left(\theta_0, \dot{\theta}_0\right) - \dot{\theta}_1\frac{\partial}{\partial \dot{x}}g\left(\theta_0, \dot{\theta}_0\right).$$

As a practical matter, it becomes very difficult to solve more than a two or three term approximation. The solutions for θ_0 and θ_1 are now obtained in sequence using the convolution integral. Proceeding for a general nonlinear function $g\left(\theta, \dot{\theta}\right)$,

$$\theta_0(t) = \int_{-\infty}^{\infty} f(t-\tau) h(\tau) d\tau$$
$$\theta_1(t) = -\int_{-\infty}^{\infty} g\left[\theta_0(t-\tau), \dot{\theta}_0(t-\tau)\right] h(\tau) d\tau$$
$$\vdots$$

and then $\theta(t) \simeq \theta_0(t) + \varepsilon\theta_1(t)$. The response is approximately given by

$$\theta(t) \simeq \int_{-\infty}^{\infty} f(t-\tau) h(\tau) d\tau$$
$$- \int_{-\infty}^{\infty} g\left[\theta_0(t-\tau), \dot{\theta}_0(t-\tau)\right] h(\tau) d\tau.$$

The above expansion searches for oscillations at the forcing frequency. One can also transform the differential equation and introduce a response frequency that is also expanded as in the Lindstedt expansion.

12.5 Subharmonic and Superharmonic Oscillations

Linearly oscillating systems will respond at the frequency of the forcing.[14] A nonlinear system will respond at subharmonic and superharmonic frequencies as well. One may ask why these harmonics exist. As Stoker[15] states, "... it is not an entirely simple matter to give a plausible physical explanation for their occurrence."

A *subharmonic* response involves oscillations of frequencies ω_m that are related to the forcing frequency Ω by the equation

$$\omega_m = \frac{\Omega}{m}, \quad m = 2, 3, \ldots.$$

Similarly, a *superharmonic* response involves oscillations of frequencies ω_n that are related to the forcing frequency Ω by the equation

$$\omega_n = n\Omega, \quad n = 2, 3, \ldots.$$

In addition, the nonlinear system loaded by a combination of forces at two distinct frequencies, say Ω_1 and Ω_2, responds at various combinations of those frequencies. Several cases are studied in this section[16] to demonstrate possible behavior.

It is noteworthy that additional periodic solutions appear for forcing frequencies *nearly equal* to a rational number, not only for forcing frequencies exactly equal to a rational number. Furthermore, not every rational number produces this effect, but only rational numbers that are ratios of small integers, for example, 1, 1/2, 1/3, 2/3.

---

[14] For an example when this is not the case, see Jordan and Smith, pages 195-196.

[15] p.103: "... it is not an entirely simple matter to give a plausible physical explanation for their occurrence. Let us recall the behavior of linear systems. If the frequency of the free oscillation of a linear system is ω_n/m, where m is an integer, say, then a periodic external force of frequency ω_n can excite the free oscillation in addition to the forced oscillation of frequency ω_n. Why should the situation be different for a nonlinear system? The explanation usually offered is as follows: Any free oscillation of a nonlinear system contains the higher harmonics in profusion, and hence it is possible that an external force with a frequency the same as one of these might be able to excite and sustain the harmonic of lower frequency. Of course that this actually should occur probably requires that the damping be not too great and that proper precautions of various kinds be taken."
One may add that a similar argument can be made for harmonics of higher frequency, but in this case, the force may have to be of larger amplitude.

[16] This section follows the outline of **Fundamentals of Vibrations** by L. Meirovitch, McGraw–Hill 2001.

12.5.1 Subharmonics

When a nonlinear equation is forced, the generation of "alien" harmonics by the nonlinear terms may cause a stable subharmonic to appear for a range of parameters, and in particular for a range of applied frequencies. The forcing amplitude also plays a part in generating and sustaining the subharmonic even in the presence of damping. Thus there will exist the tolerance of slightly varying conditions necessary for the consistent appearance of a subharmonic and its use in physical systems.

The undamped Duffing oscillator also has a periodic solution with the fundamental frequency equal to one third the driving frequency. Begin with Equation 12.16,

$$\ddot{x} + \omega_n^2 x = -\varepsilon \omega_n^2 \left(\alpha x + \beta x^3 \right) + F \cos \Omega t, \quad \varepsilon \ll 1,$$

except allow the force amplitude to be not necessarily small. Let $\omega_n = \Omega/3$, and assume the expansion in t,

$$x(t) = x_0(t) + \varepsilon x_1(t) + \varepsilon^2 x_2(t) + \cdots.$$

There is no need to expand the phase for the undamped oscillator. Following the earlier procedure yields

$$\ddot{x}_0 + \left(\frac{\Omega}{3}\right)^2 x_0 = F \cos \Omega t \quad (12.33)$$

$$\ddot{x}_1 + \left(\frac{\Omega}{3}\right)^2 x_1 = -\left(\frac{\Omega}{3}\right)^2 \left(\alpha x_0 + \beta x_0^3\right) \quad (12.34)$$

$$\ddot{x}_2 + \left(\frac{\Omega}{3}\right)^2 x_2 = -\left(\frac{\Omega}{3}\right)^2 \left(\alpha x_1 + 3\beta x_0^2 x_1\right) \quad (12.35)$$

$$\vdots$$

Equations 12.33–12.35 are solved in sequence subject to the required periodicity conditions,

$$x_i \left(\frac{\Omega}{3} t + 2\pi \right) = x_i \left(\frac{\Omega}{3} t \right),$$

and the assumed initial conditions, $x_i'(0) = 0$, for all i. The solution to Equation 12.33 is

$$x_0(t) = C_0 \cos \frac{\Omega}{3} t - \frac{9F}{8\Omega^2} \cos \Omega t. \quad (12.36)$$

This equation is substituted into the governing equation for $x_1(t)$. After expanding the x_0^3 term and simplifying, to ensure the periodic solution, the term multiplying the term $\cos(\Omega t/3)$ is set equal to zero, leading to the equation

$$C_0^2 - \frac{9F}{8\Omega^2}C_0 + 2\left(\frac{9F}{8\Omega^2}\right)^2 + \frac{4\alpha}{3\beta} = 0, \qquad (12.37)$$

a quadratic equation with the roots

$$C_0 = \frac{1}{2}\frac{9F}{8\Omega^2} \pm \frac{1}{2}\sqrt{-7\left(\frac{9F}{8\Omega^2}\right)^2 - \frac{16\alpha}{3\beta}},$$

where C_0 is a real number. The expression under the radical must be greater than or equal to zero. Additionally, introduce

$$\omega_0^2 \equiv (1+\varepsilon\alpha)\,\omega_n^2 = (1+\varepsilon\alpha)\,\frac{\Omega^2}{9}$$
$$\text{or} \quad \Omega^2 \simeq 9\omega_0^2\,(1-\varepsilon\alpha), \qquad (12.38)$$

where $(1+\varepsilon\alpha)^{-1} \simeq (1-\varepsilon\alpha)$, as per the discussion around Equation 12.24, here with $\omega_n = \Omega/3$.

Such oscillations are called subharmonic of order 3. The order coincides with the power of the nonlinearity in the spring restoring force.

12.5.2 Combination Harmonics

In linear vibration, an oscillator subjected to two forces of distinct frequencies responds in a superposition of responses, one for each loading. The nonlinear counterpart is examined here. In addition to the uncoupled responses there are responses that are integer multiples of the driving frequencies as well as linear combinations of the driving frequencies. Consider the following Duffing equation,

$$\ddot{x} + \omega_n^2 x = -\varepsilon\beta\omega_n^2 x^3 + F_1\cos\Omega_1 t + F_2\cos\Omega_2 t, \quad \varepsilon \ll 1.$$

Assume the expansion solution and equate terms of equal order to find,

$$\ddot{x}_0 + \omega_n^2 x_0 = F_1\cos\Omega_1 t + F_2\cos\Omega_2 t \qquad (12.39)$$
$$\ddot{x}_1 + \omega_n^2 x_1 = -\beta\omega_n^2 x_0^3 \qquad (12.40)$$
$$\ddot{x}_2 + \omega_n^2 x_2 = -3\beta\omega_n^2 x_0^2 x_1$$
$$\vdots$$

12.6. THE MATHIEU EQUATION

Consider only the steady state response. The solution to Equation 12.39 is

$$x_0(t) = G_1 \cos \Omega_1 t + G_2 \cos \Omega_2 t, \qquad (12.41)$$

$$G_1 = \frac{F_1}{\omega_n^2 - \Omega_1^2}$$

$$G_2 = \frac{F_2}{\omega_n^2 - \Omega_2^2}.$$

Insert Equation 12.41 into Equation 12.40, and this equation becomes[17]

$$\ddot{x}_1 + \omega_n^2 x_1 = C_1 \cos \Omega_1 t + C_2 \cos \Omega_2 t$$
$$+ C_3 \left[\cos(2\Omega_1 + \Omega_2) t + \cos(2\Omega_1 - \Omega_2) t \right]$$
$$+ C_4 \left[\cos(\Omega_1 + 2\Omega_2) t + \cos(\Omega_1 - 2\Omega_2) t \right]$$
$$+ C_5 \cos 3\Omega_1 t + C_6 \cos 3\Omega_2 t, \qquad (12.42)$$

where

$$C_1 = -\frac{3}{4}\beta \omega_n^2 G_1 \left(G_1^2 + 2G_2^2 \right), \quad C_2 = -\frac{3}{4}\beta \omega_n^2 G_2 \left(2G_1^2 + G_2^2 \right)$$

$$C_3 = -\frac{3}{4}\beta \omega_n^2 G_1^2 G_2, \quad C_4 = -\frac{3}{4}\beta \omega_n^2 G_1 G_2^2$$

$$C_5 = -\frac{1}{4}\beta \omega_n^2 G_1^3, \quad C_6 = -\frac{1}{4}\beta \omega_n^2 G_2^3.$$

It is clear from these results that the solution to the nonlinear equation consists of a linear combination of the harmonic components Ω_1, Ω_2. Since these *combination harmonics* occur in the solution to the ε^1 order term, they will generally be one order of magnitude smaller than the zero–order solution $x_0(t)$. If, however, any of the combination frequencies are near the frequency ω_n then higher, resonance–type amplitudes are possible. The expressions for G_1 and G_2 show how amplification can occur.[18]

12.6 The Mathieu Equation

Return to the problem of equations that have time–dependent coefficients. In Section 9.8, the oscillation of a pendulum suspended from a moving base was studied as an example of stability of motion. Figure 9.17 is reproduced here as Figure 12.10. This simplified model is used to represent fluid sloshing in a container, for example, fuel in an aircraft wing or in a rocket ship.

---

[17] Use $\cos a \cos b = \frac{1}{2} \left[\cos(a+b) + \cos(a-b) \right]$.
[18] A more general solution can be found in Jordan and Smith pages 196-200.

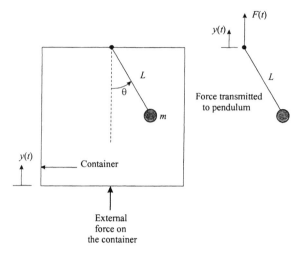

Figure 12.10: Pendulum oscillating about a vertically oscillating base.

The equations of motion were derived as

$$\ddot{y} + L\ddot{\theta}\sin\theta + L\dot{\theta}^2\cos\theta + g = \frac{F}{m}$$
$$mL^2\ddot{\theta} + mL\ddot{y}\sin\theta + mgL\sin\theta = 0.$$

Assume now that we are looking for the stability of motion about the equilibrium position $\theta = 0$. In the region about small θ, the equation in θ becomes

$$\ddot{\theta} + \frac{1}{L}(\ddot{y} + g)\theta = 0, \qquad (12.43)$$

which becomes linearized but is still coupled to the motion in the y direction. The equation in y, on the other hand, becomes

$$\ddot{y} + L\ddot{\theta}\theta + L\dot{\theta}^2 + g = \frac{F}{m}, \qquad (12.44)$$

which is simplified but still nonlinear. If θ and $\dot{\theta}$ are small, "higher order terms" such as $\dot{\theta}^2$ and $\ddot{\theta}\theta$ may be neglected. In this case, Equation 12.44 is linearized to

$$\ddot{y} + g = \frac{F}{m}. \qquad (12.45)$$

This equation is decoupled from the θ motion. Assume that the support motion is harmonic, $y = A\cos\omega t$. From Equation 12.45, the force required

12.6. THE MATHIEU EQUATION

is $F(t) = -mA\omega^2 \cos \omega t + mg$. Equation 12.43 becomes

$$\ddot{\theta} + \frac{1}{L}\left(g - A\omega^2 \cos \omega t\right)\theta = 0, \qquad (12.46)$$

which is an equation with a time–varying coefficient. An equation of this form is called the *Mathieu equation*.[19] We immediately note that $(g - A\omega^2 \cos \omega t)$ is a harmonic function that can have negative or positive values depending on the relative magnitudes of A and ω. If $g > A\omega^2$ then $g > A\omega^2 \cos \omega t$ and the coefficient of θ is always positive and response $\theta(t)$ is always stable. Otherwise, for $g < A\omega^2$, we obtain solutions with coefficients of the form $\exp(+\omega T)$ that grow without bound.

The Mathieu equation is of interest in many disciplines. Because of the possibility of instabilities, we look to delineate parameter ranges for which the solution is periodic and stable, and to define the boundaries between stability and instability. The perturbation method can be used to locate the periodic solutions as well as the stability boundaries. A general form of the Mathieu equation is

$$\ddot{\theta} + (\delta + 2\varepsilon \cos 2t)\theta = 0, \qquad (12.47)$$

where $\varepsilon \ll 1$, and the transformation from Equation 12.46 to Equation 12.47 is clear. This equation represents a quasi–harmonic system. For $\varepsilon = 0$, the linear harmonic oscillator is recovered and it is noted that parameter δ represents the natural frequency of linear oscillation squared.[20] The character of the oscillation depends on the relative magnitudes of δ and ε and the stability regions can be conveniently outlined using the parameter space δ-ε. This plane is divided into regions of stability and instability by *boundary* or *transition* curves. Any point on such a curve is characterized by periodic solutions.

Lindstedt's method can be used to obtain periodic solutions of Mathieu's equation. To this end, assume an expansion solution for $\theta(t)$ and also expand the "frequency" parameter δ,

$$\begin{aligned}\theta(\varepsilon, t) &= \theta_0(t) + \varepsilon \theta_1(t) + \varepsilon^2 \theta_2(t) + \cdots \\ \delta &= n^2 + \varepsilon \delta_1 + \varepsilon^2 \delta_2 + \cdots, \quad n = 0, 1, 2, \ldots.\end{aligned}$$

---

[19] The Mathieu equation is a special case of *Hill's* equation. Numerous specialized texts exist on this very important class of equations. The texts cited earlier on the perturbation method provide good introductions. Another useful reference is: **Nonlinear Ordinary Differential Equations**, R. Grimshaw, CRC Press, 1993. See the chapter on linear equations with periodic coefficients.

[20] The solution to the linear equation $\ddot{\theta} + \delta\theta = 0$ is $\theta = \cos\sqrt{\delta}t + \sin\sqrt{\delta}t$ or $\theta = \cos nt + \sin nt$.

The implication in the expansion for δ is that δ differs from an integer n squared by a small quantity. Again, δ represents a frequency squared term, and since we know that the behavior is quasi–oscillatory, it is found that δ does not vary much from n^2. Other possibilities exist.[21]

Substitute these expansions into Equation 12.47 and equate coefficients of like powers of ε to obtain the following set of equations,

$$\ddot{\theta}_0 + n^2 \theta_0 = 0 \qquad (12.48)$$
$$\ddot{\theta}_1 + n^2 \theta_1 = -(\delta_1 + 2\cos 2t)\theta_0 \qquad (12.49)$$
$$\ddot{\theta}_2 + n^2 \theta_2 = -(\delta_1 + 2\cos 2t)\theta_1 - \delta_2 \theta_0, \qquad n = 0, 1, 2, \ldots \quad (12.50)$$
$$\vdots$$

one set for every n. It is clear that Equations 12.48–12.50 are solved sequentially for each n. The first equation in the set provides the zero–order approximation,

$$\theta_0 = \begin{cases} \cos nt \\ \sin nt \end{cases}, \qquad n = 0, 1, 2, \ldots. \qquad (12.51)$$

The transition curves are obtained by substituting Equations 12.51 into the set of Equations 12.48–12.50, and selecting parameters so that the solutions $\theta_i(t)$, $i = 1, 2, \ldots$ are periodic. This procedure leads to an infinite number of solution *pairs*, one pair for each value of n, with the exception that, for $n = 0$, there is only one solution.

Consider first the case $n = 0$, where $\theta_0 = 1$. ($\theta_0 = 0$ is a trivial solution that does not lead to any further information.) The equation for θ_1 becomes

$$\ddot{\theta}_1 = -\delta_1 - 2\cos 2t.$$

For θ_1 to be periodic, δ_1 must equal zero, in which case

$$\theta_1 = \frac{1}{2} \cos 2t.$$

---

[21]There are other expansions for δ that lead to other "tongues" of instability, for example, where the tongues emanate from

$$\delta = \frac{(2n+1)^2}{4}, \qquad n = 0, 1, 2, \ldots.$$

The other tongues of instability emerge at higher order truncations in the various perturbation methods.

12.6. THE MATHIEU EQUATION

The equation governing θ_2, for $n = 0$, becomes

$$\ddot{\theta}_2 = -2\cos 2t \left(\frac{1}{2}\cos 2t\right) - \delta_2$$
$$= -\left(\frac{1}{2} + \delta_2\right) - \frac{1}{2}\cos 4t,$$

where use is made of the trigonometric relation $\cos^2 2t = (1 + \cos 4t)/2$. For θ_2 to be periodic, the constant term on the right hand side must be set to zero, thus, $\delta_2 = -1/2$. Hence, corresponding to $n = 0$, there is only one transition curve,

$$\delta = -\frac{1}{2}\varepsilon^2 + \cdots,$$

which, to the second-order approximation, is a parabola passing through the origin of the δ-ε parameter plane.

Next, we consider the case $n = 1$. Now there are two zero-order solutions,

$$\theta_0 = \begin{cases} \cos t \\ \sin t. \end{cases}$$

Derive the transition curve for the first of these solutions, $\theta_0 = \cos t$, for which the governing equation for θ_1 becomes

$$\ddot{\theta}_1 + \theta_1 = -(\delta_1 + 2\cos 2t)\cos t$$
$$= -(\delta_1 + 1)\cos t - \cos 3t,$$

where $2\cos 2t \cos t = \cos 3t + \cos t$ is used. We know from our vibration studies that if an undamped oscillator is forced at its natural frequency it will resonate. Here, the "natural frequency" equals 1, and the "loading" $-(\delta_1 + 1)\cos t$ is also at frequency equal to 1. Therefore, to prevent secular terms, we must set $\delta_1 = -1$. The solution to the reduced differential equation is

$$\theta_1 = \frac{1}{8}\cos 3t.$$

Inserting θ_0, θ_1 and δ_1 into the governing equation for θ_2 results in

$$\ddot{\theta}_2 + \theta_2 = -\frac{1}{8}(-1 + 2\cos 2t)\cos 3t - \delta_2 \cos t$$
$$= -\left(\frac{1}{8} + \delta_2\right)\cos t + \frac{1}{8}\cos 3t - \frac{1}{8}\cos 5t,$$

where $2\cos 2t \cos 3t = \cos 5t + \cos t$. Using the same argument as with θ_1, for θ_2 to be periodic the coefficient of $\cos t$ must be set to zero, yielding

$\delta = -1/8$. Truncating the calculations at second–order, the transition curve corresponding to $\theta_0 = \cos t$ is

$$\delta = 1 - \varepsilon - \frac{1}{8}\varepsilon^2 + \cdots.$$

Proceeding with the derivation of the transition curve corresponding to $\theta_0 = \sin t$, the equation governing θ_1 becomes

$$\begin{aligned}\ddot{\theta}_1 + \theta_1 &= -(\delta_1 + 2\cos 2t)\sin t \\ &= -(\delta_1 - 1)\sin t - \sin 3t,\end{aligned} \quad (12.52)$$

where use is made of the relation $2\cos 2t \sin t = \sin 3t - \sin t$. The solution to Equation 12.52 is periodic for $\delta_1 = 1$ and is

$$\theta_1 = \frac{1}{8}\sin 3t.$$

The equation governing θ_2 becomes

$$\begin{aligned}\ddot{\theta}_2 + \theta_2 &= -\frac{1}{8}(-1 + 2\cos 2t)\sin 3t - \delta_2 \sin t \\ &= -\left(\frac{1}{8} + \delta_2\right)\sin t - \frac{1}{8}\sin 3t + \frac{1}{8}\sin 5t,\end{aligned}$$

where $2\cos 2t \sin 3t = \sin 5t + \sin t$. For θ_2 to be periodic, it must be that $\delta_2 = -\frac{1}{8}$, and the transition curve corresponding to $\theta_0 = \sin t$ is

$$\delta = 1 + \varepsilon - \frac{1}{8}\varepsilon^2 + \cdots.$$

The procedure is now clear. Pursuing the transition curves corresponding to $n = 2$, leads to

$$\delta = 4 + \frac{5}{12}\varepsilon^2 + \cdots,$$

for $\theta_0 = \cos 2t$, and

$$\delta = 4 - \frac{1}{12}\varepsilon^2 + \cdots,$$

for $\theta_0 = \sin 2t$. Transition curves for $n = 3, 4, \ldots$ can be obtained in the same fashion. The transition curves can be plotted in the δ-ε plane, defining regions of stability and instability. See Figure 12.11 for the curves generated for $n = 0, 1, 2$. This figure is known as a *Strutt diagram*, named after Lord Rayleigh.[22] The region terminating at $(\delta, \varepsilon) = (1, 0)$ is known as the principal instability region. It is wider than the regions terminating at $\delta = n^2$,

---

[22] His birth name was John William Strutt.

12.7. THE VAN DER POL EQUATION

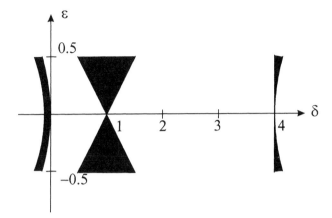

Figure 12.11: Shaded regions are unstable. The bounding curves are generated for $n = 0$, 1 and 2.

$n = 2, 3, \ldots$. The instability regions are shaded and are symmetrical about the δ axis. The stable regions are connected at the points $\delta = n^2$, $\varepsilon = 0$, $n = 0, 1, 2, \ldots$.

The analyst can use charts such as this to calculate whether the parameters used in a design are in an instability region, or near one. Recall,

$$\delta = \frac{g}{L}, \qquad \varepsilon = -\frac{A\omega^2}{2L}.$$

While δ and ε can be any values in a physical system, only certain values result in periodic solutions. The Mathieu equation with damping leads to similar results except that the shaded regions are detached from the δ axis for damping constant $c > 0$.

12.7 The van der Pol Equation

The unforced van der Pol equation is given by

$$\ddot{x} - \alpha \left(1 - x^2\right) \dot{x} + x = 0, \quad \alpha > 0. \tag{12.53}$$

If α is small in some sense, then the ε notation can be used and the van der Pol equation becomes

$$\ddot{x} + x = \varepsilon \left(1 - x^2\right) \dot{x}. \tag{12.54}$$

Expand the right hand side to order ε^2, that is, $x = x_0 + \varepsilon x_1 + \varepsilon^2 x_2$,

$$\begin{aligned}
\left(1 - x^2\right)\dot{x} &\simeq \left[1 - \left(x_0 + \varepsilon x_1 + \varepsilon^2 x_2\right)^2\right]\left(\dot{x}_0 + \varepsilon \dot{x}_1 + \varepsilon^2 \dot{x}_2\right) \\
&\simeq \left(1 - x_0^2\right)\dot{x}_0 + \varepsilon\left[-2x_0 x_1 \dot{x}_0 + \left(1 - x_0^2\right)\dot{x}_1\right] \\
&\quad + \varepsilon^2\left[-x_1^2 \dot{x}_0 - 2x_0 x_2 \dot{x}_0 - 2x_0 x_1 \dot{x}_1 + \left(1 - x_0^2\right)\dot{x}_2\right],
\end{aligned}$$

and insert into Equation 12.54. Equate terms of same order in ε to find the following perturbation equations,

$$\begin{aligned}
\varepsilon^0 &: \quad \ddot{x}_0 + x_0 = 0 \\
\varepsilon^1 &: \quad \ddot{x}_1 + x_1 = \left(1 - x_0^2\right)\dot{x}_0 \\
\varepsilon^2 &: \quad \ddot{x}_2 + x_2 = -2x_0 x_1 \dot{x}_0 + \left(1 - x_0^2\right)\dot{x}_1 \\
\varepsilon^3 &: \quad \ddot{x}_3 + x_3 = -x_1^2 \dot{x}_0 - 2x_0 x_2 \dot{x}_0 - 2x_0 x_1 \dot{x}_1 + \left(1 - x_0^2\right)\dot{x}_2 \\
&\vdots
\end{aligned}$$

The equation of $O\left(\varepsilon^3\right)$ is ignored since the expansion is to order ε^2.

Each equation is now solved sequentially and the approximate solution for $x(t)$ is given by

$$x(t) \simeq x_0 + \varepsilon x_1 + \varepsilon^2 x_2.$$

12.7.1 Limit Cycles

In the above introduction to the stability of motion, all motion either tended toward the equilibrium point, or moved out into instability. Another possibility exists, and it most closely resembles the elliptical trajectories mentioned earlier. The difference is that the periodic motion is not a closed trajectory. Rather, it is a trajectory that can be approached from the equilibrium point or from other initial conditions beyond.

In such damped vibration problems, the trajectories can start either very close to the origin or far away from the origin, and approach the same closed curve about the origin. This curve represents a periodic, but not harmonic, solution of the governing equation. This curve is called the *limit cycle*. The classical equation that has such a limit cycle is the *van der Pol* oscillator, one mathematical form of which is Equation 12.53. The van der Pol equation can only be solved numerically.

This equation is applied to a number of disciplines, including nonlinear electrical and vibratory systems. Physically, for vibratory systems, the form of the damping force, $\alpha\left(1 - x^2\right)\dot{x}$, is such that the damping is negative for small amplitudes and positive for larger amplitudes; this depends on the

12.7. THE VAN DER POL EQUATION

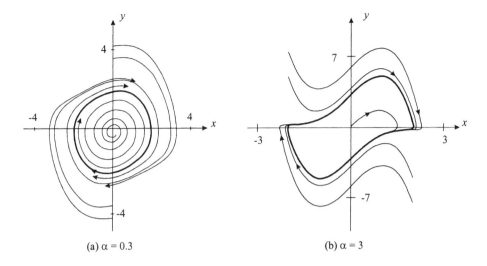

(a) α = 0.3 (b) α = 3

Figure 12.12: Trajectories for two values of α for the van der Pol equation for a number of initial conditions. The limit cycle is shown in a dark line.

sign of $\left(1 - x^2\right)$. Consider the phase plane for the van der Pol equation,

$$\dot{x} = y$$
$$\dot{y} = \alpha\left(1 - x^2\right)y - x,$$

with trajectories defined by

$$\frac{dy}{dx} = \frac{\alpha\left(1 - x^2\right)y - x}{y}.$$

Sketches of a variety of trajectories, regardless of initial conditions, approach the limit cycle asymptotically. An initial point inside the limit cycle follows an outwardly spiraling trajectory. An initial point outside the limit cycle follows an inwardly spiraling trajectory. An infinity of isoclines[23] pass through the origin, which is a singularity. The limit cycle has the interesting property that the maximum value of x is always close to 2 regardless of the value of α. Figure 12.12 shows the trajectories of two cases.

Figure 12.13 shows the time histories of this oscillator for the same two cases, for several initial conditions.

---

[23] An isocline is defined as the locus of points at which the trajectories passing through them have a constant slope. The method of isoclines is used to construct the trajectories of dynamical systems with one degree of freedom.

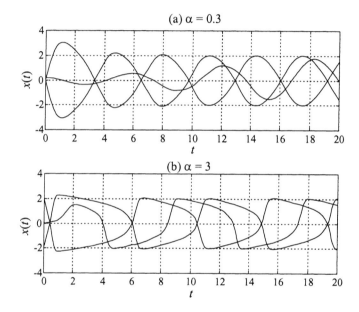

Figure 12.13: Van der Pol equation time histories for several sets of initial conditions.

12.7.2 The Forced van der Pol Equation

The forced van der Pol equation is a model for a system that is capable of self–oscillation while acted upon by another oscillator. Rand[24] points to a biological application involving the human sleep–wake cycle in which a person's biological clock is modeled by a van der Pol oscillator, and the daily night–day cycle caused by the Earth's rotation is modeled as a periodic forcing term.

The phenomenon where the period of a forced oscillation is an integer multiple of the period of the forcing is called *frequency entrainment*. The frequency of the solution, say ν, is said to be entrained by the frequency of the forcing, say ω, so that $\nu = \omega/l$ for some positive integer l called the *entrainment index*. The van der Pol equation has such characteristics.

Consider the following forced van der Pol equation,

$$\ddot{x} + \varepsilon \left(x^2 - 1\right) \dot{x} + x = F \cos \omega t. \tag{12.55}$$

The unforced case has a limit cycle with a radius of approximately 2 and a period of approximately 2π. The limit cycle is generated by the balance between the internal energy loss and energy generation. A forcing term will alter this balance.

If F is small, the excitation is called weak, and its effect depends on whether or not ω is close to the natural frequency. If it is, an oscillation will be generated that is a perturbation of the limit cycle. If F is not small, the excitation is called hard. If the natural and imposed frequencies are not close, we expect that the natural oscillation will be damped out as happens with the corresponding linear equation. Stretch time according to $\omega t = \tau$, and Equation 12.55 becomes

$$\omega^2 x'' + \varepsilon \omega \left(x^2 - 1\right) x' + x = F \cos \tau.$$

A number of cases require further examination. Suppose hard excitation far from resonance and assume that ω is not close to one. Expand as usual,

$$x(\varepsilon, \tau) = x_0(\tau) + \varepsilon x_1(\tau) + \cdots, \tag{12.56}$$

to find

$$\omega^2 x_0'' + x_0 = F \cos \tau$$
$$\omega^2 x_1'' + x_1 = -\omega \left(x_0^2 - 1\right) x_0',$$

---

[24] **Lecture Notes on Nonlinear Vibrations**, R.H. Rand. Look up the latest version of lecture notes at http://www.tam.cornell.edu/randdocs.

where $x_0(\tau)$ and $x_1(\tau)$ have period 2π. Therefore,

$$x_0(\tau) = \frac{F}{1-\omega^2} \cos \tau.$$

We know the solution to $x_1(\tau)$ is of $O(\varepsilon)$, thus,

$$x(\varepsilon, \tau) = \frac{F}{1-\omega^2} \cos \tau + O(\varepsilon).$$

The solution is a perturbation of the ordinary linear response and the limit cycle is suppressed as expected. If the excitation is soft and far from resonance, the procedure is similar as for the hard excitation but the response is usually unstable.

Suppose soft excitation near resonance. Then let $F = \varepsilon \gamma$, and near resonance $\omega = 1 + \varepsilon \omega_1$, then use expansion Equation 12.56. The resulting equations are

$$\begin{aligned} x_0'' + x_0 &= 0 \\ x_1'' + x_1 &= -2\omega_1 x_0'' - (x_0^2 - 1) x_0' + \gamma \cos \tau \\ &\vdots \end{aligned}$$

As usual, solutions are sought with period 2π, then

$$x_0(\tau) = a_0 \cos \tau + b_0 \sin \tau$$

and

$$\begin{aligned} x_1'' + x_1 &= \left[\gamma + 2\omega_1 a_0 - b_0 \left(\frac{1}{4}r_0^2 - 1\right)\right] \cos \tau \\ &+ \left[2\omega_1 b_0 + a_0 \left(\frac{1}{4}r_0^2 - 1\right)\right] \sin \tau \\ &+ \cdots \\ r_0 &= +\sqrt{a_0^2 + b_0^2}, \end{aligned}$$

where r_0 is the response amplitude for $x_0(\tau)$. For a periodic solution to exist, the coefficients of the harmonic functions must be set equal to zero,

$$\begin{aligned} 2\omega_1 a_0 - b_0 \left(\frac{1}{4}r_0^2 - 1\right) &= -\gamma \\ 2\omega_1 b_0 + a_0 \left(\frac{1}{4}r_0^2 - 1\right) &= 0. \end{aligned}$$

12.8. MOTION IN THE LARGE

These two equations can be combined into

$$r_0^2 \left[4\omega_1^2 + \left(\frac{1}{4}r_0^2 - 1 \right)^2 \right] = \gamma^2,$$

which gives possible values of r_0. Note that there may be as many as 3 real solutions for $r_0 > 0$.

Van der Pol

Balthazar van der Pol (1889-1959) was a Dutch electrical engineer who initiated modern experimental dynamics in the laboratory during the 1920's and 1930's. Van der Pol investigated electrical circuits employing vacuum tubes and found that they have stable oscillations, now called limit cycles. When these circuits are driven with a signal whose frequency is near that of the limit cycle, the resulting periodic response shifts its frequency to that of the driving signal. That is to say, the circuit becomes "entrained" to the driving signal. The waveform, or signal shape, however, can be quite complicated and contain a rich structure of harmonics and subharmonics.

In the September 1927 issue of the British journal *Nature*, he and his colleague van der Mark reported that an "irregular noise" was heard at certain driving frequencies between the natural entrainment frequencies. By reconstructing his electronic tube circuit, we now know that they had discovered deterministic chaos. Their paper is probably one of the first experimental reports of chaos—something that they failed to pursue in more detail.

Van der Pol built a number of electronic circuit models of the human heart to study the range of stability of heart dynamics. His investigations with adding an external driving signal were analogous to the situation in which a real heart is driven by a pacemaker. He was interested in finding out, using his entrainment work, how to stabilize a heart's irregular beating or "arrhythmias."

12.8 Motion in the Large

Motion in the large implies attempts to consider the complete nonlinear governing equations, rather than system properties about equilibria, or perturbation expansions that assume a smallness in some of the parameters. Nonlinear equations have no general solutions and usually require numerical solutions. However, for single degree of freedom nonlinear differential equations a somewhat general discussion is possible that sheds light on large behavior.

Consider the conservative nonlinear equation

$$\ddot{x} = f(x),$$

where $f(x)$ is a nonlinear conservative force per unit mass. It is useful to consider the energy possessed by the system. The left hand side can be written as follows,
$$\ddot{x} = \frac{d\dot{x}}{dt} = \frac{d\dot{x}}{dx}\frac{dx}{dt} = \frac{d\dot{x}}{dx}\dot{x}.$$
Then, $\ddot{x}dx = \dot{x}d\dot{x} = f(x)dx$, and
$$\int_0^x \ddot{x}dx = \int_0^{\dot{x}} \dot{x}d\dot{x} = \int_0^x f(\xi)d\xi + c, \qquad (12.57)$$
where
$$\int_0^{\dot{x}} \dot{x}d\dot{x} = \frac{1}{2}\dot{x}^2$$
is the kinetic energy per unit mass, and
$$\int_0^x f(\xi)d\xi = -V(x),$$
the negative of the potential energy per unit mass. c is a constant of integration. Equation 12.57 can be written as
$$\frac{1}{2}\dot{x}^2 + V(x) = c.$$

This equation states that the total system energy is a constant, $E = c$. It is of interest to plot the energy function in the state plane; let $y = \dot{x}$, then
$$\frac{1}{2}y^2 + V(x) = E. \qquad (12.58)$$
Equation 12.58 can be plotted as E vs. x and y, and for a particular energy level, the projection plot x vs. y is also useful. In the first instance, a three-dimensional surface can be created, one projection of which is shown in Figure 12.14 top. For lines of constant energy, it is possible to visualize the possible trajectories, as in Figure 12.14 bottom.

Example 12.6 Trajectories and Stability of a Pendulum
Identify the equilibrium points and determine their stability nature for the energy levels $E = \omega_n^2, 2\omega_n^2, 3\omega_n^2$.
Solution Example 12.2 has the equation of motion
$$\ddot{\theta} + \omega_n^2 \sin\theta = 0,$$

12.8. MOTION IN THE LARGE

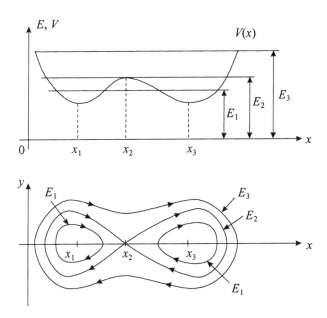

Figure 12.14: Energy curves shown in profile (top) and projection (bottom). The energy curves in the lower figure correspond to the energy "slices" in the top figure.

and the equations of the trajectories, with $\theta = x$, are

$$\dot{x} = y$$
$$\dot{y} = -\omega_n^2 \sin x.$$

The equilibrium equations are obtained for zero velocity and acceleration, that is, $\dot{x} = 0$ and $\dot{y} = 0$, then

$$y = 0, \quad \sin x = 0.$$

The equilibrium points are then given by

$$y = 0$$
$$x = \pm j\pi, \quad j = 0, 1, 2, \ldots.$$

Physically there are only two equilibrium points since the pendulum can only trace a circle of range 2π,

$$y = 0 \text{ with } x = 0 \qquad (12.59)$$
$$y = 0 \text{ with } x = \pi. \qquad (12.60)$$

The first equilibrium point, Equations 12.59, is the one usually considered in which the pendulum hangs at rest. The other equilibrium point, Equations 12.60, is where the pendulum is at rest in the upright position. If the pendulum is composed of a rigid massless rod with a point mass at one end, then it might be plausible that the upright position can be a rest equilibrium point.

Using the method of Section 12.3.1, the coefficient matrix in Equation 12.7 for the linearized state equations for the first equilibrium point $x = 0$, with $\sin x \simeq x$, becomes

$$\begin{bmatrix} d_{11} & d_{12} \\ d_{21} & d_{22} \end{bmatrix} = \begin{bmatrix} 0 & 1 \\ -\omega_n^2 & 0 \end{bmatrix}.$$

The eigenvalues of the coefficient matrix are found by setting the following determinant to zero,

$$\begin{vmatrix} -\lambda & 1 \\ -\omega_n^2 & -\lambda \end{vmatrix} = 0$$

yielding the eigenvalues $\lambda_{1,2} = \pm i\omega_n$. Both eigenvalues are pure imaginary and the equilibrium point is a center and merely stable. In the neighborhood of the second equilibrium point $x = \pi$, with[25] $\sin x \simeq -x$, the coefficient

---
[25] Expand $\sin(x + \pi)$ for small x.

12.9. ADVANCED TOPICS

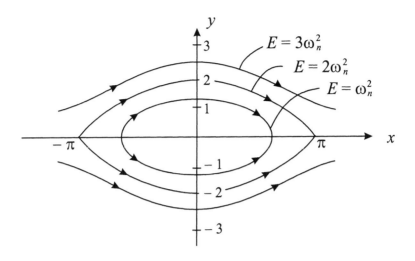

Figure 12.15: Trajectories at three energy levels.

matrix is
$$\begin{bmatrix} 0 & 1 \\ \omega_n^2 & 0 \end{bmatrix},$$
from which the eigenvalues are $\lambda_{1,2} = \pm\omega_n$. Both eigenvalues are real, with one positive and the other negative, implying that the point is a saddle and unstable.

The trajectories are shown in Figure 12.15 for the three energy levels. For $E = \omega_n^2$ the trajectory is closed, indicating periodic motion. For small $E < \omega_n^2$, the motion is harmonic. As E increases such that $\omega_n^2 < E < 2\omega_n^2$, the motion ceases to be harmonic but remains periodic. For $E \geq 2\omega_n^2$, the trajectory is open and the motion is rotary, as the mass rotates about the contact point. $E = 2\omega_n^2$ is a *separatrix*, signifying the boundary between two types of motion. The separatrix also describes rotary motion. ∎

12.9 Advanced Topics

We conclude this chapter with an examination of two more advanced topics. The first is the use of the expansion method for problems where the forcing on the system is a random process. And the second is the use of a Galerkin method to approximate the solution of a nonlinear equation.

12.9.1 Random Duffing Oscillator

Consider the Duffing equation subjected to a random force and apply the perturbation technique to derive estimates of the response correlation and spectral density. We draw on the probabilistic ideas introduced in Chapter 5. The governing equation is

$$\ddot{x} + 2\zeta\omega_n \dot{x} + \omega_n^2 \left(x + rx^3\right) = f(t).$$

Define $\varepsilon = \omega_n^2 r$. Recall that in customary oscillator notation $c/m = 2\zeta\omega_n$ and $k/m = \omega_n^2$. The excitation per unit mass $f(t)$ is a stationary Gaussian random process with power spectrum $S_{ff}(\omega)$. The equations that govern the terms in the approximation $x \simeq x_0 + \varepsilon x_1 + \varepsilon^2 x_2$ are

$$\ddot{x}_0 + 2\zeta\omega_n \dot{x}_0 + \omega_n^2 x_0 = f(t) \quad (12.61)$$
$$\ddot{x}_1 + 2\zeta\omega_n \dot{x}_1 + \omega_n^2 x_1 = -x_0^3 \quad (12.62)$$
$$\ddot{x}_2 + 2\zeta\omega_n \dot{x}_2 + \omega_n^2 x_2 = -3x_0^2 x_1. \quad (12.63)$$

Equation 12.61 is a linear oscillator driven by a random load $f(t)$. Note again that this procedure takes a nonlinear oscillator and converts it into a series of linear oscillators. The nonlinearities have been moved to the right hand side of each linear oscillator, but the system equation, that is the left hand side, is always linear and amenable to solution via the convolution integral regardless of the form of the function driving it on the right hand side. The first three terms in the infinite sequence of solutions is then,

$$x_0(t) = \int_{-\infty}^{\infty} g(t) f(t-\tau) d\tau \quad (12.64)$$

$$x_1(t) = -\int_{-\infty}^{\infty} g(t) x_0^3(t-\tau) d\tau \quad (12.65)$$

$$x_2(t) = -3 \int_{-\infty}^{\infty} g(t) x_0^2(t-\tau) x_1(t-\tau) d\tau, \quad (12.66)$$

where $g(t) = (1/\omega_d) \exp(-\zeta\omega_d t)$, $\omega_d = \omega_n \sqrt{1-\zeta^2}$. The solution to Equation 12.64 is then cubed and substituted into Equation 12.65, and that equation is integrated to find the expression $x_1(t)$. This expression and the one for $x_0^2(t)$ are then appropriately substituted into Equation 12.66 for integration. The point has been made that this is a tedious process, the accuracy of which depends on the value of ε. Since $f(t)$ is a random function of time, only the statistics of the response can be estimated. For a three–term expansion, the mean value of the response is then given by

$$E\{x\} = E\{x_0\} + \varepsilon E\{x_1\} + \varepsilon^2 E\{x_2\},$$

12.9. ADVANCED TOPICS

where

$$E\{x_0\} = \int_{-\infty}^{\infty} g(t) E\{f(t-\tau)\} d\tau$$

$$E\{x_1\} = -\int_{-\infty}^{\infty} g(t) E\{x_0^3(t-\tau)\} d\tau$$

$$E\{x_2\} = -3\int_{-\infty}^{\infty} g(t) E\{x_0^2(t-\tau) x_1(t-\tau)\} d\tau.$$

The evaluation of these expectations can be quite difficult analytically and generally requires some assumptions regarding the statistical properties of the variables.

Even more intricate is the estimation of the output correlation function $R_{xx}(\tau)$ and spectral density $S_{xx}(\omega)$,

$$\begin{aligned}R_{xx}(\tau) &= E\{x(t)x(t+\tau)\} \\ &\simeq E\left\{[x_0(t)+\varepsilon x_1(t)+\varepsilon^2 x_2(t)]\right. \\ &\quad \cdot [x_0(t+\tau)+\varepsilon x_1(t+\tau)+\varepsilon^2 x_2(t+\tau)]\} \\ &= E\left\{x_0(t)x_0(t+\tau)+\varepsilon x_0(t)x_1(t+\tau)+\varepsilon^2 x_0(t)x_2(t+\tau)\right. \\ &\quad +\varepsilon x_1(t)x_0(t+\tau)+\varepsilon^2 x_1(t)x_1(t+\tau)+\varepsilon^3 x_1(t)x_2(t+\tau) \\ &\quad \left.+\varepsilon^2 x_2(t)x_0(t+\tau)+\varepsilon^3 x_2(t)x_1(t+\tau)+\varepsilon^4 x_2(t)x_2(t+\tau)\right\}.\end{aligned}$$

To evaluate this expression we require data for numerous correlations and cross–correlations. It is reasonable to drop terms of order greater than ε^2 for consistency with the fact that the expansion for $x(t)$ is only to order ε^2, thus,

$$\begin{aligned}R_{xx}(\tau) &\simeq R_{x_0 x_0}(\tau) + \varepsilon [R_{x_0 x_1}(\tau) + R_{x_1 x_0}(\tau)] \\ &\quad + \varepsilon^2 [R_{x_1 x_1}(\tau) + R_{x_0 x_2}(\tau) + R_{x_2 x_0}(\tau)].\end{aligned}$$

The response spectral density approximation is the Fourier transform of $R_{xx}(\tau)$,

$$\begin{aligned}S_{xx}(\omega) &\simeq S_{x_0 x_0}(\omega) + \varepsilon [S_{x_0 x_1}(\omega) + S_{x_1 x_0}(\omega)] \\ &\quad + \varepsilon^2 [S_{x_1 x_1}(\omega) + S_{x_0 x_2}(\omega) + S_{x_2 x_0}(\omega)].\end{aligned}$$

While the math is relatively neat to derive, the reality of these kinds of equations is that the necessary data are rarely available to perform these computations. Usually, for an approximation the analyst might only retain terms to order ε, and even then, major simplifications are needed.

Another approach to this problem is to apply the fundamental theorem to Equation 12.61,

$$S_{x_0 x_0}(\omega) = |H(i\omega)|^2 S_{ff}(\omega).$$

Given $S_{x_0 x_0}(\omega)$, it is possible to find $S_{x_0^3 x_0^3}(\omega)$ for a Gaussian process, and then use the fundamental theorem again with Equation 12.62, and then with Equation 12.63. For non–Gaussian processes, this is more difficult.

12.9.2 The Nonlinear Pendulum Using a Galerkin Method

By way of example, a Galerkin method is used in conjunction with a perturbation approximation. Apply the Galerkin method[26] to the nonlinear pendulum. Use a one term approximation. Denote the governing equation by the shorthand function $F[\cdot]$, where θ is the exact solution and Θ is the approximate solution. The nonlinear equation is

$$F[\theta] \equiv \ddot{\theta} + \omega_n^2 \left(\theta - \frac{\theta^3}{6}\right) = 0,$$

and the assumed solution is the linear solution,

$$\Theta(t) = A \sin \omega t.$$

Substitute the assumed solution into the governing equation to find,

$$F[\Theta] \equiv -\omega^2 A \sin \omega t + \omega_n^2 \left(A \sin \omega t - \frac{1}{6} A^3 \sin^3 \omega t\right)$$
$$= \left(\omega_n^2 - \omega^2 - \frac{1}{8}\omega_n^2 A^2\right) A \sin \omega t + \frac{\omega_n^2}{24} A^3 \sin 3\omega t.$$

The Galerkin method minimizes the integral

$$\int_0^T F^2(\Theta)\, dt, \qquad (12.67)$$

with respect to A, where T is the period of the oscillation. Therefore,

$$\int_0^T F(\Theta) \frac{\partial F(\Theta)}{\partial A}\, dt = 0.$$

---

[26] See Section 11.4.3.

12.9. ADVANCED TOPICS

Then,

$$\int_0^T \left[\left(\omega_n^2 - \omega^2 - \frac{1}{8}\omega_n^2 A^2\right) A \sin\omega t + \frac{\omega_n^2}{24} A^3 \sin 3\omega t\right]$$
$$\cdot \left[\left(\omega_n^2 - \omega^2 - \frac{3}{8}A^2\omega_n^2\right) \sin\omega t + \frac{1}{8}A^2\omega_n^2 \sin 3\omega t\right] dt = 0.$$

Expand the product to yield,

$$A\left(\omega_n^2 - \omega^2 - \frac{1}{8}\omega_n^2 A^2\right)\left(\omega_n^2 - \omega^2 - \frac{3}{8}A^2\omega_n^2\right) \int_0^T \sin^2\omega t\, dt$$
$$+ \frac{\omega_n^2}{24} A^3 \left(\omega_n^2 - \omega^2 - \frac{3}{8}A^2\omega_n^2\right) \int_0^T \sin\omega t \sin 3\omega t\, dt$$
$$+ \frac{1}{8} A^3 \omega_n^2 \left(\omega_n^2 - \omega^2 - \frac{1}{8}\omega_n^2 A^2\right) \int_0^T \sin\omega t \sin 3\omega t\, dt$$
$$+ \frac{\omega_n^4}{192} A^5 \int_0^T \sin^2 3\omega t\, dt = 0.$$

Using trigonometric identities,[27] we are left with the terms

$$A\left[\left(\omega_n^2 - \omega^2 - \frac{1}{8}\omega_n^2 A^2\right)\left(\omega_n^2 - \omega^2 - \frac{3}{8}A^2\omega_n^2\right) + \frac{\omega_n^4}{192} A^4\right] = 0.$$

Assuming a nontrivial solution, $A \neq 0$, the expression in the square braces must equal zero. Expanding and combining like terms results in

$$\omega^4 + \omega^2\omega_n^2 \left(\frac{1}{2}A^2 - 2\right) + \omega_n^4 \left(1 - \frac{1}{2}A^2 + \frac{5}{96}A^4\right) = 0.$$

This is a quadratic in ω^2 that can be solved as

$$\omega_{1,2}^2 = \begin{cases} \omega_n^2 \left(1 - 0.145 A^2\right) \\ \omega_n^2 \left(1 - 0.352 A^2\right). \end{cases}$$

---

[27] The integrals are over the period $T = 2\pi/\omega$. Replacing the functions under the integrals as follows, many integrate to zero,

$$\int_0^T \sin^2 \omega t\, dt = \int_0^T \left(\frac{1}{2} - \frac{1}{2}\cos 2\omega t\right) dt$$
$$\int_0^T \sin\omega t \sin 3\omega t\, dt = \int_0^T \left(\frac{1}{2}\cos 2\omega t - \frac{1}{2}\cos 4\omega t\right) dt$$
$$\int_0^T \sin^2 3\omega t\, dt = \int_0^T \left(\frac{1}{2} - \frac{1}{2}\cos 6\omega t\right) dt.$$

Replace T by $2\pi/\omega$ in each trig function so that the only nonzero integrals result from the integrals of $1/2$.

The first root ω_1^2 is the value that minimizes Equation 12.67. Lindstedt's solution is $\omega_n^2 \left(1 - 0.375A^2\right)$. The one-term Galerkin approximate solution is

$$\Theta(t) = A \sin \omega_n^2 \left(1 - 0.145A^2\right) t.$$

Since a one-term solution was assumed, only one initial condition is needed. The initial angle $\Theta(0)$ does not affect this solution, but the initial velocity $\dot\Theta(0)$ and the value of A are related by

$$A\left(1 - 0.145A^2\right) = \frac{\dot\Theta(0)}{\omega_n^2}.$$

12.10 Concluding Summary

In this final chapter of our text, a very basic introduction is developed on the kinds of analysis and behavior one can expect if there are even small nonlinearities in an oscillating system. The phase plane is used to graphically represent the oscillatory character of autonomous systems. The way in which the stability characteristics of a nonlinear oscillator are quantified are introduced. The perturbation method is developed as a way to approximate the effects of small nonlinearities, whether in damping or stiffness, on the response and period of oscillation.

Of course, many important topics are not included here and are left to the more specialized books. Numerous references are footnoted to texts that are completely devoted to nonlinear methods.

12.11 Problems

1. Construct the phase diagram for the simple harmonic oscillator $\ddot{x} + \omega_n^2 x = 0$.

2. Construct the phase diagram for the equation $\ddot{x} - \omega_n^2 x = 0$.

3. Examine the equation $\ddot{x} + (x^2 + \dot{x}^2 - 1)\dot{x} + x = 0$ for how damping adds and removes energy from the system depending on the values of x and $\dot{x}$. Define $\dot{x} = y$.

4. Obtain an approximation to the forced response, of period 2π, for the equation

$$x'' + \frac{1}{4}x + 0.1x^3 = \cos \tau.$$

5. Obtain an approximation to the forced response, of period 2π, for the equation
$$x'' + \frac{1}{2}x + 0.1x^3 = \cos\tau.$$

6. For the forced, quasi-harmonic system, *(i)* derive Equation 12.23, *(ii)* derive Equation 12.26.

7. [Jordan and Smith, pp. 132.] Obtain approximately the solutions of period 2π of the equation
$$\ddot{x} + \Omega^2 x - \varepsilon x^2 = \Gamma\cos t, \quad \varepsilon > 0.$$

8. Derive the forced periodic response of the equation
$$x'' + (9 + \varepsilon\beta)x - \varepsilon x^3 = \Gamma\cos\tau.$$

9. Solve Equation 12.27 for $\Omega \simeq 1$ using the approach suggested in the footnoted Equation 12.32.

10. For the subharmonic response, *(i)* derive Equations 12.33–12.35, *(ii)* derive Equation 12.36, *(iii)* derive Equation 12.37, *(iv)* derive Equation 12.38.

11. [Jordan and Smith, pp. 195.] Find the subharmonic response of order $1/n$ for the linear equation
$$\ddot{x} + \frac{1}{n^2}x = \Gamma\cos t.$$

12. Consider the following Duffing equation,
$$\ddot{x} + \alpha x + \varepsilon x^3 = \Gamma\cos\omega t,$$
$\alpha, \Gamma, \omega > 0$. The values of ω for which the subharmonics occur are unknown. Expand x and ω in a perturbation solution.

13. For the combination harmonic response, derive Equation 12.42.

12.12 Mini–Projects

Consider some of these papers as possible mini–projects or supplementary readings.

1. *Vibration Analysis of a Rotating Timoshenko Beam*, S.C. Lin, K.M. Hsiao, *J. Sound and Vibration* (2001) **240**(2), 303–322. The governing equations for linear vibration of a rotating Timoshenko beam are derived by d'Alembert's principle and the principle of virtual work. In order to capture all inertia effects and the coupling between extensional and flexural deformation, the consistent linearization of the fully geometrically nonlinear beam theory is used.

2. *Multiple Resonant or Non–resonant Parametric Excitations for Nonlinear Oscillators*, A. Maccari, *J. Sound and Vibration* (2001) **242**(5), 855-866. The transient and steady state response of a very general nonlinear oscillator subject to a finite number of parametric excitations is considered by the asymptotic perturbation method.

3. *New Nonlinear Modelling for Vibration Analysis of a Straight Pipe Conveying Fluid*, S.I. Lee, J. Chung, *J. Sound and Vibration* (2002) **254**(2), 313–325. A new nonlinear model of a straight pipe conveying fluid is presented for vibration analysis when the pipe is fixed at both ends.

4. *Statistical Seismic Response Analysis and reliability Design of Nonlinear Structure System*, B.-Y. Moon, B.-S. Kang, *J. Sound and Vibration* (2002) **258**(2), 269–285. This paper proposes a method to analyze seismic response and reliability design of a complex nonlinear structure under random excitation.

5. *Nonlinear statics and dynamics of a simply supported nonuniform tube conveying an incompressible inviscid fluid*, S.V. Sorokin, A.V. Terentiev, *J. Fluids and Structures* **17** (2003) 415–431. The nonlinear governing equations of motion are derived using Hamilton's principle.

6. *Dynamical analysis of two coupled parametrically excited van der Pol oscillators*, Q. Bi, *Int. J. of Nonlinear Mechanics*, **39** (2004) 33–54. Transition boundaries are sought to divide the parameter space into a set of regions. Stabilities and chaos is studied.

7. *Nonlinear longitudinal vibrations of non–slender piezoceramic rods*, U. von Wagner, *Int. J. Nonlinear Mechanics* 39 (2004) 673–688. Perturbation techniques are used to solve nonlinear equations of motion of piezoceramics excited by weak electrical fields.

8. *Quadratic nonlinear oscillators*, R.E. Mickens, *J. Sound and Vibration* 270 (2004) 427–432.

9. *Large multi–hinged space systems: a parametric stability analysis*, M. Lavagna, A.E. Finzi, *Acta Astronautica* 54 (2004) 295–305. The paper presents the

12.12. MINI–PROJECTS

parametric stability analysis of a set of equilibrium configurations with respect to an orbiting reference frame for a three–hinged space system.

Index

Accelerometer, mechanical, transfer function for, 265-267
Aerodynamic loads, 200-201
Analysis, design, interrelationship of, 6
Approximations, deterministic, 13
Arbitrary constraints, evaluation of, 538-540
Arbitrary loading
  convolution, 148-155, 166-168
  Laplace transform, 135-143, 165
Axial vibration, beams, 505-514, 552-554
  Hamilton's approach to governing equation, 507-508
  Newton, Sir Isaac, approach to governing equation, 505-507
  orthogonality of normal modes, 512-514
  second order Galerkin's equations for, 610-612
  simplified eigenvalue problem, 509-512
  uniform, with elastic boundary, 510-512

Base excitation, 115-120
  structure with rotating unbalance, 128-129
Beam vibration, 541-540, 559-560
  response of beam to traveling force, 549-550
  response of beam with moving support, 548-549
  transverse vibration of beam on elastic foundation, 546-548
  transverse vibration of beam with axial force, 541-544
  transverse vibration of beam with elastic restraints, 544-546
Bernoulli, Daniel, 535-537
Bessel, Friedrich Wilhelm, 587-588
Broad band processes, 239-242

Cantilever beam, transverse time-dependent displacement at free end, 620-624
Circular membranes, 582-588
Colored noise, response to, 248
Component modeling, 11
Computational modeling, 6-7
Constraint equation, 301-302
Containers, sloshing of fluids in, 454-457
Continuous limit of discrete formulation, 487-489
Continuous models for vibration, 16-17, 487-573, 575-639
  approximate methods, 600-612
  Galerkin method, 608-612

[Continuous models for vibration, approximate methods]
  Rayleigh-Ritz method, 602-608
  Rayleigh's quotient, 600-602
 beam vibration, 541-540
  response of beam to traveling force, 549-550
  response of beam with moving support, 548-549
  transverse vibration of beam on elastic foundation, 546-548
  transverse vibration of beam with axial force, 541-544
  transverse vibration of beam with elastic restraints, 544-546
 continuous limit of discrete formulation, 487-489
 longitudinal (axial) vibration of beams, 505-514
  Hamilton's approach to governing equation, 507-508
  Newton, Sir Isaac, approach to governing equation, 505-507
  orthogonality of normal modes, 512-514
  simplified eigenvalue problem, 509-512
 random vibration of continuous structures, 596-600
 torsional vibration of shafts, 514-518
  torsion of shaft with rigid disk at end, 516-518
 transverse vibration of beams, 518-541
  boundary conditions, 523-524
  derivation of equations of motion for beam with shear distortion, Timoshenko beam, 519-523
  orthogonality of normal modes, 531-541
  simplified eigenvalue problem, 524-531
  Timoshenko beam, 519-523
 variables not separate, 613-620
  flow in pipe with constant tension, 624-630
  response to nonharmonic, time-dependent boundary conditions, 613-624
 vibration of membranes, 575-588
  circular membranes, 582-588
  rectangular membranes, 575-582
 vibration of plates, 588-596
  derivation of equation of motion for rectangular plates, 589-593
  eigenvalue problem, 593-596
 vibration of strings, 489-505
  boundary value problem for string, 498-499
  modal solution for fixed-fixed boundary conditions, 499-505
  wave equation via Hamilton's principle, 495-498
  wave propagation solution, 491-495
Control, qualitative vibration and, 5-6
Convolution for general loading, 152-154
Correlation coefficient, reliability, 226-228
Cramer's rule, 37

d'Alembert, Jean Le Rond, 308
d'Alembert's principle, 303-308
 equations of motion by virtual work and, 306-308
Damping, 94
 discrete models with, single degree of freedom vibrations, 93-168
  arbitrary loading, 135-143, 148-155
  damped response, 108-129
  damping, 94
  forced vibration with damping, 104-108
  free vibration with damping, 94-104
  harmonic excitation, 108-129
  impulsive excitation, 146-148
  periodic excitation, not harmonic, 129-135

INDEX

[Damping, discrete models with, single degree of freedom vibrations]
 step loading, 143-146
 inertia, 11-13
 Lagrange's equation with, 322-323
Density function, use of, 211
Design
 analysis, interrelationship of, 6
 computational, 6-7
Determinant matrices, matrix inverse, 35-37
Deterministic approximations, 13
Deterministic mathematical modeling, 81-86
Differential equations, 23-32
 homogeneous solution, 25-29
 assumed solution, 28-29
 linear equations, solution of, 25
 superposition, principle of, 25
 particular solution, 29-32
Dimensionality, 15
Dirac, Paul Adrien Maurice, 148
Discrete models, 15-16
 with damping, single degree of freedom vibrations, 93-168
 arbitrary loading, 135-143, 148-155
 damped response, 108-129
 damping, 94
 forced vibration with damping, 104-108
 free vibration with damping, 94-104
 harmonic excitation, 108-129
 impulsive excitation, 146-148
 periodic but not harmonic excitation, 129-135
 step loading, 143-146
 single degree of freedom vibration, 39-92
 free vibration with no damping, 67-71, 86-90
 harmonic forced vibration with no damping, 72-81, 90-91
 mathematical modeling, deterministic, 44-67, 81-86
 rocket ship, 40-44

rotating vector approach, equation of motion, 64-67
 satellite transport, 40
Duhamel, Jean Marie Constant, 155

Earthquake spectra, 238
Eigenfunctions, mode shapes, 501
Eigenvalues
 geometry of, 435-439
 natural frequencies, 500-501
 repeated frequencies, 438-439
 square matrix, 37-38
Eigenvectors, square matrix, 37-38
Elastic pendulum, Hamilton's principle for derivation of equation of motion, 320-322
Elevator cable system, 359-363
Energy formulation, 59-64
Ensemble averaging, 229-233
Equation of motion, d'Alembert's principle for derivation of, 304-305
Equations, differential, 23-32
 homogeneous solution, 25-29
 assumed solution, 28-29
 linear equations, solution of, 25
 superposition, principle of, 25
 particular solution, 29-32
Equations of motion, 14
 derivation of, 343-351, 411-413
 [k] matrix properties, 349-351
 [m] matrix properties, 349-351
Ergodicity, 235
Euler, Leonhard, 99-100
Excitation due to rotating unbalance, 120-122
Expansion methods, 654-671
Exponential density, 217-218

Fatigue life, 201-203
Feedback control, 263-267, 287
 performance of systems, 267-272, 287-289
 gain factor, 270-271
 poles, second order system, 269-270
 stability of response, 272

[Feedback control, performance of systems]
zeros of second order system, 269-270
Flexibility, 334-343
equations of motion via, 339-343
matrix, 337-338
preliminary concepts of, 410-411
Flow in pipe with constant tension, 624-630
Fluids in containers, sloshing of, 454-457
Forced pendulum, 667-670
Forced van der Pol equation, 685-687
Forced vibration
with damping, 104-108
via modal analysis, 397-402
Forcing, harmonic, 154-155
Fourier, Jean Baptiste Joseph, 134-135
Fourier series representative, application of, 131-134
Free vibration via modal analysis, 391-393
Free vibration with damping, 94-104
direct method, 415-416
by direct method, 380-386
phase plane, 104
time constraints, 103-104
Free vibration with no damping, 67-71, 86-90
alternate formulation, 69-70
initial displacement, 68-69
initial velocity, 68-69
phase plane, 70-71
Frequency ratio, vs. peak amplitude, 110-114
Fundamental nonlinear equations, 645-646

Galerkin, Boris Grigorievich, 612-613
Galerkin method, 608-612
Gaussian density, 218-220

Hamilton, Sir William Rowan, 322

Hamilton's principle, 318-322, 345-347, 507-508
Harmonic excitation
in complex notation, 114-129
damped response, 108-129, 158-162
Harmonic force with different frequency, 63-64
Harmonic forced vibration with no damping, 72-81, 90-91
free plus forced vibration response, 73-74
resonance, 74-79
l'Hôpital, Guillaume Francois Antione Marquis, 79
one degree of freedom beating, 79
resonance response, direct approach to, 77-79
water, vibration of structure in, 79-81
Harmonic loading, 63
Harmonic loads, as boundary conditions, 549
Homogeneous solution to equation, 25-29
assumed solution, 28-29

Impulsive excitation, 146-148
Inertia, 11-13
springs with, 61-62
Influence coefficients, 335-343
Inverse, matrix, 35-37
Inverse problem, 175, 334
Inverse vibration, 443-454
deterministic inverse vibration problem, 446-449
effect of uncertain data, 449-454
mass, estimation of, 481
stiffness, estimation of, 481

Jointly distributed variables, 225-226
Jump phenomenon, 666
Lagrange, Joseph Louis, 172
Lagrange's equation, 169-172, 188-189, 308-318, 345-347
application to single degree of freedom oscillator, 170-172

[Lagrange's equation]
  with damping, 322-323
  for small oscillations, 317-318
Laplace, Pierre-Simon, 142-143
Laplace transform, arbitrary loading, 135-143
Leibnitz, Gottfried, 43-44
l'Hôpital, Guillaume Francois Antione Marquis, 79
Lindstedt-Poincare method, 658-662
Linear equations, solution of, 25
  superposition, principle of, 25
Linear models approximating nonlinear behavior, 14-15
Linearization, Taylor series, 21-23
Logarithmic decrement, 100-103
Lognormal density, 220-221
Longitudinal vibration of beams, 505-514, 552-554
  Hamilton's approach to governing equation, 507-508
  Newton, Sir Isaac, approach to governing equation, 505-507
  orthogonality of normal modes, 512-514
  simplified eigenvalue problem, 509-512

Mass coupling, equations of motion with, 347-349
Mathematical modeling, deterministic, 44-67, 81-86
  damping, 46-49
  deterministic approximation, sources of, 49
  dimensional analysis, 49-52
  dimensions of vibration, 50-52
  energy formulation, 59-64
  harmonic force with different frequency, 63-64
  harmonic loading, 63
  mass, 46-49
  Newton, Sir Isaac, second law, 52-59
  pendulum, nonlinear, linearized, 55-56
  problem idealization, formulation, 44-46

electrical analogy, 48
electrical-mechanical analogy, 48
linear approximation, 45-46
Taylor series, 46
springs with inertia, 61-62
stiffness, 46-49
torsional vibration, 54-55
variable mass system, 56-59
vertical motion, pendulum motion, compared, 61
Mathematics, 21-38
  differential equations, 23-32
    homogeneous solution, 25-29
    linear equations, solution of, 25
    particular solution, 29-32
  linearization, 21-23
  matrices, 32-38
    Cramer's rule, 37
    determinant, matrix inverse, 35-37
    operations, 34-35
    square matrix, eigenvalues, 37-38
  Taylor series, 21-23
  transition, 38
Mathieu equation, 675-681
Matrices, 32-38
  Cramer's rule, 37
  determinant, matrix inverse, 35-37
  operations, 34-35
  square matrix
    eigenvalues, 37-38
    eigenvectors, 37-38
Matrix inverse, 35-37
  determinant matrices, 35-37
Mechanical accelerometer for rapidly accelerating vehicle, 139-142
Membranes, vibration of, 575-588
  circular membranes, 582-588
  rectangular membranes, 575-582
Miner's rule for fatigue damage, 202-203
Modal analysis, 416-419
  compared to direct method, 407-410

[Modal analysis]
  with forcing, 393-402
  with proportional damping, 402-407
Modal matrix [P], 425-427
Modal orthogonality, 387-393
Modeling
  component, 11
  computational, 6-7
  continuous, 16-17
  discrete, 15-16
  multi-degree of freedom, 16
  nonlinear, 17
  single degree of freedom, 16
  system, 11
  types of, 14-17
  for vibration, 7-17
Monte Carlo simulation, 470-476
  generation of random variates, 473-474
  power spectral density, generating time-history for random process defined by, 475-476
  random number generation, 472-473
Motion
  equations of, 14
  in large, 687-691
  second law of, Newton, Sir Isaac, 17-18
Multi degree of freedom vibration, 332-486
  direct method, free vibration with damping, 376-386
  eigenvalue problem
    geometry of, 435-439
    repeated frequencies, 438-439
  equations of motion, derivation of, 343-351
    [k] matrix properties, 349-351
    [m] matrix properties, 349-351
  flexibility, 334-343
  free vibration with damping, 376-386
  generalization to n degrees of freedom, 423-427
  influence coefficients, 335-343
  inverse problems, 334
  inverse vibration, 443-454

  deterministic inverse vibration problem, 446-449
  effect of uncertain data, 449-454
  modal analysis, 386-410
  modal analysis compared to direct method, 407-410
  modal analysis with forcing, 393-402
  modal analysis with proportional damping, 402-407
  modal orthogonality, 387-393
  modal matrix [P], 425-427
  Monte Carlo simulation, 470-476
    generation of random variates, 473-474
    power spectral density, generating time-history for random process defined by, 475-476
    random number generation, 472-473
  periodic structures, 333-334, 439-443
    effects of imperfection, 442-443
    perfect lattice models, 440-442
  Rayleigh's quotient, 465-470
  sloshing of fluids in containers, 454-457
  stability of motion, 458-461
  stiffness, 334-343
  stochastic response of linear MDOF system, 461-465
  undamped vibration, 351-370
    coupled pendula--beating, 368-376
    forced vibration by direct method, 363-368
    two degree of freedom motion, solution by direct method, 351-363
  unrestrained systems, 427-435
    repeated frequencies, 435
Multi-degree of freedom vibration, damping in, 405-406

Narrow band processes, 239-242
Newton, Sir Isaac
  approach to governing equation, 505-507

INDEX

[Newton, Sir Isaac]
  second law of motion, 17-18, 52-59
Non-autonomous systems, periodic solutions of, 666-671
Nonlinear behavior, linear models approximating, 14-15
Nonlinear models, 17
  stability and, 11
Nonlinear oscillation, 641-699
  expansion methods, 654-671
  fundamental nonlinear equations, 645-646
  jump phenomenon, 666
  Lindstedt-Poincare method, 658-662
  Mathieu equation, 675-681
  motion in large, 687-691
  non-autonomous systems, periodic solutions of, 666-671
  nonlinear pendulum, using Galerkin method, 694-696
  nonlinear vibration, examples of, 642-645
  perturbation methods, 654-671
  phase plane, 646-654
    stability of equilibria, 651-654
  quasi-harmonic systems, forced oscillations of, 662-666
  random duffing oscillator, 692-694
  subharmonic oscillations, 672-675
  superharmonic oscillations, 672-675
  van der Pol equation, 681-687
    forced van der Pol equation, 685-687
    limit cycles, 682-684
Nonlinear pendulum, using Galerkin method, 694-696
Nonlinear vibration, examples of, 642-645
Normal (Gaussian) density, 218-220

Ocean wave forces, 203-205
Ocean wave spectra, 237
One degree of freedom beating, 79

Oscillation, nonlinear, 641-699
  expansion methods, 654-671
  fundamental nonlinear equations, 645-646
  jump phenomenon, 666
  Lindstedt-Poincare method, 658-662
  Mathieu equation, 675-681
  motion in large, 687-691
  non-autonomous systems, periodic solutions of, 666-671
  nonlinear pendulum, using Galerkin method, 694-696
  nonlinear vibration, examples of, 642-645
  perturbation methods, 654-671
  phase plane, 646-654
    stability of equilibria, 651-654
  quasi-harmonic systems, forced oscillations of, 662-666
  random duffing oscillator, 692-694
  subharmonic oscillations, 672-675
  superharmonic oscillations, 672-675
  van der Pol equation, 681-687
    forced van der Pol equation, 685-687
    limit cycles, 682-684
Oscillations, Lagrange's equation, 317-318
Oscillator, variables for, 282-283
Oscillator control, 174

Parameter variations, sensitivity, 279-282, 290-291
Particular solution to equation, 29-32
Peak amplitude, *vs.* frequency ratio, 110-114
Pendulum
  with base motion, 458-461
  nonlinear, 656-657
  linearized, 55-56
  simple, 643-645
  trajectories, stability of, 688-691
Periodic excitation
  not harmonic, 129-135

[Periodic excitation]
  not harmonic excitation, 162-164
Periodic structures, 333-334, 439-443, 480-481
  effects of imperfection, 442-443
  perfect lattice models, 440-442
Perturbation methods, 654-671
Phase plane, 646-654
  stability of equilibria, 651-654
Poincare, Jules Henri, 661-662
Power spectrum, 236-242
Probabilistic forces, single degree of freedom vibration, 195-260
  fatigue life, 201-203
  material properties, 206
  mathematical expectation, 211-214
  variance, 213-214
  ocean wave forces, 203-205
  probability densities useful in applications, 215-222
    exponential density, 217-218
    lognormal density, 220-221
    normal (Gaussian) density, 218-220
    Rayleigh density, 221-222
    uniform density, 215-217
  random processes, 228-242
    basic random process descriptors, 228-229
    ensemble averaging, 229-233
    power spectrum, 236-242
    stationarity, 233-235
  random variables, 208-211, 222-228
    correlation, 224-228
    covariance, 224-228
    probability density function, 209-211
    probability distribution, 208-209
  random vibration, 200-201, 242-249
    derivation of equations, 244
    formulation, 243-244
    response correlations, 245-246
    response spectral densities, 246-247
  statistics, 207
  wind forces, 205
Probability, definition for, 197-199
Probability densities, 215-222, 253-254
Problem idealization, formulation, 7-11
Proportional damping via modal analysis, forced vibration with, 404-405

Quadratic density, 217
Qualitative probability, 5
Qualitative structures, 4
Qualitative systems, 4
Qualitative vibration, 5-6
Quasi-harmonic systems, forced oscillations of, 662-666

Random duffing oscillator, 692-694
Random loading, 173
Random processes, 228-242
  basic random process descriptors, 228-229
  ensemble averaging, 229-233
  power spectrum, 236-242
  stationarity, 233-235
Random variables, 208-211, 222-228
  correlation, 224-228
  covariance, 224-228
  probability density function, 209-211
  probability distribution, 208-209
Random vibration, 200-201, 242-249, 257-258
  continuous structures, 596-600
  derivation of equations, 244
  formulation, 243-244
  response correlations, 245-246
  response spectral densities, 246-247
Randomness, 172-174
Ransient responses, 137-139
Rayleigh, John William Strutt, 404
Rayleigh density, 221-222
Rayleigh-Ritz method, 602-608

INDEX

Rayleigh's quotient, 465-470, 482, 600-602
  for three degree of freedom system, 468-470
  for two degree of freedom system, 466-468
Reciprocity, 339
Rectangular membranes, vibration, 575-582
Rectangular plates
  derivation of equation of motion for, 589-593
  forced, simply-supported, 595-596
Resonance in electric motors, 126-128
Resonance response, direct approach to, 77-79
Response mean, variance, 246
Response to nonharmonic, time-dependent boundary conditions, 613-624
Rocket ship, 40-44
  Leibnitz, Gottfried, 43-44
  Newton, Sir Isaac, 41-43
Rotating shafts, 122-126
Rotating unbalance, excitation due to, 120-122
Rotating vector, equation of motion, 64-66
  solution, 66-67

Satellite in shipping container, vibration analysis of, 182-186
Satellite transport, 40
Second law of motion of Sir Isaac Newton, 17-18
Self-excited system, stability, 175-176
Sensitivity analysis, as part of control design, 281-282
Short duration harmonic forcing, 145-146
Single degree of freedom vibration, 39-92, 169-194
  control, 174-175
  discrete models, 39-92

  free vibration with no damping, 67-71, 86-90
    alternate formulation, 69-70
    initial displacement, 68-69
    initial velocity, 68-69
    phase plane, 70-71
  harmonic forced vibration with no damping, 72-81, 90-91
    free plus forced vibration response, 73-74
    resonance, 74-79
    water, vibration of structure in, 79-81
  inverse problem, 175
  Lagrange's equation, 169-172
  mathematical modeling, deterministic, 44-67, 81-86
    damping, 46-49
    deterministic approximation, sources of, 49
    dimensional analysis, 49-52
    energy formulation, 59-64
    harmonic force with different frequency, 63-64
    harmonic loading, 63
    mass, 46-49
    Newton, Sir Isaac, second law, 52-59
    pendulum, nonlinear, linearized, 55-56
    problem idealization, formulation, 44-46, 48
    springs with inertia, 61-62
    stiffness, 46-49
    torsional vibration, 54-55
    variable mass system, 56-59
    vertical motion, pendulum motion, compared, 61
  probabilistic forces, 195-260
    fatigue life, 201-203
    material properties, 206
    mathematical expectation, 211-214
    ocean wave forces, 203-205
    probability, 207
    probability densities useful in applications, 215-222

[Single degree of freedom vibration, probabilistic forces]
random processes, 228-242
random variables, 208-211, 222-228
random vibration, 200-201, 242-249
statistics, 207
wind forces, 205
randomness, 172-174
rocket ship, 40-44
Leibnitz, Gottfried, 43-44
Newton, Sir Isaac, 41-43
rotating vector approach, equation of motion, 64-66
solution, 66-67
satellite transport, 40
self-excited system, stability, 175-176
Single degree of freedom vibrations, discrete models with damping, 93-168
arbitrary loading
convolution, 148-155
Laplace transform, 135-143
damped response, 108-129
damping, 94
forced vibration with damping, 104-108
free vibration with damping, 94-104
phase plane, 104
time constraints, 103-104
harmonic excitation, 108-129
in complex notation, 114-129
impulsive excitation, 146-148
periodic but not harmonic excitation, 129-135
step loading, 143-146
Sloshing of fluids in containers, 454-457
Small oscillations, Lagrange's equation, 317-318
Springs with inertia, 61-62
Square matrix
eigenvalues, 37-38
eigenvectors, 37-38
Stability of motion, 458-461

State space, base excitation in, 283-286
Statics, equilibrium and, 13
Stationarity, 233-235
Statistics, 207
Step loading, 143-146, 165-166
Stiffness, 11-13, 334-343
preliminary concepts of, 410-411
Stiffness matrix, 338-339
potential energy and, 350-351
Stochastic response, linear multidegree of freedom system, 481-482
Stochastic response of linear MDOF system, 461-465
Strings, vibration of, 551-560
Subharmonic oscillations, 672-675
Superharmonic oscillations, 672-675
System modeling, 11
System models, types of, 14-17
System uncertainty, 12-13

Taylor, Brook, 21-23
Taylor series, 21-23
linearization and, 21-23
Time constraints, free vibration with damping, 103-104
Time scales, 104-104
Timoshenko, Stephen P., 540-541
Timoshenko beam, 519-523
Torsional vibration, 54-55
shafts, 554
Torsional vibration of shafts, 514-518
torsion of shaft with rigid disk at end, 516-518
Transient response, automatic control of, 272-278, 289-290
control actions, 272-274
control of transient response, 274-278
Transverse vibration, beams, 518-541, 554-559
boundary conditions, 523-524
derivation of equations of motion for beam with shear distortion, 519-523

[Vibration]
nonlinear, examples of, 642-645
nonlinear oscillation, 641-699
qualitative, 5-6
  control, 5-6
  uncertainties, 5-6
random, 200-201, 242-249, 257-258
satellite, in shipping container, analysis of, 182-186
shafts, torsional, 554
single degree of freedom vibration
  advanced topics, 169-194
  with damping, discrete models, 93-168
  discrete models, 39-92
  probabilistic forces, 195-260
stresses induced, evaluation of, 538-540
strings, 489-505, 551-560
structure in water, 79-81
torsional, 54-55
  shafts, 514-518
transverse
  beam, 518-548, 554-559
  response of forced, damped beam, 533-535
undamped, 351-370, 413-415
variational principles, 295-331
vibration control, 261-294
  feedback control, 263-267
  performance of, 267-272
  parameter variations, sensitivity of, 279-282
  state variable models, 282-286
    integrals, 286
    matrix derivatives, 286
  transient response, automatic control of, 272-278
    control actions, 272-274
    control of transient response, 274-278
Vibration of plates, 588-596
  derivation of equation of motion for, rectangular plates, 589-593
  eigenvalue problem, 593-596
Vibration of strings, 489-505
  boundary value problem for string, 498-499
  modal solution for fixed-fixed boundary conditions, 499-505
  wave equation via Hamilton's principle, 495-498
  wave propagation solution, 491-495
Vibration-induced stresses, evaluation of, 538-540
Virtual work, 297-308
  energy, 297-300
  virtual work, 300-303
  work, 297-300

Water, vibration of structure in, 79-81
Water landing, space module, 177-182
Wave forces, on oil drilling platform, 203-204
Whirling, 122-126
White noise, 242
  oscillator response to, 247-248
Wind forces, 205
Wind spectra, 237-238

INDEX

[Transverse vibration, beams]
  orthogonality of normal modes, 531-541
  simplified eigenvalue problem, 524-531
  Timoshenko beam, 519-523
Transverse vibration response, forced, damped beam, 533-535
Traveling string, 628
Two body single degree of freedom system, Lagrange's, 315-317
Two degree of freedom inverse vibration problem, 448-449
Two degrees of freedom, no damping, 356-359
Types of system models, 14-17

Uncertain two degree of freedom system, 452-454
Uncertainties, qualitative vibration, 5-6
Undamped nonlinear oscillator, phase plane of, 648-651
Undamped nonlinear pendulum, phase plane for, 647-648
Undamped vibration, 351-370, 413-415
  coupled pendula, beating, 368-376
  forced vibration by direct method, 363-368
  two degree of freedom motion, solution by direct method, 351-363
Undamped vibration absorber, 366-368
Uniform density, 215-217
Unrestrained rotation, 430-434
Unrestrained systems, 427-435, 479-480
  repeated frequencies, 435
Unrestrained translation, 434-435

van der Pol, Balthazar, 687
van der Pol equation, 681-687
  forced van der Pol equation, 685-687
  limit cycles, 682-684

Variable mass system
Variational principle
  d'Alembert's princip
  Hamilton's principl
  Lagrange's equatio
    with damping, 32!
    for small oscillatic
  virtual work, 297-3
    energy, 297-300
    virtual work, 300-
    work, 297-300
Vector, rotating, equ: tion, 64-66
Vertical motion, pen(tion, compared, (
Vibration
  absorber, undampe
  beam, 541-540, 559
  axial, 505-514
  continuous models,
    advanced models,
  continuous structur 596-600
  dimensions of, 50-5:
  forced
    with damping, 104
    by direct method, :
    modal analysis, 39
    proportional damp analysis, 404-40£
  free
    with damping, 94-376-386, 415-416
    modal analysis, 39
    with no damping, (
  free plus forced resp
  harmonic forced, wi ing, 72-81, 90-91
  inverse, 446-449
  mass, 443-454, 48]
  stiffness, estimatic
  two degree freedor
  longitudinal, beam, 552-554
  mathematics, 21-38
  membranes, 575-58
  modeling, 7-17
  multi degree of free(
    advanced topics, 4!